AF768837

Plant-Animal Interactions: An Ecological Approach

Plant-Animal Interactions: An Ecological Approach

Edited by Isaac Clements

SYRAWOOD
PUBLISHING HOUSE
New York

Published by Syrawood Publishing House,
750 Third Avenue, 9th Floor,
New York, NY 10017, USA
www.syrawoodpublishinghouse.com

Plant-Animal Interactions: An Ecological Approach
Edited by Isaac Clements

© 2018 Syrawood Publishing House

International Standard Book Number: 978-1-68286-536-1 (Hardback)

This book contains information obtained from authentic and highly regarded sources. Copyright for all individual chapters remain with the respective authors as indicated. All chapters are published with permission under the Creative Commons Attribution License or equivalent. A wide variety of references are listed. Permission and sources are indicated; for detailed attributions, please refer to the permissions page and list of contributors. Reasonable efforts have been made to publish reliable data and information, but the authors, editors and publisher cannot assume any responsibility for the validity of all materials or the consequences of their use.

The publisher's policy is to use permanent paper from mills that operate a sustainable forestry policy. Furthermore, the publisher ensures that the text paper and cover boards used have met acceptable environmental accreditation standards.

Trademark Notice: Registered trademark of products or corporate names are used only for explanation and identification without intent to infringe.

Cataloging-in-Publication Data

Plant-animal interactions : an ecological approach / edited by Isaac Clements.
p. cm.
Includes bibliographical references and index.
ISBN 978-1-68286-536-1
1. Animal-plant relationships. 2. Ecology. I. Clements, Isaac.
QH549.5 .P53 2018
591.54--dc23

Printed in China.

TABLE OF CONTENTS

PREFACE

The main aim of this book is to educate learners and enhance their research focus by presenting diverse topics covering this vast field. This is an advanced book which compiles significant studies by distinguished experts in the area of analysis. This book addresses successive solutions to the challenges arising in the area of application, along with it; the book provides scope for future developments.

Plant-animal interactions are biological processes that occur between plants and animals in an ecosystem. Biodiversity is a key factor in the propagation and sustenance of an ecosystem. The symbiotic relationships that are formed by different species contribute to vital environmental processes such as pollination, seed dispersal, etc. Species evolution also depends on the surrounding environment. Natural resources are common for all species and cooperation and competition depends on the availability of food and water. From theories to research to practical applications, case studies related to all contemporary topics of relevance to this field have been included herein. With state-of-the-art inputs by acclaimed experts of this field, this book targets students and professionals.

It was a great honour to edit this book, though there were challenges, as it involved a lot of communication and networking between me and the editorial team. However, the end result was this all-inclusive book covering diverse themes in the field.

Finally, it is important to acknowledge the efforts of the contributors for their excellent chapters, through which a wide variety of issues have been addressed. I would also like to thank my colleagues for their valuable feedback during the making of this book.

Editor

1

Ant-Plant Interaction in a Tropical Savanna: May the Network Structure Vary over Time and Influence on the Outcomes of Associations?

Denise Lange, Kleber Del-Claro*

Instituto de Biologia, Universidade Federal de Uberlândia, Uberlândia, Minas Gerais, Brazil

Abstract

Plant-animal interactions occur in a community context of dynamic and complex ecological interactive networks. The understanding of who interacts with whom is a basic information, but the outcomes of interactions among associates are fundamental to draw valid conclusions about the functional structure of the network. Ecological networks studies in general gave little importance to know the true outcomes of interactions and how they may change over time. We evaluate the dynamic of an interaction network between ants and plants with extrafloral nectaries, by verifying the temporal variation in structure and outcomes of mutualism for the plant community (leaf herbivory). To reach this goal, we used two tools: bipartite network analysis and experimental manipulation. The networks exhibited the same general pattern as other mutualistic networks: nestedness, asymmetry and low specialization and this pattern was maintained over time, but with internal changes (species degree, connectance and ant abundance). These changes influenced the protection effectiveness of plants by ants, which varied over time. Our study shows that interaction networks between ants and plants are dynamic over time, and that these alterations affect the outcomes of mutualisms. In addition, our study proposes that the set of single systems that shape ecological networks can be manipulated for a greater understanding of the entire system.

Editor: Fabio S. Nascimento, Universidade de São Paulo, Faculdade de Filosofia Ciências e Letras de Ribeirão Preto, Brazil

Funding: The authors had financial support of the Conselho Nacional de Desenvolvimento Científico e Tecnológico (CNPq - for K Del Claro 301605/2013-0; 473055/2012-0) and Coordenação de Aperfeiçoamento de Pessoal de Nível Superior (CAPES - dor D. Lange PNPD). The authors ensure that the funders had no role in study design, data collection and analysis, decision to publish, or preparation of the manuscript.

Competing Interests: The authors have declared that no competing interests exist.

* Email: kdelclaro@gmail.com

Introduction

Plant-animal interactions occur in a community context of dynamic and complex ecological interactive networks and the understanding of who interacts with whom is a basic information, but the outcomes of interactions among associates are fundamental to draw valid conclusions about the functional structure of the network [1–3].

Evolutionary ecologists are becoming increasingly interested in bipartite network analysis depicting interspecific interactions, as a tool for studies within an ecological context [4–6]. Bipartite network analysis includes several metrics, such as connectance, nestedness, cluster coefficients, web asymmetry, number of compartments, species degree, among others (see more details in [7]) and enables conclusions to be drawn on structure, stability and the robustness of interactions involving two groups of organisms [8–13] According to Hagen and co-workers [6], new data analytical tools such as network analysis, now form an essential ingredient in the study of complex systems.

In addition, experimental manipulations are widely used to explore the outcomes of interactions, mainly insect-plant interactions. The experimental methods vary according to the studied system. In a review [14], the authors argue that experimental manipulations also help to explore the structures that maintain preserved viable communities in an interactive way. In addition, some authors claim that long-term studies (e.g. intergenerational and/or temporal variation studies) and manipulative experiments within networks are fruitful avenues for future research on the mechanics of observed patterns in ecological networks [1,15].

Although common and widespread in terrestrial ecosystems, extrafloral nectar-mediated ant-plant mutualisms have only recently been the focus of network analysis [16–29]. Ants are the main organisms linked to plant biotic defenses against herbivores and the most common resource that plants offer to attract ants is nectar produced by extrafloral nectaries (EFNs), a liquid substance rich in carbohydrates with dilute concentration of amino acids, lipids, phenols, alkaloids and volatile organic compounds [30]. The attracted ants might provide plants with effective protection against natural enemies [31], and carbohydrates of extrafloral nectar have been suggested to be a key resource for arboreal ants [32]. Ants feeding on extrafloral nectar increased their survivorship, colony growth and reproduction [33]. These mutualistic ant-plant interactions are highly facultative and vary in time and space, [34–35] mainly depending on the characteristics of the ant species [36–38], ant density [39–40], herbivore defensive strategies [37], as well as the plant species [41], and its phenology [29].

Although studies using experimental manipulation to explore the outcomes of this type of mutualistic interactions are relatively common in the literature (see [31,42]), most of them have analyzed interactions between small subsets of ants and plants over a brief time interval [43–45]. The few studies that have directly investigated the effects of ants and EFN-bearing plant associations on communities, suggest that EFNs might influence species composition, abundance, and interactions on the community scale [41–42,46–47].

Here, we evaluate the dynamic of an interaction network between ants and plants with extrafloral nectaries, by verifying the temporal variation in structure and outcomes of mutualism for the plant community (leaf herbivory). To reach this goal, we used two tools: bipartite network analysis and experimental manipulation. Our main hypothesis is that ant-plant interactions are dynamic and exhibit temporal variation in their structure. This variation will influence the outcomes of mutualism, interfering within the effectiveness of ants as biotic control agents of plants against leaf herbivory. This type of study highlights the importance of considering the effects of variation in species composition, as well as their characteristics (i.e. natural history, morphology and behavior) in evolutionary ecology, which attempts to understand the patterns behind the topological features and functional structure of mutualistic networks.

Materials and Methods

Study site and plant characterization

Field study was conducted between September 2008 and April 2010 in a private natural Savanna reserve (Clube Caça e Pesca Itororó de Uberlândia/CCPIU - 48°17′ W; 18°58′ S) in Uberlândia, Minas Gerais State, south-eastern Brazil. The Biology Institute of Universidade Federal de Uberlândia has a memorandum of understanding with CCPIU, an agreement between Mr. Nilson Dias, head of CCPIU, and Dr. Kleber Del Claro, director of Biology Institute that enables ecological studies in the area. The vegetation is dominated by cerrado *strictu sensu*, consisting of trees 2–8 m in height, with an understory dominated by shrubs, grasses and scattered perennial herbs. As in other Cerrado areas, the climate in the region is rainy from October to April and dry from May to September (autumm-winter) [48].

Extrafloral nectaried plants are present in at least 25% of species and 31% of individuals in a Cerrado area [49] however, as verification of the interaction outcomes requires experimental manipulation, we chose a subset of plant species with extrafloral nectaries. The plant community chosen for this study corresponded to the most abundant EFN-tree species present in the study area, such as: *Caryocar brasiliense* (Cambess) (Caryocaraceae), *Lafoensia pacari* (A. St.-Hil.) (Lythraceae), *Ouratea spectabilis* (Mart.) Engl. (Ochnaceae), *O. hexasperma* (A. St.-Hil.) Baill (Ochnaceae), *Qualea grandiflora* (Mart.) (Vochysiaceae), *Q. multiflora* (Mart.) (Vochysiaceae), *Q. parviflora* (Mart.) (Vochysiaceae), *Stryphnodendron adstringens* (Mart.) Coville (Fabaceae) and *S. polyphyllum* (Mart.) (Fabaceae) (see [50]). The description of the location of EFNs in each species, as well as their morphology is found in [49,51], respectively.

Experimental manipulation and data collection for network construction

To identify which ant species visited the EFNs of each plant species and the effect of these ants on the leaf herbivory of the species, we conducted a long-term experimental manipulation. In August 2008, 30 similar individuals of each species (2–3 m tall, with a similar number of stems and phenology –without young leaves, buds, flowers and/or fruits) were selected. From each individual, we selected two similar stems (80 cm long) that were randomly designated as treatment or control. The treatment stems received a band of adhesive paper covering the whole circumference of the stem at its insertion with the trunk, and a layer of non-toxic ant repellent resin (Tanglefoot, Rapids, Michigan) was laid over strip. Ants already present were manually removed, as well as all structures that could act as bridges to provide access to the treatment stems. The control stems also received a band of adhesive paper with resin, however, it covered only half of the stem diameter, enabling ants to continue to climb the stem. In September 2008 and 2009, we tagged nine leaves on each stem of each individual. Three leaves were randomly chosen from each part of the stem: apex, middle and near the trunk insertion. The leaf area loss (leaf herbivory) of each leaf was examined in April (2009 and 2010), because after this month the most species lose its leaves resprouting in September or October. Therefore, in this study the data were collected into two periods: from September 2008 to April 2009 (corresponding to period 2009), and from September 2009 to April 2010 (corresponding to period 2010). Measurements of herbivory rates on leaves were assessed by placing the leaves on a transparent grid divided into millimeters. An index of herbivory from each leaf was calculated as the proportion of points in the grid falling within damaged and undamaged areas of the leaf blade [52]. The herbivory rate corresponded to the mean of the ratios of leaf damage assessed in the nine leaves. The proportions were evaluated in the field without leaf removal.

Fortnightly (from September to April), initially between 8:30 h and 10:30 h and subsequently between 14:00 h and 16:30 h, treatment stems were monitored for the integrity of the resin barrier and visiting ants from control stems were counted and identified. For the identification of unknown ant species, one individual of each species was placed in 70% alcohol. Ant species were identified at the Museu de Zoologia da Universidade de São Paulo (MZUSP) and voucher specimens were deposited at the Museu da Biodiversidade do Cerrado at the Universidade Federal de Uberlândia.

Data analyses

To understand the network structure evaluated in this study, for each year, we carried out two incidence matrices (ants per plants), a qualitative matrix (considering the presence or absence of each ant species on each plant species) and other quantitative measurements (with frequency of interactions). We used the following metrics to check the properties of network interactions: connectance (sensu [53]), degree of species (k) and average degree (for plants and ants) (both sensu [54]), specialization index of the network (sensu [55]), nestedness index (sensu [56]), and web asymmetry (sensu [57]). We compared the frequency distribution of species per degree for ant and plant between years via a Kolmogorov-Smirnov two-sample test with the D statistic and we tested the differences in frequency distribution of species per degree between networks using XL-Stat Pro 7.52.

To measure the nestedness we used the *NODF* index (*Nestedness metric based on Overlap and Decreasing Fill*, see [56]), calculated by the software Aninhado 3.0 [58]. We estimated the significance of *NODF* using a Monte Carlo procedure with 1,000 randomizations, using a null model *Ce* (type II), in which the interaction probability between an ant and a plant is proportional to their total number of interactions. The *NODF* index is strongly recommended, due to its theoretical and statistical consistency [56].

Table 1. Ant species associated with plant species studied.

Code	Subfamilies/Ant Species	Plant Species 1	2	3	4	5	6	7	8	9	Total (*n*=270)
	Formicinae										
1	*Brachymyrmex* sp.1	3	7	5	8	4	10	2	9	5	**53**
2	*Camponotus* sp.1	1	0	2	1	0	0	0	0	0	**4**
3	*C. crassus* Mayr, 1862	15	18	17	12	19	20	15	16	20	**152**
4	*C. blandus* (Smith, F., 1858)	5	8	5	10	4	2	9	6	6	**55**
5	*C. trapeziceps* Forel, 1908	1	2	1	3	3	5	2	2	0	**19**
6	*C. leydigi* Forel, 1886	1	1	0	3	0	0	0	1	0	**6**
7	*C. lespesii* Forel, 1886	1	0	0	0	0	1	0	0	0	**2**
8	*C. vittatus* Forel, 1904	0	0	0	0	2	1	0	0	0	**3**
	Myrmicinae										
9	*Crematogaster erecta* Mayr, 1866	1	2	2	2	1	0	3	5	0	**16**
10	*C. bruchi* Forel, 1912	1	1	2	1	3	1	1	0	0	**10**
11	*Pheidole* sp.1	1	1	1	0	0	1	1	0	1	**6**
12	*P.* sp.2	0	2	1	1	1	0	0	0	0	**5**
13	*Solenopsis* sp.1	0	2	3	1	0	0	3	2	1	**12**
14	*S.* sp.2	0	0	1	1	1	0	4	0	1	**8**
15	*Cephalotes* sp.1	1	0	0	0	0	0	0	0	0	**1**
16	*C. pusillus* (Klug, 1824)	10	17	16	13	17	11	15	17	6	**122**
17	*C. bruchi* (Forel, 1912)	0	1	0	0	0	0	0	0	1	**2**
18	*C. atratus* (Linnaeus, 1758)	1	0	0	0	0	0	0	0	0	**1**
19	*Nesomyrmex spininodis* (Mayr, 1887)	2	0	1	2	1	0	0	3	0	**9**
	Dolichoderinae										
20	*Azteca* sp.1	2	6	1	3	4	2	0	5	4	**27**
21	*Linepithema aztecoides* Wild, 2007	0	0	0	0	0	0	0	1	0	**1**
22	*Forelius brasiliensis* (Forel, 1908)	0	0	0	0	0	1	0	0	0	**1**
	Pseudomyrmicinae										
23	*Pseudomyrmex* sp.1	0	1	1	0	0	1	0	0	0	**3**
24	*P.* sp.2	0	0	0	0	0	0	1	0	0	**1**
25	*P. gracilis* (Fabricius, 1804)	4	9	13	5	3	9	7	10	3	**63**
26	*P. flavidulus* (Smith, F., 1858)	6	14	4	5	3	5	5	18	1	**61**
	Ectatomminae										
27	*Gnamptogenys semiferox* Brown, 1958	0	1	0	0	1	0	1	0	0	**3**
28	*Ectatomma tuberculatum* (Olivier, 1792)	9	4	7	5	2	6	1	1	1	**36**
29	*E. edentatum* Forel, 1912	0	0	0	0	0	2	0	0	0	**2**

Table 1. Cont.

Code	Subfamilies/Ant Species	Plant Species									
		1	**2**	**3**	**4**	**5**	**6**	**7**	**8**	**9**	**Total ($n=270$)**
30	*E. planidens* Borgmeier, 1939	0	0	0	1	0	0	1	0	0	**2**
	Ponerinae										
31	*Pachycondyla villosa* (Fabricius, 1804)	1	1	3	0	1	0	0	0	1	**7**

Column numbers correspond to the total of individual plants ($n=30$ trees per species) on which ant species were observed. Code numbers of plant species are as follows: (1) *Stryphnodendron adstringens*, (2) *S. polyphyllum*, (3) *Qualea grandiflora*, (4) *Ouratea hexasperma*, (5) *Q. parviflora*, (6) *Q. multiflora*, (7) *O. spectabilis*, (8) *Caryocar brasiliense* and (9) *Lafoensia pacari*. The code numbers of ant and plant species are the same as in Figure 1.

We calculated the balance between the numbers of plant "*I*" and ant "*J*" species in each network using the following equation: $W = (I-J)/(I+J)$, where "*W*" is the web asymmetry. Values equal zero for balanced webs; positive numbers indicate more plant species and negative values, more ant species; rescaled from −1 to 1 [57]. In addition, we calculated the network specializations using the H_2' index (sensu [55]), which is a dimensional measure derived from Shannon's index and range from zero (extreme generalization) to one (extreme specialization) [55]. To calculate the network specialization index (H_2'), we used the software R 2.13.2 (bipartite package) from the quantitative matrices. Finally, a bipartite graph of ant-plant interactions was constructed using the *Program for Analysis and Visualization of Large Networks* – Pajek 1.27 [59].

Since data from leaf loss area exhibited a normal distribution, they were analyzed using factorial ANOVA and Tukey post hoc comparisons. To compare herbivory between control (with ants) and treatment (excluding ants) stems in each plant species, we used the Student's *t*-test adjusted with the Bonferroni correction. Data for ant abundance were log(x+1) transformed and compared using a paired *t*-test. These statistical analyses were performed in Systat 12.0. All data are included within the manuscript.

Results

Ant diversity associated with plants

A total of 31 ant species from six subfamilies and 14 genera were associated with the studied EFN-bearing plants between 2009 and 2010 (Table 1). The most common genera were *Camponotus*, *Cephalotes* and *Pseudomyrmex*. Seven species [*Brachymyrmex* sp.1, *Camponotus crassus* Mayr, 1862, *Camponotus blandus* (Smith, F., 1858), *Cephalotes pusillus* (Klug, 1824), *Pseudomyrmex gracilis* (Fabricius, 1804), *Pseudomyrmex flavidulus* (Smith, F., 1858), *Ectatomma tuberculatum* (Olivier, 1792)] were recorded on all plant species, and two of these, *C. crassus* (present in more than 56% of plant individuals) and *C. pusillus* (in more than 45% of plant individuals), were ant species with a higher abundance and frequency (Table 1). The highest richness of associated ant species was observed on the plants: *S. adstringens*, *S. polyphyllum* and *Q. grandiflora*, which exhibited 61.3% of all linked ant species ($n=19$ species); followed by *O. hexasperma* (58%, $n=18$ species), *Q. parviflora* (54.8%, $n=17$ species), *Q. multiflora* and *O. spectabilis* (51.6%, $n=16$ species); *C. brasiliense* (45.1%, $n=14$ species) and *L. pacari* (41.9%, $n=13$ species).

There was a change in the network interaction over time; in 2009, the network was composed of 26 ant species and two species: *Crematogaster bruchi* Forel, 1912 and *Linepithema aztecoides* Wild, 2007 were observed only in 2009. However, in 2010 the network was enlarged and comprise 29 ant species. Five species [*Camponotus lespesii* Forel, 1886, *Camponotus vittatus* Forel, 1904, *Cephalotes atratus* (Linnaeus, 1758), *Forelius brasiliensis* (Forel, 1908) and *Ectatomma planidens* Borgmeier, 1939] were recorded only in 2010 (Figure 1). Another five ant species were considered the most generalist, interacting with all plant species in both years: *C. crassus, C. blandus, C. pusillus, P. gracilis* and *P. flavidulus*. With almost equal importance and perhaps as generalist as the other species, *E. tuberculatum* and *Camponotus trapeziceps* Forel, 1908 were associated with eight out of nine plants in both years. These seven species maintained the same degree (number of links) over time and were also the most frequent according to their presence in the tagged individuals ($n=30$) of each plant species (Figure 1). The ants *Crematogaster erecta* Mayr, 1866, *Nesomyrmex spininolis* (Mayr, 1887), *Solenopsis* sp.2, *Pachycondyla villosa* (Fabricius, 1804), *Pheidole* sp.2, *Pheidole* sp.1, *Camponotus* sp.1, *Gnamptogenys semiferox* Brown, 1958,

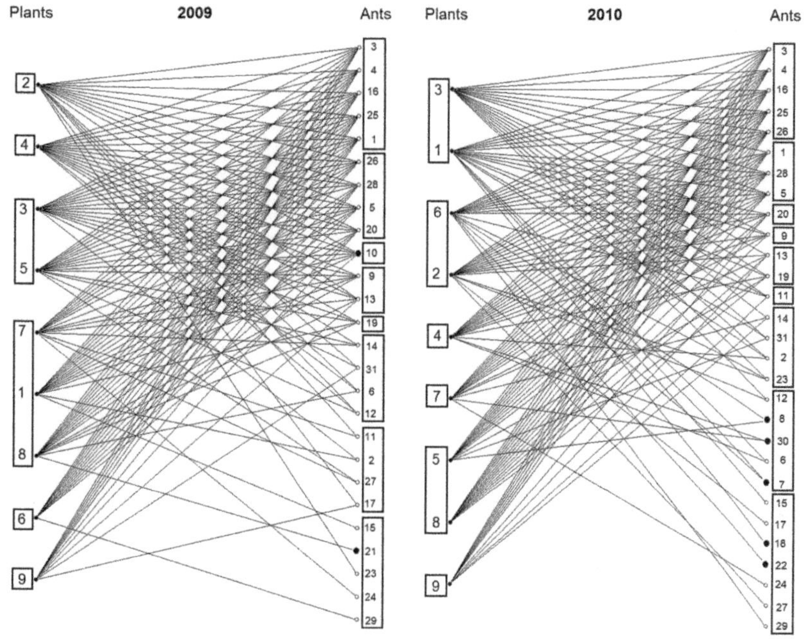

Figure 1. Ant-plant interaction networks in Brazilian Savanna studied in 2009 and 2010. Circles represent species, and links indicate extrafloral nectar feeding associations among ants and plants. Network 2009: 26 ant species and nine plant species; Network 2010: 29 ant species and nine plant species (see Table 1 for species identities). Bold circles in ant columns denote species that occurred in only one year. Rectangles denote species with the same degree.

Cephalotes bruchi (Forel, 1912) and *Pseudomyrmex* sp.1, also maintained their degree. The other 14 species demonstrated degree variation between 2009 and 2010. Some ants were considered as specialists, interacting with only one plant species: *Cephalotes* sp.1, *Pseudomyrmex* sp.2 and *Ectatomma edentatum* Forel, 1912 in both years; *L. aztecoides* and *Pseudomyrmex* sp.1 in 2009; and *Cep. bruchi, C. atratus, F. brasiliense* and *G. semiferox* in 2010. *Stryphnodendron polyphyllum* in 2009 and *Q. grandiflora* and *S. adstringens* in 2010 were the plant species that showed the greatest degree, with 17 links in each year. The plant with the lowest number of interactions was *L. pacari* with 10 links for each year.

The structure and dynamic of the interaction network

The properties of the interaction network varied over time, but the pattern of nestedness and asymmetry was maintained. In 2010, there was a greater number of ant species interacting with plants and a greater number of interactions (Table 2). However, in 2009, 53.84% (*C*) of potential links were realized, whereas in 2010, this was reduced to 47.89%. Although the number of interactions varied between species and years (see Figure 1), the average degree was similar for plants (Network $2009 = 14 \pm 0.745$; Network $2010 = 13.88 \pm 0.790$, mean ± SE) and ant species (Network $2009 = 4.84 \pm 0.62$; Network $2010 = 4.31 \pm 0.579$). For ants, the degree varied between one and nine, but the more common degrees between species were 1, 2, 3, 8 e 9 in both years. For plants, the degree varied between 10 and 17, with a similar distribution between years. Following this tendency, the frequency distribution of degree did not vary between years either for ants (Kolmogorov-Smirnov, $D = 0.111$; $P > 0.05$) or for plants (Kolmogorov-Smirnov, $D = 0.059$; $P > 0.05$). Both webs showed asymmetry (2009 = 0.48; 2010 = 0.52), nestedness (2009 = 49.14; $P <$

Table 2. The metrics of the interaction network for the association of ant-extrafloral-bearing trees.

Network metrics	Network 2009	Network 2010
Number of ant species	26	29
Number of associations	1195	1271
Degree of plant species (Average degree ± SE)	4.84±0.62	4.31±0.579
Degree of ant species (Average degree ± SE)	14±0.745	13.88±0.790
Network connectance	53.84%	47.89%
H_2' index	0.104	0.128
Web asymmetry	0.48	0.52
Specialization asymmetry	−0.48	−0.52
Nestedness value (NODF)	49.14; $P<0.001$	45.11; $P<0.001$

Table 3. Results of ant-exclusion experiments.

Variables	df	Mean Square	*F*-ratio	*P*-value
Species	8	1056.642	53.636	0.000*
Stems	1	591.317	30.016	0.000*
Years	1	8472.147	430.056	0.000*
Species and Stems	8	24.820	1.260	0.261
Species and Years	8	60.391	3.066	0.002*
Stems and Years	1	0.000	0.000	0.997
Species, Stems and Years	8	13.696	0.695	0.696
Error	1016	19.700		

Factorial ANOVA comparing the herbivory (% leaf area loss) and variables: plant species, stem type (with and without ants), and year.
* statistical difference.

0.001; and 2010 = 45.11; $P<0.001$), and a low degree of specialization (0.104 and 0.128 for 2009 and 2010, respectively).

The outcomes of interaction

Leaf herbivory varied temporally and according to plant species and treatment (Table 3 and Table S1). Herbivory was higher in 2009 (Figure 2) and ant abundance on plants was higher in 2010 ($t = -2.55$; $P = 0.01$; Figure 3). In general, there was significantly greater herbivory in ant-excluded stems in both years (Figure 2). However, only four plant species (*Q. multiflora* and *S. polyphyllum* in 2009; *Q. multiflora*, *Q. grandiflora* and *O. spectabilis* in 2010) had a lower leaf area loss in stems visited by ants than ant-excluded stems (see Table S2). Independently of stem treatment, leaf area loss in both years was significantly higher in *Q. parviflora* and *C. brasiliense* than in other plant species (Figure 4 and Table S3).

Discussion

The structure and dynamic of the interaction network

The networks studied here, exhibited the same general pattern as other mutualistic networks: nestedness, asymmetry and low specialization (see Table 2 and Figure 1), and this pattern was maintained over time. Our data and those of others [9,20–21,29,42], showed that interaction networks between ants and plants with EFNs are shaped by a few generalist species and by multiple species that are linked to one or a few species. Specialization in this type of interaction is rare [42] however, as we observed, due to occasional ant or plant distribution (differential abundance, capacity of dispersion and colonization), species with low links or only one link are common, and might act, at least temporally, as specialists. Furthermore, seasonal variations might occur on the ant fauna associated with liquid resources (i.e. extrafloral nectar; honeydew) found on plants [41,60], which can

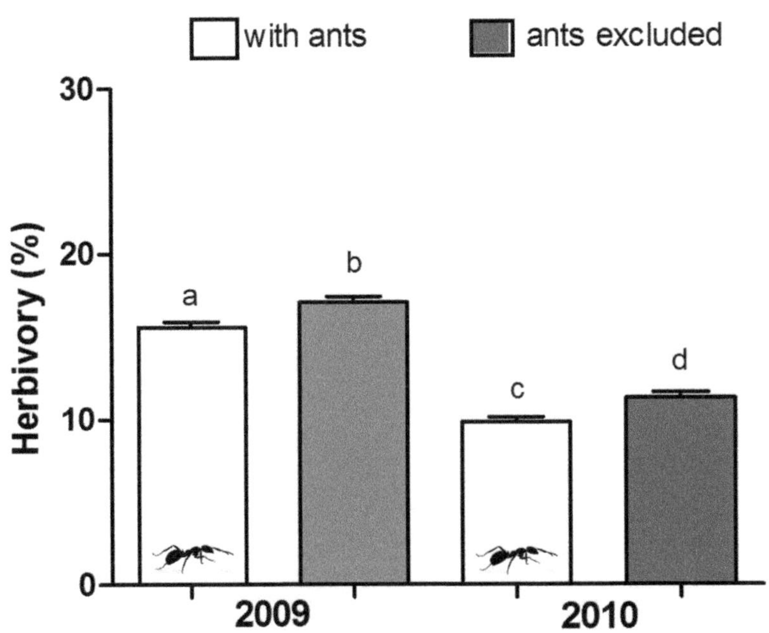

Figure 2. Comparison of herbivory between plants with or without ant exclusion. Different letters represent a statistical difference ($P<0.05$; ANOVA and Tukey post hoc comparison; Table S1). The horizontal line represents the mean, boxes represent standard error and whiskers, the standard deviation.

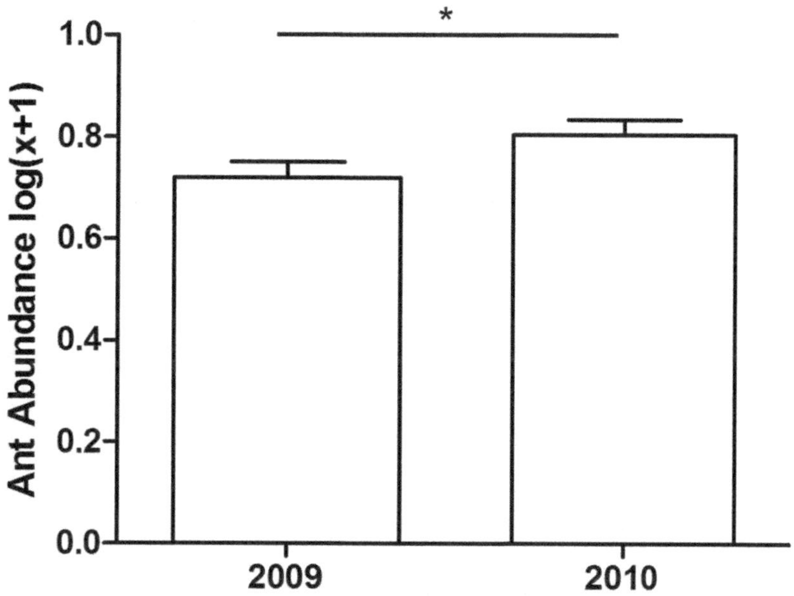

Figure 3. Ant abundance in stems with ants in the core of extrafloral nectaried trees. The asterisks represent statistical difference (paired *t*-test, $P<0.05$). The horizontal line represents the mean, boxes represent standard error and whiskers, the standard deviation.

also explain the observed high amount of a few connected species [20–21,29].

We also showed that not only the degrees of ant species changed over the time, but also those of plant species. A major number of ant species foraging on plants also means an increase in predation strategies against herbivores, what to the plant community means complementary effects (functional complementarity) of ant services (e.g. [60]). In plant-pollinator networks, from the consumer's viewpoint, differences in flowering phenology and/or nutritional variation in floral resources might explain a complementary role of different flower species [61]. Knowing that temporal and phenological variation also occurs in extrafloral nectar production [29] the same might be expected from studies of ant-plant systems.

In our data, the connectance decreased from 2009 to 2010; this was probably due to the effect of the absence in 2010 of one highly connected ant species observed in 2009 (*Crematogaster bruchi*) and the occurrence in 2010 of five different and less connected species in 2010 (Figure 1 and Table 1). Other studies in mutualistic networks [21,62] also demonstrated that the incorporation of new species into the network did not interfere with the web structure, thus, new species interact with generalists and maintain the nested aspect of the network. Therefore, most generalist species in the network (hubs, those with a large number of interactions) become

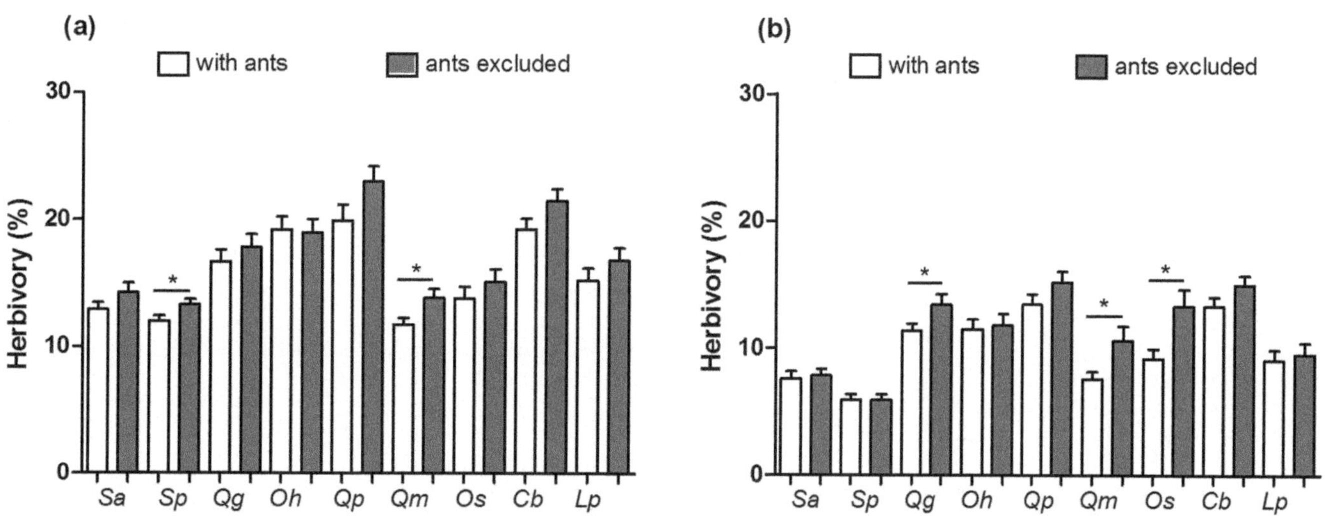

Figure 4. Herbivory between stems with ants or ants excluded in the extrafloral nectaried trees. Abbreviations: (Sa) *Stryphnodendron adstringens*, (Sp) *S. polyphyllum*, (Qg) *Qualea grandiflora*, (Oh) *Ouratea hexasperma*, (Qp) *Q. parviflora*, (Qm) *Q. multiflora*, (Os) *O. spectabilis*, (Cb) *Caryocar brasiliense* and (Lp) *Lafoensia pacari*; (a) 2009 and (b) 2010. The asterisks represent a statistical difference (*t*-test, $P<0.05$). The horizontal line represents the mean, boxes represent standard error and whiskers depict 95% intervals.

responsible for the maintenance of the network properties (see also [9,11,27]. This fact was demonstrated in this study by permanency of the species *C. crassus, C. blandus, C. pusillus, P. gracillis* on top of the network from year to year.

The existence of a few dominant species and many non-dominant species (subordinate and opportunistic species - [63]) in the community, contributes to the increase in the coexistence of these species in plants and, consequently, to the nestedness pattern of the network. Conversely, hierarchically superior species tend to be more abundant and frequent, as the model predicts the dominance hierarchy [47].

Although the default network has been maintained over time, about 50% of plant species and ants had a degree that was unchanged. This can be explained by the variation in the attractiveness of plant species as well as by the dynamics of the ant community. Several authors have observed variation in EFN-bearing plant attractiveness to ants and have related this variation to soil quality [26], plant biology and herbivory level [64], and abiotic factors [25].

Although variation in the degree of ant and plant species between the years studied was observed, the asymmetry, nestedness and specialization degree of the network maintained similar rates (Table 2). This result reinforces the idea that the general pattern of the network between ants and plants with EFNs does not vary over time, despite changes in the internal network (e.g. species degree and connectance) [29,65]. The degree of specialization of the two networks studied was similar to that seen in other networks involving interactions between ants and plants with EFNs (see [18]). Despite the existence of interspecific competition among ants for the best resources (plants that produce nectar with higher quality), each ant visits such plants only for short periods of time, and does not necessarily nest on the plants, making interactions between ants and EFN-bearing plants non-specialized (a low H_2') [18].

The outcomes of interaction

Most ant-plant experimental manipulative studies have historically been directed toward demonstrating ant benefits to plants and were carried out in systems exhibiting a unique producer (reviewed by [31]). Only recently has the viewpoint of ant fitness adequately been explored [33,66], and the present study is the first to experimentally investigate the beneficial effect of ants on the ant-plant interaction network. Our results indicate that some general patterns viewed in singular systems are also present in a network subset. For example: a) the variation over time in the outcomes of benefits observed on the plant community (Figure 2) also occur when single species are analyzed (Figure 4; e.g. [34–35]); b) the assembly of ant species linked to plant species varied between years in a network (Figure 1) as in single ant-plant associations (e.g. [60]). We must consider that this depends on ant species features such as abundance (Figure 3), dispersion capacity, body size, aggressive behavior (e.g. [36,41,67]) and also on herbivore defensive strategies [37,68].

Several authors have associated the outcome of mutualism between ants and plants with the behavior of the ant species involved [33,36,67]. As pointed out by other studies in the Cerrado biome [68–70], *Camponotus* species are the main defenders in these ant–plant interactions. Species of this genus are very agile and aggressive, feeding not only on EFNs, but also preying upon small arthropods [45]. Some ant species, such as those in the genus *Cephalotes*, despite being abundant in tropical systems, do not benefit the plants and are considered parasitic ant-plant systems [38,68].

Differences in the use of arboreal sugars by dominant and non-dominant ants (discussed in the previous section) suggest that arboreal sugar composition might affect ant assemblage composition and thus the range of ecological functions performed by those ants [71]. The variation in the attractiveness of plant species for the ants together with the physical and chemical defenses of plants, might explain the observed variation in the percentage herbivory between species recorded over time.

As mentioned earlier, attractiveness is related to biotic and abiotic factors, which vary temporally. This variation was evident in both degrees of plant species and herbivory between years. According to Del-Claro and Oliveira [60], the temporal variation in biotic and abiotic conditions of the Cerrado environment can greatly influence the results of mutualistic interactions. In this context, we showed that ant presence had a strong positive effect on the herbivory of the plant community. In this facultative mutualistic interaction (sensu [72]), several ant species are involved in the relationship and the outcomes of the interaction can be related to several factors, including the behavior of associated ant species [38], as well as temporal variation [73].

Conclusion

In this study, we showed that ant presence had a strong positive effect on the herbivory of the plant community during the two periods evaluated. However, this effect was smaller in 2009 than in 2010. Additionally, in 2009 we observed lesser ant abundance and degree of specialization (H_2' index) compared to 2010. These results show the importance of greater abundance and specialization of ants in ant-plant protection system.

In conclusion, the internal changes in the ant-plant network influenced the variation of the outcomes of these interactions to the plant species. We also showed that ant-plant interactions can be manipulated as in single systems, as subsets of networks, which provide a more general and broader vision of the processes occurring in the entire community. We suggest that studies such as this one should be encouraged, because they can increase knowledge concerning the importance of spatial and temporal variation in ecology of interactions.

Supporting Information

Table S1 Tukey test comparing the leaf area loss between stems with and without ants for the years 2009 and 2010.

Table S2 Comparison of leaf area loss between stems with and without ants for all plant species separately in 2009 and 2010.

Table S3 Tukey test comparing the leaf area loss of plant species between stems with and without ants over 2009 and 2010.

Acknowledgments

We wish to thank the CCPIU for the use of the Cerrado reserve, R. Feitosa for his support in the identification of ant-species, and H.M. Torezan-Silingardi, N. Leiner, J.C. Santos, E. Tizo-Pedroso, W.D. Fernandes, M.O. Gonzaga, N. Blüthgen, and A. Aguirre for their comments and suggestions on the manuscript.

Author Contributions

Conceived and designed the experiments: DL KD-C. Performed the experiments: DL KD-C. Analyzed the data: DL KD-C. Contributed reagents/materials/analysis tools: DL KD-C. Contributed to the writing of the manuscript: DL KD-C.

References

1. Ings TC, Montoya JM, Bascompte J, Blüthgen N, Brown L, et al. (2009) Ecological networks – beyond food webs. J Anim Ecol 78: 253–269. DOI: 10.1111/j.1365-2656.2008.01460.x
2. Dáttilo W, Sánchez-Galván I, Lange D, Del-Claro K, Rico-Gray V (2014) Importance of interaction frequency in analysis of ant-plant networks in tropical environments. J Trop Ecol 30: 165–168. DOI: 10.1017/S0266467413000813
3. Thompson JN (2013) Relentless Evolution. University of Chicago Press, Chicago. 499 p.
4. Bascompte J, Jordano P (2007) Plant–animal mutualistic networks: the architecture of biodiversity. Annu Rev Ecol Evol S 38: 567–593. DOI: 10.1146/annurev.ecolsys.38.091206.095818
5. Vázquez DP, Blüthgen N, Cagnolo L, Chacoff NP (2009) Uniting pattern and process in plant–animal mutualistic networks: a review. Ann Bot 103: 1445–1457. DOI: 10.1093/aob/mcp057
6. Hagen M, Kissling WD, Rasmussen C, de Aguiar MAM, Brown LE, et al. (2012) Biodiversity, species interactions and ecological networks in a fragmented world. Adv Ecol Res 46: 89–210. DOI: 10.1016/B978-0-12-396992-7.00002-2
7. Dormann CF, Fründ J, Blüthgen N, Gruber B (2009) Indices, graphs and null models: analyzing bipartite ecological networks. The Open Ecol J 2: 7–24.
8. Montoya JM, Pimm SL, Solé RV (2006) Ecological networks and their fragility. Nature 442: 259–264. DOI: 10.1038/nature04927
9. Bastolla U, Fortuna MA, Pascual-Garcia A, Ferrera A, Luque B, et al. (2009) The architecture of mutualistic networks minimizes competition and increases biodiversity. Nature 458: 1018–1021. DOI: 10.1038/nature07950
10. Thébault E, Fontaine C (2010) Stability of ecological communities and the architecture of mutualistic and trophic networks. Science 329: 853–856. DOI: 10.1126/science.1188321
11. Mello MAR, Santos GMM, Mechi MR, Hermes MG (2011) High generalization in flower-visiting networks of social wasps. Acta Oecol 167: 131–140. DOI: 10.1016/j.actao.2010.11.004
12. Saavedra S, Stouffer DB, Uzzi B, Bascompte J (2011) Strong contributors to network persistence are the most vulnerable to extinction. Nature 478: 233–235. DOI: 10.1038/nature10433
13. Benadi G, Blüthgen N, Hovestadta T, Poethke HJ (2013) Contrasting specialization–stability relationships in plant–animal mutualistic systems. Ecol Model 258: 65–73. DOI: 10.1016/j.ecolmodel.2013.03.002
14. Del-Claro K, Torezan-Silingardi HM (2009) Insect-plant interactions: new pathways to a better comprehension of ecological communities in neotropical savannas. Neotropic Entomol 38(2): 159–164. DOI: 10.1590/S1519-566X2009000200001
15. Del-Claro K, Stefani V, Lange D, Vilela AA, Nahas L, et al. (2013) The importance of natural history studies for a better comprehension of animal-plant interactions networks. Bioscence J 29(2): 439–448.
16. Guimarães PR, Rico-Gray V, Reis SF, Thompson JN (2006) Asymmetries in specialization in ant–plant mutualistic networks. Proc R Soc B 273: 2041–2047. DOI: 10.1098/rspb.2006.3548
17. Guimarães PR, Rico-Gray V, Oliveira PS, Izzo TJ, dos Reis SF, et al. (2007) Interaction intimacy affects structure and coevolutionary dynamics in mutualistic networks. Curr Biol 17: 1–7. DOI: 10.1016/j.cub.2007.09.059
18. Blüthgen N, Menzel F, Hovestadt T, Fiala B, Blüthgen N (2007) Specialization, constraints, and conflicting interests in mutualistic networks. Curr Biol 17: 341–346. DOI: 10.1016/j.cub.2006.12.039
19. Chamberlain SA, Holland JN (2008) Density-mediated, context-dependent consumer–resource interactions between ants and extrafloral nectar plants. Ecology 89: 1364–1374. DOI: 10.1890/07-1139.1
20. Chamberlain SA, Kilpatrick JR, Holland JN (2010) Do extrafloral nectar resources, species abundances and body sizes contribute to the structure of ant-plant mutualistic networks? Oecologia 164: 741–750. DOI: 10.1007/s00442-010-1673-6
21. Díaz-Castelazo C, Guimarães PR, Jordano P, Thompson JN, Marquis RJ, et al. (2010) Changes of a mutualistic network over time: reanalysis over a 10-year period. Ecology 91(3): 793–801.
22. Díaz-Castelazo C, Sánchez-Galván IR, Guimarães PR, Raimundo RLG, Rico-Gray V (2013) Long-term temporal variation in the organization of an ant–plant network. Ann Bot 111: 1285–1293. DOI: 10.1093/aob/mct071
23. Sugiura S (2010) Species interactions-area relationships: biological invasions and network structure in relation to island area. Proc Roy Soc February: 1–9.
24. Holland JN, Chamberlain SA, Miller TEX (2011) Consequences of ants and extrafloral nectar for a pollinating seed-consuming mutualism: ant satiation, floral distraction, or plant defense? Oikos 120: 381–388. DOI: 10.1111/j.1600-0706.2010.18958.x
25. Rico-Gray V, Díaz-Castelazo C, Ramírez-Hernández A, Guimaraes PR Jr, Holland JN (2012) Abiotic factors shape temporal variation in the structure of an ant–plant network. Arthropod-Plant Interac 6(2): 289–295.
26. Dáttilo W, Rico-Gray V, Rodrigues DJ, Izzo T (2013) Soil and vegetation features determine the nested pattern of ant–plant networks in a tropical rainforest. Ecol Entomol 38(4): 74–380. DOI: 10.1111/een.12029
27. Dáttilo W, Guimarães PR, Izzo T (2013) Spatial structure of ant-plant mutualistic networks. Oikos 122(11): 1643–1648. DOI: 10.1111/j.1600-0706.2013.00562.x
28. Dáttilo W, Fagundes R, Gurka CAQ, Silva MSA, Vieira MCL, et al. (2014) Individual-based ant-plant networks: diurnal-nocturnal structure and species-area relationship. PLoS One (in press).
29. Lange D, Dáttilo W, Del-Claro K (2013) Influence of extrafloral nectary phenology on ant-plant mutualistic networks in a Neotropical Savanna. Ecol Entomol 38: 463–469. DOI: 10.1111/een.12036
30. González-Teuber M, Heil M (2009) Nectar chemistry is tailored for both attraction of mutualists and protection from exploiters. Plant Signal & Behav 4: 809–813.
31. Rosumek FB, Silveira FAO, Neves FS, Barbosa NP, Diniz L, et al. (2009) Ants on plants: a meta-analysis of the role of ants as plant biotic defenses. Oecologia 160: 537–549. DOI: 10.1007/s00442-009-1309-x
32. Davidson DW, Cook SC, Snelling RR (2003) Explaining the abundance of ants in lowland tropical rainforest canopies. Science 300: 969–972. DOI: 10.1126/science.1082074
33. Byk J, Del-Claro K (2011) Ant-plant interaction in the Neotropical savanna: direct beneficial effects of extrafloral nectar on ant colony fitness. Popul Ecol 53(2): 327–332. DOI: 10.1007/s10144-010-0240-7
34. Rico-Gray V (1993) Use of plant-derived food resources by ants in the dry tropical lowlands of coastal Veracruz, Mexico. Biotropica 25: 301–315.
35. Bronstein JL, Alarcón R, Geber M (2006) The evolution of plant-insect mutualisms. New Phytol 172: 412–428. DOI: 10.1111/j.1469-8137.2006.01864.x
36. Davidson DW, McKey D (1993) The evolutionary ecology of symbiotic ant-plant relationship. J Hymenopt Res 2: 13–83.
37. Floren A, Biun A, Linsenmair EK (2002) Arboreal ants as key predators in tropical lowland forest trees. Oecologia 131: 137–144. DOI: 10.1007/s00442-002-0874-z
38. Byk J, Del-Claro K (2010) Nectar- and pollen–gathering *Cephalotes* ants provide no protection against herbivory: a new manipulative experiment to test ant protective capabilities. Acta Ethol 13: 33–38. DOI: 10.1007/s10211-010-0071-8
39. Horvitz CC, Schemske DW (1984) Effects of ants and attended herbivore on seed production of a neotropical herb. Ecology 65: 1369–1378. DOI: 10.2307/1939117
40. Rashbrook VK, Compton SG, Lawton JH (1992) Ant herbivore interactions: reasons for the absence of benefits to a fern with foliar nectaries. Ecology 73(6): 2167–2174. DOI: 10.2307/1941464
41. Blüthgen N, Verhaagh M, Goitía W, Jaffé K, Morawetz W, et al. (2000) How plants shape the ant community in the Amazonian rainforest canopy: the key role of extrafloral nectaries and homopteran honeydew. Oecologia 125: 229–240. DOI: 10.1007/2004420000449
42. Rico-Gray V, Oliveira PS (2007) The ecology and evolution of ant-plant interactions. The University of Chicago Press, Chicago. 331 p.
43. Bentley BL (1977) The protective function of ants visiting the extrafloral nectaries of *Bixa orellana* L. (Bixaceae). J Ecol 65: 27–38.
44. Suzuki N, Ogura K, Katayama N (2004) Efficiency of herbivore exclusion by ants attracted to aphids on the vetch *Vicia angustifollia* L. (Leguminosae). Ecol Rev 19: 275–282. DOI: 10.1111/j.1440-1703.2004.00636.x
45. Nascimento EA, Del-Claro K (2010) Ant visitation to extrafloral nectaries decreases herbivory and increases fruit set in *Chamaecrista debilis* (Fabaceae) in a Neotropical savanna. Flora 205: 754–756. DOI: 10.1016/j.flora.2009.12.040
46. Rico-Gray V, Palacios-Rios M, Garcia-Franco JG, Mackay WP (1998) Richness and seasonal variation of ant– plant associations mediated by plant-derived food resources in the semiarid Zapotitlan Valley, Mexico. Am Midl Nat 140: 21–26. DOI: 10.1674/0003-0031(1998)140[0021:RASVOA]2.0.CO;2
47. Blüthgen N, Stork NE, Fiedle K (2004) Bottom-up control and co-occurrence in complex communities: honeydew and nectar determine rainforest ant mosaic. Oikos 106: 344–358. DOI: 10.1111/j.0030-1299.2004.12687.x
48. Oliveira PS, Marquis RJ (2002) The Cerrados of Brazil. Ecology and natural history of a Neotropical Savanna. Columbia University Press, New York. 398 p.
49. Oliveira PS, Leitão-Filho HF (1987) Extrafloral nectaries: Their taxonomic distribution and abundance in the woody flora of cerrado vegetation in Southeast Brazil. Biotropica 19: 140–148.
50. Appolinario V, Shiavini I (2002) Levantamento fitossociológico de espécies arbóreas de cerrado (stricto sensu) em Uberlândia, Minas Gerais. Bol do Herbário Ezechias Paulo Heringer 10: 57–75.
51. Machado SR, Morellato LPC, Sajo MG, Oliveira PS (2008) Morphological patterns of extrafloral nectaries in woody plant species of the Brazilian cerrado. Plant Biol 10: 660–673. DOI: 10.1111/j.1438-8677.2008.00068.x.

52. Dirzo R, Domingues C (1995) Plant-animal interactions in Mesoamerican tropical dry forest. In: Bullock SH, Money SH, Medina HA, editors. Seasonally dry tropical forests. Cambridge University Press, Cambridge. pp. 304–325.
53. Jordano P (1987) Patterns of mutualistic interactions in pollination and seed dispersal: connectance, dependence, and coevolution. Amer Nat 129: 657–677.
54. Dunne JA (2006) The Network Structure of Food Webs. In: Pascual M, Dunne JA, editors. Ecological Networks: linking structure to dynamic in food webs. Oxford University Press, Oxford, UK. pp. 325–347.
55. Blüthgen N, Menzel F, Blüthgen N (2006) Measuring specialization in species interactions networks. BMC. Ecology 6: 9–12. DOI: 10.1186/1472-6785-6-9
56. Almeida-Neto M, Guimarães PR, Guimarães P, Loyola RD, Ulrich W (2008) A consistent metric for nestedness analysis in ecological systems: reconciling concept and measurement. Oikos 117: 1227–1239. DOI: 10.1111/j.2008.0030-1299.16644.x
57. Vázquez DP, Melián CJ, Williams NM, Blüthgen N, Krasnov BR, et al. (2007) Species abundance and asymmetric interaction strength in ecological networks. Oikos 16: 1120–1127. DOI: 10.1111/j.0030-1299.2007.15828.x
58. Guimarães PR, Guimarães P (2006) Improving the analyses of nestedness for large sets of matrices. Environ Model Softw 21: 1512–1513. DOI: 10.1016/j.envsoft.2006.04.002
59. Batagelj V, Mrvar A (1998) Pajek - Program for Large Network Analysis. Connections 21(2): 47–57.
60. Del-Claro K, Oliveira PS (2000) Conditional outcomes in a neotropical treehopper-ant association: temporal and species-specific variation in ant protection and homopteran fecundity. Oecologia 124: 156–165. DOI: 10.1007/s004420050002
61. Blüthgen N, Klein AM (2011) Functional complementarity and specialization: Why biodiversity is important in plant-pollinator interactions. Basic Appl Ecol 12(4): 282–291. DOI: 10.1016/j.baae.2010.11.001
62. Olesen JM, Eskildsen LI, Venkatasamy S (2002) Invasion of pollination networks on oceanic islands: importance of invader complexes and endemic super generalists. Divers Distrib 8: 181–192.
63. Hölldobler B, Wilson EO (1990) The ants. Harvard University Press, Cambridge, MA. 746 p.
64. Heil M, Koch T, Hilpert A, Fiala B, Boland W, et al. (2001) Extrafloral nectar production of the ant-associated plant, *Macaranga tanarius*, is an induced, indirect, defensive response elicited by jasmonic acid. Proc Natl Acad Sci of the USA 98: 1083–1088. DOI: 10.1073/pnas.98.3.1083
65. Jordano P, Bascompte J, Olesen J (2003) Invariant properties in coevolutionary networks of plant–animal interactions. Ecol Lett 6: 69–81. DOI: 10.1046/j.1461-0248.2003.00403.x
66. Lach L, Hobbs ER, Majer EJD (2009) Herbivory-induced extrafloral nectar increases native and invasive ant worker survival. Pop Ecol 51: 237–243. DOI: 10.1007/s10144-008-0132-2
67. Stanton ML, Palmer TM (2011) The high cost of mutualism: effects of four species of East African ant symbionts on their myrmecophyte host tree. Ecology 92(5): 1073–1082. DOI: 10.1890/10-1239.1
68. Sendoya SF, Freitas AVL, Oliveira PS (2009) Egg-laying butterflies distinguish predaceous ants by sight. Am Nat 174: 134–140. DOI: 10.1086/599302
69. Del-Claro K, Berto V, Réu WF (1996) Effect of herbivore deterrence by ants on the fruit set of an extrafloral nectary plant, *Qualea multiflora* (Vochysiaceae). J Trop Ecol (12): 887–892.
70. Oliveira PS, Silva AF da, Martins AB (1987) Ant foraging on extrafloral nectaries of *Qualea grandiflora* (Vochysiaceae) in cerrado vegetation: ants as potential antiherbivore agents. Oecologia 74: 228–230.
71. Gibb H, Cunningham SA (2009) Arboreal sugars affect ecological dominance in ant assemblages across a restoration chronosequence. Insect Soc 56: 405–412. DOI: 10.1007/s00040-009-0038-9
72. Schemske DW (1982) Ecological correlates of a neotropical mutualism: Ant assemblages at *Costus* extrafloral nectaries. Ecology 63: 932–941. DOI: 10.2307/1937233
73. Heil M, McKey D (2003) Protective ant-plant interactions as model systems in ecological and evolutionary research. Ann Rev Ecol Syst 34: 425–453. DOI: 10.1146/annurev.ecolsys.34.011802.132410

2

Overexpression of the PAP1 Transcription Factor Reveals a Complex Regulation of Flavonoid and Phenylpropanoid Metabolism in *Nicotiana tabacum* Plants Attacked by *Spodoptera litura*

Tomoko Mitsunami[1], Masahiro Nishihara[2], Ivan Galis[3], Kabir Md Alamgir[3], Yuko Hojo[3], Kohei Fujita[2], Nobuhiro Sasaki[2], Keichiro Nemoto[4], Tatsuya Sawasaki[4], Gen-ichiro Arimura[1]*

1 Department of Biological Science & Technology, Faculty of Industrial Science & Technology, Tokyo University of Science, Tokyo, Japan, **2** Iwate Biotechnology Research Center, Kitakami, Japan, **3** Institute of Plant Science and Resources, Okayama University, Kurashiki, Japan, **4** Proteo-Science Center, Ehime University, Matsuyama, Japan

Abstract

Anthocyanin pigments and associated flavonoids have demonstrated antioxidant properties and benefits for human health. Consequently, current plant bioengineers have focused on how to modify flavonoid metabolism in plants. Most of that research, however, does not consider the role of natural biotic stresses (e.g., herbivore attack). To understand the influence of herbivore attack on the metabolic engineering of flavonoids, we examined tobacco plants overexpressing the *Arabidopsis PAP1* gene (encoding an MYB transcription factor), which accumulated anthocyanin pigments and other flavonoids/ phenylpropanoids. In comparison to wild-type and control plants, transgenic plants exhibited greater resistance to *Spodoptera litura*. Moreover, herbivory suppressed the PAP1-induced increase of transcripts of flavonoid/phenylpropanoid biosynthetic genes (e.g., *F3H*) and the subsequent accumulation of these genes' metabolites, despite the unaltered *PAP1* mRNA levels after herbivory. The instances of down-regulation were independent of the signaling pathways mediated by defense-related jasmonates but were relevant to the levels of PAP1-induced and herbivory-suppressed transcription factors, An1a and An1b. Although initially *F3H* transcripts were suppressed by herbivory, after the *S. litura* feeding was interrupted, *F3H* transcripts increased. We hypothesize that in transgenic plants responding to herbivory, there is a complex mechanism regulating enriched flavonoid/phenylpropanoid compounds, via biotic stress signals.

Editor: Jin-Song Zhang, Institute of Genetics and Developmental Biology, Chinese Academy of Sciences, China

Funding: This work was financially supported in part by the MEXT Grants for Excellent Graduate Schools program of Kyoto University; the Research Fund of Tokyo University of Science; and a Grant-in-Aid for Scientific Research from the Japan Society for the Promotion of Science to G.A. (No. 24770019). The LC-MS/MS analyses were supported by the Japan Advanced Plant Science Network. The funders had no role in study design, data collection and analysis, decision to publish, or preparation of the manuscript.

Competing Interests: The authors have declared that no competing interests exist.

* Email: garimura@rs.tus.ac.jp

Introduction

The production of specialized metabolites in plants has attracted great interest as a result of the successful development of mechanisms that enable metabolic enzymes to be produced [1–3]. To synchronously activate metabolic genes, plants use a large variety of transcription factors. These factors may control multiple genes or orchestrate entire metabolic pathways or their specific branches. PRODUCTION OF ANTHOCYANIN PIGMENT1 (PAP1) is a typical R2R3 MYB-type transcription factor in *Arabidopsis thaliana* that is able to ectopically activate the entire array of genes involved in the biosynthesis of phenolic acids and flavonoid compounds (including anthocyanins) in the flavonoid biosynthetic pathway in several plant species, including tobacco, salvia, petunia and rose [4–8]. The overexpression of *Arabidopsis PAP1* in transgenic tobacco plants under the control of a constitutive promoter led to the induction of anthocyanins in the leaves, flowers, stems and roots; the deep red or purple appearance of the pants parts was evidence [5]. Moreover, *PAP1*-overexpressing petunia flowers have been shown to increase not only levels of anthocyanins but also emissions of floral volatiles (benzenoids) [7]. Gene expression profiling and measurements of the levels of pathway intermediates suggest that both increased metabolic flux and the transcriptional activation of scent and color genes underlie the enhancement of petunia flower color and scent production [7]. The coordinated regulation of metabolic steps within or between pathways involved in vital plant functions – for instance, pathways involved in floral display determining plant-pollinator interactions – provides plants with a clear advantage. Indeed, bees and humans can discriminate between the floral scents of *PAP1*-overexpressing and control rose flowers [6]. Therefore, PAP1 overexpression may find applications in horticulture and human health.

Manipulating flavonoid metabolism, especially in response to biotic or abiotic stresses, may have unpredicted side effects, as these compounds belong to the largest group of specialized metabolites produced by plants [9]. Flavonoids serve as essential components of a number of structural polymers that provide protection; lignin, for instance, due to its antioxidant and free radical scavenging properties, protects plants from ultraviolet light and defends plants against herbivores and pathogens [10]. Anthocyanin accumulation in *Arabidopsis* is enhanced in response to a number of environmental stress factors, such as cold, drought, pathogen attack and nutrient depletion [11]. Moreover, it appears that transgenic tobacco expresses structural genes and regulatory genes involved in flavonoid biosynthesis, and these genes defend the plant against herbivores [12,13]. Therefore, *PAP1*-overexpressing plants that typically accumulate large amounts of flavonoids might possess increased plant resistance to biotic stresses, including herbivores, but to date no practical evidence of this prediction has been provided. Whether the *PAP1* transgene interferes with endogenous TFs and processes involved in plants' defense and development has not been extensively investigated.

A separate branch of flavonoid catabolism controlled by other MYB transcription factors (e.g., MYBJS1) was previously shown to control the accumulation of anti-herbivore compounds, phenolamides (PAs), in *Nicotiana* species [14–17]. In addition, CoA-activated phenolic acids with polyamines in PA production is controlled by the MYBJS1 homologue MYB8 in a herbivory-inducible fashion in wild tobacco, *N. attenuata* [18]. MYB8 (and MYBJS1) also likely contributes to lignin accumulation by targeting the control of the hydroxycinnamoyl-CoA:shikimate/quinate hydroxycinnamoyl transferase (HCT) gene in tobacco [19–21].

Because *PAP1*-overexpressing plants constitutively produce large amount of anthocyanins, we asked if the accumulation of other inducible flavonoids and phenylpropanoids was affected. PAP1 and MYBJS1 seem to share common gene targets in the pathways they regulate, and so these two transcription factors are likely to engage in complex interactions when *PAP1* is introduced into plants under a constitutive promoter and also, in response to herbivory, when other MYBs are additionally activated upon. It is likely that when *PAP1*-overexpressing plants are damaged by herbivores, the dynamic activation of the metabolic flow via other MYB members competes with activated coumaric acid units used in the constitutive synthesis of anthocyanins by PAP1-promoted enzymes. Both transcriptional and post-transcriptional regulations of PAP1 have been shown to control anthocyanin levels in ubiquitous light/dark rhythms [22]. Moreover, when anthocyanin production is regulated by independent PAP-transcription and environmental stress, anthocyanin levels are reduced [23]. Given that the stress of herbivory could affect PAP1 function in *PAP1*-overexpressing plants, we asked how efficiently transgenes such as PAP1 could be expected to perform under natural, uncontrolled environmental conditions.

In the current study, we found that the defensive properties of *PAP1*-overexpressing plants increased upon herbivore attack, and addressed the issue of metabolic fluxes and transcriptional regulation in the flavonoid/phenylpropanoid pathways that are co-regulated by powerful PAP1-activators and herbivory. Since jasmonic acid (JA) and its derivatives (referred as jasmonate [JAs]) coordinate the expression of regulatory and structural genes in the flavonoid/phenylpropanoid pathways under stress, we also assessed the role of JAs (and other phytohormones) in the herbivory- and PAP1-mediated regulation of phenylpropanoid metabolism. Finally, we discuss the implications of enhancing the production of specialized metabolites in plants under natural environmental conditions.

Materials and Methods

Production of transgenic tobacco plants

The full-length coding region of *Arabidopsis PAP1* (At1g56650) cDNA was inserted into the binary vector pSMABR35SsGFP by replacing the GFP reporter gene of the vector [24]. The resulting plasmid, pSMABR-35SPAP1, was transformed into *Agrobacterium tumefaciens* strain EHA101 by electroporation. *PAP1*-overexpressing tobacco (*N. tabacum* cv. SR1) was produced by *Agrobacterium*-mediated transformation as described previously [24]. After rooting and acclimatization, the regenerated plants were grown in a controlled greenhouse to set seeds. Nine lines of transgenic T_1 seeds were tested for germination on Murashige and Skoog medium supplemented with 5 mg l^{-1} bialaphos at 25°C under a 16-h photoperiod at a light intensity of ca. 30 μE m^{-2} s^{-1}. T_2 seeds harvested from each T_1 individual plant that showed ca. 3:1 segregation ratio were tested for bialaphos-resistance again. Finally, T_2 and T_3 homozygous plant lines were used for further analyses. A homozygous tobacco T_2 line transformed with the binary plasmid pSMABR-35SpGUS (in which the reporter *GFP* gene of pSMABR35SsGFP was replaced by β-glucuronidase [GUS] gene), expressing bialaphos resistance [bar] and GUS genes, was used as a control.

Three representative *PAP1*-overexpressing plants (*PAP1* lines), constitutively expressing *Arabidopsis PAP1* under the control of the cauliflower mosaic virus (CaMV) 35S promoter, were used. The *PAP1* lines exhibited substantial expression of *PAP1* based on their increased pigmentation throughout their development, in comparison to wild-type (WT) and transgenic plants expressing *uid*A (coding for the GUS reporter gene and serving as a control) (Figures S1 and S2). Except for having increased pigmentation, *PAP1* lines developed and matured in a similar manner to their control plants, as shown in previous studies [5,6] (Figure S2).

Plants and caterpillars

Tobacco plants were grown in plastic pots in a growth chamber at 25°C (16 h photoperiod at a light intensity of 80 μE m^{-2} s^{-1}) for about 5 weeks. Eggs of *Spodoptera litura* (Fabricius) (Lepidoptera: Noctuidae) were obtained from Sumika Technoservice Co. Ltd. (Takarazuka, Japan), and the hatched larvae were reared on artificial diet (Insecta LF, Nihon Nosan Kogyo Ltd., Tokyo, Japan) in the laboratory at 25°C.

Chemical and herbivore treatments and leaf sample preparation

For chemical treatment, methyl jasmonate (MeJA, 0.1 mM, Wako Pure Chemical Industrials, Ltd., Osaka, Japan) in 2 mL of 0.1% ethanol solution was sprayed onto intact plants in plastic pots. Ethanol solution (0.1%) was used as the control. For herbivore treatment, four third-instar larvae were incubated with a whole tobacco plant in a pot (Figure S3). All herbivory bioassays were carried out in a climate-controlled chamber at 25°C (16 h photoperiod). For the determination of phytochemical concentrations and RNA extractions, leaf tissue samples were harvested into the liquid nitrogen, flash-frozen, and kept at −80°C prior to extraction.

Performance of *S. litura* larvae

Larvae were reared on artificial diet until the assay started, as described above. Second-instar *S. litura* larvae were weighed prior to release onto a potted plant (ranging from 0.7 to 1.2 mg) and released one at a time. Each larva was allowed to move on the whole plant in a climate-controlled room at 25°C under a 16-h photoperiod. Twenty-six to thirty-two larvae were analyzed for each line.

Analysis of phenolic compounds

Phenylpropanoids and flavonoids in tobacco leaves were determined using high-performance liquid chromatography (HPLC) with a slight modification of the method described by Galis et al. (2013) and Matsuba et al. (2010) [25,26]. Replicated analyses were conducted with five independent leaf samples.

For anthocyanin (cyanidin-3-rutinoside) determination, the ground leaf samples (300 mg each) were further homogenized in the liquid nitrogen and extracted with 3 ml of 80% methanol containing 1% trifluoroacetic acid (TFA) at 4°C overnight. After passing through a 0.22 μm syringe filter (Millipore, Billerica, MA, USA), filtrates (10 μl) were subjected to an HPLC (L-6320 Intelligent pump, L-4200 UV-VIS detector, AS-4010 Intelligent Autosampler, L-5025 column oven, Hitachi, Tokyo, Japan) equipped with a COSMOSIL $5C_{18}$-MS-II packed column (4.6 mm×50 mm; Nacalai Tesque, Kyoto, Japan). The separation was performed by the linear gradient elution of 10–70% (v/v) methanol in 1.5% aqueous phosphoric acid for 5 min at a flow rate of 1.5 ml min^{-1} at 35°C. The elution profiles were monitored at 528 nm and recorded using the Chromato-Pro (Runtime Instruments, Kanagawa, Japan). The authentic compound used for quantification was cyanidin-3-rutinoside (Funakoshi Co. Ltd., Tokyo, Japan).

To determine phenylpropanoids (chlorogenic acid) and flavonoids (kaempferol 3-*O*-rutinoside and rutin), the ground leaf samples (100 mg each) were briefly homogenized in the liquid nitrogen at room temperature in 0.75 ml of 84 mM acetate buffer (adjusted to pH 4.8 with ammonia) in 40% methanol (v/v) using a FastPrep FP120 instrument (Thermo Savant, Thermo Fisher Scientific, Pittsburgh, PA, USA) in the presence of 2.3-mm diameter zirconia beads. Samples were centrifuged for 20 min at 16,100 g, 4°C, and supernatants were collected. Pellets were re-extracted in 0.75 ml buffer containing 80% (v/v) methanol by shaking at RT for 15 min and centrifuged as before; supernatants were pooled with the first extracts. Aliquots (20 μl) of each sample were analyzed on a Shimadzu Prominence HPLC system equipped with a COSMOSIL $5C_{18}$-MS-II packed column (4.6 mm×50 mm; Nacalai Tesque, Kyoto, Japan) in a gradient setup: solvent A (0.1% formic acid, 0.1% ammonia in water, pH 3.6) and solvent B (methanol); time (min)/B (%) gradient conditions: 0/2, 10/80, 15/80, 16/95, 21/95, 24/2, followed by a 6-min equilibration step. Flow rate was set to 1 ml/min and the column was maintained at 35°C. Chromatographic peaks of chlorogenic acid isomers and rutin were detected by an SPD-M20A diode array detector and compared with external calibration curves of authentic chlorogenic acid (Sigma-Aldrich) and rutin (Ishizu Pharmaceuticals Co. Ltd., Osaka, Japan) standards, respectively. Kaempferol 3-*O*-rutinoside content was estimated based on rutin external calibration.

Identification of phenolic acid and flavonoid compounds

Leaf extracts of *PAP1* lines originally prepared for HPLC analyses were used to identify the major UV-absorbing peaks in chromatograms. A 20-μl aliquot of leaf sample was first separated on HPLC as above, and peak fractions of major UV-absorbing compounds were collected using an automated fraction collector (FRC-10A, Shimadzu, Kyoto, Japan). Fractions from 10 runs were pooled, evaporated, re-dissolved in 70% methanol in water (v/v) and spun shortly before analysis. A 10-μl aliquot of each fraction was subjected to measurement on a triple quadrupole liquid chromatography-tandem mass spectrometry (MS) system (LC-MS/MS 6410, Agilent Technologies, Santa Clara, CA, USA) equipped with a Zorbax SB-C18 column (2.1 mm id×50 mm, (1.8 μm), Agilent Technologies), maintained at 40°C and coupled to an MS detector operating in positive scan mode (m/z 100–1000; fragmentor at 135 V). Samples were passed through a UV-PDA detector before entering the MS electrospray ion source for reference and later identification of the peaks. Solvent A (0.1% formic acid in water) and solvent B (0.1% formic acid in acetonitrile) were used in time (min)/B (%) gradient mode: 0/5, 0.5/5, 2/40, 6/40, 10/95, 15/95, 16/5, 20/5 at a constant flow rate of 0.4 ml/min. The mass spectra corresponding to UV-absorbing peaks in the chromatograms were analyzed and used to identify each compound. Only the rutin peak could be assigned against an authentic rutin standard (Ishizu Pharmaceutical Co., Ltd., Osaka, Japan), while the remaining peaks were putatively assigned based on their m/z, fragmentation properties and characteristic UV absorbance spectra.

Phytohormone analysis

Leaf hormone concentrations were determined according to the method described by Fukumoto et al. (2013) [27]. Approximately 100 mg of the ground leaf samples was further homogenized in liquid nitrogen and extracted in 1 ml of ethylacetate spiked with internal standards (IS): 25 ng d3-JA; 5 ng d3-JA-Ile; 10 ng d6-ABA and 10 ng d4-SA. After brief homogenization in a FastPrep instrument as above, samples were centrifuged at 4°C, 20 min at 16,100 *g*. Supernatants were stored and pellets re-extracted in 0.5 ml of ethylacetate without IS. After centrifugation, pooled supernatants were evaporated under reduced pressure in a rotary vacuum concentrator and re-suspended in 250 μl of 70% methanol/water (v/v). Ten-microliter aliquots were analyzed on a triple quadrupole LC-MS/MS 6410 equipped with a Zorbax SB-C18 chromatographic column [2.1 mm id×50 mm, (1.8 μm), Agilent Technologies] protected by a narrow bore guard column Zorbax SB-C8 [2.1 mm id×12.5 mm, (5 μm)]. Samples were chromatographically resolved using solvents A [0.1% formic acid in water (v/v)] and B [0.1% formic acid in acetonitrile (v/v)] used in the following gradient: time (min)/B (%) 0/15, 4.5/98, 12/98, 12.1/15, 18/15. Flow rate was set to 0.4 ml/min and column temperature maintained at 40°C. Mass transitions for each compound [hormone/Q1 precursor ion (m/z)/Q3 product ion (m/z)] were detected in the ESI-negative mode of the mass spectrometer: JA/209/59; JA-Ile/322/130; ABA/263/153; SA/137/93; d3-JA/212/59; d3-JA-Ile/325/130; d6-ABA/269/159; d4-SA/141/97. Hormone amounts were calculated from the ratio of endogenous hormone peak and known amount of IS spike, and adjusted to the exact fresh mass (FM) of the sample used for extraction. Replicated analyses were conducted with five independent leaf samples.

cDNA synthesis and quantitative PCR

Approximately 100 mg of the ground leaf samples was homogenized in liquid nitrogen, and total RNA was isolated from leaf tissues using NucleoSpin RNA Plant (Machery-Nargel, Düren, Germany) following the manufacturer's protocol. First-strand cDNA was synthesized using a ReverTra Ace qPCR RT Master mix (Toyobo, Osaka, Japan) and 0.5 μg of total RNA at 37°C for 15 min. Real-time PCR was performed on a Light Cycler Nano system (Roche Applied Science, Indianapolis, IN, USA) using FastStart Essential DNA Green Master (Roche Applied Science) and gene-specific primers (Table S1). The following protocol was used: initial polymerase activation: 60 s at 95°C; 40 cycles of 15 s at 95°C and 45 s at 60°C; and then melting curve analysis preset by the instrument. Relative transcript abundances were determined after normalization of raw signals with housekeeping transcript abundances of *N. tabacum ELONGATION FACTOR 1α* (*NtEF1α*; GenBank D63396). Replicated analyses were conducted with four to five independent leaf samples.

Statistical analysis

All experiments were repeated at least four times, and statistical analyses were performed using a Tukey's HSD test after a one-way ANOVA. A P value<0.05 was considered statistically significant.

Results

Transgenic tobacco plants constitutively expressing *PAP1* genes are better able to defend themselves against herbivores than are non-transgenic tobacco plants

We evaluated the resistance of *PAP1* lines, constitutively expressing *Arabidopsis PAP1*, to the arthropod herbivore *S. litura*. Larvae were applied onto the leaves of potted WT, GUS and *PAP1* lines for 2 days (Figure 1). Larvae on the PAP-5 and PAP-8 lines had markedly lower weight gain compared to those on WT or GUS lines ($P<0.05$). Larvae on another *PAP1* line (PAP-2) exhibited slightly, but not significantly, less weight gain compared to those on WT or GUS lines ($P>0.05$).

Biosynthesis of flavonoids/phenylpropanoids in *PAP1* lines

Since PAP1 has been shown to activate the phenylpropanoid/flavonoid pathway [28], we measured endogenous levels of their metabolites in *PAP1* lines. According to HPLC analysis supported by identification of the most abundant compounds by LC-MS/MS, there was increased accumulation of an anthocyanin compound (cyanidin-3-*O*-rutinoside) as well as of two other flavonoid compounds (kaempferol 3-*O*-rutinoside and rutin) and of phenylpropanoid compounds (chlorogenic acid isomers 1 and 2) in leaves of *PAP1* lines, relative to WT and GUS lines (Figure 2). PAP-8 leaves accumulated all of the above five compounds at the highest levels, unsurprisingly as this line was best able to defend itself against *S. litura* larvae (see Figure 1). Interestingly, *S. litura* infestation substantially diminished the increase of the concentration of these compounds in *PAP1* lines, whereas no significant change (either increase or decreased) of those compounds was observed in both WT and GUS lines (Figure 2).

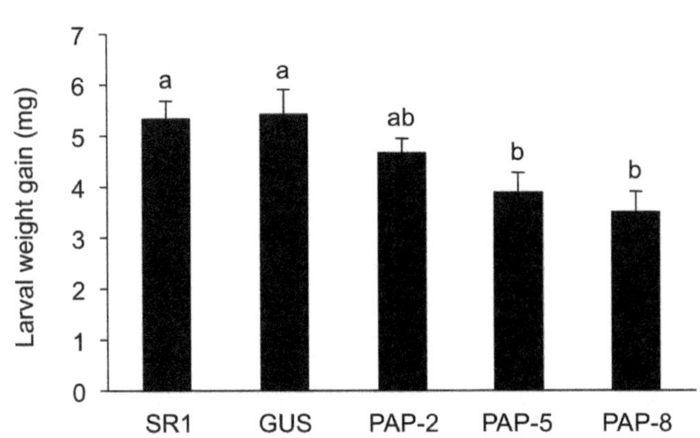

Figure 1. The net body weight that *Spodoptera litura* larvae gained during 4 days on wild-type (WT), GUS control and *PAP1* lines. Data are shown as the mean+standard errors ($n=26$–32). Means followed by different small letters are significantly different ($P<0.05$).

We therefore assessed the transcriptional pattern of flavonoid/phenylpropanoid pathway genes in leaves of *PAP1* lines not infested or infested with *S. litura* for 2 days (Figure 3). In comparison to GUS lines, PAP-8 showed significantly increased transcript levels of eight genes (*PALs*, *CHS*, *CHI*, *F3H*, *F3'H*, *ANS* and *DFR*) involved in phenylpropanoid and flavonoid biosynthesis. The increase of the transcript level was, however, diminished by 34, 60, 29, 39 and 29% of the initial increase in five cases, *CHS*, *CHI*, *F3H*, *ANS* and *DFR*, respectively, in *PAP1* lines in response to herbivory. *PAP1* was expressed similarly between uninfested and infested PAP-8 leaves. Moreover, we also determined the transcript levels of four genes responsible for producing endogenous transcription factors: MYBJS1 (another MYB transcription factor sharing transcriptional targets with PAP1 in the phenylpropanoid pathway [18]), An2 (PAP1 homologue MYB transcription factor [29]), and An1a and An1b (NtAn2-interacting, basic helix-loop-helix (bHLH) transcription factors [30]). An1a, An1b and An2 are known to be responsible for anthocyanin pigmentation in tobacco flowers [29,30]. The transcript of *MYBJS1* was induced similarly between GUS and PAP-8 leaves indicates that the *MYB* gene was not under the direct control of PAP1. *An2* was not expressed in either GUS or PAP-8 leaves, confirming what was previously known, namely, that this transcription factor is abundantly expressed in flowers but not in leaves [29]. On the other hand, although PAP-8 increased transcript levels of *An1a* and *An1b* transcript levels of GUS lines, the increase of the transcript levels was diminished by herbivory; this result was also observed in the structural genes mentioned above (see Figures 3 and S4 for 2 days and 4 days of exposure, respectively).

Upstream and downstream JA signaling for *PAP1*-activated genes

Since JA signaling along with other phytohormone signaling plays a central role in defense responses against chewing herbivores [31], it might be expected that induced JA production and signaling would be relevant to the low concentration of flavonoid/phenylpropanoid pathway products in *PAP1* lines attacked by *S. litura*. To assess whether *PAP1* overexpression affects JA biosynthesis and signaling, we analyzed endogenous levels of jasmonoyl-L-isoleucine (JA-Ile, an active form of JAs [32]) and of two other phytohormones: abscisic acid (ABA, involved in protective wound-healing processes [33]) and salicylic acid (SA, involved in pathogenesis and herbivore responses [34–36]). As shown in Figure 4A, JA and JA-Ile increased and accumulated at similar rates in the infested PAP-8 leaves compared to in the leaves of infested GUS lines. Those data indicate that PAP1 is not an upstream transcription regulator for the biosynthesis of JA. Moreover, neither ABA nor SA was increased in leaves of GUS or PAP-8 lines upon infestation, indicating that neither of those two hormones is directly linked by PAP1 actions.

Next, we investigated the gene expression levels in leaves of GUS and PAP-8 lines by applying an exogenous solution of MeJA (the commonly used active form of JAs [37]: 0.1 mM). Even though the PAP1-regulated transcript levels of *An1a*, *An1b* and *F3H* are negatively affected by herbivory (as shown in Figure 3), MeJA did not reduce their transcript levels to the same extent that it reduced transcript levels of *PAP1* (Figure 4B). The MeJA concentration used for our treatment was undoubtedly sufficient to induce defense responses in leaf tissues, as the transcript levels of a typical MeJA-inducible protease inhibitor (PIs [38]) and *MYBJS1* genes were substantially up-regulated in the same treated samples.

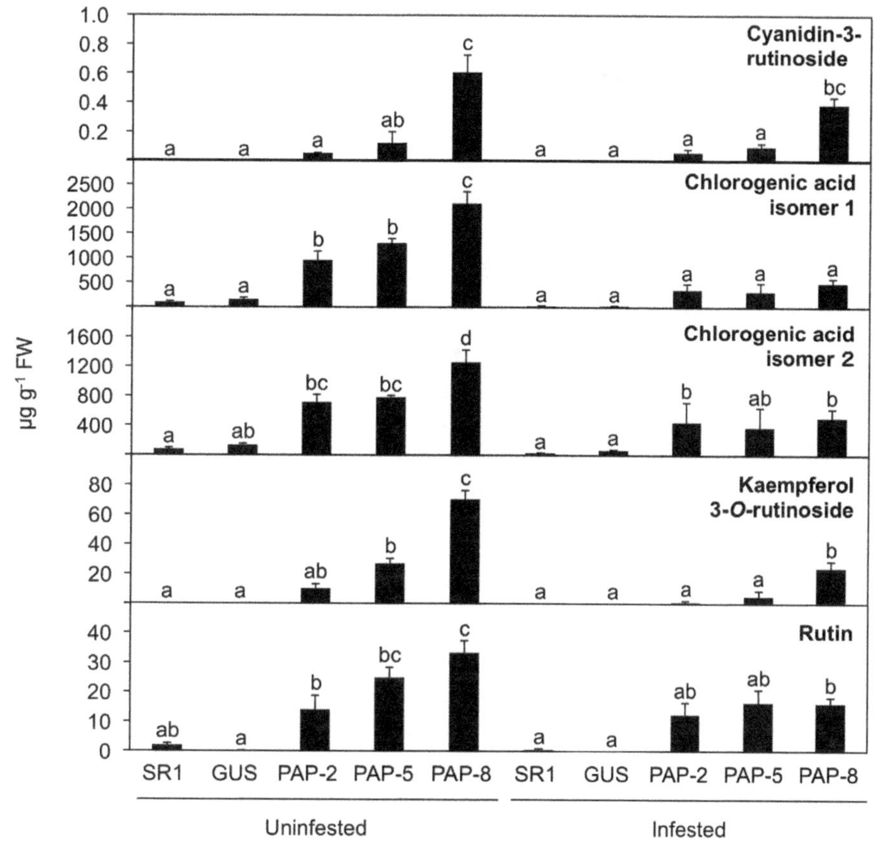

Figure 2. Endogenous accumulation of flavonoids/phenylpropanoids. Values for the major anthocyanin compound (cyanidin-3-*O*-rutinoside), other flavonoid (kaempferol 3-*O*-rutinoside and rutin) and phenylpropanoid (chlorogenic acid) compounds were determined in leaves of wild-type (WT), GUS control and *PAP1* lines not infested or infested with *Spodoptera litura* for 2 days. Data are shown as the mean+standard error ($n = 5$). Means followed by different small letters are significantly different ($P < 0.05$).

We therefore conclude that the down-regulation of PAP1-activated genes after herbivory is closely linked not to the JA signaling and biosynthesis cascades but, rather, to other herbivory-associated signals discussed below.

Reversible and transient regulatory patterns of *F3H* and *PI2* during herbivory and post-herbivory periods

To clarify the temporal regulation of PAP1-regulated genes, we analyzed the temporal patterns of the expression levels of *F3H* (PAP1-regulated) and *PI2* (JA-inducible) not only during 48 h after the first exposure to *S. litura* but also for up to 48 h after the larvae were removed (post-herbivory recovery; Figure 5). Consistent with the data shown in Figure 3, the regulation of the expression levels of *An1a*, *An1b* and *F3H* by PAP1 was suppressed 48 h after the first exposure to *S. litura*. After the larvae were removed, the expression levels remained low during the first 8 h but then increased 24 h and 48 h after the damage had ended. In contrast, *PI2* and *PAP1* displayed different expression profiles: the expression level of *PI2* reached a maximum at 48 h after first exposure to *S. litura* but dropped to the basal level 8 h, 24 h and 48 h after the damage had ended, consistent with the defense function of the PI2 protein; the expression level of *PAP1* was unchanged through the experimental period. Collectively, these results suggest that *F3H* might be regulated positively by an additional, JA-independent regulator(s) during the post-herbivory period.

Discussion

To date, light stimuli and abiotic stress have been reported to modulate anthocyanin concentrations in several plant taxa. For instance, in grape berries, and flowers of rose and *Brunsfelsia calycina*, anthocyanin accumulation appears to be suppressed in response to high temperature [39–41]. Most impressively, Rowan et al. (2009) reported that abiotic stress treatments of high temperature and low light cause reductions in the concentrations of anthocyanins and down-regulation of the genes involved in anthocyanin biosynthesis in leaves of *PAP1*-overexpressing *Arabidopsis* plants [23]. The inhibition of transcription occurs independently of *PAP1*, and the loss of anthocyanins is not the result of an obvious catabolic enzyme regulated at the transcription level. The authors addressed whether the decreased flux of anthocyanins into the vacuole occurred as a consequence of a decline in the expression of biosynthetic genes by reducing the expression of three genes (*TT8*, *TTG1* and *EGL3*) of the PAP1 transcriptional complex and enhancing the expression of the potential transcriptional repressors *AtMYB3*, *AtMYB6* and *AtMYBL2*. Those results especially reflect to our finding that transcript levels of both biosynthetic genes and endogenous transcription factor (positive regulator) genes were decreased in infested PAP1 plants compared to those in the uninfested PAP1 plants (Figure 3). *An1a* and *An1b* function by interacting with *An2* (PAP1 homologue) to regulate anthocyanin biosynthesis [30], thus implying that those two factors can also interact with *Arabidopsis* PAP1. In fact, *An1a* and *An1b* transcripts were

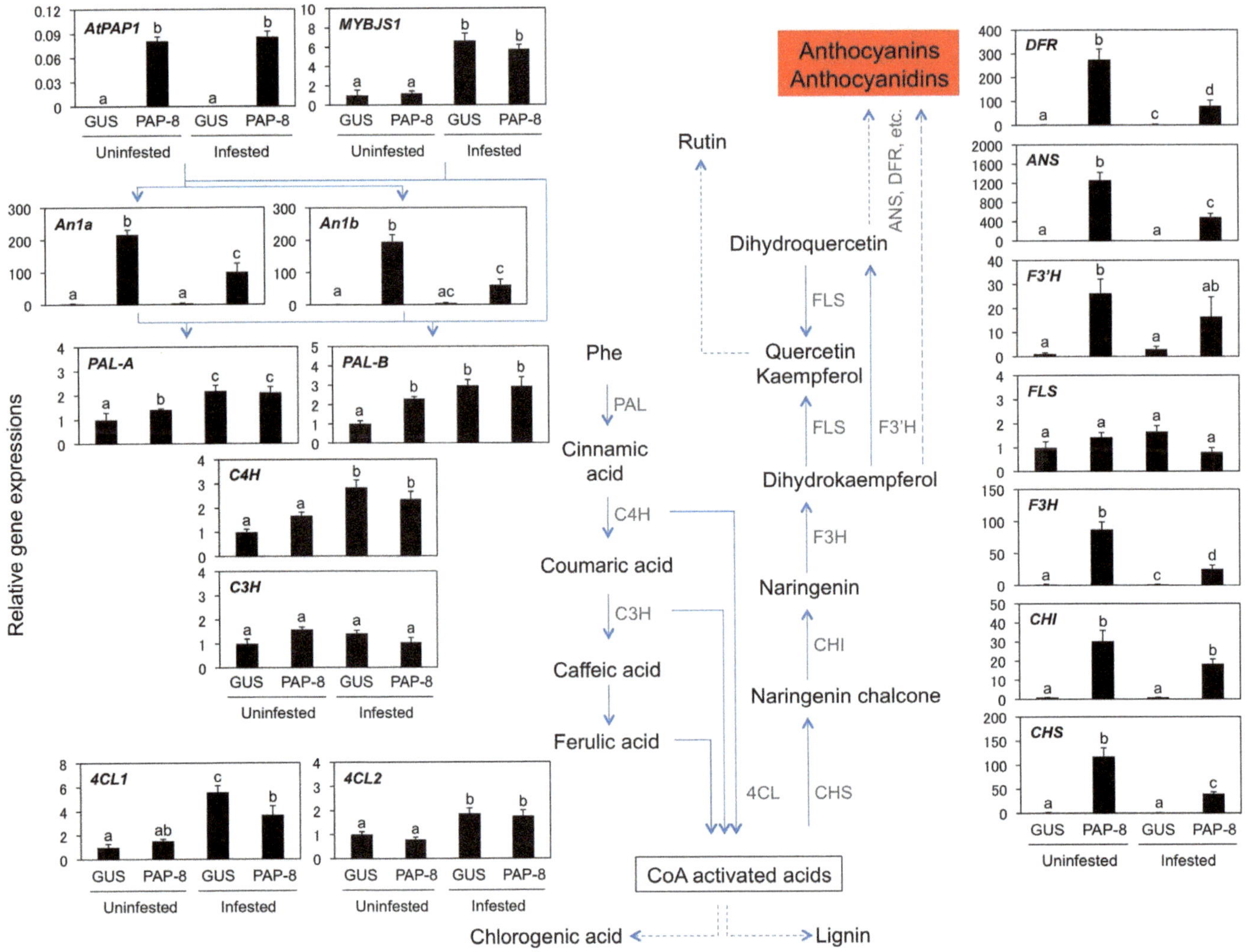

Figure 3. Expression of genes for transcription factors PAP1 and MYBJS1, An1a, An1b and structural genes involved in the flavonoid/phenylpropanoid pathway. Relative transcript levels of genes were determined in leaves of GUS control and *PAP1* lines (PAP-8) not infested or infested with *Spodoptera litura* for 2 days. Transcript levels of genes were normalized by those of *NtEF1α*. Data are shown as the mean+ standard errors ($n = 4$–5). Means followed by different small letters are significantly different ($P<0.05$). Expression of genes in GUS and PAP-8 leaves after 4 days of damage are presented in Figure S4. ANS, anthocyanidin synthase; C3H, cinnamate 3-hydroxylase; C4H, cinnamate 4-hydroxylase; CHI, chalcone isomerase; CHS, chalcone synthase; 4CL, 4-coumarate coenzyme A ligase; DFR, dihydroflavonol 4-reductase; F3H, flavanone 3-hydroxylase; F3′H, flavonoid 3′-hydroxylase; FLS, flavonol synthase; PAL, phenylalanine ammonia-lyase.

enhanced in the overexpression of *PAP1*, suggesting that those two bHLH proteins are able to interact with foreign PAP1 genes and are involved in the feedback regulation by the MYB/bHLH complexes as reported in *Arabidopsis* [42]. The diminished expression of those two genes suppress the expression of structural genes – genes that are also suppressed by herbivory – and regulates the flux of flavonoid/phenylpropanoid (Figures 2 and 3). Presumably, in return, plants gain energy and the resources required for induced defense responses (so-called homeostasis and feedback regulation), which might eventually make plants better able to withstand to generalist herbivores.

We present here two important findings, namely, that *PAP1* lines that accumulated great amounts of flavonoid/phenylpropanoid pathway products defend against the generalist herbivore *S. litura*, and that the *PAP1*-stimulated accumulations of those compounds are attenuated by herbivory (Figure 6). The first finding is in line with a previous report that transgenic tobacco plants expressing *Arabidopsis* transcription factor *AtMYB12* exhibit more resistance to *S. litura* and *Helicoverpa armigera* than do non-transgenic plants, most probably due to these plants' enhanced accumulation of rutin [13], a chemical that is toxic for those herbivores [43]. As rutin is one of predominant products in *PAP1* lines (Figure 2), this flavonoid compound is predicted to be a major anti-herbivore agent in our case as well.

The suppression of flavonoid/phenylpropanoid pathway products by herbivory is likely to reflect the phenomena described above, such as the response of *Arabidopsis* plants to abiotic stress, namely, the overexpression of *PAP1* [23]. In addition to the proposed reason for suppression (see above), we should consider an alternative (or additional) explanation, such as that PAP1 protein could be post-transcriptionally degraded under herbivory even though normal transcript levels of *PAP1* in PAP-8 leaves are maintained during herbivory (Figure 3). A similar mechanism is consistent with the repression of anthocyanin in the dark; in this case, the COP1/SPA ubiquitin ligase plays a critical role in the degradation of PAP1 proteins [22]. Although we conducted

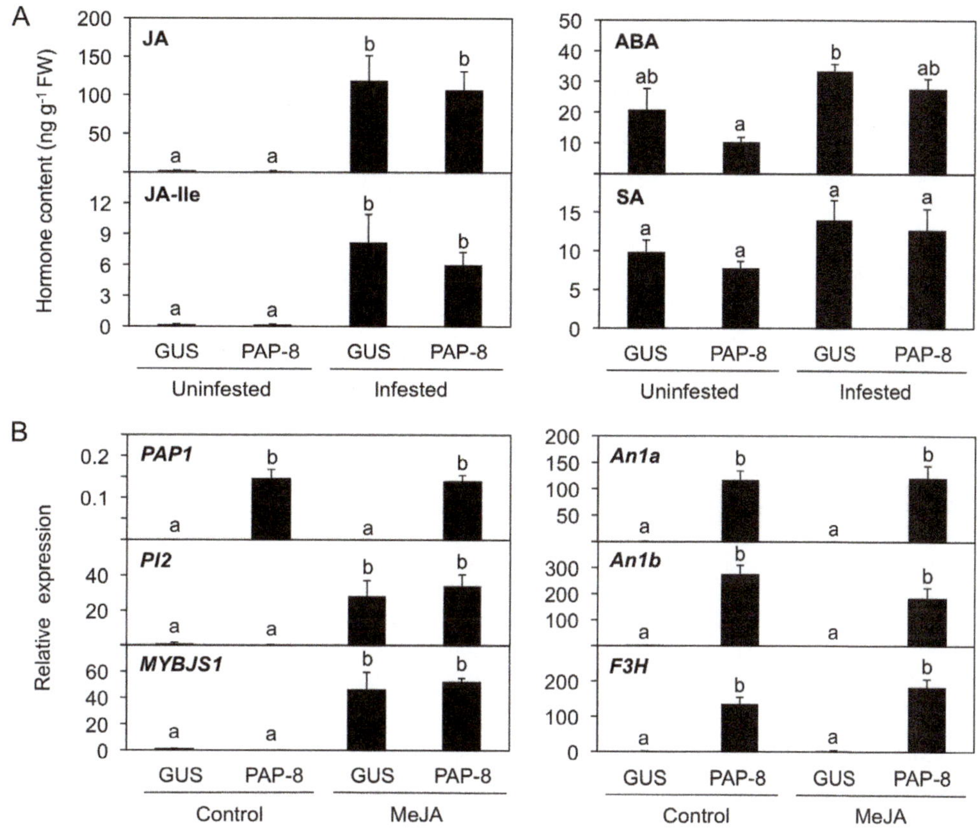

Figure 4. Upstream and downstream transcription regulation of PAP1-activated genes. A, Endogenous phytohormone levels were determined in leaves of GUS control and *PAP1* lines (PAP-8) not infested or infested with *Spodoptera litura* for 2 days. ABA, abscisic acid; JA, jasmonic acid; JA-Ile, jasmonoyl-L-isoleucine; SA, salicylic acid. B, The effect of treatment with methyl jasmonate (MeJA) on gene expression was determined in leaves of GUS and PAP-8 lines treated with 0.1 mM MeJA (in 0.1% ethanol solution) for 24 h. Plants treated with 0.1% ethanol solution for 24 h served as controls. Transcript levels of genes were normalized by comparing them to those of *NtEF1α*. Data are shown as the mean+standard error ($n = 4$–5). Means followed by different small letters are significantly different ($P < 0.05$). F3H, flavanone 3-hydroxylase; PI2, proteinase inhibitor 2.

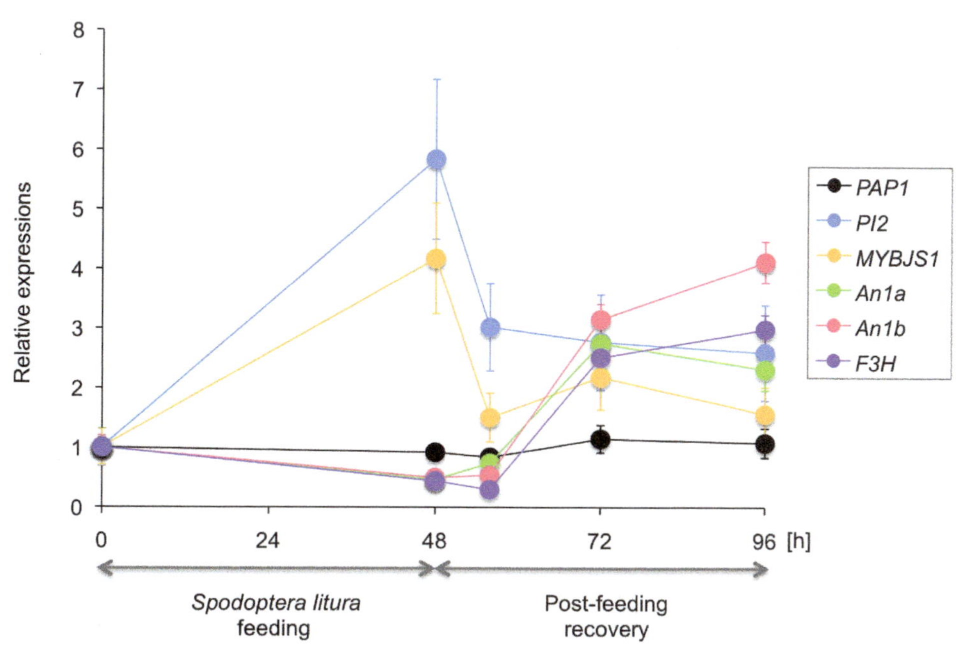

Figure 5. Temporal patterns of *F3H* and *PI2* gene regulation in leaves of *PAP1* lines (PAP-8) plants during exposure to *Spodoptera litura* for 48 h and for up to another 48 h after removal of the larvae (recovery period). Transcript levels of genes were normalized by comparing them to those of *NtEF1α*. Data are shown as the means ± standard errors ($n = 4$–5).

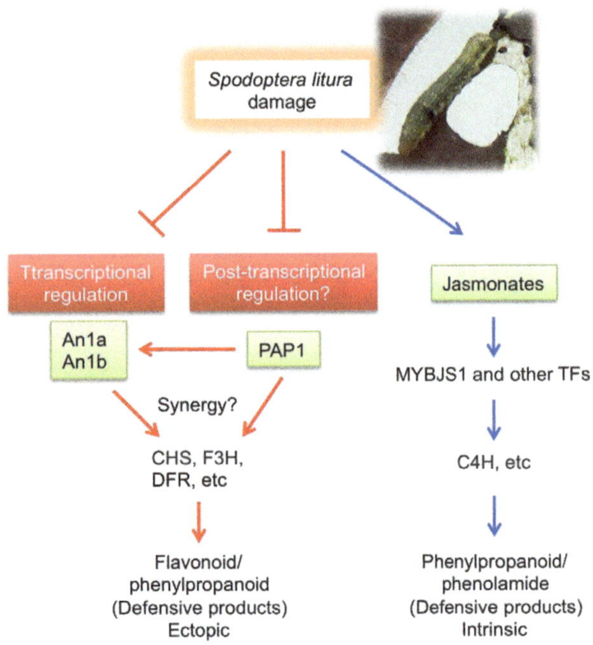

Figure 6. Schematic presentation of effect of *Spodoptera litura* feeding on the PAP1-regulated phenylpropanoid biosynthetic genes in a manner independent of jasmonate signaling. Arrows and bars indicate positive and negative interactions, respectively. TFs, transcription factors.

immunoblot analysis using polyclonal antibodies against partial PAP1 peptide residues ([H]CKIKMKKRDITP[OH]) (produced by Sigma-Aldrich, Ishikari, Japan), our efforts unfortunately failed to detect sufficient PAP1 protein signals extracted from uninfested and infested PAP-8 leaves. Therefore, it remains to be clarified whether PAP1 degradation in response to herbivory is subject to similar mechanisms for light/dark responses observed in *Arabidopsis*.

Alternatively, the possibility that the general decrease observed in the concentration of phenolic compounds after two days of herbivory could be a result of an altered metabolic precursor allocation (competition) cannot be excluded. For example, in transgenic *N. attenuata* plants, the transient silencing of the expression of a hydroxycinnamoyl-CoA:shikimate/quinate hydroxycinnamoyl transferase (HCT) gene involved in lignin biosynthesis dramatically increased herbivory-induced phenolamides produced in the parallel metabolic branch [21]. It was proposed that a strong metabolic tension in the plant, exacerbated during herbivory, exists over the allocation of coumaroyl-CoA units among lignin and other phenolic compounds. This tension could be extended to anthocyanins and flavonoids, as shown in our experiments (see Figure 2), although the branches that are simultaneously enhanced have not yet been identified. In addition, the enhanced catabolism of some phenolics during herbivory stress could be responsible. Unfortunately, very little is known about the anthocyanins, kaempferol-3-*O*-rutinoside and rutin catabolism in plants, and the hypothesis that there is competition among metabolic precursors remains to be tested.

Generally, since JA-mediated defense signaling is predominantly activated when plants are attacked by chewing herbivores [31], it might be expected that induced JA production and signaling would be relevant to the defense responses to *S. litura* attack. Even though *PI2* expression was highly induced during herbivory and exogenous MeJA application (see Figures 3 and 4), it was likely that the *PAP1*-activated genes were suppressed independently of the induction of JA and JA-Ile (an active form of JA) (Figure 4A). Moreover, since MeJA application also failed to suppress PAP1-activated genes in PAP-8 plants (Figure 4B), JA signaling is not likely a part of the down-regulation pathway. Also, in our case SA and ABA signaling pathways seem to be irrelevant in our case (see Figure 4A). Those facts may reflect a recent report that a mitogen-activated protein kinase (MPK4) suppresses a JA-independent defense pathway in *N. attenuata* in response to *Manduca sexta* feeding [44].

Moreover, the reverse mechanism was also likely used, given that PAP1-regulated *F3H* expression was dramatically increased in PAP-8 leaves during the post-herbivory period, when JA signaling was already switched off (Figure 5). This up-regulation might be caused by the fine-tuning of the PAP1 protein level after the cessation of herbivory, resulting in increased PAP1 levels beyond the basal level in the undamaged condition. Alternatively, additional signal(s) might be facilitated only when JA signaling is turned off, although these speculations require further investigation.

Horticultural ecosystems are very complicated and flexible. Hence, prior to setting genetically modified plants in the field, a wide range of systematic studies should be conducted to understand realistic horticultural ecosystems, where a myriad of plants, animals, and microorganisms interact and coevolve. However, it still remains to be elucidated how other multiple and simultaneous stresses can potentially influence flavonoid/phenylpropanoid catabolism in real agricultural settings.

Supporting Information

Figure S1 Relative mRNA levels of *PAP1* in leaves of wild-type (WT), GUS and *PAP1* lines. Transcript levels of genes were normalized by those of *NtEF1α*. Data are shown as the mean+standard errors ($n = 5$). ND, not detected.

Figure S2 Morphology of *PAP1* lines. A, 4-week-old plants; B, 8-week-old plants; C, flowers.

Figure S3 GUS and PAP-8 plants infested with *Spodoptera litura* for 2 days (A) and 4 days (B).

Figure S4 Expression of genes in GUS and PAP-8 leaves infested with *Spodoptera litura* for 4 days. Relative transcript levels of genes were determined in leaves of GUS control and *PAP1* lines (PAP-8) not infested or infested with *S. litura* for 4 days. Transcript levels of genes were normalized by comparing them to those of *NtEF1α*. Data are shown as the mean+standard errors ($n = 4$–5). Means followed by different small letters are significantly different ($P<0.05$).

Table S1 RT-PCR primers used for this study.

Acknowledgments

We thank Ms. Emily Wheeler for editorial assistance.

Author Contributions

Conceived and designed the experiments: TM MN IG KF NS KN GA. Performed the experiments: TM MN KA YH KF NS KN GA. Analyzed the data: TM MN IG KA YH KF NS KN GA. Contributed reagents/materials/analysis tools: MN IG TS GA. Wrote the paper: MN IG GA.

References

1. Dixon RA, Liu C, Jun JH (2013) Metabolic engineering of anthocyanins and condensed tannins in plants. Curr Opin Biotechnol 24: 329–335.
2. Glenn WS, Runguphan W, O'Connor SE (2013) Recent progress in the metabolic engineering of alkaloids in plant systems. Curr Opin Biotechnol 24: 354–365.
3. Wang Y, Chen S, Yu O (2011) Metabolic engineering of flavonoids in plants and microorganisms. Appl Microbiol Biotechnol 91: 949–956.
4. Borevitz JO, Xia Y, Blount J, Dixon RA, Lamb C (2000) Activation tagging identifies a conserved MYB regulator of phenylpropanoid biosynthesis. Plant Cell 12: 2383–2394.
5. Xie DY, Sharma SB, Wright E, Wang ZY, Dixon RA (2006) Metabolic engineering of proanthocyanidins through co-expression of anthocyanidin reductase and the PAP1 MYB transcription factor. Plant J 45: 895–907.
6. Ben Zvi MM, Shklarman E, Masci T, Kalev H, Debener T, et al. (2012) *PAP1* transcription factor enhances production of phenylpropanoid and terpenoid scent compounds in rose flowers. New Phytol 194: 430–439.
7. Ben Zvi MM, Negre-Zakharov F, Masci T, Ovadis M, Shklarman E, et al. (2008) Interlinking showy traits: co-engineering of scent and colour biosynthesis in flowers. Plant Biotechnol J 6: 403–415.
8. Zhang Y, Yan YP, Wang ZZ (2010) The *Arabidopsis* PAP1 transcription factor plays an important role in the enrichment of phenolic acids in *Salvia miltiorrhiza*. J Agric Food Chem 58: 12168–12175.
9. Mellway RD, Tran LT, Prouse MB, Campbell MM, Constabel CP (2009) The wound-, pathogen-, and ultraviolet B-responsive *MYB134* gene encodes an R2R3 MYB transcription factor that regulates proanthocyanidin synthesis in poplar. Plant Physiol 150: 924–941.
10. Korkina LG (2007) Phenylpropanoids as naturally occurring antioxidants: from plant defense to human health. Cell Mol Biol 53: 15–25.
11. Chalker-Scott L (1999) Environmental significance of anthocyanins in plant stress responses. Photochem Photobiol 70: 1–9.
12. Kumar V, Nadda G, Kumar S, Yadav SK (2013) Transgenic tobacco overexpressing tea cDNA encoding dihydroflavonol 4-reductase and anthocyanidin reductase induces early flowering and provides biotic stress tolerance. PLoS ONE 8: e65535.
13. Misra P, Pandey A, Tiwari M, Chandrashekar K, Sidhu OP, et al. (2010) Modulation of transcriptome and metabolome of tobacco by Arabidopsis transcription factor, *AtMYB12*, leads to insect resistance. Plant Physiol 152: 2258–2268.
14. Gális I, Simek P, Narisawa T, Sasaki M, Horiguchi T, et al. (2006) A novel R2R3 MYB transcription factor NtMYBJS1 is a methyl jasmonate-dependent regulator of phenylpropanoid-conjugate biosynthesis in tobacco. Plant J 46: 573–592.
15. Kaur H, Heinzel N, Schöttner M, Baldwin IT, Gális I (2010) R2R3-NaMYB8 regulates the accumulation of phenylpropanoid-polyamine conjugates, which are essential for local and systemic defense against insect herbivores in *Nicotiana attenuata*. Plant Physiol 152: 1731–1747.
16. Bassard JE, Ullmann P, Bernier F, Werck-Reichhart D (2010) Phenolamides: bridging polyamines to the phenolic metabolism. Phytochemistry 71: 1808–1824.
17. Gulati J, Baldwin IT, Gaquerel E (2014) The roots of plant defenses: Integrative multivariate analyses uncover dynamic behaviors of roots' gene and metabolic networks elicited by leaf herbivory. Plant J 77: 880–892.
18. Onkokesung N, Gaquerel E, Kotkar H, Kaur H, Baldwin IT, et al. (2012) MYB8 controls inducible phenolamide levels by activating three novel hydroxycinnamoyl-coenzyme A:polyamine transferases in *Nicotiana attenuata*. Plant Physiol 158: 389–407.
19. Hoffmann L, Besseau S, Geoffroy P, Ritzenthaler C, Meyer D, et al. (2004) Silencing of hydroxycinnamoyl-coenzyme A shikimate/quinate hydroxycinnamoyltransferase affects phenylpropanoid biosynthesis. Plant Cell 16: 1446–1465.
20. Hoffmann L, Maury S, Martz F, Geoffroy P, Legrand M (2003) Purification, cloning, and properties of an acyltransferase controlling shikimate and quinate ester intermediates in phenylpropanoid metabolism. J Biol Chem 278: 95–103.
21. Gaquerel E, Kotkar H, Onkokesung N, Galis I, Baldwin IT (2013) Silencing an *N*-acyltransferase-like involved in lignin biosynthesis in *Nicotiana attenuata* dramatically alters herbivory-induced phenolamide metabolism. PLoS ONE 8: e62336.
22. Maier A, Schrader A, Kokkelink L, Falke C, Welter B, et al. (2013) Light and the E3 ubiquitin ligase COP1/SPA control the protein stability of the MYB transcription factors PAP1 and PAP2 involved in anthocyanin accumulation in Arabidopsis. Plant J 74: 638–651.
23. Rowan DD, Cao M, Lin-Wang K, Cooney JM, Jensen DJ, et al. (2009) Environmental regulation of leaf colour in red 35S:*PAP1 Arabidopsis thaliana*. New Phytol 182: 102–115.
24. Mishiba K, Yamasaki S, Nakatsuka T, Abe Y, Daimon H, et al. (2010) Strict *de novo* methylation of the 35S enhancer sequence in gentian. PLoS One 5: e9670.
25. Galis I, Schuman MC, Gase K, Hettenhausen C, Hartl M, et al. (2013) The use of VIGS technology to study plant-herbivore interactions. In: Becker A, editor. Methods in Molecular Biology - Virus-induced gene silencing: Methods and protocols. Totowa, NJ: Humana Press Inc. pp. 109–138.
26. Matsuba Y, Sasaki N, Tera M, Okamura M, Abe Y, et al. (2010) A novel glucosylation reaction on anthocyanins catalyzed by acyl-glucose-dependent glucosyltransferase in the petals of carnation and delphinium. Plant cell 22: 3374–3389.
27. Fukumoto K, Alamgir K, Yamashita Y, Mori IC, Matsuura H, et al. (2013) Response of rice to insect elicitors and the role of OsJAR1 in wound and herbivory-induced JA-Ile accumulation. J Integr Plant Biol 55: 775–784.
28. Tohge T, Nishiyama Y, Hirai MY, Yano M, Nakajima J, et al. (2005) Functional genomics by integrated analysis of metabolome and transcriptome of Arabidopsis plants over-expressing an MYB transcription factor. Plant J 42: 218–235.
29. Pattanaik S, Kong Q, Zaitlin D, Werkman JR, Xie CH, et al. (2010) Isolation and functional characterization of a floral tissue-specific R2R3 MYB regulator from tobacco. Planta 231: 1061–1076.
30. Bai Y, Pattanaik S, Patra B, Werkman JR, Xie CH, et al. (2011) Flavonoid-related basic helix-loop-helix regulators, NtAn1a and NtAn1b, of tobacco have originated from two ancestors and are functionally active. Planta 234: 363–375.
31. Arimura G, Köpke S, Kunert M, Volpe V, David A, et al. (2008) Effects of feeding *Spodoptera littoralis* on Lima bean leaves: IV. Diurnal and nocturnal damage differentially initiate plant volatile emission. Plant Physiol 146: 965–973.
32. Staswick PE, Tiryaki I (2004) The oxylipin signal jasmonic acid is activated by an enzyme that conjugates it to isoleucine in Arabidopsis. Plant Cell 16: 2117–2127.
33. Lulai EC, Suttle JC, Pederson SM (2008) Regulatory involvement of abscisic acid in potato tuber wound-healing. J Exp Bot 59: 1175–1186.
34. Boatwright JL, Pajerowska-Mukhtar K (2013) Salicylic acid: an old hormone up to new tricks. Mol Plant Pathol 14: 623–634.
35. Zhang PJ, Zheng SJ, van Loon JJ, Boland W, David A, et al. (2009) Whiteflies interfere with indirect plant defense against spider mites in Lima bean. Proc Natl Acad Sci USA 106: 21202–21207.
36. Chung SH, Rosa C, Scully ED, Peiffer M, Tooker JF, et al. (2013) Herbivore exploits orally secreted bacteria to suppress plant defenses. Proc Natl Acad Sci USA 110: 15728–15733.
37. Farmer EE, Ryan CA (1990) Interplant communication: airborne methyl jasmonate induces synthesis of proteinase inhibitors in plant leaves. Proc Natl Acad Sci USA 87: 7713–7716.
38. Srinivasan T, Kumar KR, Kirti PB (2009) Constitutive expression of a trypsin protease inhibitor confers multiple stress tolerance in transgenic tobacco. Plant Cell Physiol 50: 541–553.
39. Mori K, Goto-Yamamoto N, Kitayama M, Hashizume K (2007) Loss of anthocyanins in red-wine grape under high temperature. J Exp Bot 58: 1935–1945.
40. Dela G, Or E, Ovadia R, Nissim-Levi A, Weiss D, et al. (2003) Changes in anthocyanin concentration and composition in 'Jaguar' rose flowers due to transient high-temperature conditions. Plant Sci 164: 333–340.
41. Vaknin H, Bar-Akiva A, Ovadia R, Nissim-Levi A, Forer I, et al. (2005) Active anthocyanin degradation in *Brunfelsia calycina* (yesterday-today-tomorrow) flowers. Planta 222: 19–26.
42. Baudry A, Caboche M, Lepiniec L (2006) TT8 controls its own expression in a feedback regulation involving TTG1 and homologous MYB and bHLH factors, allowing a strong and cell-specific accumulation of flavonoids in *Arabidopsis thaliana*. Plant J 46: 768–779.
43. Pandey A, Misra P, Chandrashekar K, Trivedi PK (2012) Development of *AtMYB12*-expressing transgenic tobacco callus culture for production of rutin with biopesticidal potential. Plant Cell Rep 31: 1867–1876.
44. Hettenhausen C, Baldwin IT, Wu J (2013) *Nicotiana attenuata* MPK4 suppresses a novel jasmonic acid (JA) signaling-independent defense pathway against the specialist insect *Manduca sexta*, but is not required for the resistance to the generalist *Spodoptera littoralis*. New Phytol 199: 787–799.

3

Resource-Mediated Indirect Effects of Grassland Management on Arthropod Diversity

Nadja K. Simons[1]*, Martin M. Gossner[1], Thomas M. Lewinsohn[2], Steffen Boch[3], Markus Lange[4], Jörg Müller[5], Esther Pašalić[1], Stephanie A. Socher[3], Manfred Türke[1], Markus Fischer[3,6], Wolfgang W. Weisser[1]

1 Terrestrial Ecology Research Group, Department of Ecology and Ecosystem Management, School of Life Sciences Weihenstephan, Technische Universität München, Freising, Germany, **2** Department of Animal Biology, Institute of Biology, University of Campinas, Campinas, Sao Paulo, Brazil, **3** Institute of Plant Sciences, University of Bern, Bern, Switzerland, **4** Max-Planck-Institute for Biogeochemistry, Jena, Germany, **5** Institute of Biochemistry and Biology, University of Potsdam, Potsdam, Germany, **6** Biodiversity and Climate Research Centre, Senckenberg Gesellschaft für Naturforschung, Frankfurt/Main, Germany

Abstract

Intensive land use is a driving force for biodiversity decline in many ecosystems. In semi-natural grasslands, land-use activities such as mowing, grazing and fertilization affect the diversity of plants and arthropods, but the combined effects of different drivers and the chain of effects are largely unknown. In this study we used structural equation modelling to analyse how the arthropod communities in managed grasslands respond to land use and whether these responses are mediated through changes in resource diversity or resource quantity (biomass). Plants were considered resources for herbivores which themselves were considered resources for predators. Plant and arthropod (herbivores and predators) communities were sampled on 141 meadows, pastures and mown pastures within three regions in Germany in 2008 and 2009. Increasing land-use intensity generally increased plant biomass and decreased plant diversity, mainly through increasing fertilization. Herbivore diversity decreased together with plant diversity but showed no response to changes in plant biomass. Hence, land-use effects on herbivore diversity were mediated through resource diversity rather than quantity. Land-use effects on predator diversity were mediated by both herbivore diversity (resource diversity) and herbivore quantity (herbivore biomass), but indirect effects through resource quantity were stronger. Our findings highlight the importance of assessing both direct and indirect effects of land-use intensity and mode on different trophic levels. In addition to the overall effects, there were subtle differences between the different regions, pointing to the importance of regional land-use specificities. Our study underlines the commonly observed strong effect of grassland land use on biodiversity. It also highlights that mechanistic approaches help us to understand how different land-use modes affect biodiversity.

Editor: Christian Rixen, WSL Institute for Snow and Avalanche Research SLF, Switzerland

Funding: The work was funded by the DFG Priority Program 1374 "Infrastructure-Biodiversity-Exploratories" (DFG-WE 3081/21-1.). www.dfg.de/spp/en. Additional funds for exchange visits to analyze data were provided by Unicamp (PRP/Faepex) for TML and MMG and by DAAD (Project TUMBRA) for NKH. www.toek.wzw.tum.de/index.php?id = 117. This work was supported by the German Research Foundation (DFG) and the Technische Universität München within the funding programme Open Access Publishing. The funders had no role in study design, data collection and analysis, decision to publish, or preparation of the manuscript.

Competing Interests: The authors have declared that no competing interests exist.

* Email: nadja.simons@tum.de

Introduction

Negative effects of intensive grassland land use on biodiversity have been found for many taxa including plants [1], herbivorous and carnivorous arthropods [2–4], pollinating insects [5] and birds [6]. Despite the growing consensus that intensive land use has generally negative effects on biodiversity [7], the particular mechanisms that lead to a decrease in biodiversity are often not fully understood [8], because land use consists of various modes that can have opposing or additive effects on biodiversity. In semi-natural grasslands important land-use modes are mowing, grazing and fertilization. Several observational and experimental studies found decreasing diversities of plants and arthropods with increasing frequency of mowing events e.g. [4,9–11], with increasing fertilization intensity e.g. [12,13] or with increasing grazing intensity [14–16]. Whereas effects of land-use modes on plants are often direct, e.g. when mowing hinders seed set of late-flowering plants, effects on higher trophic levels such as insect herbivores or carnivores may be either direct or mediated by changes in the plant community.

Several hypotheses have been proposed to describe the effects of differently diverse plant communities on the diversity and abundance of herbivores and predators. The 'Resource Heterogeneity Hypothesis' (RHH) predicts that more diverse resources provide more niches for a greater number of specialized species at higher trophic levels [17–20], i.e. herbivore diversity is promoted by increased plant diversity and predator diversity increases in response to herbivore diversity – a positive trophic cascade. In contrast to the RHH, the 'More Individuals Hypothesis' (MIH) proposes that diversity of consumers increases when resource quantity increases, i.e. plant biomass for herbivores and herbivore biomass for predators [21]. According to the MIH, this positive effect of resource abundance (or biomass) on consumer diversity is mediated by an increase in total consumer abundance. Borer et al. [22] studied plant diversity effects on arthropod diversity in experimental plant communities and found that more diverse plant communities hosted more diverse herbivore communities, but that this effect was mediated by higher overall plant and herbivore biomass. While the RHH assumes a direct link between resource and consumer diversity, the MIH assumes an indirect link through resource and consumer abundance, i.e. resource abundance increases consumer abundance which in turn increases consumer diversity. Thus, arthropod biomass and diversity may be differently affected by increasing land-use intensity depending on the mechanistic relationships between land use and the herbivore and carnivore communities.

We combined detailed information on grassland land-use modes (fertilization, grazing and mowing) with biomass and species richness data of plants and arthropods to analyse effects of land-use intensity on the arthropod community. Although experimental short-term manipulation of land use can elucidate immediate effects of the land-use modes (such as mowing), it is not clear whether these mechanisms operate similarly when different land-use practices are combined in agricultural grasslands and under long term conditions. This can only be assessed by studying grasslands which have been used as meadows or pastures continuously for several years or decades. In our study, we build on the work of Socher et al. [1] who tested for direct and indirect effects of grassland land use on plant diversity and biomass. We studied effects of land use on the arthropod community including both direct and indirect (via the plant community) chains of effects. The study system includes grasslands which have been managed for at least 20 years and comprise a wide range of land-use intensities. By including three different regions in Germany and by sampling in two consecutive years, we were able to consider the generality of observed patterns.

Based on the two hypotheses mentioned above and previous knowledge on mechanisms of grassland land use, we defined two alternative models. In the first model ('Resource Heterogeneity Model') we tested whether land-use intensity affects herbivore and predator diversity via changes in the diversity of their respective resources (RHH). According to the RHH we expected positive effects of plant diversity on herbivore diversity and of herbivore diversity on predator diversity (Figure 1 A). In the second model ('Resource Abundance Model'), we tested whether effects of land-use intensity affect arthropods through changes in resource quantity (MIH) and added herbivore and predator biomass to the model (Figure 1 B). According to the MIH, we expected plant biomass to have a positive effect on herbivore diversity through herbivore biomass and positive effects of herbivore biomass on predator diversity, respectively.

Within both models (Figure 1 A & B), we expected mowing to decrease plant diversity (by a loss of disturbance-sensitive species) and to decrease plant biomass (through mechanical disturbances during the growing period). Grazing was expected to increase plant diversity (increasing number of niches for plants or preventing competitive exclusion) and decrease plant biomass (recurrent disturbance of plant growth). Another possible plant species response to grazing or mowing is overcompensating for tissue loss leading to the opposite expectation (i.e. increase of plant biomass following grazing or mowing). However, overcompensation has usually been demonstrated for particular species, not at the community level e.g. [23,24], where a decrease of plant biomass is more likely. Fertilization was expected to decrease plant diversity (dominance of fast-growing species) and increase plant biomass (increased nutrient input). Based on the findings of Socher et al. [1] we expected plant biomass to be negatively correlated with plant diversity.

In our study we asked the following questions: 1) How do land-use modes, singly or in combination, affect arthropod diversity? 2) How are effects of land-use intensity on herbivore diversity mediated by the responses of plants to land use (do they follow the 'Resource Heterogeneity Hypothesis' or the 'More Individuals Hypothesis')? 3) Are the responses of predators to land use governed by the same mechanisms as responses of herbivores?

Materials and Methods

Study sites and land use

The Biodiversity Exploratory research program (www.biodiversity-exploratories.de) was established in 2006 within three regions in Germany: (1) Schorfheide-Chorin (SCH) in north-east Germany (3–140 m a.s.l., 53°02′ N 13°83′ E, annual mean precipitation 500–600 mm, mean temperature 8–8.5°C). (2) Hainich-Dün (HAI) in central Germany (285–550 m a.s.l., 51°20′ N 10°41′ E, 500–800 mm, 6.5–8°C). (3) Swabian Alb (ALB) in south-west Germany (460–860 m a.s.l., 48°43′ N 9°37′ E, 700–1000 mm, 6–7°C). Within each region, 50 plots of 50 m × 50 m size serve as basis for surveys of biodiversity or ecosystem processes. Those plots were chosen from a total of 500 candidate plots in each region on which initial vegetation and land-use surveys were conducted. The 50 plots per region cover the whole regional gradient of land-use modes and intensity [25].

Land-use modes on the studied grasslands include mowing with different frequency (meadows), grazing by different kinds of livestock (cattle and sheep pastures) or both mowing and grazing (mown pastures). Grasslands are either unfertilized or fertilized with different amounts of fertilizer. During the study years all grassland plots continued to be managed by farmers in the same way as the surrounding grasslands. Grassland land use was assessed each year (since 2006) by standardized interviews with farmers and land-owners. From these interviews, we derived information on fertilization, grazing and mowing: Fertilization intensity was calculated as the total amount of nitrogen applied per hectare and year, in the form of organic or chemical fertilizer. For grazing, information on livestock units and the duration of grazing periods were combined as a measure of grazing intensity. Mowing intensity was calculated as the number of mowing events per year.

For our analyses we used land-use information from the two years prior to sampling and the sampling year. Mowing intensity included mowing events in the sampling year only up to the sampling event (e.g. for samples taken in June, mowing events later than June in the sampling year were not included). As mowing not only has long-term but also short-term effects on arthropods (e.g. Humbert et al. [26] showed that mowing with machinery leads to high mortality in Orthoptera), we included the number of days between the arthropod sampling day and the last mowing event prior to the respective sampling as an additional land-use variable ('Time after mowing'). As two arthropod samplings were conducted per year (see below), we used the mean number of days for each year. For a detailed description of land-use intensity calculations and an overview of land-use information see Appendix S1 as well as Table S1 & Table S2.

Plant and arthropod sampling and plot selection

Plant diversity and plant biomass on the plots were assessed between mid-May and mid-June in 2008 & 2009 using vegetation surveys and aboveground biomass harvests following the methods

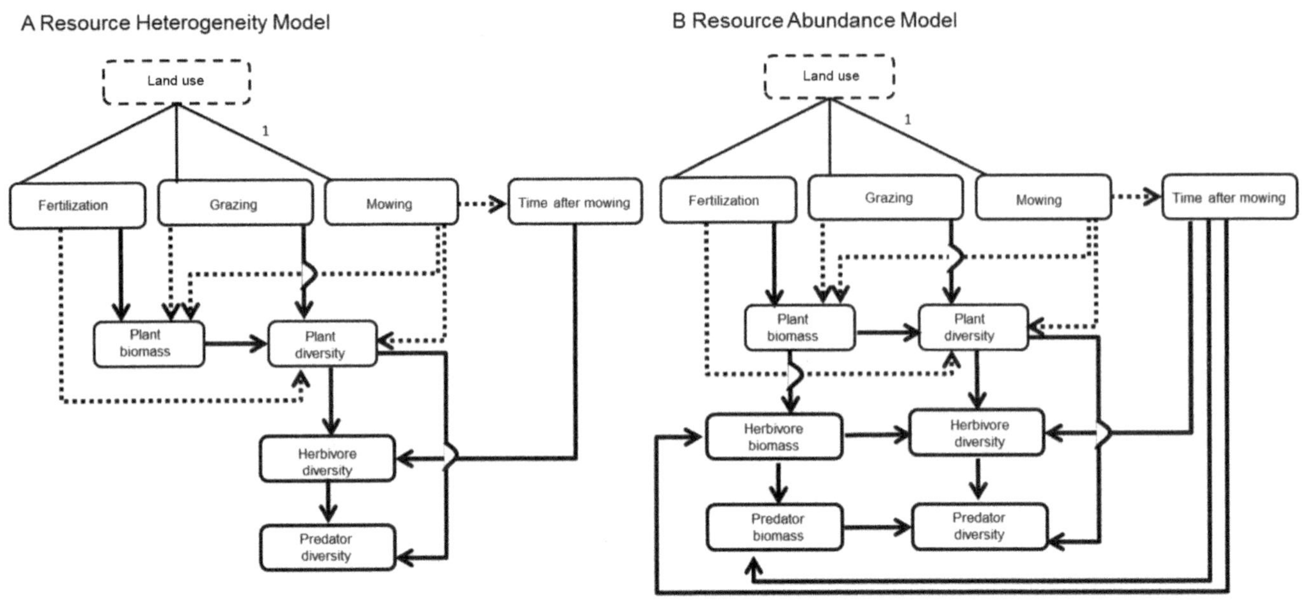

Figure 1. Concepts for structural equation models. 'Land use' is a latent (i.e. unmeasured) variable influencing the management components. The path coefficient for the correlation between 'Land use' and cutting frequency was a priori set to 1. The arrows indicate expected causal effects between measured variables; solid arrows represent expected positive effects and dotted arrows expected negative effects. For more detailed information on model definition and variable calculation see the Method section and Appendix S1.

described by Socher et al. [1]: On a 4 m × 4 m subplot in each plot all vascular plant species were identified in the field following the nomenclature of Wisskirchen & Häupler [27], categorized into functional groups (grasses, herbs, legumes) and their cover (%) was estimated. As vegetation was generally relatively uniform across the plot, we consider our subplot sample as representative for the 50 m × 50 m plot. Plant biomass was clipped at a height of 2 to 3 cm above ground and dried for 48 hours at 80°C before weighing. Occasionally occurring shrubs and litter were excluded before drying. In 2008, biomass samples were taken in two 25 cm × 50 cm subplots next to the vegetation survey plot. In 2009, biomass was harvested on eight subplots of 50 cm × 50 cm size adjacent to one side of the vegetation survey plot to better reflect small-scale variability. In both years, total plant biomass [g/m^2] was recorded as the mean from the subplots. In mown and grazed plots, the subplots were fenced until biomass harvest to obtain productivity estimates unaffected by current year grazing or mowing. For the analyses we used plant species richness per plot as measure for plant diversity.

We sampled arthropods by sweep-netting with a total of 60 double-sweeps (one double-sweep is defined as moving the net from the left to the right and back perpendicular to the walking direction) along three plot borders (150 m in total) in June and August 2008 & 2009 during periods of dry weather conditions. The two sampling months were chosen because they cover the main activity period of arthropods and were found to represent the variation in diversity among plots equally well as sampling in five months (tested with a subset of plots in 2009, data not shown). We chose sweep-netting over suction sampling because of logistic difficulties with suction sampling on our large number of plots and because suction sampling was found to underrepresent Heteroptera (Gossner et al., unpublished data from our plots and [28]).

Sampling was conducted within seven days or fewer per region and within two weeks in all regions. Arthropods were preserved in 70% ethanol, sorted by taxonomic group and identified to species level by taxonomic experts. We focused on Araneae, Hemiptera (Cicadomorpha, Fulgoromorpha and Heteroptera), Coleoptera and Orthoptera because of their numerical dominance in the studied grasslands. Only adult individuals were considered for the analysis, because identification of juveniles is often difficult. Samples from both months within one year were pooled for analysis. Because sampling effort was standardized there was no need to adjust for differences in the number of sampled individuals. Accordingly, we used species richness as diversity measure.

Due to external influences, such as aggressive cows preventing us from entering several plots, the number of plots with a complete dataset (plant and arthropod data) differed somewhat between years: in 2008 we considered 124 plots and in 2009 we considered 141 plots (see Table 1 for the number of plots per region). Of those plots, 117 plots were considered in both years.

Ethics Statement

Field work permits were issued by the responsible state environmental offices of Baden-Württemberg, Thüringen and Brandenburg (according to § 72 BbgNatSchG, i.e. nature conservation law of Brandenburg). Out of the 150 initially sampled plots, 17 are part of grassland areas that are protected under the Habitats Directive (FFH), 58 are protected as LSG (landscape conservation site), 17 are protected as NSG (nature conservation site) and 5 are situated within a former military training area. Except the 5 plots which are situated in the former military training area and 6 plots that are situated on state property, all other grasslands are privately owned by the farmers. One individual of *Phytoecia cylindrica* (Coleoptera, Linnaeus, 1758), which is protected under BArtSchV (Federal Protection of Species Order) was sampled in 2008.

Classification of trophic guilds and arthropod biomass assessment

All identified arthropod species were assigned to one of three trophic guilds (herbivores, predators and decomposers) based on

Table 1. Mean ± standard error for the sampled variables within each of the three regions.

	Vascular plants		Herbivores			Predators		
	No. species	Biomass (g)	No. species	Abundance	Biomass (g)	No. species	Abundance	Biomass (g)
2008								
Swabian Alb N = 43	33.3±10.3	598.1±302.9	33.3±10.6	628.6±570.9	15.3±14.3	5.4±3.5	10.6±8.3	0.4±0.4
Hainich-Dün N = 38	25.9±9.6	592.6±228.8	31.4±12.3	307.4±205.0	9.6±8.6	5.6±2.8	15.8±10.0	0.5±0.5
Schorfheide-Chorin N = 43	16.0±3.7	752.0±281.7	26.7±7.7	281.1±145.0	8.8±4.9	5.1±2.7	13.2±9.5	0.3±0.3
F-value$_{2,121}$	46.72	4.56	4.64	11.89	5.09	0.29	3.25	0.91
P	<0.001	0.012	0.011	<0.001	0.008	0.752	0.042	0.404
Means over all regions	25.0±2.2	648.0±11.0	30.0±1.9	410.0±19.6	11.2±3.1	5.0±1.3	13.0±2.6	0.4±0.02
2009								
Swabian Alb N = 47	32.1±11.2	272.1±123.1	20.5±5.6	180.9±144.5	6.7±4.8	3.8±2.0	4.6±2.8	0.3±0.3
Hainich-Dün N = 48	33.6±12.1	299.7±128.7	16.6±7.1	73.0±93.4	2.8±4.4	4.3±2.5	8.5±7.2	0.2±0.2
Schorfheide-Chorin N = 46	19.8±4.2	331.2±130.5	22.1±8.0	163.2±105.1	4.7±4.0	5.2±2.9	9.8±7.2	0.3±0.6
F-value$_{2,140}$	27.25	2.49	7.648	11.76	9.25	4.1	9.0	0.5
P	<0.001	0.086	<0.001	<0.001	<0.001	0.019	<0.001	0.61
Means over all regions	29.0±2.2	301.0±7	20.0±1.6	139.0±10.6	4.7±2.2	4.0±1.2	8.0±2.3	0.3±0.7

N = number of plots sampled in the respective year. Arthropod biomass was estimated from the species' mean length (see text). Means were tested for differences between regions with ANOVA. Degrees of freedom are indicated with the F-value$_{df\ Region,\ df\ Residuals}$. Abundances are expresses as number of adult individuals whereas biomass is given in grams. Species lists of herbivores and predators can be found in Appendix S5 and Appendix S6, original datasets are available from the Dryad Digital Repository: http://doi.org/10.5061/dryad.f3b77.

their known main food resource as adults (Appendix S2). Because decomposers made up less than 2% of individuals in 90% of our samples, they were not included in the analysis. Arthropod biomass per plot was estimated by applying the general power function developed by Rogers et al. [29] to each sampled species:

$$biomass[mg] = 0.305 * L^{2.62}$$

Where L is the mean length (in mm) of a species, derived from the same sources as given for the identification of feeding guilds (Appendix S2). The contribution of a species to total biomass was then calculated as: species-specific biomass × abundance of the species. Biomass for each plot was calculated for herbivores and predators separately by summing over all species. We used biomass rather than abundances as measure for resource abundance to be consistent with plant assessments.

Statistical analysis

To compare means of the response variables among regions, we used ANOVA (aov) with 'Region' as explanatory variable in R [30].

For structural equation modelling, there are, in principle three main approaches [31]: 1) Strictly confirmatory, where a single model is defined and tested against the data. 2) Alternative modelling, where similar models are tested and compared by model selection for best fit. 3) Model generating, where one tentative initial model is simplified until the model fit cannot be improved. We used a combination of approaches 1 and 3 by defining two models based on the 'Resource Heterogeneity' (RHH) and 'More Individuals' (MIH) hypotheses (confirmatory approach, Figure 1) but using step-wise deletion of paths within these models when the model structure did not fit the data (model generating). The second step was included to assess whether the discrepancies between model and data structure were due to erroneous assumptions about interactions among variables (in this case model fit would be improved by step-wise deletion) or due to missing (i.e. not measured) variables (in that case model fit would not be improved).

Grassland land use was included in the models by the variables fertilization, grazing and mowing as well as 'Time after mowing' (as described in the first section of Materials and Methods). In addition, we added a latent (i.e. unmeasured) variable 'Land use' to describe the combined effect of land use. This latent variable accounts for correlations between the different land-use modes. As latent variables have no underlying data which the model algorithm can use to calculate its variance, either the latent's variance or one regression weight between the latent and one of the connected variables has to be fixed to 1. We chose to fix the regression weight between the latent variable and mowing frequency because mowing is a good descriptor of land-use intensity and it is correlated with both fertilization and grazing intensity [32]. The two other regression weights (between 'Land use' and grazing or fertilization) and the latent's variance can then be estimated by the model algorithm. To test for a non-linear effect of grazing intensity, we tried a quadratic transformation of grazing intensity, but because it did not improve model fit we kept the linear relationship. As the correlations between the land-use modes varied between regions, the model sometimes did not converge; in these cases, we included correlations between mowing and fertilization and between mowing and grazing instead of the latent variable. Except for the additional variables and pathways that were added in the 'Resource Abundance Model', all pathways were expected to be identical in both models (Fig. 1 A & B), as motivated in the Introduction.

We used the software package 'sem' in R for structural equation modelling [30] which fits models using Maximum Likelihood based on observed and expected covariance matrices. The model fit is estimated as the overall model p-value which indicates if the two covariance matrices are significantly different from each other ($p<0.05$, bad model fit) or not ($p>0.05$, good model fit [31], p.128f). A step-wise selection procedure was applied by hand using the corrected Akaike's Information Criterion (AICc) for models without appropriate fit (model p-value <0.05). If a second measure of model quality, the Goodness-of-fit (GoF) index, was above 0.75 for the final model, models without adequate fit were still used for comparison among regions.

To calculate the total effect of 'Land use' on the plant and arthropod variables, the standardized path coefficients within each possible pathway between 'Land use' and the respective variable were multiplied and all resulting products were summed (e.g. coefficients from 'Land use' to fertilization and from fertilization to plant biomass were multiplied and added to the products of pathways including grazing and mowing). Total effects among plant and arthropod variables were calculated accordingly. Data was transformed as necessary (further details in Appendix S3).

Results

Over all plots we recorded 271 vascular plant species in 2008 and 281 vascular plant species in 2009; 54,660 herbivore individuals from 392 species were sampled in 2008 and 19,507 herbivores from 335 species in 2009; 1,823 predator individuals from 162 species were sampled in 2008 and 1,075 predators from 154 species in 2009. The average plant biomass per plot was 648 g/m^2 in 2008 and 301 g/m^2 in 2009, mean estimated herbivore biomass per plot was 11.2 g and 4.7 g (2008 and 2009 respectively) and mean estimated predator biomass per plot was 0.4 g and 0.3 g (including 1 and 5 plots with no predators in 2008 and 2009 respectively). The lower plant and herbivore biomass in 2009 compared to 2008 can be attributed to a shorter time-span between the onset of the vegetation period and sampling (as calculated for 'Time after mowing', see Appendix S1). Increasing plant biomass was correlated with a decrease in plant species diversity (see below) and with an increase in relative cover of grass species (compared to herbaceous species) ($R^2_{adj} = 0.269$ in 2008 and $R^2_{adj} = 0.081$ in 2009, $P<0.001$ in both years and all regions; Spearman's correlation).

In 2008 and 2009, plant and herbivore species richness as well as herbivore abundance and herbivore biomass differed significantly between regions and were highest in ALB (Table 1). Plant biomass also differed significantly between regions in 2008 and was highest in SCH while in 2009 it was similar among regions (Table 1). Predator species richness differed between regions only in 2009 (being highest in SCH) and was very similar among regions in 2008. Predator abundances differed between regions both in 2008 and 2009 and were lowest in ALB. Predator biomass did not differ between regions in both years (Table 1).

Resource Heterogeneity Model

The Resource Heterogeneity Model fitted the 2008 and 2009 data well for each region (Figure 2). Except for the omission of the latent variable 'Land use' in three cases, no further path-selection procedure was applied. Despite this omission, correlations among the land-use modes were consistent between years and regions, where fertilization was positively correlated with 'Land use' (or

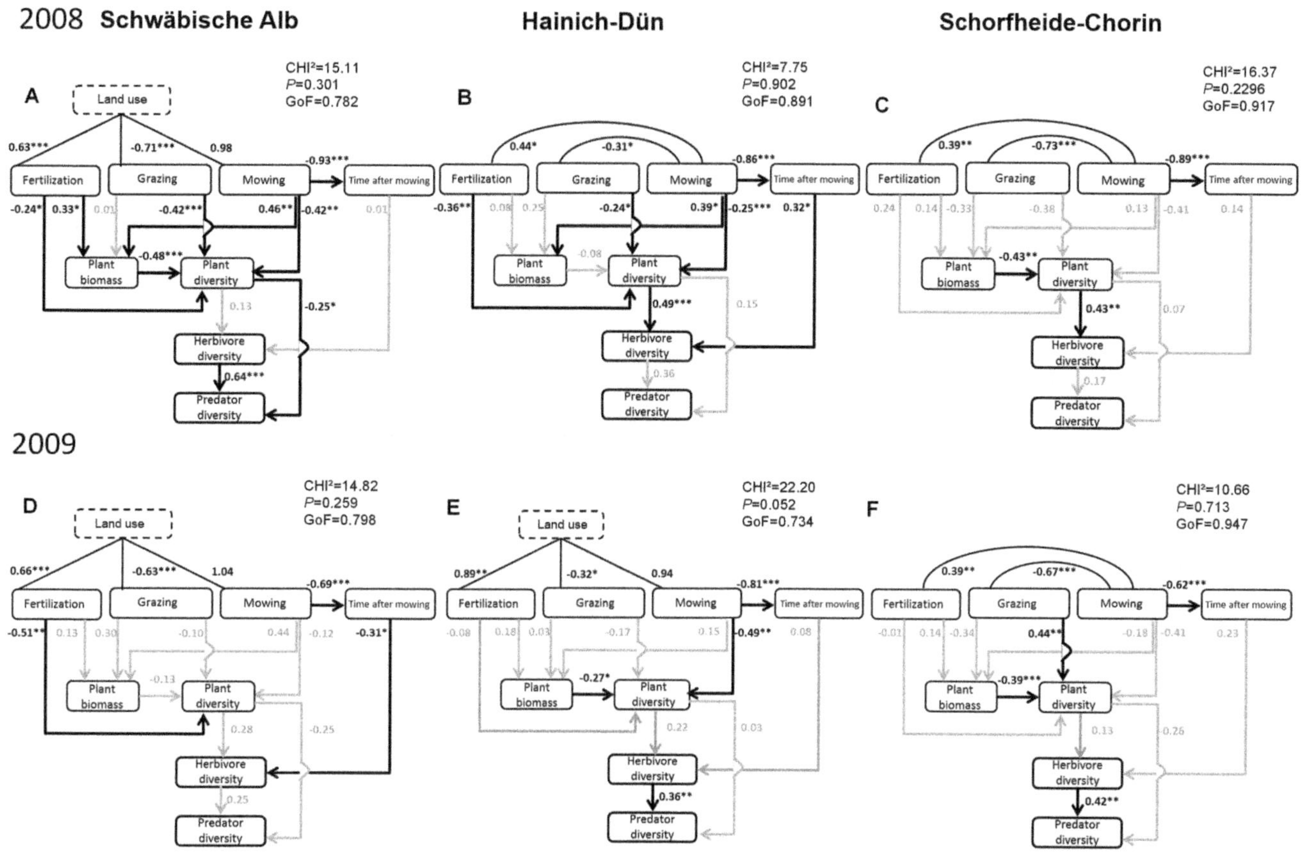

Figure 2. Standardized regression weights and significance levels from the Resource Heterogeneity Model. Models are shown for 2008 (A, B, C) and 2009 (D, E, F) for the Swabian Alb (A, D), Hainich-Dün (B, E) and Schorfheide-Chorin (C, F). Black solid lines and numbers indicate significant paths, grey arrows and numbers indicate non-significant paths. Correlation graphs for plant and arthropod measures can be found in Appendix S4. GoF = Goodness of fit. Significance levels: P<0.05: */P<0.01: **/P<0.001: ***. Total effects in Tables 2 & 3 were calculated using the standardized regression weights, e.g. plant diversity effects on predator diversity in Swabian Alb 2008 were calculated as 0.13 (coefficient from plant diversity to herbivore diversity)*0.64 (herbivore diversity to predator diversity) *−0.25 (plant diversity to predator diversity) = −0.17.

mowing) and grazing was negatively correlated with 'Land use' (or mowing).

Fertilization generally had the predicted negative effect on plant diversity and a positive effect on plant biomass, although effects were significant only in ALB (for diversity in 2008 & 2009 and for biomass only in 2008) and HAI (for diversity in 2008). We found positive but not significant effects of grazing intensity on plant biomass (Figure 2) but grazing intensity consistently reduced plant diversity (except for a significant increase in SCH in 2009). Mowing frequency consistently increased plant biomass and decreased plant diversity (although only significant in ALB and HAI and mostly in 2008). Differences in effect direction were found in SCH (positive effect of fertilization on plant diversity, and negative effects of grazing and mowing on plant biomass) but all of those effects were not significant.

As expected, plant biomass was always negatively related to plant diversity even though the paths for HAI 2008 and ALB 2009 were not significant. Effects of plant diversity on herbivore diversity were positive in all regions and years, as predicted by the RHH (although only significant in HAI and SCH). Similarly, we found a positive effect of herbivore diversity on predator diversity in all regions and years (here it was significant in 3 of 6 cases).

When summing the direct and indirect effects of 'Land use' on the target variables, we found total effects to be positive for plant biomass and negative for plant diversity (Table 2). The only exception was SCH, due to the strong influence of grazing (negative on plant biomass in 2008 and positive on plant diversity in 2009). The total effect of 'Land use' on herbivore diversity was negative in all regions and both years, whereas total effects on predator diversity were generally very weak (<0.1). The total effects of plant diversity on herbivore diversity were all positive as they are identical to the direct effects in this model. Total effects of plant diversity on predator diversity were, however, generally weak. Overall, 'Land use' decreased plant diversity directly and indirectly through an increase in plant biomass, which led to a decrease in herbivore diversity and a following decrease in predator diversity.

Resource Abundance Model

The Resource Abundance Model fitted the 2008 data adequately well for HAI and SCH, but not for ALB, even after applying path-selection procedures. For 2009, we found adequate model fits for all three regions after applying path-selection to HAI (Figure 3). Compared with the Resource Heterogeneity Model, the relationships among land-use modes remained the same, except that the latent variable could be kept in the model for HAI in 2008 (because the model converged in the first run). The direct effects of the land-use modes on plant diversity and biomass changed slightly in magnitude but not in direction compared with

Table 2. Standardized total effects derived from the Resource Heterogeneity Model for both years.

		Land use (mode) effects on				Plant diversity effects on	
		Plant biomass	Plant diversity	Herbivore diversity	Predator diversity	Herbivore diversity	Predator diversity
Swabian Alb	2008	0.65	−0.59	−0.09	0.09	0.13	−0.17
	2009	0.35	−0.44	0.10	0.03	0.28	0.06
Hainich-Dün	2008					0.49	0.32
Fertilization	2008	0.08	−0.37	−0.18	−0.06		
Grazing	2008	0.25	−0.26	−0.13	−0.05		
Mowing	2008	0.39	−0.56	−0.54	−0.19		
	2009	0.28	−0.56	−0.19	−0.08	0.22	0.11
Schorfheide-Chorin	2008					0.43	0.15
Fertilization	2008	0.41	0.18	0.10	0.01		
Grazing	2008	−0.33	−0.24	−0.10	0.0		
Mowing	2008	0.17	0.01	−0.01	0.0		
	2009					0.13	−0.20
Fertilization	2009	0.14	−0.06	0.0	0.02		
Grazing	2009	−0.03	0.46	0.06	0.02		
Mowing	2009	−0.19	−0.10	−0.16	−0.09		

Total effects were calculated by multiplying the standardized path coefficients on the single pathways between two variables and summing up those values for all possible pathways. Standardized total effects can range between −1 and 1. Effects are shown from the first row on the second row. For a description on how total effects were calculated see last paragraph in Material and Methods and refer to legend of Figure 2 for an example. Original datasets are available from the Dryad Digital Repository: http://doi.org/10.5061/dryad.f3b77.

the first model and between years. The change in magnitude is due to the fact that effects were standardized (i.e. values are relative to each other) which is influenced by the change in number of variables included in the model. As in the first model, plant biomass was negatively related to plant diversity (although not significant in HAI 2008) and plant diversity had a positive effect on herbivore diversity (although only significant in ALB & HAI 2008 and SCH 2009; Figure 3).

We did not find a significant direct effect of plant biomass on herbivore biomass as predicted by the MIH in any region or year. As the relationship between plant and herbivore biomass may be weakened if there are shifts in the relative abundances of differently sized herbivore species, we also tested the Resource Abundance Model using arthropod abundances instead of biomass but again did not find any effect of plant biomass on herbivore abundances, neither in 2008 or 2009 (Figure S1: Standardized regression weights and significance levels from the resource abundance model including arthropod abundances). Herbivore biomass had a significant positive effect on herbivore diversity in all regions and both years. Herbivore biomass also had a positive effect on predator biomass, significant in three out of six cases (ALB 2008, 2009, SCH 2009). Predator biomass itself had a significant positive effect on predator diversity in all regions and both years.

Total effects of 'Land use' on plant and arthropod diversity as well as on plant biomass were identical in direction and similar in effect strength to the Resource Heterogeneity Model for both years, including the reverse effects in SCH due to grazing. Total effects of plant diversity on herbivore diversity were positive, but total effects of plant biomass on herbivore diversity were absent or very weak. Total effects of herbivore biomass on predator diversity were positive (Table 3).

In summary, 'Land use' increased plant biomass, but this effect was not leading to an increase in herbivore biomass. Nevertheless, increasing herbivore biomass generally increased predator biomass which in turn increased predator diversity.

Discussion

We tested whether effects of land-use intensity on the diversity of arthropod herbivores and predators are mediated by changes in the diversity or biomass of their respective resources, i.e. plants and herbivores. Although the strength of effects varied between regions and years, we found negative effects of land-use intensity on the diversity of both arthropod groups which were mediated by different pathways. For herbivores, the results were consistent with the 'Resource Heterogeneity Hypothesis', as we found significant effects of plant diversity on herbivore diversity with no evidence for effects mediated by plant biomass. For predators, results were more consistent with the 'More Individuals Hypothesis', i.e. predator diversity and biomass were more strongly affected

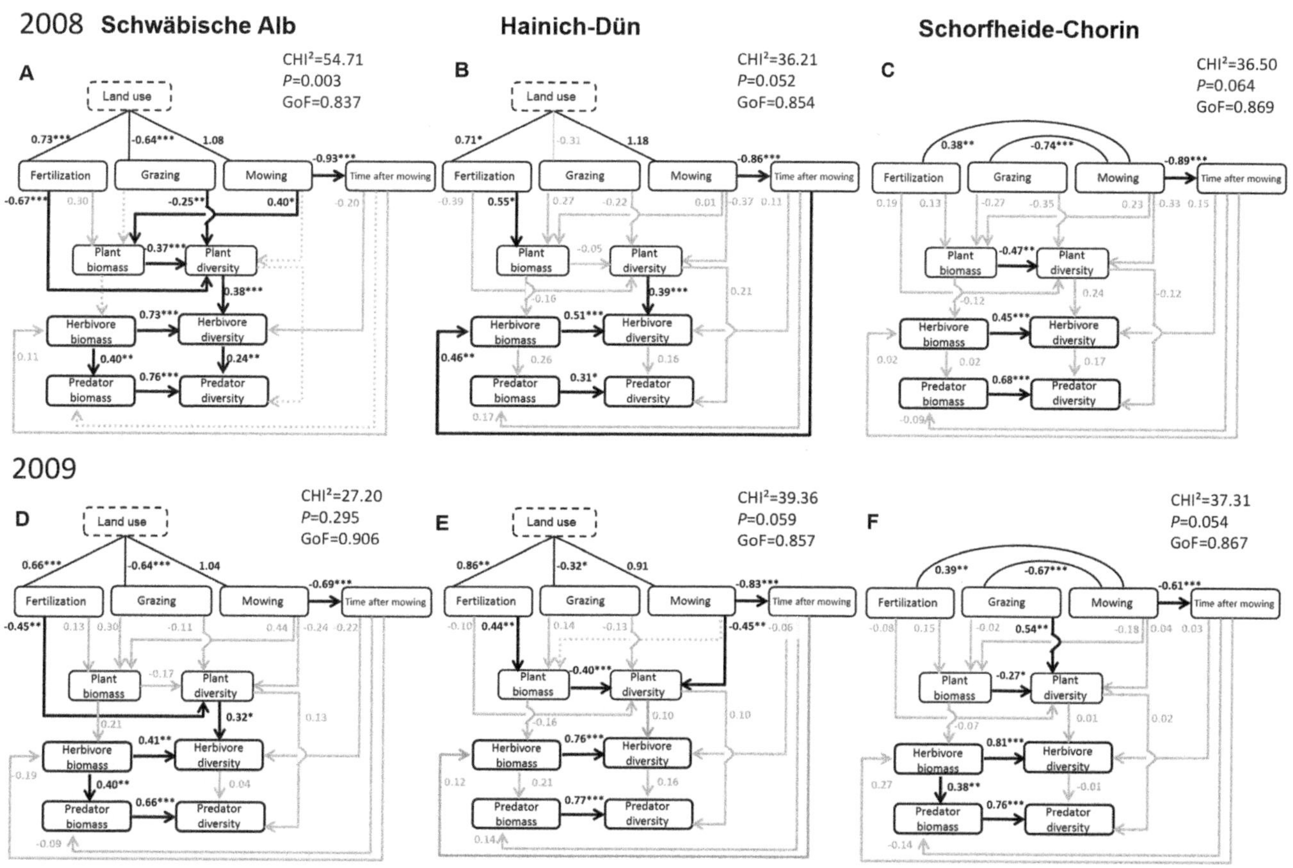

Figure 3. Standardized regression weights and significance levels from the Resource Abundance Model. Models are shown for 2008 (A, B, C) and 2009 (D, E, F) for the Swabian Alb (A, D), Hainich-Dün (B, E) and Schorfheide-Chorin (C, F). Black solid lines and numbers indicate significant paths, grey arrows and numbers indicate non-significant paths. Grey dotted paths were excluded during the step-wise selection procedure. Correlation graphs for plant and arthropod measures can be found in Appendix S4. GoF = Goodness of fit. Significance levels: $P<0.05$: */$P<0.01$: **/ $P<0.001$: ***.

Table 3. Standardized total effects derived from the Resource Abundance Model for both years.

		Land use (mode) effects on				Plant diversity effects on		Plant biomass effects on	Herbivore biomass effects on
		Plant biomass	Plant diversity	Herbivore diversity	Predator diversity	Herbivore diversity	Predator diversity	Herbivore diversity	Predator diversity
Swabian Alb	2008	0.65	−0.56	−0.09	−0.06	0.38	0.09	−0.14	0.48
	2009	0.05	−0.48	0.07	0.02	0.32	0.15	0.04	0.28
Hainich-Dün	2008	0.84	−0.71	−0.70	−0.36	0.39	0.27	−0.10	0.16
	2009	−0.37	−0.43	−0.53	−0.38	0.77	0.52	0.03	0.66
Schorfheide-Chorin	2008					0.24	−0.08	0.17	0.09
Fertilization	2008	−0.28	0.32	0.09	−0.02				
Grazing	2008	0.23	−0.46	−0.12	0.03				
Mowing	2008	0.13	−0.39	−0.25	0.06				
	2009					0.01	0.03	−0.06	0.28
Fertilization	2009	−0.02	−0.08	0.0	0.0				
Grazing	2009	−0.18	0.59	0.01	0.02				
Mowing	2009	0.15	0.0	0.01	0.02				

Total effects were calculated by multiplying the standardized path coefficients on the single pathways between two variables and summing up those values for all possible pathways. Standardized total effects can range between −1 and 1. Effects are shown from the first row on second row. For an example on how effects were calculated see legend of Figure 2.

indirectly by changes in herbivore biomass than by direct effects of herbivore diversity.

Total land-use effects on arthropods and differences between regions

The negative total effect of land-use intensity on herbivore and predator diversity is consistent with previous reports on the consequences of grassland land use on arthropods [2–4], but the effects of the individual land-use modes (especially grazing) differed between regions and years and were only partly consistent with expectations. For instance, moderate grazing was found to have a positive effect on arthropod diversity in several studies [33] but we found only weak total effects of grazing intensity on arthropod diversity (<0.2). A wider range of grazing intensities in our study system compared to other studies might explain the absence of a positive effect in our case; e.g. Dennis et al. [34] found higher abundance and diversity of arthropods under moderate grazing intensity of sheep compared to high grazing intensity. The different types of livestock in our study system likely changed effects of grazing as well because herbivore diversity in 2008 was significantly lower on cattle-grazed plots compared to sheep-grazed plots and mixed grazing by cattle and horses had a significant positive effect on predator diversity compared with cattle or sheep grazing (Figure S2: Effects of livestock type on plant and insect species richness in 2008). Differences in the grazing gradients between regions and changes in grazing practices between years can also explain the change from a negative effect of grazing to a positive effect of grazing on plant diversity in SCH.

The different correlations among the land-use modes in the different regions are, in fact, a striking result of our study. Whereas all modes were significantly correlated with the latent variable 'Land use' in ALB, fertilization and grazing were sometimes (HAI) or always (SCH) independent of each other in the other regions. The discrepancies were probably caused by the range of land-use options realized in the different regions. In ALB, where most of the grasslands are managed by small farming enterprises or farming families, we find extensively grazed, unfertilized plots, e.g. sheep pastures on nutrient-poor hillsides, as well as intensively grazed, fertilized plots. Thus, grazing and fertilization are closely linked in this region. In the other two regions, plots with low grazing intensity sometimes receive low fertilizer input (for instance, in organic farming practices) but in other cases they are highly fertilized mown pastures that are only grazed at the end of the plant growth period. This weakened the correlation between grazing and fertilization.

Effects of land-use intensity on plant biomass and diversity

Socher et al. [1] extensively discussed the effects of fertilization, grazing and mowing on plant biomass and diversity using the same plots as the present study, thus here we only summarize the main points: Increasing fertilization intensity generally decreased plant diversity and increased biomass as found in many preceding studies. Negative effects of high mowing frequency on plant diversity support previous findings from different types of grasslands and various regions [35]. The negative effect of grazing on plant diversity appears to contrast the general finding that grazing increases plant diversity [36]. However, most previous studies compared grazed with ungrazed sites and thus found that grazing increased plant diversity via increased sward heterogeneity [11,37]. In our case, grazing ranged from no grazing to very high grazing intensities, which could result in a non-linear or hump-shaped effect of grazing on plant diversity. As including a non-linear effect of grazing did not improve our model (see Method section) it seems likely that the negative effect of very high grazing intensities is exceeding the positive effect moderate grazing has compared to non-grazed sites.

Effects of changes in the plant community on herbivores

Our results showed clear evidence that land-use effects on herbivore diversity are mediated by plant diversity as predicted by the 'Resource Heterogeneity Hypothesis'. The fact that we did not find effects on herbivore diversity mediated by plant biomass is in contrast to findings from experimental plant communities [22], which showed that arthropod diversity was only indirectly affected by plant diversity through increased plant biomass. This indirect effect on arthropod diversity was additionally mediated by arthropod biomass (measured as biovolume) and therefore followed the 'More Individuals Hypothesis'[22]. The differences between results from our study in managed grasslands and results from experimental plant communities may be due to the absence of correlations between the proportion of particular plant functional groups and plant diversity in biodiversity experiment, as both variables are manipulated similarly. In contrast, the grasslands in our study system showed an increasing cover proportion of grasses with decreasing plant diversity. As found by Haddad et al. [38] the presence of grass species (which are productive but have low nutritional quality for herbivores) led to a decrease in total insect abundance by 25% even though total plant biomass increased. Only when grass species were absent and all plants were of higher nutritional quality, insect abundance was best explained by plant biomass. Hence, the higher cover of grass species on grasslands with low plant diversity in our study system led to a higher plant biomass but at the same time could not sustain higher herbivore biomass possibly because the overall nutritional value for the arthropods did not increase together with plant biomass.

Effects of changes in the herbivore community on predators

We found direct and indirect effects between herbivore and predator diversity, indicating mechanisms in accordance with both the 'Resource Heterogeneity' and the 'More Individuals Hypothesis'. One example would be the ALB 2008 where both direct and indirect effects were significant and strong. This indicates a complementary role of both mechanisms which might be the result of different predator groups reacting to either one of the mechanisms. In our study, the total effect of herbivore biomass on predator diversity was stronger in five out of six cases than the direct effect of herbivore diversity (compare Figure 3 and Table 3). This agrees with results from a plant diversity experiment, where effects of plants on herbivores were consistent with the 'Resource Heterogeneity Hypothesis' but effects of herbivores on predators were more in agreement with the 'More Individuals Hypothesis' [12]. Further research is needed to understand how predator diversity is affected by land use, as herbivore biomass was not affected by any of the included factors. This is relevant for sustainable land use of grasslands in agricultural dominated landscapes, because high predator abundance and diversity enhances biocontrol potential e.g. [39].

Where to go from here

We tested the effect of land-use intensity on important grassland herb-dwelling arthropods over a wide geographic range. To achieve standardized sampling in the vegetation layer across a large number of plots, sweep-net sampling was used which is a suitable method to representatively sample important herbivores (e.g. Heteroptera) as well as predators (e.g. Araneae) among arthropods (e.g. [40–42]). Nevertheless, it is a less well-performing method to sample other functional groups such as pollinators (butterflies and bees) [40] or ground-dwelling species [43]. Additional methods (such as suction sampling, pitfall traps or pan traps) might therefore be advised if a study's focus is not only on herb-dwelling species. Disentangling the differences which might apply to different functional groups within herbivores and predators (e.g. sucking vs. chewing herbivores or predators with different hunting strategies) will further increase our understanding of land-use effects. One promising approach was recently proposed by Lavorel et al. [44] who included producer and consumer traits in structural equation models to understand how land use affects ecosystem services through changes in the trait composition of the groups which provide the services.

Conclusions

Our results emphasize the importance of studying indirect effects of land-use intensity on the arthropod community, as they showed that herbivores and predators respond to changes of different aspects of their resources. We confirm that herbivore diversity is responding positively to higher plant diversity in grasslands, whereas herbivore biomass matters more than diversity for how predator diversity is affected by land use. By including different regions we showed on the one hand that the negative effects of high fertilization intensity and high mowing frequency on arthropod diversity are consistent over large scales; but on the other hand the variability of land-use traditions clearly indicates that findings cannot be easily extended to a wider geographical context. Our results thus not only emphasize the importance of land use for biodiversity changes, but also the need for more differentiated approaches to disentangle how different land-use modes have different effects on biodiversity, and how chains of effects differ for different aspects of biodiversity.

Supporting Information

Figure S1 Standardized regression weights and significance levels from the resource abundance model including arthropod abundances. Models are shown after step-wise deletion of non-significant paths. Black solid lines and numbers indicate significant paths; grey arrows indicate non-significant paths. Grey, dotted paths were excluded during the step-wise selection procedure. Significance level: $p<0.05$: */$p<0.01$: **/$p<0.001$: ***.

Figure S2 Effects of livestock type on plant and insect species richness in 2008. Means per plot and standard errors are shown. Horizontal lines indicate significant differences based on Tukey's HSD test. Significance levels: $p<0.05$: */$p<0.01$: **/$p<0.001$: ***.

Table S1 Mean and range of land-use activities in the three regions and during the years considered for the analysis with samplings from 2008.

Table S2 Mean and range of land-use activities in the three regions and during the years considered for the analysis with samplings from 2009.

Appendix S1 Assessment and calculation of land-use information.

Appendix S2 Classification of trophic guilds.

Appendix S3 Structural equation model setup and path-selection procedure.

Appendix S4 Bivariate correlations between model variables.

Appendix S5 List of arthropod species sampled in 2008

Appendix S6 List of arthropod species sampled in 2009.

Acknowledgments

Special thanks go to Leonardo Ré Jorge, who gave valuable comments during the discussion on the model structure. We thank Luis Sikora for his help with sweep-netting the grasslands and Roland Achtziger, Theo Blick, Boris Büche, Michael-Andreas Fritze, Günter Köhler, Frank Köhler, Franz Schmolke and Thomas Wagner for identifying the arthropods and providing data on trophic guilds and body size. A database on Orthoptera was kindly provided by Frank Dziock. We thank the managers of the three exploratories, Swen Renner, Sonja Gockel, Kerstin Wiesner and Martin Gorke for their work in maintaining the plot and project infrastructure; Andreas Hemp and Uta Schumacher for insuring successful field work in the Schorfheide; Simone Pfeiffer and Christiane Fischer giving support through the central office, Michael Owonibi for managing the central data base and Eduard Linsenmair, Dominik Hessenmöller, Jens Nieschulze, Ingo Schöning, François Buscot, Ernst-Detlef Schulze and the late Elisabeth Kalko for their role in setting up the Biodiversity Exploratories project.

Author Contributions

Conceived and designed the experiments: MF MMG WWW. Performed the experiments: SB ML JM EP SAS MT. Analyzed the data: NKS MMG TML. Contributed to the writing of the manuscript: NKS MMG TML SB ML JM EP SAS MT MF WWW.

References

1. Socher SA, Prati D, Boch S, Müller J, Klaus VH, et al. (2012) Direct and productivity-mediated indirect effects of fertilization, mowing and grazing on grassland species richness. Journal of Ecology 100: 1391–1399.
2. Bell JR, Wheater CP, Cullen WR (2001) The implications of grassland and heathland management for the conservation of spider communities: a review. Journal of Zoology 255: 377–387.
3. Di Giulio M, Edwards PJ, Meister E (2001) Enhancing insect diversity in agricultural grasslands: the roles of management and landscape structure. Journal of Applied Ecology 38: 310–319.
4. Nickel H, Hildebrandt J (2003) Auchenorrhyncha communities as indicators of disturbance in grasslands (Insecta, Hemiptera) - a case study from the Elbe flood plains (northern Germany). Agriculture Ecosystems & Environment 98: 183–199.
5. Weiner CN, Werner M, Linsenmair KE, Bluthgen N (2011) Land use intensity in grasslands: Changes in biodiversity, species composition and specialisation in flower visitor networks. Basic and Applied Ecology 12: 292–299.
6. Chamberlain DE, Fuller RJ, Bunce RGH, Duckworth JC, Shrubb M (2000) Changes in the abundance of farmland birds in relation to the timing of agricultural intensification in England and Wales. Journal of Applied Ecology 37: 771–788.
7. Allan E, Bossdorf O, Dormann CF, Prati D, Gossner MM, et al. (2014) Interannual variation in land-use intensity enhances grassland multidiversity. Proceedings of the National Academy of Sciences of the United States of America 111: 308–313.

8. Littlewood NA, Stewart AJA, Woodcock BA (2012) Science into practice - how can fundamental science contribute to better management of grasslands for invertebrates? Insect Conservation and Diversity 5: 1–8.
9. Marini L, Fontana P, Scotton M, Klimek S (2008) Vascular plant and Orthoptera diversity in relation to grassland management and landscape composition in the European Alps. Journal of Applied Ecology 45: 361–370.
10. Morris MG, Lakhani KH (1979) Responses of grassland invertebrates to management by cutting. 1. Species-diversity of Hemiptera. Journal of Applied Ecology 16: 77–98.
11. Woodcock B, Potts S, Tscheulin T, Pilgrim E, Ramsey A, et al. (2009) Responses of invertebrate trophic level, feeding guild and body size to the management of improved grassland field margins. Journal of Applied Ecology 46: 920–929.
12. Haddad NM, Crutsinger GM, Gross K, Haarstad J, Knops JMH, et al. (2009) Plant species loss decreases arthropod diversity and shifts trophic structure. Ecology Letters 12: 1029–1039.
13. van den Berg LJL, Vergeer P, Rich TCG, Smart SM, Guest D, et al. (2011) Direct and indirect effects of nitrogen deposition on species composition change in calcareous grasslands. Global Change Biology 17: 1871–1883.
14. Ryder C, Moran J, Mc Donnell R, Gormally M (2005) Conservation implications of grazing practices on the plant and dipteran communities of a turlough in Co. Mayo, Ireland. Biodiversity and Conservation 14: 187–204.
15. Sjodin NE, Bengtsson J, Ekbom B (2008) The influence of grazing intensity and landscape composition on the diversity and abundance of flower-visiting insects. Journal of Applied Ecology 45: 763–772.
16. Watkinson AR, Ormerod SJ (2001) Grasslands, grazing and biodiversity: editors' introduction. Journal of Applied Ecology 38: 233–237.
17. Hutchinson GE (1959) Homage to Santa Rosalia or Why are there so many kinds of animals? American Naturalist 93: 145–159.
18. Lewinsohn TM, Roslin T (2008) Four ways towards tropical herbivore megadiversity. Ecology Letters 11: 398–416.
19. Southwood TRE, Brown VK, Reader PM (1979) The relationships of plant and insect diversities in succession. Biological Journal of the Linnean Society 12: 327–348.
20. Strong DR, Jr., Lawton JH, Southwood TRE (1984) Insects on Plants: Community Patterns and Mechanisms. Cambridge, MA: Harvard University Press.
21. Srivastava DS, Lawton JH (1998) Why more productive sites have more species: An experimental test of theory using tree-hole communities. American Naturalist 152: 510–529.
22. Borer ET, Seabloom EW, Tilman D, Novotny V (2012) Plant diversity controls arthropod biomass and temporal stability. Ecology Letters 15: 1457–1464.
23. Andreasen C, Hansen CH, Moller C, Kjaer-Pedersen NK (2002) Regrowth of weed species after cutting. Weed Technology 16: 873–879.
24. Becklin KM, Kirkpatrick HE (2006) Compensation through rosette formation: the response of scarlet gilia (Ipomopsis aggregata: Polemoniaceae) to mammalian herbivory. Canadian Journal of Botany-Revue Canadienne De Botanique 84: 1298–1303.
25. Fischer M, Bossdorf O, Gockel S, Hansel F, Hemp A, et al. (2010) Implementing large-scale and long-term functional biodiversity research: The Biodiversity Exploratories. Basic and Applied Ecology 11: 473–485.
26. Humbert JY, Ghazoul J, Walter T (2009) Meadow harvesting techniques and their impacts on field fauna. Agriculture Ecosystems & Environment 130: 1–8.
27. Wisskirchen R, Haeupler H (1998) Standardliste der Farn- und Blütenpflanzen Deutschlands. Stuttgart (Hohenheim): Eugen Ulmer.
28. Brook A, Woodcock B, Sinka M, Vanbergen A (2008) Experimental verification of suction sampler capture efficiency in grasslands of differing vegetation height and structure. Journal of Applied Ecology 45: 1357–1363.
29. Rogers LE, Hinds WT, Buschbom RL (1976) A general weight versus length relationshop for insects. Annals of the Entomological Society of America 69: 387–389.
30. R Core Team (2013) R: A language and environment for statistical computing. 3.0.2 ed. Vienna, Austria: R Foundation for Statistical Computing.
31. Grace JB (2006) Structural Equation Modeling and Natural Systems. Cambridge, UK: Cambridge University Press.
32. Blüthgen N, Dormann CF, Prati D, Klaus VH, Kleinebecker T, et al. (2012) A quantitative index of land-use intensity in grasslands: Integrating mowing, grazing and fertilization. Basic and Applied Ecology 13: 207–220.
33. Scohier A, Dumont B (2012) How do sheep affect plant communities and arthropod populations in temperate grasslands? Animal 6: 1129–1138.
34. Dennis P, Skartveit J, McCracken DI, Pakeman RJ, Beaton K, et al. (2008) The effects of livestock grazing on foliar arthropods associated with bird diet in upland grasslands of Scotland. Journal of Applied Ecology 45: 279–287.
35. Hopkins A, Wilkins RJ (2006) Temperate grassland: key developments in the last century and future perspectives. Journal of Agricultural Science 144: 503–523.
36. Marion B, Bonis A, Bouzille JB (2010) How much does grazing-induced heterogeneity impact plant diversity in wet grasslands? Ecoscience 17: 229–239.
37. Rook AJ, Dumont B, Isselstein J, Osoro K, WallisDeVries MF, et al. (2004) Matching type of livestock to desired biodiversity outcomes in pastures - a review. Biological Conservation 119: 137–150.
38. Haddad NM, Tilman D, Haarstad J, Ritchie M, Knops JMH (2001) Contrasting effects of plant richness and composition on insect communities: A field experiment. American Naturalist 158: 17–35.
39. Geiger F, Bengtsson J, Berendse F, Weisser WW, Emmerson M, et al. (2010) Persistent negative effects of pesticides on biodiversity and biological control potential on European farmland. Basic and Applied Ecology 11: 97–105.
40. Buffington ML, Redak RA (1998) A comparison of vacuum sampling versus sweep-netting for arthropod biodiversity measurements in California coastal sage scrub. Journal of Insect Conservation 2: 99–106.
41. Doxon E, Davis C, Fuhlendorf S (2011) Comparison of two methods for sampling invertebrates: vacuum and sweep-net sampling. Journal of Field Ornithology 82: 60–67.
42. Spafford R, Lortie C (2013) Sweeping beauty: is grassland arthropod community composition effectively estimated by sweep netting? Ecology and Evolution 3: 3347–3358.
43. Standen V (2000) The adequacy of collecting techniques for estimating species richness of grassland invertebrates. Journal of Applied Ecology 37: 884–893.
44. Lavorel S, Storkey J, Bardgett R, Bello F, Berg M, et al. (2013) A novel framework for linking functional diversity of plants with other trophic levels for the quantification of ecosystem services. Journal of Vegetation Science 24: 942–948.

4

Aphid Honeydew Quality as a Food Source for Parasitoids Is Maintained in Bt Cotton

Steffen Hagenbucher[1¤], Felix L. Wäckers[2], Jörg Romeis[1*]

1 Agroscope, Institute for Sustainability Sciences (ISS), Zurich, Switzerland, 2 Centre for Sustainable Agriculture, Lancaster University, Lancaster, United Kingdom

Abstract

Bt-transgenic cotton has proven to be highly efficient in controlling key lepidopteran pests. One concern with the deployment of Bt cotton varieties is the potential proliferation of non-target pests. We previously showed that Bt cotton contained lower concentrations of insecticidal terpenoids as a result of reduced caterpillar damage, which benefited the aphid *Aphis gossypii*. It is thus important that non-target herbivores are under biological control in Bt cotton fields. The induction or lack of induction of terpenoids could also influence the quality of aphid honeydew, an important food source for beneficial insects. We therefore screened *A. gossypii* honeydew for cotton terpenoids, that are induced by caterpillars but not the aphids. We then tested the influence of induced insect-resistance of cotton on honeydew nutritional quality for the aphid parasitoid *Lysiphlebus testaceipes* and the whitefly parasitoid *Eretmocerus eremicus*. We detected the cotton terpenoids gossypol and hemigossypolone in *A. gossypii* honeydew. Although a feeding assay demonstrated that gossypol reduced the longevity of both parasitoid species in a non-linear, dose-dependent manner, the honeydew was capable of sustaining parasitoid longevity and reproduction. The level of caterpillar damage to Bt and non-Bt cotton had no impact on the quality of honeydew for the parasitoids.These results indicate that the nutritional quality of honeydew is maintained in Bt cotton and is not influenced by induced insect resistance.

Editor: Nicolas Desneux, French National Institute for Agricultural Research (INRA), France

Funding: This project was supported by the Swiss National Science Foundation (SNF grant no. 31003A-120477). The funders had no role in study design, data collection and analysis, decision to publish, or preparation of the manuscript.

Competing Interests: The authors have declared that no competing interests exist.

* Email: joerg.romeis@agroscope.admin.ch

¤ Current address: Department of Plant Protection Biology, Swedish University of Agricultural Sciences, Alnarp, Sweden

Introduction

In 2012, insect-resistant, genetically engineered (GE) cotton expressing Cry toxins from *Bacillus thuringiensis* (Bt) was grown on 22.5 million hectares world-wide [1]. These Bt cotton plants are highly resistant to major lepidopteran pests [2,3], and their large-scale adoption has led to area-wide population declines of some of these target pests [4,5]. The high specificity of Cry toxins makes them a valuable tool for integrated pest management (IPM) programs as the use of Bt-transgenic varieties can lead to reductions in insecticide use with benefits for biological control [3,6–8].

Research in different regions of the world has indicated that the abundance of mirid (Hemiptera: Miridae) and stink bug (Hemiptera: Pentatomidae) herbivores that are not suppressed by the insecticidal trait has increased on Bt cotton [3,9]. The reasons for these population increases are complex and include the reduction in insecticide use and a decrease in resource competition [3,9–11]. Another possible factor that could lead to the emergence of non-target pests is reduced indirect, plant-mediated competition between herbivores, which can be mediated by secondary plant metabolites [12]. Cotton plants possess a range of defence compounds that belong to a group of closely related terpenoids: gossypol, hemigossypolone, and the heliocides 1–4 [13–15]. The cotton defence system is inducible, and damage by some herbivores results in a systemic increase in terpenoid production [14,16–19].

Because Bt cotton plants are less damaged by tissue-feeding caterpillars than conventional cotton plants [20], the overall amount of terpenoids could be lower and the plants thereby more susceptible to insects that do not induce terpenoid production and are also not affected by the Cry toxins. We previously reported that the cotton aphid *Aphis gossypii* (Glover) (Hemiptera: Aphididae) was indeed more abundant on Bt plants than on non-Bt plants infested with *Heliothis virescens* (Fabricius) (Lepidoptera: Noctuidae) larvae, and that the increase in *A. gossypii* abundance was correlated with lower levels of caterpillar-induced terpenoids [15]. We have also investigated whether this positive effect on aphids affects the next higher trophic level, i.e., aphid parasitoids. While fewer parasitoid mummies were found on caterpillar-infested non-Bt than on caterpillar-infested Bt cotton, aphid quality as hosts for *Lysiphlebus testaceipes* (Cresson) (Hymenoptera: Braconidae) was not affected by the changes in terpenoid production of the plants even though the aphids contained cotton terpenoids [21].

Adult parasitoids, however, could also be exposed to cotton terpenoids when they consume honeydew on cotton plants. Honeydew is an important carbohydrate source for parasitoids, especially in agro-ecosystems that often lack alternative sugar sources [22–25]. Honeydew consumption increases parasitoid longevity [24,26,27] and fecundity [28], and honeydew can serve as a host-finding cue for parasitoids of honeydew producers [29,30]. However, honeydew is often inferior to nectar or nectar–sugar solutions as a food source for parasitoids [22,24,26,31–34]. One possible reason for this is that honeydew contains insect-derived sugars that have a lower nutritional value for insects than plant-derived sugars and that may even be toxic [33,35]. Additionally, secondary plant compounds can pass through the insect gut unaltered and can thus be present in the honeydew, which may further reduce its suitability as food [24,34,36]. Aphids feeding on caterpillar-infested cotton plants in which plant defenses have been induced may consume elevated levels of induced secondary metabolites. When they excrete these metabolites in their honeydew, this would in turn expose the honeydew feeding insects to higher levels of secondary metabolite in their food.

We here assessed the impact of caterpillar-induced cotton terpenoids on the quality of *A. gossypii* honeydew as a food source for two parasitoid species common in USA cotton fields, i.e., *L. testaceipes* and *Eretmocerus eremicus* (Rose and Zolnerowich) (Hymenoptera: Aphelinidae). While *L. testaceipes* is one of the principal parasitoids of *A. gossypii* [37], *E. eremicus* is an important parasitoid of *Bemisia tabaci* (Gennadius) (Hemiptera: Aleyrodidae) (Gerling et al., 2001). Although the latter species is more likely to feed on the honeydew of its host than on that of *A. gossypii*, it would still encounter aphid honeydew in its natural habitat. For example, in Tifton, GA (USA), *A. gossypii* invades cotton in early June and peaks in mid-July before populations start to collapse [38]. Aphids are, however, still present during the remaining season and often reach a second population peak in late July/August [38], which is also the time were whiteflies start to appear in the cotton fields (personal observation). In this case aphid honeydew could be an important source of nutrition when whitefly numbers are still low. Our hypothesis was that *A. gossypii* honeydew would contain lower levels of terpenoids on Bt cotton than on non-Bt cotton, such that honeydew from Bt cotton would be superior to honeydew from non-Bt cotton as a food source for the parasitoids.

Materials and Methods

Plant material

Two varieties of GE cotton plants provided by Monsanto Company (St. Louis, USA) were used for the experiments. One variety ("Bt cotton"), Deltapine DPL143B2 RF (event: MON15985×MON88913), expresses two Bt toxins (Cry1Ac and Cry2Ab, Bollgard II) and carries a herbicide-tolerance trait. The second variety ("non-Bt cotton"), Deltapine DPL147 RF (event: MON88913), has a similar genetic background but contains only the herbicide-tolerance trait. For insect rearing, the closely related non-GE variety Deltapine DPL491 was used. Plants were grown in 3-l plastic pots (one plant per pot) containing heat-sterilized humus-rich soil. At planting, 15 mg of the slow-release fertilizer Osmocote (16% N, 11% P_2O_5, 11% K_2O; Scotts UK Professional, Bramford, UK) was added to each pot. Subsequently, the plants were fertilized weekly using 10N:10P:8K at 20 ml/l and were watered daily. Plants were enclosed in gauze cages (height: 71 cm, diameter: 35 cm, mesh-width: 0.264 mm) to protect them from glasshouse pests.

Insect material

A colony of *A. gossypii* was obtained from Syngenta (Stein, Switzerland) and reared permanently on 4- to 8-week-old non-Bt cotton plants (DPL491). Larvae of *H. virescens* and adult *B. tabaci* were regularly obtained from Syngenta. For the experiments, *B. tabaci* colonies were established by enclosing five adults on the youngest fully developed leaf of a 4- to 8-week-old cotton plant (DPL 491) in a small plastic cage (diameter: 35 mm). After 2 weeks, the colonies were used for the experiments.

Mummies of *L. testaceipes* were obtained from Katz Biotech AG (Germany), and mummies of *E. eremicus* were obtained from Andermatt Biocontrol AG (Switzerland). All insects were kept at 25±5°C and 70±10% RH and under long-day conditions (16 h of light:8 h of dark).

Honeydew collection from induced and uninduced cotton plants

In the glasshouse, Bt and non-Bt cotton plants were used when they had four fully expanded true leaves. Plants were either infested by a single *H. virescens* larva (3rd instar) that was caged on the youngest fully developed leaf for 7 days in a gauze bag or plants were left uninfested (control). *Heliothis virescens* larval weight was recorded before the larva was placed on the plant and after 7 days to calculate the weight gain.

After the larva was removed, 100 *A. gossypii* (mixed stages) were transferred to the youngest fully developed leaf of each plant (not identical to the leaf on which the larva was released). Uninfested plants were treated in the same way. The aphids were enclosed in a clip-cage built from 9-cm-diameter Petri dishes. After 7 days, honeydew collection was initiated by placing glass plates (15×15 mm) on the bottom of each Petri dish. During the following 7 days, the glass plates were collected every 24 h, and the bottom of the Petri dish was replaced to prevent growth of fungi. The honeydew-sprinkled glass plates were stored at −80°C for later use.

After the 7 days of honeydew collection, damage caused by *H. virescens* larvae was recorded. For this, the damaged leaf from each plant was collected and photographed, and the damaged leaf area was measured with ImageJ v1.42 (National Institutes of Health, USA) software. Additionally, the youngest fully developed leaf of each plant was collected for terpenoid quantification.

For analysis of the terpenoid content of honeydew, we used a different set of plants that were treated as described in the previous paragraphs. Honeydew was collect from 7–10 plants per treatment. However, collection was only conducted during the first day and not during the entire seven day period, as in the previous experiment. The glass plates used were pre-weighed before the honeydew collection. After 24 h, the plates were removed and stored at −80°C until later analysis by high-performance liquid chromatography (HPLC).

Parasitoid longevity as affected by gossypol

Females of *L. testaceipes* and *E. eremicus* were between 1 and 14 h old when used for these experiments. To assume mating, every female parasitoid was kept in a plastic vial with one male for 4 h. Subsequently, females were kept individually in glass vials (60×11.7 mm) that were closed with Parafilm and contained a piece of cotton wool soaked with water to provide a RH>95% to minimize evaporation of water from the food sources (sucrose solutions containing gossypol and *A. gossypii* honeydew, as described in the next paragraphs). The cotton wool was replaced every other day. Survival of the parasitoids was recorded twice each day (at 9–10 am and at 5–6 pm). Experiments were conducted at 25±5°C and 85±10% RH under long-day conditions (16 h of light:8 h of dark). Thirty females were tested per treatment and parasitoid species. Females lost or killed during handling were removed and were not included in the data analysis, leading to a total of 27–30 replications for *E. eremicus* and 28–30 replications for *L. testaceipes*.

Female parasitoids of both species where fed different concentrations of the cotton terpenoid gossypol (≥95%, Sigma, St. Louis, USA). Because gossypol is not soluble in water, it was first dissolved in 1% dimethyl sulfoxide (DMSO). Subsequently, different concentrations of gossypol (0.000001, 0.00001, 0.0001, and 0.001%) were dissolved in a 1 M sucrose solution. A 1 M sucrose solution with 1% DMSO, a pure 1 M sucrose solution, and pure water served as controls. A single drop (2 µl) of the test solution was placed on the inner surface of the Parafilm that was used to seal the vial. The Parafilm and the test solution were changed every other day.

Parasitoid longevity as affected *A. gossypii* honeydew

Female parasitoids of both species where fed *A. gossypii* honeydew from *H. virescens*-infested and uninfested Bt and non-Bt cotton plants. For details on the experimental set-up and environmental conditions see section "Parasitoid longevity as affected by gossypol". The following treatments were compared: water only (control), 1 M sucrose solution, honeydew from uninfested Bt cotton, honeydew from infested Bt cotton, honeydew from uninfested non-Bt cotton, and honeydew from infested non-Bt cotton. Thirty females were tested per treatment and parasitoid species. Females lost or killed during handling were removed and were not included in the data analysis, leading to a total of 24–30 replications for *E. eremicus* and 27–30 replications for *L. testaceipes*.

Honeydew was provided *ad libitum* on the glass plates, on which it was collected and was replaced daily. The plates were carefully broken in two pieces to fit them into the vial, one halve per vial. The wet cotton wool was replaced every other day.

Parasitoid reproductive capacity as affected by *A. gossypii* honeydew

The parasitoids *L. testaceipes* and *E. eremicus* were prepared and fed with honeydew as described for the longevity experiment. After 24 h of feeding, each *L. testaceipes* female was removed from the glass tube and placed in a Petri dish (diameter: 50 mm) containing parts of a non-Bt (DPL491) cotton leaf with >100 *A. gossypii* of mixed stages. After 4 h, the parasitoid was removed, and the cotton leaf was placed on a non-transgenic cotton seedlings to allow the aphids to settle on the plant. The plants were checked daily for parasitoid mummies. Mummies were collected, placed in plastic containers (17×17×8 mm), and checked daily for emerging parasitoids. The number of mummies produced per female, parasitoid emergence rate, and the sex ratio of parasitoid offspring were recorded.

After 2, 4, and 6 days of feeding on honeydew, each *E. eremicus* female was removed from the glass tube and placed in a plastic container (17×17 mm). The container was placed over a colony of *B. tabaci* nymphs on a non-Bt cotton seedling (see section "Insect material" for details). After 4 h, the container was removed and the parasitoid was transferred back into the glass tube. The number of parasitoids emerging from parasitized *B. tabaci* was recorded.

Both experiments were conducted at 25±5°C and 75±10% RH. In the case of *E. eretmocerus* each treatment was represented by 23–26 parasitoid females. For *L. testaceipes* there were 26–31 females per treatment.

Parasitoid gustatory response to gossypol

Starved, water-satiated female parasitoids (≤14 h old, mated) were used for this experiment. To ensure that females were water-satiated, they were individually kept in a plastic vial (17×17×8 mm) containing a piece of cotton wool soaked with water for a minimum of 1 h before they were used in the experiment. At the start of the experiment, a droplet of the test solution (1 µl for *L. testaceipes* and 0.3 µl for *E. eremicus*) was placed on the lid of the plastic vial. The parasitoid's tendency to walk upwards soon brought it into contact with the test solution. The parasitoids were observed until they made contact with the droplet, which happened within 60 sec in all cases. After the parasitoid contacted the test solution, its behavior was recorded. The reaction was either scored as acceptance (feeding for a minimum of 5 s) or rejection (less than 5 s of contact). When the test solution caused a feeding response, the duration of the feeding event (time during which the mouthparts were in contact with the test solution) was recorded. Observations were made with a dissecting microscope.

The following test solutions were used: 1 M sucrose solution, 1 M sucrose+1% DMSO, and sucrose/DMSO solutions containing 0.000001, 0.00001, 0.0001, and 0.001% gossypol. In total, 28–33 parasitoids of each species were observed per treatment.

HPLC analyses of terpenoids in plant material and honeydew

Pre-weighed honeydew-covered glass plates (15×15 mm), obtained as described in the "Honeydew collection" section, were placed in a Petri dish that contained a water-saturated piece of cotton wool. The Petri dishes were closed with Parafilm and kept in a climatic chamber (25°C and 85% RH) for 30 min to saturate the honeydew with water. Subsequently, each glass plate was weighed on a microbalance (Mettler Toledo MX5; division d = 1 µg; tolerance ±2 µg). To adjust for the weight loss of honeydew through evaporation of water during the measurement, the weight of the samples was recorded after 10 s on the balance.

Samples were placed in a 200-ml Erlenmeyer flask, and 5 ml of water was added. The flasks were shaken for 30 minutes (in darkness) to dissolve the honeydew in the water, and the solution was transferred into 10-ml plastic tubes. To extract the gossypol from the solution, 1 ml of heptane was added, and the heptane phase was transferred to a new vial. This was repeated three times. Heptane was evaporated with compressed air at 50°C. The gossypol residue was redissolved in 100 µl of acetonitirile (Multi-solvent HPLC grade, Scharlau, Sentmenat, Spain), water (purified by a Gradient A10, Millipore, Billerica, USA), and 85% phosphoric acid (Fluka, Buchs, Switzerland) (80:20:0.1) and used for HPLC. A 50-µl volume of each sample was injected into the HPLC.

Terpenoid content (gossypol, hemigossypolone, heliocides 1 and 4, and heliocides 2 and 3) was analyzed by HPLC using the protocol previously described in detail [15]. Because the terpenoid concentrations in the leaves were high, 1000 µl of the extraction solution was used, and only 10 µl was injected into the HPLC.

Statistical analysis

All analyses were conducted using R 2.13.2.

Plant damage and weight gain by *H. virescens* were analyzed using the Welch *t*-test for non-homogenous variances. This test was chosen after detecting a non-homogenous variance using Levene's Test for Homogeneity of Variance. The effect of gossypol on the acceptance rate of sucrose solution by parasitoids was analyzed by comparing all treatments against DMSO using

Pearson's Chi-squared test. Significance levels were adjusted with Bonferroni-correction, resulting in an adjusted $\alpha = 0.01$.

One-way analyses of variance (ANOVA) was used to analyze data sets with homogenous variance (tested with Levene's Test for Homogeneity of Variance) and normal distribution (analyzes using Q-Q Plots). This was the case for the terpenoid contents of plants and of honeydew, the effect of gossypol on the feeding time of parasitoids, longevity of both parasitoids and the assessed parameters of the *L. testaceipes* reproductive capacity after honeydew feeding (number of mummies, offspring, emergence rate, and the sex ratio). Some data sets were transformed to meet ANOVA requirements (log transformation: terpenoids content of plants and of honeydew, feeding time on gossypol; arc sin transformation: emergence rate and sex ratio after honeydew feeding). Means were subsequently separated using the Tukey HSD-test. Data from parasitoids that were provided only with water were omitted from the analysis, to reduce Heterogeneity of Variance.

The total number of offspring produced by *E. eremicus* females after honeydew feeding was analyzed using Kruskal-Wallis rank sum test.

Results

Heliothis virescens performance, plant damage, and terpenoid concentration in leaves

When infested with a single *H. virescens* larva for 7 days, Bt cotton plants remained nearly undamaged (mean ± SE of the area consumed on the infested leaf: 0.1 ± 0.03 cm^2) compared to non-Bt cotton plants (38.3 ± 5.01 cm^2; *t*-test: $t = -7.52$; $P<0.001$). No larva on Bt cotton (n = 26) was alive after 7 days, but all larvae on the non-Bt cotton survived and gained considerable weight (mean ± SE weight gain per larva: 177.7 ± 20.22 mg; n = 26).

The constitutive expression of foliar terpenoids in the youngest fully developed leaf did not differ between uninfested Bt and non-Bt cotton plants. Total terpenoid levels were significantly increased in the youngest leaves of herbivore-infested non-Bt plants 7 days after caterpillar damage had ceased (ANOVA, $F_{3,78} = 19.5$, $P < 0.001$) (Table 1). The terpenoid concentration increased 186% relative to the concentration in uninfested non-Bt cotton. The concentration of all terpenoids except hemigossypolone increased significantly in response to herbivore damage. Terpenoid levels did not differ between infested and uninfested Bt cotton (Table 1).

Terpenoid concentration in *A. gossypii* honeydew

Gossypol and hemigossypolone were detected in the honeydew of *A. gossypii* feeding on cotton (Table 2). Terpenoid concentrations did not differ between the honeydew produced by aphids fed on Bt and non-Bt cotton plants or between *H. virescens*-infested and uninfested plants (ANOVA, $F_{3,32} = 2.02$, $P = 0.13$). Terpenoid concentration in honeydew was <1% of that in the leaf tissue.

Parasitoid longevity as affected by gossypol and *A. gossypii* honeydew

The longevity of *E. eremicus* females significantly differed when the females were fed with sucrose, sucrose+DMSO, or sucrose+DMSO+gossypol solutions (ANOVA; $F_{5,164} = 6.00$, $P<0.001$) (Figure 1A). *Eretmocerus eremicus* longevity was significantly shorter with gossypol at 0.00001 and 0.0001% than with the DMSO control solution. Feeding on gossypol at a lower (0.000001%) and a higher (0.001%) concentration did not significantly affect *E. eremicus* longevity. The longevity of *L. testaceipes* females was also significantly affected by feeding on the gossypol and control solutions (ANOVA, $F_{5,170} = 4.76$; $p<0.001$)

Table 1. Terpenoid concentration (ng/mg dw) in leaves of Bt and non-Bt cotton as affected by prior infestation with *Heliothis virescens* in a glasshouse experiment.

Terpenoids	ANOVA			Concentration (ng/mg dw ± SE)			
	df	F	P	Bt uninfested	Non-Bt uninfested	Bt infested	Non-Bt infested
HGQ	3,78	0.57	0.64	4078±374.6 a	4105±410.1 a	4615±443.7 a	6936±587.1 a
G	3,78	21.6	<0.001	1353±212.2 b	1685±173.6 b	2141±307.6 b	5774±697.8 a
H1/H4	3,78	12.5	<0.001	335±27.4 b	314±28.7 b	296±23.4 b	591±46.6 a
H2/H3	3,78	10.4	<0.001	702±52.4 bc	612±60.2 c	964±74.6 ab	1303.7±161.2 a
Total	3,78	19.6	<0.001	6470±567.6 b	6717±581.5 b	8118±745.3 b	12614±1541.6 a

Terpenoid concentration (mean ± SE; n = 10) was measured in the youngest fully developed leaf of each Bt and non-Bt cotton plant that was uninfested or infested with *H. virescens* (3rd instar) for 7 days. HGQ: hemigossypolone; G: gossypol; H1/H4: heliocide 1 and 4; H2/H3: heliocide 2 and 3.
Means in a row followed by different letters are significantly different ($P\leq0.05$; Tukey HSD test).

Table 2. Terpenoid concentrations (ng/mg fw) in the honeydew of *Aphis gossypii* feeding on cotton.

Terpenoids	ANOVA			Concentration (ng/mg fw)			
	df	*F*	*P*	Bt uninfested	Non-Bt uninfested	Bt infested	Non-Bt infested
HGQ	3,17	0.38	0.77	10.4±4.90	30.3±13.33	12.6±4.31	39.2±16.58
G	3,26	1.1	0.37	4.2±1.07	7.4±1.72	6.3±2.13	40.4±33.75
H1/H4	ND	ND	ND	ND	ND	ND	ND
H2/H3	ND	ND	ND	ND	ND	ND	ND
Total	3,32	2.02	0.13	11.1±3.37	35.9±13.40	15.4±3.71	71.6±37.76

Values are means ± SE, n = 7–10. Honeydew was collected from aphids on Bt and non-Bt cotton plants that were uninfested or infested with larvae of *Heliothis virescens* (3rd instar) for 7 days. HGQ: hemigossypolone; G: gossypol; H1/H4: heliocide 1 and 4; H2/H3: heliocide 2 and 3.
ND- not detectable.

(Figure 1B). *Lysiphlebus testaceipes* longevity was significantly shorter with gossypol at 0.000001 and 0.00001% than with the DMSO control solution (Tukey HSD test, $p<0.05$), but was unaffected by higher gossypol concentrations. For both parasitoid species, longevity was similar with the DMSO control solution and the pure sucrose solution ($p>0.05$).

Parasitoid longevity as affected by *A. gossypii* honeydew

Eretmocerus eremicus longevity significantly differed between females fed with honeydew from different sources and the sucrose control solution (ANOVA, $F_{4,128}=4.28$; $P<0.001$) (Figure 1C). Parasitoid longevity was shorter when the females were fed with honeydew rather than with the sucrose solution, except when the honeydew was obtained from uninfested Bt cotton. Among the honeydew treatments, *E. eremicus* longevity did not differ significantly among honeydews collected from Bt or non-Bt cotton plants that were infested or uninfested with an *H. virescens* larva. Similarly, the longevity of *L. testaceipes* females was shorter when the females were fed with *A. gossypii* honeydew rather than with the sucrose control solution (Figure 1D), but longevity did not differ among the four honeydew treatments (ANOVA, $F_{4,138}=13.04$; $P<0.001$).

Parasitoid reproductive capacity as affected by *A. gossypii* honeydew

The food source did not affect the number of mummies produced, the number of emerged adults, adult emergence rate (%), or the progeny sex ratio for *L. testaceipes* (Table 3). The reproductive capacity of *E. eremicus* females on days 2, 4, and 6 after feeding on honeydew was low (it ranged from 1.2±0.25 to 1.9±0.45 progeny per female) and did not differ among food treatments (Kruskal-Wallis rank sum test: df = 4, $\chi^2=1.87$; $P=0.76$).

Parasitoid gustatory response to gossypol

The percentage of *L. testaceipes* females that accepted the test solutions ranged from 61.3 to 84.4% and did not differ between the gossypol treatments and the DMSO control (Pearson's χ^2-test with Bonferroni-correction: $P>0.01$). The average (± SE) feeding time ranged from 25.5±3.8 to 32.4±5.53 s and did not significantly differ among the treatments (ANOVA, $F_{5,130}=0.79$, $P=0.56$).

The percentage of *E. eremicus* females that accepted the test solutions ranged from 78.6 to 100% and did not differ between the gossypol treatments and the DMSO control (Pearson's χ^2-test with Bonferroni-correction: $P>0.01$). The average feeding time ranged from 57.3±9.07 to 89.8±14.93 s and did not significantly differ among the treatments (ANOVA, $F_{5,156}=0.64$, $P=0.67$).

Discussion

Gossypol and other cotton terpenoids that are produced as a response to caterpillar damage are ingested by *A. gossypii* and excreted in honeydew. Because Bt cotton suffers much less caterpillar damage than non-Bt cotton, we expected that the terpenoid concentration in *A. gossypii* honeydew would be lower on Bt cotton than on non-Bt cotton and that parasitoid fitness might be greater when feeding on honeydew from Bt cotton. Although gossypol at specific concentrations reduced the longevity of two hymenopteran parasitoid species parasitoid fitness was unaffected by honeydew source.

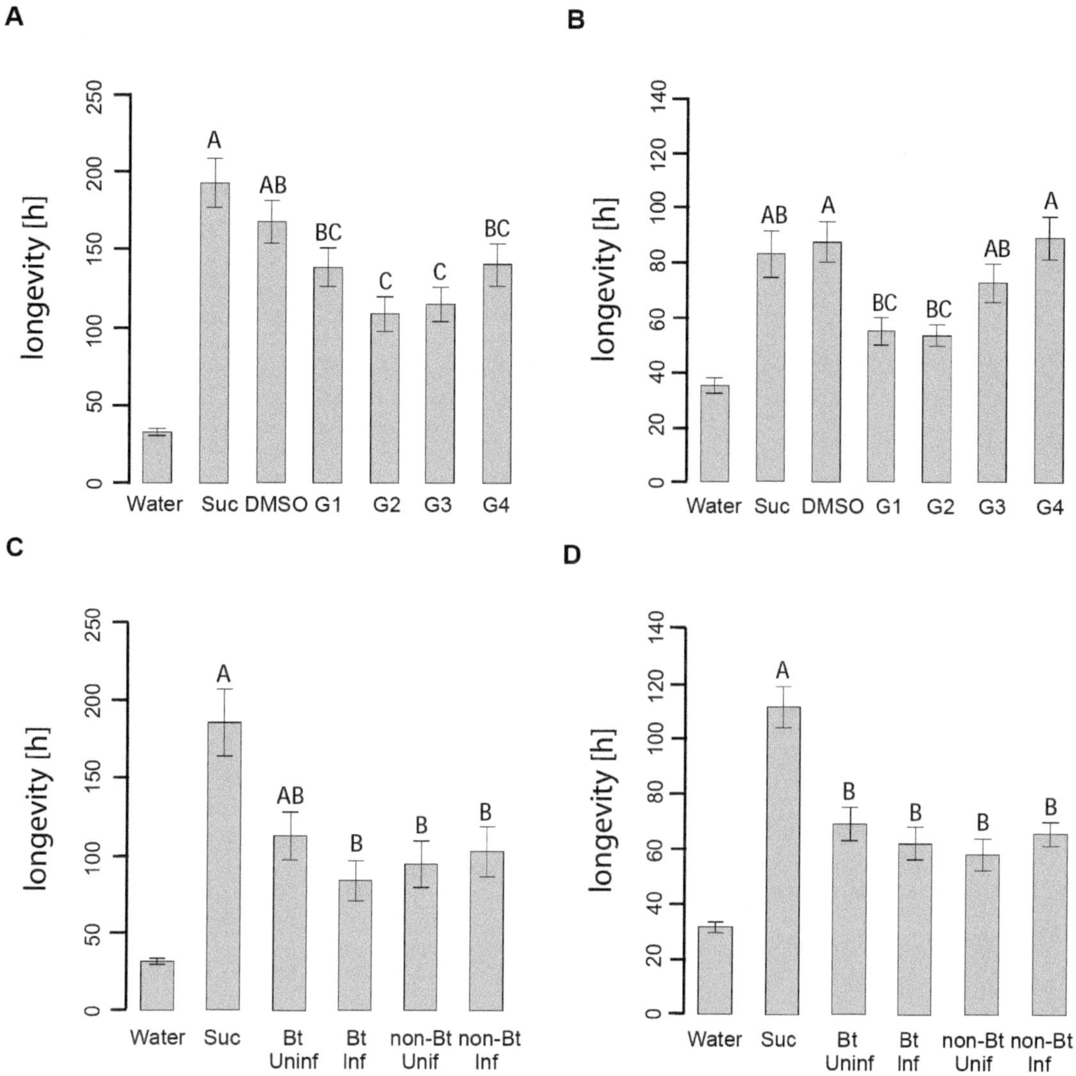

Figure 1. Parasitoid longevity when fed with gossypol or honeydew from *Aphis gossypii*. Longevity (hours ± SE) of (A) female *Eretmocerus eremicus* (n = 27–30) and (B) female *Lysiphlebus testaceipes* (n = 28–30) when fed with: Water, 1 M sucrose solution (Suc), 1 M sucrose+1% DMSO (DMSO), or sucrose/DMSO solutions containing different concentrations of gossypol (G1: 0.000001% gossypol; G2: 0.00001%, G3: 0.0001%; G4: 0.001%). Longevity (hours ± SE) of (C) female *Eretmocerus eremicus* (n = 24–30) and (D) female *Lysiphlebus testaceipes* (n = 27–30) when fed with: Water, 1 M sucrose solution (Suc), or honeydew from *Aphis gossypii* that were kept on Bt or non-Bt cotton that were uninfested (Uninf) or infested (Inf) with a *Heliothis virescens* larva. Within each panel, different letters above bars indicate statistically significant differences (Tukey HSD test, $P < 0.05$). The water control was not included in the statistical analysis.

Gossypol concentration in honeydew

Phloem-feeders that consume insecticidal. plant compounds must protect themselves against these toxic compounds. Certain species like *Brevicoryne brassicae* (L.) (Hemiptera: Aphididae) are known to detoxify or sequester secondary plant compounds [39,40]. Besides metabolizing a toxin, herbivores can excrete it either with their feces, or in the case of phloem-feeders, in their honeydew. The glucosinolate sinigrin was found in the honeydew of *Myzus persicae* (Sulzer) (Hemiptera: Aphididae) feeding on *Brassica nigra* (L.) (Brassicales: Brassicaceae) [41], cardenolides were present in *Aphis nerii* (Boyer de Fonscolombe) (Hemiptera: Aphididae) honeydew from milkweed [42], and alkaloids have been reported from honeydew of various species including *Aphis jacobaeae* (Schrank) (Hemiptera: Aphididae) feeding on milkweed [36,43].

Although cotton terpenoids have been detected in the feces of *H. virescens* [44], the current report provides the first evidence for the presence of terpenoids, i.e., gossypol and hemigossypolone, in the honeydew of a phloem-feeder. The heliocides 1–4 were not detected. The latter result may be attributed to the fact that heliocides 1–4 occur only at relatively low levels in the plant or that they are digested in the aphid gut. Compared to the concentrations of secondary metabolites in the honeydew of other plant–aphid systems, the concentrations of cotton terpenoids in *A. gossypii* honeydew were very low. Based on honeydew fresh weight, concentrations of cotton terpenoids in honeydew never exceeded 0.01% and were usually closer to 0.001%. While we could detect the terpenoids, the quantification of such low amounts reached the limits of our analytical methods (as is indicated by the high variability of terpenoid concentrations) and thus needs to be taken with caution. Although our obtained data

Table 3. *Lysiphlebus testaceipes* reproduction as affected by *Aphis gossypii* honeydew.

Food source	N	No. of mummies produced per plant	No. of emerged adults	Emergence rate (%)	Females (%)
1 M Sucrose	27	16.4±3.32	9.3±2.03	56.6	52.4
Bt unifested	26	14.6±2.88	6.8±1.32	46.2	54.6
Bt infested	30	13.9±2.69	6.7±1.21	48.5	51.7
non-Bt uninfested	31	19.0±2.75	9.1±0.63	47.5	50.2
non-Bt infested	27	18.8±3.56	9.2±1.53	48.7	53.6
ANOVA					
df		4,137	4,137	4,134	4,114
F		0.63	0.72	0.52	0.53
P		0.64	0.58	0.72	0.71

Values are means ± SE, n = 26–31. Female parasitoids were fed for 24 h with 1 M sucrose or with honeydew produced by *Aphis gossypii* feeding on Bt and non-Bt cotton plants that were uninfested or infested with larvae of *Heliothis virescens* (3rd instar). The females were then allowed to parasitize *Aphis gossypii* for 4 h.

suggest an increased concentration of terpenoids in herbivore-damaged non-Bt cotton, this could not be supported by the statistical analysis. For comparison, previous studies reported cardenolide concentrations as high as 46% in the honeydew of *A. nerii* [42] and pyrrolizidine alkaloid concentrations as high as 0.76% of in the honeydew of *A. jacobaeae* [43]. The data are, however, not directly comparable because the latter two studies presented the toxin concentration based on honeydew dry weight. The relatively low concentration of cotton terpenoids in honeydew in the present study may be explained by an efficient detoxification by the cotton aphids. Alternatively, the concentration of terpenoids in phloem sap may be low, given their hydrophobic properties.

Impact of gossypol on parasitoids

While the impact of gossypol on lepidopteran herbivores is well known [14,45], little information is available on its effect on other arthropods and especially on predators and parasitoids. We found that certain gossypol concentrations reduced the longevity of females of the parasitoids *L. testaceipes* and *E. eremicus*. The effect was non-linear in that the maximum effect occurred at intermediate (*E. eremicus*) or low (*L. testaceipes*) concentrations. This lack of a linear relation between gossypol dose and parasitoid longevity cannot be explained by parasitoid deterrence at high gossypol concentrations because no such deterrence was evident in our gustatory response experiment. A similar lack of a linear dose-response was detected with female spined soldier bugs *Podisus nigrispinus* (Dallas) (Hemiptera: Pentatomidae) [46]. The results of earlier tri-trophic studies were mixed. The fitness of the parasitoid *Campoletis sonorensis* (Cameron) (Hymenoptera: Ichneumonidae) was reduced when its hosts had fed on diets with high concentrations of gossypol but was enhanced when its hosts fed on diets with low concentrations of gossypol [47]. In contrast, the aphid parasitoid *Lysiphlebus japonica* (Stary & Schlinger) (Hymenoptera: Braconidae) did not develop differently in hosts from plants with high or low gossypol concentrations [48]. In our previous study, *L. testaceipes* fitness was similar when its hosts developed on caterpillar-induced and uninduced cotton plants [21].

Impact of honeydew on parasitoids

Although the parasitoids consumed *A. gossypii* honeydew, their longevity was much shorter when fed on honeydew rather than on a sucrose solution. The aphid host plant, i.e., Bt or non-Bt plants that were infested or uninfested with a *H. virescens*, did not influence honeydew quality in terms of parasitoid fitness. The reproductive capacity did not differ between parasitoids that had fed on honeydew compared to those that had fed on sucrose, and again, the response to honeydew was not influenced by plant type (Bt or non-Bt) or prior exposure to a chewing herbivore (a caterpillar). The observed fecundity of *E. eremicus* was relatively low, which was most likely caused by the short exposure time of the parasitoid to its host. For example, Soler & van Lenteren [49] reported a total fecundity for this species of around 160 eggs per female over a 14 day lifespan, corresponding to about 10 eggs per day. This ratio compares well with the observed fecundity of *L. testaceipes* which has been reported to be about double to that of *E. eremicus* in a much shorter life-time [50].

We thus conclude that neither the genetic transformation nor the induced resistance of Bt cotton against caterpillar damage greatly affects honeydew quality. The overall concentration of cotton terpenoids in honeydew was probably too low to affect the parasitoids. Furthermore, the ingestion of the terpenoids by *A. gossypii* apparently did not result in a change in honeydew composition. A change in honeydew sugar and amino acid composition has been documented as a response to the ingestion of *Galanthus nivalis* agglutinin by *Rhopalosiphum padi* (L.) (Hemiptera: Aphididae), and such changes affected the nutritional quality of the food source for the aphid parasitoid *Aphidius ervi* (Haliday) (Hymenoptera: Braconide) [34]. In the case of *E. eremicus*, it should be noted that *Eretmocerus* species are host feeders [51]. It is possible that females fed on their host during parasitization and thereby obtained an alternate source of energy that might have countered any negative effects of the honeydew diet on parasitoid reproductive capacity. Whether gossypol or other terpenoids are present in the hemolymph of the hosts is unknown.

Conclusions

Honeydew from *A. gossypii* feeding on cotton can be utilized as a carbohydrate source by the parasitoids *L. testaceipes* and *E. eremicus*. While cotton terpenoids were detected in the honeydew, the presence of these insecticidal compounds did not adversely affect the adult parasitoids. Overall, honeydew quality as a food source did not differ between Bt-transgenic and non-transgenic cotton plants. Similarly, it was not altered when plants had been attacked by caterpillars and terpenoid production was induced.

Consequently, honeydew quality and utilization by natural enemies is unlikely to differ between Bt and non-Bt cotton fields.

Author Contributions

Conceived and designed the experiments: SH JR FLW. Performed the experiments: SH. Analyzed the data: SH JR. Contributed to the writing of the manuscript: SH JR FLW.

References

1. James C (2012) Global status of commercialized biotech/GM crops: 2012. ISAAA Brief No. 44. ISAAA, Ithaca, NY.
2. Sivasupramaniam S, Moar W, Ruschke L, Osborn J, Jiang C, et al. (2008) Toxicity and characterization of cotton expressing *Bacillus thuringiensis* Cry1Ac and Cry2Ab2 proteins for control of lepidopteran pests. Journal of Economic Entomology 101: 546–554.
3. Naranjo SE (2011) Impacts of Bt transgenic cotton on integrated pest management. Journal of agricultural and food chemistry 59: 5842–5851.
4. Carriere Y, Ellers-Kirk C, Sisterson M, Antilla L, Whitlow M, et al. (2003) Long-term regional suppression of pink bollworm by *Bacillus thuringiensis* cotton. Proceedings of the National Academy of Sciences USA 100: 1519–1523.
5. Wu K-M, Lu Y-H, Feng H-Q, Jiang Y-Y, Zhao J-Z (2008) Suppression of cotton bollworm in multiple crops in China in areas with Bt toxin–containing cotton. Science 321: 1676–1678.
6. Romeis J, Meissle M, Bigler F (2006) Transgenic crops expressing *Bacillus thuringiensis* toxins and biological control. Nature Biotechnology 24: 63–71.
7. Romeis J, Shelton A, Kennedy GG, Eds(2008) Integration of Insect-Resistant Genetically Modified Crops within IPM Programs. Springer Science+Business Media B.V.
8. Lu Y, Wu K, Jiang Y, Guo Y, Desneux N (2012) Widespread adoption of Bt cotton and insecticide decrease promotes biocontrol services. Nature 487: 362–365.
9. Lu Y, Wu K, Jiang Y, Xia B, Li P, et al. (2010) Mirid bug outbreaks in multiple crops correlated with wide-scale adoption of Bt cotton in China. Science 328: 1151–1154.
10. Whitehouse M, Wilson L, Constable G (2007) Corrigendum to: Target and non-target effects on the invertebrate community of Vip cotton, a new insecticidal transgenic. Crop and Pasture Science 58: 383–383.
11. Zeilinger AR, Olson DM, Andow DA (2011) Competition between stink bug and heliothine caterpillar pests on cotton at within-plant spatial scales. Entomologia Experimentalis et Applicata 141: 59–70.
12. Kaplan I, Denno RF (2007) Interspecific interactions in phytophagous insects revisited: a quantitative assessment of competition theory. Ecology Letters 10: 977–994.
13. Stipanovic RD, Altman DW, Begin DL, Greenblatt GA, Benedict JH (1988) Terpenoid aldehydes in upland cottons: analysis by aniline and HPLC methods. Journal of Agricultural and Food Chemistry 36: 509–515.
14. Hagenbucher S, Olson D, Ruberson J, Wäckers F, Romeis J (2013) Resistance mechanisms against arthropod herbivores in cotton and their interactions with natural enemies. Critical Reviews in Plant Sciences 32: 458–482.
15. Hagenbucher S, Wäckers FL, Wettstein FE, Olson DM, Ruberson JR, et al. (2013) Pest trade-offs in technology: reduced damage by caterpillars in Bt cotton benefits aphids. Proceedings of the Royal Society B: Biological Sciences 280: 20130042.
16. McAuslane HJ, Alborn HT, Toth JP (1997) Systemic induction of terpenoid aldehydes in cotton pigment glands by feeding of larval *Spodoptera exigua*. Journal of Chemical Ecology 23: 2861–2879.
17. Bezemer T, Wagenaar R, Van Dam N, Van Der Putten W, Wäckers F (2004) Above-and below-ground terpenoid aldehyde induction in cotton, *Gossypium herbaceum*, following root and leaf injury. Journal of Chemical Ecology 30: 53–67.
18. Anderson P and Agrell J (2005) Within-plant variation in induced defence in developing leaves of cotton plants. Oecologia 144: 427–434.
19. Olson DM, Davis R, Wäckers FL, Rains G, Potter T (2008) Plant–herbivore–carnivore interactions in cotton, *Gossypium hirsutum*: linking belowground and aboveground. Journal of Chemical Ecology 34: 1341–1348.
20. Adamczyk Jr J, Adams L, Hardee D (2001) Field efficacy and seasonal expression profiles for terminal leaves of single and double *Bacillus thuringiensis* toxin cotton genotypes. Journal of Economic Entomology 94: 1589–1593.
21. Hagenbucher S, Wäckers F, Romeis J (2014) Indirect multi-trophic interactions mediated by induced plant resistance: impact of caterpillar feeding on aphid parasitoids. Biology Letters 10: 20130795.
22. Wäckers F (2005) Suitability of (extra-) floral nectar, pollen, and honeydew as insect food sources. In: Wäckers FL, van Rijn PCJ, Bruin J, eds. Plant-provided Food for Carnivorous Insects: A Protective Mutualism and its Applications. Cambridge: Cambridge University Press, pp 17–74.
23. Hogervorst PA, Wäckers FL, Romeis J (2007) Detecting nutritional state and food source use in field-collected insects that synthesize honeydew oligosaccharides. Functional Ecology 21: 936–946.
24. Wäckers FL, Van Rijn PC, Heimpel GE (2008) Honeydew as a food source for natural enemies: Making the best of a bad meal? Biological Control 45: 176–184.
25. Tena A, Pekas A, Waeckers FL, Urbaneja A (2013) Energy reserves of parasitoids depend on honeydew from non-hosts. Ecological Entomology 38: 278–289.
26. Hogervorst PA, Wäckers FL, Romeis J (2007) Effects of honeydew sugar composition on the longevity of Aphidius ervi. Entomologia Experimentalis et Applicata 122: 223–232.
27. Dulaurent AM, Rossi JP, Deborde C, Moing A, Menassieu P, et al. (2011) Honeydew feeding increased the longevity of two egg parasitoids of the pine processionary moth. Journal of Applied Entomology 135: 184–194.
28. Tena A, Llácer E, Urbaneja A (2013) Biological control of a non-honeydew producer mediated by a distinct hierarchy of honeydew quality. Biological Control 67: 117–122.
29. Bouchard Y, Cloutier C (1984) Honeydew as a source of host-searching kairomones for the aphid parasitoid *Aphidius nigripes* (Hymenoptera: Aphidiidae). Canadian Journal of Zoology 62: 1513–1520.
30. Budenberg W (1990) Honeydew as a contact kairomone for aphid parasitoids. Entomologia Experimentalis et Applicata 55: 139–148.
31. Wäckers FL (2000) Do oligosaccharides reduce the suitability of honeydew for predators and parasitoids? A further facet to the function of insect-synthesized honeydew sugars. Oikos 90: 197–201.
32. Lee JC, Heimpel GE, Leibee GL (2004) Comparing floral nectar and aphid honeydew diets on the longevity and nutrient levels of a parasitoid wasp. Entomologia Experimentalis et Applicata 111: 189–199.
33. Mandour N, Ren S, Qiu B (2007) Effect of *Bemisia tabaci* honeydew and its carbohydrates on search time and parasitization of *Encarsia bimaculata*. Journal of Applied Entomology 131: 645–651.
34. Hogervorst PA, Wäckers FL, Woodring J, Romeis J (2009) Snowdrop lectin (*Galanthus nivalis* agglutinin) in aphid honeydew negatively affects survival of a honeydew-consuming parasitoid. Agricultural and Forest Entomology 11: 161–173.
35. Wäckers F (2001) A comparison of nectar-and honeydew sugars with respect to their utilization by the hymenopteran parasitoid *Cotesia glomerata*. Journal of Insect Physiology 47: 1077–1084.
36. Dreyer DL, Jones KC, Molyneux RJ (1985) Feeding deterrency of some pyrrolizidine, indolizidine, and quinolizidine alkaloids towards pea aphid (*Acyrthosiphon pisum*) and evidence for phloem transport of indolizidine alkaloid swainsonine. Journal of Chemical Ecology 11: 1045–1051.
37. Colfer R, Rosenheim J (2001) Predation on immature parasitoids and its impact on aphid suppression. Oecologia 126: 292–304.
38. Abney MR, Ruberson JR, Herzog GA, Kring TJ, Steinkraus DC, et al. (2008) Rise and fall of cotton aphid (Hemiptera: Aphididae) populations in southeastern cotton production systems. Journal of Economic Entomology 101: 23–35.
39. Kazana E, Pope TW, Tibbles L, Bridges M, Pickett JA, et al. (2007) The cabbage aphid: a walking mustard oil bomb. Proceedings of the Royal Society B: Biological Sciences 274: 2271–2277.
40. Ramsey JS, Rider DS, Walsh TK, De Vos M, Gordon K, et al. (2010) Comparative analysis of detoxification enzymes in *Acyrthosiphon pisum* and *Myzus persicae*. Insect Molecular Biology 19: 155–164.
41. Merritt SZ (1996) Within-plant variation in concentrations of amino acids, sugar, and sinigrin in phloem sap of black mustard, *Brassica nigra* (L.) Koch (Cruciferae). Journal of Chemical Ecology 22: 1133–1145.
42. Malcolm SB (1990) Chemical defence in chewing and sucking insect herbivores: plant-derived cardenolides in the monarch butterfly and oleander aphid. Chemoecology 1: 12–21.
43. Vrieling K, Smit W, van der Meijden E (1991) Tritrophic interactions between aphids (*Aphis jacobaeae* Schrank), ant species, *Tyria jacobaeae* L., and *Senecio jacobaea* L. lead to maintenance of genetic variation in pyrrolizidine alkaloid concentration. Oecologia 86: 177–182.
44. Rojas MG, Stipanovic R, Williams H, Vinson S (1992) Metabolism of gossypol by *Heliothis virescens* (F.) (Lepidoptera: Noctuidae). Environmental Entomology 21: 518–526.
45. Bottger G, Shehaan ET, Lukefahr M (1964) Relation of gossypol content of cotton plants to insect resistance. Journal of Economic Entomology 57: 283–285.
46. Evangelista Junior WS, Santos RL, Torres JB, Zanuncio JC (2011) Effect of gossypol on survival and reproduction of the zoophytophagous stinkbug *Podisus nigrispinus* (Dallas). Revista Brasileira de Entomologia 55: 267–271.
47. Gunasena G, Vinson S, Williams H, Stipanovic R (1989) Development and survival of the endoparasitoid *Campoletis sonorensis* (Hymenoptera: Ichneumonidae) reared from gossypol-exposed *Heliothis virescens* (F.) (Lepidoptera: Noctuidae). Environmental Entomology 18: 886–891.
48. Sun YC, Feng L, Gao F, Ge F (2011) Effects of elevated CO_2 and plant genotype on interactions among cotton, aphids and parasitoids. Insect Science 18: 451–461.
49. Soler R, van Lenteren J (2004) Reproduction and development of *Eretmocerus eremicus* (Hymenoptera: Aphelinidae) on *Trialeurodes vaporariorum* (Homoptera: Aleyrodidae. Proceedings of the Netherlands Entomological Society 15: 111–117.

50. Silva RJd, Bueno VHP, Silva DB, Sampaio MV (2008) Fertility life table of *Lysiphlebus testaceipes* (Cresson) (Hymenoptera, Braconidae, Aphidiinae) in *Rhopalosiphum maidis* (Fitch) and *Aphis gossypii* Glover (Hemiptera, Aphididae). Revista Brasileira de Entomologia 52: 124–130.

51. Jervis MA, Kidd NA (1986) Host-feeding strategies in hymenopteran parasitoids. Biological Reviews 61: 395–434.

5

Migratory Herbivorous Waterfowl Track Satellite-Derived Green Wave Index

Mitra Shariatinajafabadi[1]*, Tiejun Wang[1], Andrew K. Skidmore[1], Albertus G. Toxopeus[1], Andrea Kölzsch[2,3], Bart A. Nolet[3], Klaus-Michael Exo[4], Larry Griffin[5], Julia Stahl[6], David Cabot[7]

1 Faculty of Geo-Information Science and Earth Observation (ITC), University of Twente, Enschede, The Netherlands, **2** Max Planck Institute for Ornithology, Department of Migration and Immuno-Ecology, Vogelwarte Radolfzell, Radolfzell, Germany, **3** Department of Animal Ecology and Project group Movement Ecology, Netherlands Institute of Ecology (NIOO-KNAW), Wageningen, The Netherlands, **4** Institute of Avian Research, Wilhelmshaven, Germany, **5** Wildfowl & Wetlands Trust, Slimbridge, Gloucestershire, United Kingdom, **6** Sovon Dutch Centre for Field Ornithology, Nijmegen, The Netherlands, **7** Environmental Consultancy Services, Carrigskeewaun, Carrowniskey, Westport, Co. Mayo, Ireland

Abstract

Many migrating herbivores rely on plant biomass to fuel their life cycles and have adapted to following changes in plant quality through time. The green wave hypothesis predicts that herbivorous waterfowl will follow the wave of food availability and quality during their spring migration. However, testing this hypothesis is hampered by the large geographical range these birds cover. The satellite-derived normalized difference vegetation index (NDVI) time series is an ideal proxy indicator for the development of plant biomass and quality across a broad spatial area. A derived index, the green wave index (GWI), has been successfully used to link altitudinal and latitudinal migration of mammals to spatio-temporal variations in food quality and quantity. To date, this index has not been used to test the green wave hypothesis for individual avian herbivores. Here, we use the satellite-derived GWI to examine the green wave hypothesis with respect to GPS-tracked individual barnacle geese from three flyway populations (Russian n = 12, Svalbard n = 8, and Greenland n = 7). Data were collected over three years (2008–2010). Our results showed that the Russian and Svalbard barnacle geese followed the middle stage of the green wave (GWI 40–60%), while the Greenland geese followed an earlier stage (GWI 20–40%). Despite these differences among geese populations, the phase of vegetation greenness encountered by the GPS-tracked geese was close to the 50% GWI (i.e. the assumed date of peak nitrogen concentration), thereby implying that barnacle geese track high quality food during their spring migration. To our knowledge, this is the first time that the migration of individual avian herbivores has been successfully studied with respect to vegetation phenology using the satellite-derived GWI. Our results offer further support for the green wave hypothesis applying to long-distance migrants on a larger scale.

Editor: Gil Bohrer, The Ohio State University, United States of America

Funding: This research received financial support from the EU Erasmus Mundus External Cooperation Window (EM8) Action 2 project, http://www.erasmusmundus8.net/ (project number 10438223). Tracking devices for the Greenland barnacle geese were provided by Dr. David Cabot (5) and the WWT (2). Tracking devices for the Svalbard barnacle geese were provided through grants from Scottish Natural Heritage, Solway Coast Area of Outstanding Natural Beauty Sustainable Development Fund, the BBC, and the Heritage Lottery Fund Awards for All. The funders had no role in study design, data collection and analysis, decision to publish, or preparation of the manuscript.

Competing Interests: David Cabot is an employee of Environmental Consultancy Services. There are no patents, products in development, or marketed products to declare.

* Email: m.shariatinajafabadi@utwente.nl

Introduction

Satellite remote sensing is increasingly being used in ecological studies [1–4] and some new systems are facilitating the use of satellite data in ecological studies. For example, the Environmental-Data Automated Track Annotation (Env-DATA) System enables the processing of a large array of remote sensing weather and geographical data to analyze spatio-temporal patterns of animal movement tracks [5]. The integration of passive acoustic monitoring (PAM), visual sighting surveys, satellite telemetry records, and photo-identification catalogs in a biogeographic database (OBIS-SEAMAP) is another example of a system that provides new views and tools for assessing the ecology of marine mammals and biodiversity on a global scale [6].

The normalized difference vegetation index (NDVI) is a global vegetation indicator derived from remote sensors that integrate signals from the red (RED) and near-infrared (NIR) reflectance of Earth's objects, according to the equation: NDVI = (NIRRED)/(NIR+RED) [7,8]. NDVI calculations are based on the principle that actively growing green plants strongly absorb radiation in the visible region of the spectrum, while strongly reflecting radiation in the near-infrared region. NDVI is therefore interpreted as a measure of green leaf biomass [9]. Since the plant biomass trends generally correspond to the trend in NDVI [10] and the NDVI is closely related to net primary productivity [11], the NDVI derived from multispectral satellite data is commonly used by ecologists to estimate vegetation biomass (e.g. food quantity) as well as to assess seasonal changes in plant biomass over large regions [1,12].

Satellite NDVI time-series data has also been widely adopted as a proxy for plant phenology in ecological studies [13–16]. The plant phenology itself has been recognized as a good proxy for plant quality, as young plants are generally of high nutritional value, with low levels of secondary plant chemicals [17]. The nutritional quality declines with maturation stage (or vegetative biomass) [18]. Forage quality is highest during the early phenological stages (young growing plants) and then declines rapidly as the vegetation matures over the growing season [19]. Recent studies in the Arctic tundra using plant data [20] have shown that three NDVI metrics are significantly related to the date of peak nitrogen concentration. The strongest relationship was found with the date at which NDVI values reached 50% of their annual maximum ($R^2 = 0.87$).

NDVI has been employed as a proxy for the forage quality and timing of the availability of high-quality vegetation in studies of herbivore behavior and habitat use. For example, Mueller et al. [21] examined the relationship between vegetation productivity and animal habitat utilization, and they found that the intermediate range of NDVI was significantly associated with the highest food quality and resource availability for herbivores like Mongolian gazelles (*Procapra gutturosa*). Hamel et al. [22] assessed the relationship between two NDVI indices and the date of peaks in fecal crude protein, which represents temporal variability in the high-quality vegetation available for alpine ungulates. They concluded that NDVI can reliably be used to measure the yearly changes in the timing of the availability of high-quality vegetation for temperate herbivores.

Further support for the use of NDVI is provided by several more examples: Ryan et al. [23] studied the relationship between NDVI and forage nutrient indicators in a free-ranging African herbivore ecosystem. They suggested that NDVI can be used to index the nitrogen content of forage and that this is correlated with improved physical condition in African buffalo (*Syncerus caffer*). An individual-based movement modeling approach has been used to investigate how changes in NDVI, i.e. spatio-temporal variability in vegetation productivity, affected the migratory movements and their timing for zebra [24] and elephants [25]. Stoner et al. [26] used NDVI to evaluate the relative differences in habitat quality between the home ranges of natal and adult cougars (*Puma concolor*).

It has been hypothesized that movements of migratory herbivores are linked to plant phenology. This so-called green wave hypothesis states that herbivores time their spring migration to take advantge of successive peaks of nutrition and digestibility of plant growth as they migrate toward their breeding destination [27]. A space-time-time matrix of greenness is a tool for relating instantaneous green-up (or any other resource state) to animal movement [28]. It was calculated from satellite NDVI time-series data, and used by Bischof et al. [28] to study the relationship between plant phenology and the use of space by migratory and resident red deer (*Cervus elaphus*). They found that migrants had much greater access to early plant phenology than the residents. Deer were also more likely to migrate to areas that provided greater gains in instantaneous rate of green-up, which was interpreted as "springness" [28]. Rather than "surfing the green wave" during their migration, the red deer moved rapidly from the winter to the summer range, thereby "jumping the green wave". The space-time-time matrix of greenness was also defined as the relative phenological development. It has been successfully used to explain the difference in altitudinal migration between giant pandas (*Ailuropoda melanoleuca*) and golden takin (*Budorcas taxicolor bedfordi*) in relation to spatio-temporal variations in food quality and quantity [29]; the indicator of greeness was called the satellite-derived green wave index (GWI) in our study. Although the satellite-derived GWI has been proved to be a useful tool to study the migration of herbivorous mammals with respect to vegetation phenology, it has never been tested for migrating avian herbivores. We therefore set out to investigate the satellite-derived GWI for three different populations of barnacle geese (*Branta leucopsis*).

Barnacle geese are highly selective herbivores [30], and they prefer to eat the parts of a plant with the highest nutritional quality [31]. The green wave hypothesis has been successfully tested for this species using direct field measurements of plant biomass and quality at selected field sites [19]. Moreover, the timing of the spring migration in European greater white-fronted geese (*Anser albifrons*) in relation to the green wave has been well predicted using peaks in the acceleration of temperature (GDDjerk), which seem to be closely related to the onset of spring [32].

Our aim was to test if the satellite-derived GWI can be used for studying the green wave hypothesis with respect to avian herbivore migrants. We therefore examined a prediction based on the green wave hypothesis: if barnacle geese are surfing the green wave, then the phase of vegetation greenness they encounter will closely match the 50% GWI (i.e. the assumed date of peak nitrogen concentration).

Materials and Methods

Study Area and Barnacle Goose Populations

There are five separate populations of barnacle geese in the Western Palearctic, including three Arctic and two temperate breeders [31,33]. We studied the three long-distance migratory populations, from Russia, Svalbard (Norway), and Greenland, which use different wintering sites but breed in the Arctic (Figure 1).

In order to catch and fix transmitters on barnacle geese, we obtained a license under the Wild Flora and Fauna Protection Act (*Flora en Fauna Wet*), number FF75A/2007/056, and approval from the Dutch Ethical Committee, under protocol number CL 0703. A license to conduct this study in the Natura2000 area "*Waddenzee*" was obtained from the Province of Friesland, number 00692701. In the UK, permission to fit satellite tags was granted by the British Trust for Ornithology Unconventional Marks Panel. The Greenland barnacle geese were caught and fitted with transmitters under a license issued by the National Parks and Wildlife Service, Dublin, under the Wildlife Act, 1976, section 32.

The Russian barnacle goose population overwinters on the Wadden Sea coast of Denmark, Germany and the Netherlands. The geese leave this area in April-May and migrate via stopovers in the Baltic Sea (most notably on the Swedish island of Gotland and in western Estonia), the White Sea, and the Kanin Peninsula. They arrive at their breeding sites on the Arctic coast of Russia in early June, after a flight of 3000–3700 km [34,35]. The Svalbard population overwinters on the Solway Firth, UK. Birds leave from mid-April onwards, and typically have stopovers on the coastal islands of Norway (Helgeland in mid-Norway, and Vesterålen in northern Norway) for two to three weeks. They arrive at their breeding sites in Svalbard from mid-May onwards, after flying some 3100 km [31,36]. The Greenland population leaves its overwintering sites on islands off the north and west coasts of Scotland and Ireland in the second half of April. They migrate via stopovers in Iceland and arrive at their breeding sites on northeast Greenland in late May [37].

MODIS NDVI Data

We used the 16-day composite MODIS NDVI data (MOD13A2) (http://glovis.usgs.gov/), collected by NASA's MODIS Terra satellite at a 1-km resolution and spanning the

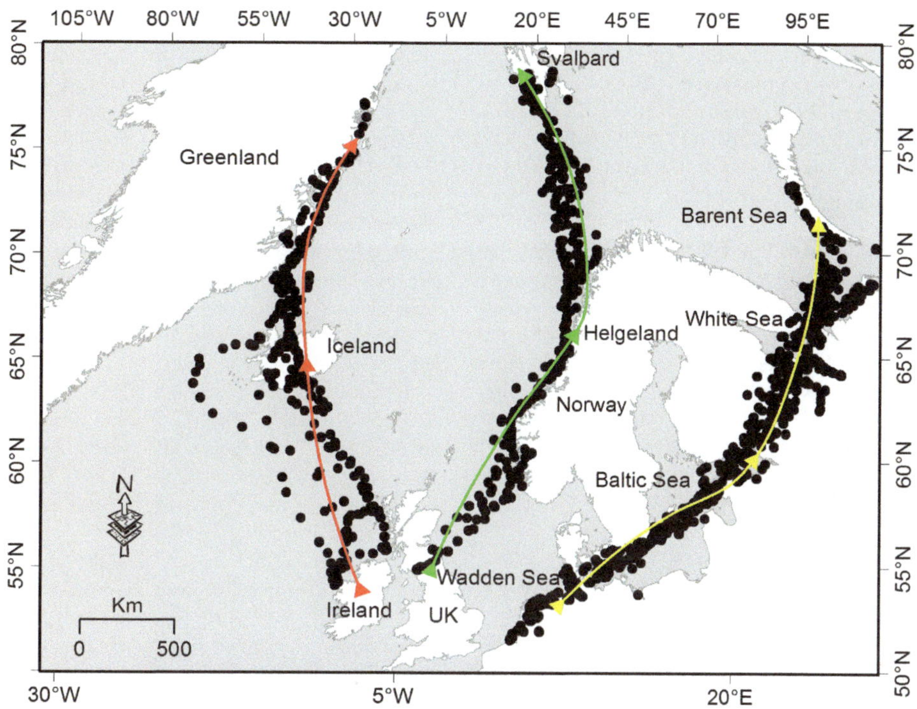

Figure 1. Spring migration route for three barnacle goose populations from their wintering to their breeding sites. The yellow, green and red arrows indicate the Russian, Svalbard and Greenland flyways, respectively. In each flyway, the dots show examples of the spatial distribution of GPS locations recorded for the 12 Russian, 8 Svalbard and 7 Greenland barnacle geese, from 2008 to 2010.

period from 2008 to 2010. This is useful for continental and global ecological studies [8,38]. The MODIS NDVI product is given in the sinusoidal projection system that ensures consistency of the size of the sites, independently of their latitude. The composition methods that are used to produce the MOD13A2 products reduce artifacts due to clouds, aerosols and satellite-view zenith angle [8]. However, some noise from residual cloud and aerosol contamination, as well as sensor problems, remain in the data, which causes misclassification of phenological parameters [8]. In order to minimize the overall noise in the NDVI time series, a Savitzky-Golay filter was applied to each annual NDVI cycle. In the next step, double logistic function-fitting, suitable for modeling the yearly NDVI time series of boreal and arctic-alpine vegetation, was applied to maintain the integrity of the time series data [14,39].

The effects of snow and large solar zenith angles at high latitudes cause a dramatic decrease in the NDVI during the winter [40]. Since snow cover negatively affects the NDVI, the melting snow at the end of winter allows the NDVI to rise, although the rise is not necessarily related to increased vegetation activity [41]. To reduce the effect of snow in high latitudes, the winter NDVI (i.e. the NDVI of any snow-affected pixel during the winter season from October until February) was therefore estimated using a method proposed by Beck et al. [14].

For our next analysis we aimed at a temporal resolution of 1 day rather than that of the 16-day composite, so the 23 NDVI images were interpolated to 365 images for each year using simple linear regression.

Satellite-Derived Green Wave Index (GWI)

The satellite-derived green wave index (GWI) is a transformation of the interpolated NDVI and has a ratio output ranging from 0–100% for each cell and indicating the annual minimum and maximum NDVI, respectively. The greenness of two pixels at a given time can be compared by looking at the GWI irrespective of their absolute NDVI, because the GWI is normalized to account for differences such as land cover variances [13,38]. The GWI were calculated following the method proposed by White et al. [13] and Beck et al. [38]:

$$GWI_t = (NDVI_t - NDVI_{min})/(NDVI_{max} - NDVI_{min}) \times 100 \quad (1)$$

where for each pixel $NDVI_{min}$ is the annual minimum NDVI, $NDVI_{max}$ is the annual maximum NDVI, and $NDVI_t$ and GWI_t are the NDVI and green wave index at time t, respectively [13,38]. The pixels with GWI = 0, or near 0%, appear in areas that are at, or near, their minimum greenness. The pixels with GWI of 100%, or near 100%, indicate areas that are at, or near, their maximum greenness [42]. A GWI of 50% indicates the intermediate stage of the greenness and incorporates a quality versus quantity trade-off (i.e. an area with high quality forage) [20,43].

GPS Tracking Data of Barnacle Geese

The geese were captured on their overwintering sites in the Netherlands, Solway Firth, and Ireland, and fitted with solar GPS/ARGOS transmitters (Solar GPS 100 PTT; PTT-platform transmitter terminal; Microwave Telemetry, Inc., Columbia, MD, USA). The Russian and Svalbard barnacle geese were equipped with 30 g transmitters (except for the individuals with ID 78198, 78378 and 178199 in the Svalbard population, which were equipped with 45 g transmitters). The Greenland barnacle geese were equipped with 45 g transmitters (except for the individuals with ID 65698 and 70563, which were equipped with 30 g transmitters). The PTTs were programmed to record the position of the individual goose four times per day for the Russian population, and every two hours for the Svalbard and Greenland populations, from dawn to dusk. The data collected included the goose ID, date, time, longitude, latitude, speed, course, and altitude. The GPS locations were uploaded to ARGOS satellites every four days [44–46]. From the Russian population, 12 females were tagged, whereas from the Svalbard and Greenland population, 15 males were tagged in total. However, the barnacle goose is a monogamous species and pair bonds persist during migration and for a long period, so the data sets were comparable [27].

For each of the three years (2008–2010), GPS tracks of incomplete spring migrations were removed from our analysis, resulting in 26 full data tracks for 12 female birds of the Russian population, 9 full data tracks for 8 male birds of the Svalbard population, and 7 full data tracks for 7 male birds of the Greenland population (see Table 1). The barnacle geese tracking data of all three populations can be viewed at movebank.org:

(i) Russian population: "Migration timing in barnacle geese (Barents Sea), data from Kölzsch et al. and Shariatinajafabadi et al. 2014", DOI:10.5441/001/1.ps244r11
(ii) Svalbard population: "Migration timing in barnacle geese (Svalbard), data from Kölzsch et al. and Shariatinajafabadi et al. 2014", DOI:10.5441/001/1.5k6b1364
(iii) Greenland population: "Migration timing in barnacle geese (Greenland), data from Kölzsch et al. and Shariatinajafabadi et al. 2014", DOI:10.5441/001/1.5d3f0664.

Delineation of Stopover Sites

During their spring migration, the geese stop at several sites along the way to rest, refuel or await better weather conditions [36]. To delineate stopover sites for each individual, groups of continuous GPS positions were identified where the movements of individuals between two positions in a cluster were no greater than 30 km, which is the maximum distance between resting and foraging grounds at wintering sites [32]. The stopover sites were selected where the birds remained for at least 48 h in such a GPS cluster [47]. The location of each site was defined as the center of each selected group, by taking the average of the latitudes and longitudes of the GPS positions [32]. In total, for 2008 to 2010, we recognized 57 stopover sites along the Russian flyway, 18 along the Svalbard flyway, and 14 along the Greenland flyway (for 12, 8 and 7 geese, respectively) (see Table 1).

Relating Satellite-Derived Green Wave Index to Barnacle Goose Migration

We used two approaches to test whether barnacle geese 'surf' along the green wave. One approach was a visualization method to identify correlations between barnacle goose movements during the spring migration and vegetation phenology. For the visualization method, first we divided the study area into three flyways, i.e. Russian, Svalbard and Greenland. Then we used the GPS-tracking data of migrating barnacle geese and related these to the spatio-temporal pattern in GWI (i.e. the vegetation phenology). In this regard, the annual GWI trajectories were stratified for each flyway separately by latitude, plotted along axes of time and latitude, and colored according to GWI value. Thus, each cell in the stratified image represented the average of the actual GWI values in each latitudinal band at a certain time.

The timing of 50% NDVI correlates with the peak in food quality [20]. So, our second approach was to define the date at which the actual GWI value reached 50% of its annual maximum at each of the stopover sites, and compare that to the date on which the geese arrived at that site using regression analysis. To perform the analysis, data from different stopover sites were combined from the three years for each population, leading to 57 stopover sites for the Russian population, 18 for the Svalbard population, and 14 for the Greenland population.

To predict the geese arrival dates from three populations at each stopover site, we used a linear, mixed-effect model, with a fixed effect for the date of 50% GWI, as well as considering the random effect of individual geese within different tracking years and the random effect of each tracking year.

A slope approximately equal to 1 and an intercept near 0 represents surfing the green wave (i.e. where the date of 50% GWI at a given stopover site was also the date on which that stopover was occupied by the geese). The coefficient of determination, R^2, was used to assess the strength of the relation.

In addition to regression analysis, we calculated the root-mean-square deviation (RMSD) to measure how well the observed arrival dates at stopover sites fitted with arrival dates predicted from the satellite-derived GWI. We defined RMSD values <10 days as a good fit, 10–15 days as moderate, and >15 days as poor, based on Duriez et al. [48].

The effect of tracking year and flyway on the actual GWI values was tested using a two-way factorial ANOVA, with year (three levels) and flyways (three levels) as well as their interaction. Where a significant effect was found, we used a Bonferroni correction at $p = 0.0167$ to compare means within each factor level.

Barnacle geese forage on food patches with the highest grass density [31] and they also forage on agricultural fields in temperate regions [19,34]. We therefore extracted the actual GWI values only from grassland and cropland land cover types in a 15-km radius around each of the 57, 18, and 14 stopover sites for the Russian, Svalbard, and Greenland populations respectively (Appendix S1). This distance is based on the core foraging range for barnacle geese [49]. In order to do the statistical analysis (i.e. regression and ANOVA), the actual GWI values were extracted from the real stopover site locations.

Results

Visualization of Barnacle Goose Migration against Satellite-Derived GWI

The northward migration of barnacle geese correlated well with the plant phenology (Figure 2). Their spring migration during the study period fell within the early stage (GWI 20–40%), middle stage (40–60%), or late stage greenness (60–80%) based on the GWI values.

In two years, 2008 and 2009, Russian barnacle geese left the lower latitudes in late-April, when the GWI values were near to 70%. For a one-month period (late-April to late-May), the geese migrated to higher latitudes, following a mid-range of GWI values (GWI 40–60%). They arrived at the breeding sites, where the GWI values were close to 20%, at the end of May and beginning of June. The Svalbard geese followed the same phenological stage of the vegetation as the Russian geese, but stayed closer to 40% GWI during their migration to higher latitude.

In contrast, the spring migration of the Greenland geese and their response to the plant phenology was different to the other two populations. The Greenland geese left the lower latitudes around the start of April, when the GWI was about 40%. During their migration to higher latitudes, they tracked a constant but lower range of GWI values (20–40%) than the Russian and Svalbard geese, i.e. the Greenland geese followed an earlier stage of the GWI than the Russian and Svalbard geese (2008 and 2009 in Figure 2). However, in 2010, we observed that the geese from all three populations tracked a higher range of GWI during their northward migration. The GWI range was 60–80% for the

Table 1. Tag ID, year of tracking, and number of stopover sites for each barnacle goose.

Russian population (n = 12)			Svalbard population (n = 8)			Greenland population (n = 7)		
Bird ID	Track year	No. of stopover sites	Bird ID	Track year	No. of stopover sites	Bird ID	Track year	No. of stopover sites
78033	2009–2010	2	33953	2010	2	65698	2009	2
78034	2009–2010	2	33954	2010	1	70563	2010	2
78035	2009–2010	2	78198	2008	5	78199	2010	2
78036	2009–2010	3	78378	2008–2009	3	78207	2008	2
78037	2009	2	86824	2009	1	78208	2008	2
78039	2009–2010	4	86828	2009	1	78209	2008	1
78041	2008–2010	6	178199	2008	3	78210	2008	3
78043	2008–2010	10	186827	2009	2			
78044	2008–2010	10						
78045	2008	4						
78046	2008–2009	2						
78047	2008–2010	10						

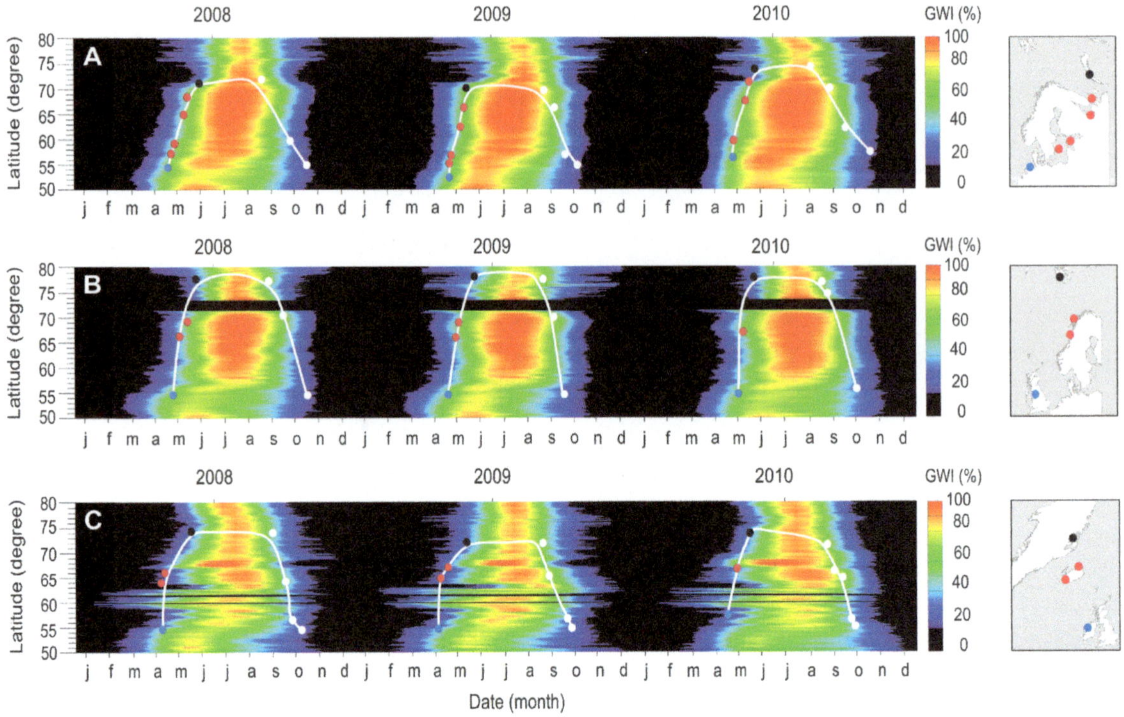

Figure 2. The GWI summary plots showing plant phenology over three years (2008–2010). The Russian (A), Svalbard (B) and Greenland (C) flyways are indicated. The GWI is estimated from MODIS NDVI and ranges from 0% (minimum greenness) to 100% (maximum greenness). The northward spring migration has been shown on the GWI background, as well as the return movement throughout the year. Each dot in the figure represents the average of both the latitude of the site locations and the time for 12 Russian, 8 Svalbard and 7 Greenland barnacle geese, from 2008 to 2010. The site locations include breeding (black dots), overwintering (blue dots), and stopover (red dots) sites for the spring migration and white dots for the autumn migration. The map of each flyway with the site locations overlaid is shown in the right-hand column. The white smoothed line shows the general migration pattern of the geese with respect to the vegetation phenology. The black bands on the western flyways (Svalbard and Greenland) indicate areas with no NDVI information (i.e. ocean).

Russian and Svalbard geese, whereas it was 40–60% for the Greenland geese. Indeed, in 2010, the GWI values showed that all the tracked geese migrated northward when the vegetation was in a later phenological stage than the two preceding years (Figure 2). In all three years, the maximum greenness was rarely attained for the habitats between 50–55 latitude in each of the flyways (Figure 2). Unlike the spring migration, the autumn migration of barnacle geese did not fall in a specific GWI stage but instead they followed a rather wide range of GWI (Figure 2).

In order to further illustrate how barnacle geese follow the phenological development of the vegetation, the GWI was mapped during the spring migration in 2008 and showed the barnacle goose locations for the corresponding time periods (Figure 3). This map strongly supports the hypothesis that phenological development drives barnacle goose movement during the spring migration.

Correlation between Barnacle Goose Spring Migration and Date of 50% GWI

For individuals from the Russian flyways, the residual variance estimate ($\hat{\sigma} = 27.55$) was larger than the random effect variance estimates of individual geese within different tracking years ($\hat{\sigma} = 0.37$) and given the random effect of a tracking year ($\hat{\sigma} = 5.87$). Moreover, for individuals on the Svalbard and Greenland flyways, we determined an estimate of zero for the random effect variance; this simply indicated that the level of "between-group" and "within-group" variability is not sufficient to warrant incorporating a random effect in the model. We therefore eliminated the random effect from the model and fitted an OLS regression to individuals on the Russian, Svalbard and Greenland flyways.

In all three flyways, we found a significant relationship between the arrival dates at the stopover sites and the date of 50% GWI at that specific stopover (Table 2). However, the relationship was stronger for the Russian ($R^2 = 0.71$, $p < 0.001$, $n = 57$) and Svalbard geese ($R^2 = 0.70$, $p < 0.001$, $n = 18$) than for the Greenland geese ($R^2 = 0.31$, $p < 0.05$, $n = 14$) (Table 2, Figure 4). Furthermore, there was a good fit between observed arrival dates at stopover sites and arrival dates predicted using the GWI index for the Russian (RMSD of 6.21), Svalbard (RMSD of 8.82) and Greenland geese (RMSD of 8.83) (Figure 4).

Comparison of GWI at Spring Stopover Sites for the Three Flyway Populations

A factorial ANOVA revealed a significant main effect of flyway on GWI values at stopover sites (Table 3). It suggested that the GWI values at the stopover sites in the Russian and Svalbard flyways were significantly higher than at the stopover sites in the Greenland flyway (Figure 5A). Moreover, the GWI was affected by year and it was significantly higher in 2010 than in the other years (Table 3, Figure 5B). The difference in GWI values between the Russian and Svalbard flyways and between the years 2008 and 2009 was not significant (Figure 5A, and 5B). We could not find a significant interaction effect between the year and flyway on the GWI values at stopover sites (Table 3).

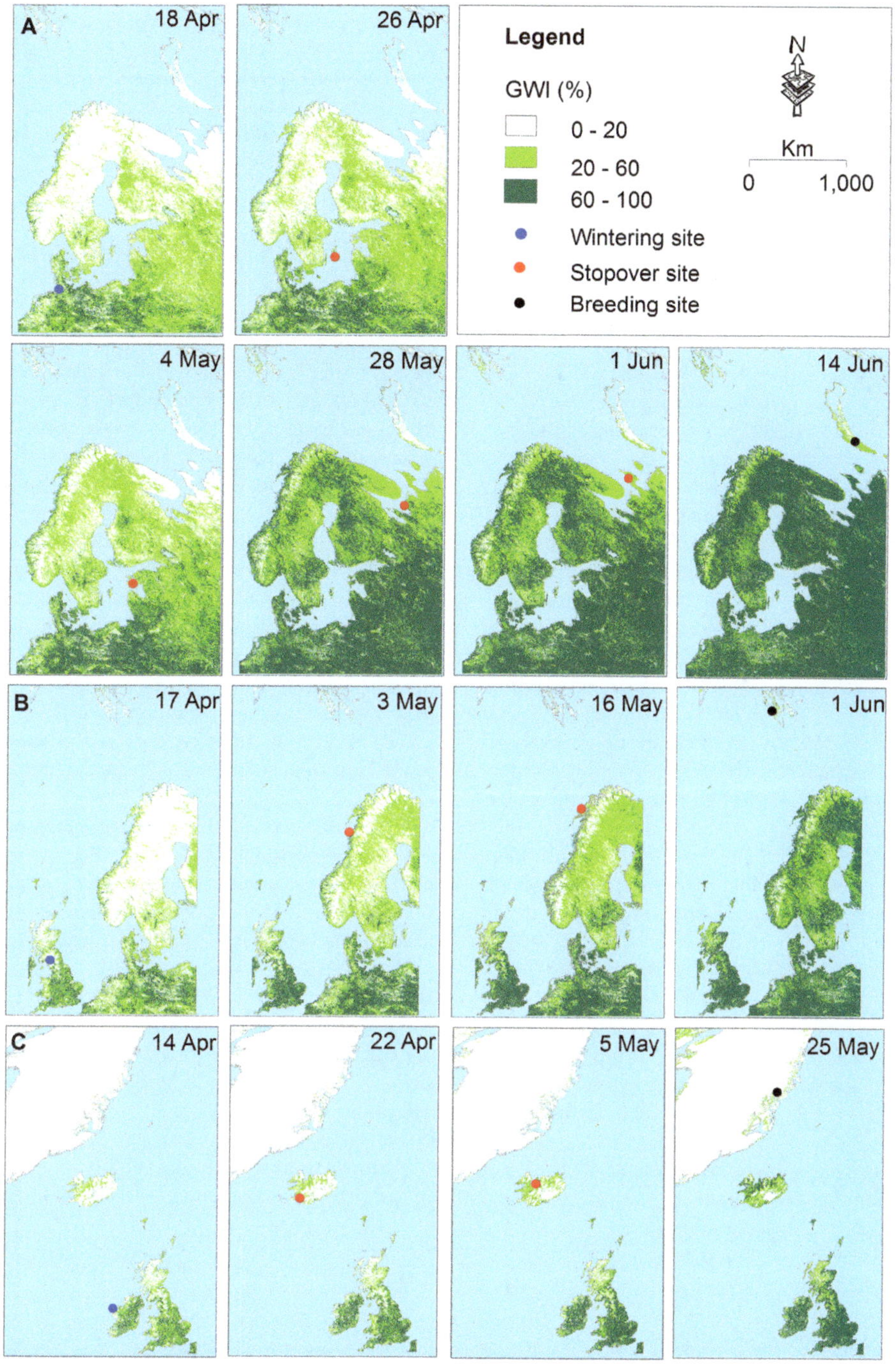

Figure 3. The northward movement of three individual barnacle geese in relation to the green wave. The map indicates the Russian (A), Svalbard (B), and Greenland flyways (C). The individuals' IDs were: 78045, 178199, and 78207 for birds on the Russian, Svalbard and Greenland flyways, respectively, in 2008.

Discussion

Migratory Barnacle Geese Track Satellite-Derived Green Wave Index

Using the satellite-derived green wave index (GWI), we have shown how strongly the spring migration of barnacle geese is correlated with the "green wave" of vegetation phenology. To our knowledge, this is the first time that the migration of individual avian herbivores has been successfully studied with respect to vegetation phenology by using the satellite-derived GWI and GPS tracking of individual birds. Our results revealed that, over a three-year period, their arrival time at the stopover sites during their spring migration coincided well with a specific range of GWI. This range is referred to as the "green wave" and we divided it into three stages (early, middle, and late) in this study. The GWI values selected at the habitat indicate that barnacle geese do not select areas with maximum plant biomass. They preferred areas with an intermediate range of plant biomass, and thereby made a trade-off

Table 2. Results of ordinary least squares regression between the arrival date of the barnacle geese at the stopover sites and the date of 50% GWI, for three different flyways, from 2008 to 2010.

Flyway	d.f.	R^2	*p*-value	Coefficient	Intercept
Russia (n = 57)	55	0.71	<0.001	0.86	20.31
Svalbard (n = 18)	16	0.70	<0.001	0.90	11.96
Greenland (n = 14)	12	0.31	<0.05	0.38	79.20

d.f. degree of freedom, R^2 coefficient of determination.

between forage quality and quantity. Areas with a low GWI (< 20%), where the ingestion rate is limited, and with a high GWI (> 80%), where the energy intake rate decreases because of the low nutritional value and digestibility of mature forage [21,50], were both avoided by the barnacle geese during their spring migration. Thus, their migratory behavior was consistent with the prediction derived from the green wave hypothesis – that avian herbivores follow the successive spring flushes of plants along their northward migration route. The decrease of the GWI values from June–July onwards, and thus the lack of maximum greenness for some areas of the northern mid-latitudes is presumably due to harvesting and also to the ripening and senescence of other crops in agricultural areas [51].

As Figure 2 shows, in contrast to their spring migration, barnacle geese do not appear to follow the green wave during their autumn migration. The geese are not as tied to tracking the green wave during the autumn migration because they have other constraints, such as the need to build up as good physical condition as possible after the energy stresses of the moult period. Moreover, the timing of arrival at the destination is not important in the autumn as it is in spring. They therefore tend to remain in the Arctic and accumulate fat reserves until the autumn snow forces them to migrate southwards [52] and they wait for the best weather before departing, for example to make use of tailwinds [53]. Although the geese took rests on their southward migration, they could not refuel enough during the resting periods and still depended on the energy stores they had accumulated in the Arctic before departure [54].

For the tracked barnacle geese from the Russian and Svalbard flyway populations, we found a strong significant relationship and a good fit between the arrival date at stopover sites and the dates of 50% GWI at that specific stopover (see Figure 4 and Table 2). Moreover, data points were dispersed around the 1:1 line, and the slope of the regression line was close 1. This suggests that the Russian and Svalbard geese were able to surf the green wave and that they benefited from having access to early vegetation phenology by closely tracking the 50% GWI. However, for the Greenland geese, we observed a relatively weak relationship between arrival date and 50% GWI. Furthermore, the dispersion of data points was mostly below the 1:1 line. This indicates that the Greenland geese arrived earlier at the stopover sites with respect to the green wave. However, their early arrival at the stopover sites may still have an advantage even if there is a lag between their arrival time and the peak in food quality. For instance, it was found that the rate of fat deposition of geese is influenced by their knowledge and experience of feeding at the same foraging sites over several years [55]. Thus, the early arrival of the geese can reduce the competition for food by deterring other birds from occupying the same foraging sites. In addition, individuals who are unable to follow the green wave properly, and thus unable to accumulate large fat reserves, would still benefit from the opportunity to breed successfully by arriving early at the breeding sites [56]. The early arrivals would have less competition for food there, and they could occupy the best nesting sites [57]. Moreover, an early start to breeding means the goslings hatch early and benefit from the longest period of high food quality and pre-migratory fettering [52].

For selective avian herbivores, such as geese, the higher nutritional quality and digestibility of plants occurs at the start of the growing season, when there is an intermediate plant biomass

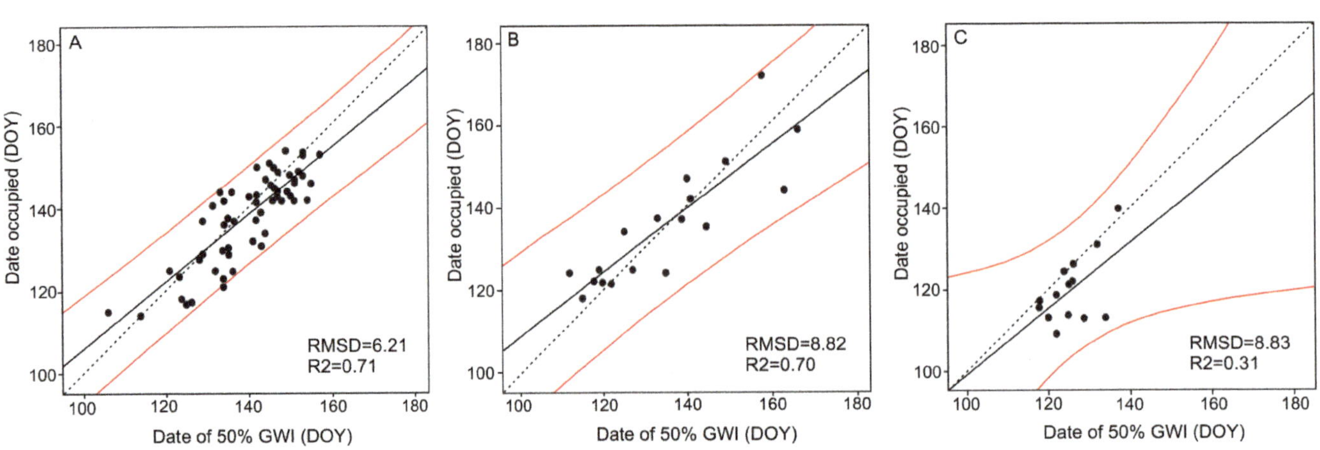

Figure 4. The relationship between date of 50% GWI and arrival date at stopover sites during migration. The Russian (A), Svalbard (B) and Greenland (C) barnacle goose populations are indicated. The solid black line shows the OLS regression line, while the dotted line is the 1:1 line. The red line shows the 95% confidence interval. GWI = green wave index, DOY = day of the year counting from 1st January.

Table 3. Summary statistics of a factorial ANOVA examining the effects of flyway, year and their interaction on GWI values at stopover sites.

Source of variation	d.f.	*F*-value	*p*-value
Flyway	2	12.68	<0.001
Year	2	14.1	<0.001
Flyway*year	4	0.96	0.43

p-value<0.001, n = 89, R^2 = 0.44.
d.f. degree of freedom, R^2 coefficient of determination.

[31,32]. It has been demonstrated that there was a successive wave of nutrient biomass along the spring migration route of Russian barnacle geese [19]. Moreover, along the Russian flyway, the maximum value of nutrient biomass was also found to occur at each stopover site when it was occupied [19]. The peak of forage biomass quality for Russian barnacle geese in the Baltic Sea and Barents Sea, sampled from leaf tips, was around 20th April and 20th June, respectively [58]. These two periods are almost similar to the arrival time of Russian barnacle geese at the Baltic Sea and Barents Sea coast seen in our study. Thus, the most plausible explanation for the association between the 50% GWI and the observed dates of geese occupying the stopover sites is that the GWI reflects the forage quality.

Our research and that done by Van der Graaf et al. [19] led to the same conclusions for Russian barnacle geese and their following of the green wave, despite using methods with very different scales. We used satellite imagery to cover the complete geographical range without any field data, while Van der Graaf et al. [19] used only field data from a limited number of sites. It is clear that using satellite imagery, such as NASA's MODIS NDVI data which are freely available, saves a lot of time and cost for this kind of research covering vast geographical areas. Moreover, satellite imagery is available for any time period, and makes this kind of research possible in very remote areas. The satellite-derived GWI has also been successfully used to correlate the altitudes of movements of ground animals, like giant pandas and golden takin, with phenological development of the vegetation [29,38]. Our results show that this index can also be applied to the movement of avian herbivores that move comparatively faster and cover larger distances with respect to vegetation phenology.

Differences in the Satellite-Derived GWI at Spring Stopover Sites

The comparison of the satellite-derived GWI values at spring stopover sites between the three flyway populations showed a significant effect for the flyway. Our results showed that the Russian and Svalbard barnacle geese are more similar in terms of how they track vegetation phenology, as there was no significant difference in the GWI values between these two flyways. On the other hand, the Greenland geese were significantly ahead of the other two flyways with respect to following the green wave.

Based on the deposition rate hypothesis, birds decided to migrate when foraging conditions start to deteriorate and staying is no longer profitable [56]. The Greenland geese probably need to leave Ireland earlier because spring occurs earlier there than on the other flyways and the grass quality is assumed to decline due to maturation. These geese have no mid-point to migrate to which would be ideal in terms of surfing the green wave; instead only

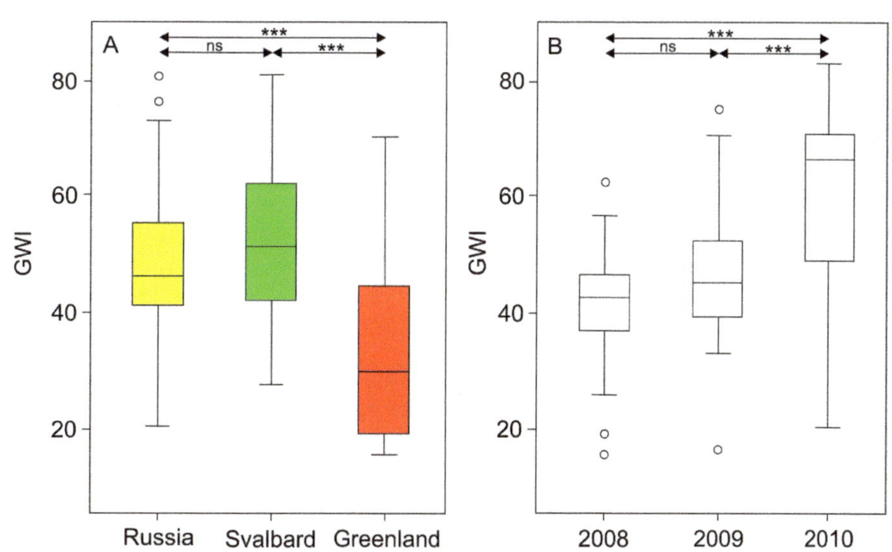

Figure 5. Box plots showing the development of the green wave index (GWI) at stopover sites. The range of GWI values is shown for the three flyways (A), and for the three different years (2008–2010) (B). Each box plot shows the median (line within the box), the 25th percentile (lower end of the box), the 75th percentile (upper end of the box), and 10th to 90th percentile (solid lines). The open circles show the outliers. The significant differences in GWI at the stopover sites between the three different flyways and the three different years seen in an ANOVA analysis using a Bonferroni correction are indicated (here *p*-value = 0.05/3). ***$p \leq 0.001$, ns = non-significant.

Iceland is available as a stopover and they must arrive there earlier in terms of spring's progress than they would perhaps choose under more ideal circumstances.

Besides the flyways, our results showed the significant effect of the type of year on how barnacle geese follow the green wave. The Russian and Svalbard geese followed the middle stage of the green wave in 2008 and 2009, but a later stage in 2010. In contrast, the Greenland geese followed the earlier stage of the green wave in 2008 and 2009, but the middle stage in 2010. In other words, the geese we tracked followed the markedly higher value of the satellite-derived GWI in 2010 in all three populations. We think this was due to the extreme weather in northern and western Europe in 2010. The continental temperate climate zone in western Europe was particularly dry for the spring season of 2010, certainly compared with the two previous years [59]. This could have led to an earlier start to the growing season at higher latitudes because an increase in the mean annual air temperature in early spring corresponds to an advance in leafing [60]. An earlier start to the growing season at higher latitudes would have meant that the geese were more likely to catch the later phenological stages of plant growth along their flyway in 2010 if they had started migrating at their normal time. As shown by Tombre et al. [16], if the geese cannot predict the conditions they might encounter at the next stopover, they are unable to respond quickly to the advancing spring. For instance, the lack of correlation in the onset of spring between the Solway Firth and Helgeland stopovers meant the geese were unable to migrate earlier if spring was early at both sites [16]. Moreover, the timing of the Russian geese migration from the Baltic Sea was not linked to the advancement of plant growth, most likely because of the low correlation in the weather patterns between the Baltic Sea and White Sea [33]. Using the third derivative of daily temperature sums (GDDjerk), Kölzsch et al. [61] showed that the geese are able to closely follow the green wave during their spring migration if predictability of climatic conditions was high between stopovers. Therefore, in the case that predictability is low, the geese might rely more on fixed cues such as photoperiod (length of daylight hours), and do not migrate earlier in the year if spring is early.

Conclusions

By using the satellite-derived green wave index, we have shown that individual barnacle geese surf the wave of high-nutrition plants. Remote sensing tools provide the opportunity to predict plant biomass and to study plant phenology in remote areas such as the Arctic, where it is difficult to collect plant data on a large spatial and temporal scale. In addition, by applying GWI (a metric derived from the NDVI time series) as a remote sensing tool to determine accurately the timing of high quality vegetation for herbivores (i.e. the date at which GWI reaches 50% of its maximum value), we were able to investigate how the geese from the three populations made use of the green wave during the three years studied. Remote sensing data, and NDVI in particular, are among the technological advances that are proving useful in studying large-scale movement ecology, and they have helped us gain a better understanding of how vegetation dynamics and distribution affect movement patterns in animal populations. To our knowledge, this is the first time that the migration of individual avian herbivores has been successfully studied with respect to vegetation phenology by using the satellite-derived green wave index.

Acknowledgments

The barnacle goose data for the Russian population were provided by FlySafe, a project initiated by the Integrated Applications Promotion (IAP) Programme of the European Space Agency. The study was carried out in cooperation with the Institute of Avian Research, Germany, the Dutch Centre of Field Ornithology (Sovon), and the University of Amsterdam from 2007 onwards. GPS tracks of geese from the Svalbard and Greenland populations were collected by the Wildfowl & Wetlands Trust (UK). The Svalbard barnacle geese were caught with the help of the North Solway Ringing Group, and the Greenland barnacle geese were caught with the help of the National Parks and Wildlife Service, Dublin, Ireland. We thank Willem Nieuwenhuis for his help with the programming, Dr. Roshanak Darvishzadeh for her comments on the revised manuscript, and Jackie Senior for editing the text.

Author Contributions

Conceived and designed the experiments: MS TW AS AT AK BN. Performed the experiments: MS TW AS. Analyzed the data: MS TW AS. Contributed reagents/materials/analysis tools: KE LG DC. Wrote the paper: MS TW AS AT KE AK BN LG JS DC.

References

1. Pettorelli N, Vik JO, Mysterud A, Gaillard J-M, Tucker CJ, et al. (2005) Using the satellite-derived NDVI to assess ecological responses to environmental change. Trends Ecol Evol 20: 503–510.
2. St-Louis V, Pidgeon AM, Kuemmerle T, Sonnenschein R, Radeloff VC, et al. (2014) Modelling avian biodiversity using raw, unclassified satellite imagery. Philos T Roy Soc B 369: 20130197.
3. Madritch MD, Kingdon CC, Singh A, Mock KE, Lindroth RL, et al. (2014) Imaging spectroscopy links aspen genotype with below-ground processes at landscape scales. Philos T Roy Soc B 369: 20130194.
4. Di Marco M, Buchanan GM, Szantoi Z, Holmgren M, Marasini GG, et al. (2014) Drivers of extinction risk in African mammals: the interplay of distribution state, human pressure, conservation response and species biology. Philos T Roy Soc B 369: 20130198.
5. Dodge S, Bohrer G, Weinzierl R, Davidson SC, Kays R, et al. (2013) The environmental-data automated track annotation (Env-DATA) system: linking animal tracks with environmental data. Mov Ecol 1: 3.
6. Fujioka E, Soldevilla MS, Read AJ, Halpin PN (2013) Integration of passive acoustic monitoring data into OBIS-SEAMAP, a global biogeographic database, to advance spatially-explicit ecological assessments. Ecol Inform 21: 59–73.
7. Myneni RB, Hall FG, Sellers PJ, Marshak AL (1995) The interpretation of spectral vegetation indexes. IEEE Trans Geosci Remote Sens 33: 481–486.
8. Huete A, Didan K, Miura T, Rodriguez EP, Gao X, et al. (2002) Overview of the radiometric and biophysical performance of the MODIS vegetation indices. Remote Sens Environ 83: 195–213.
9. Tucker CJ, Vanpraet CL, Sharman MJ, Vanittersum G (1985) Satellite remote-sensing of total herbaceous biomass production in the Senegalese Sahel - 1980–1984. Remote Sens Environ 17: 233–249.
10. Walker DA, Auerbach NA, Shippert MM (1995) NDVI, biomass, and landscape evolution of glaciated terrain in northern Alaska. Polar Rec 31: 169–178.
11. Box E, Holben B, Kalb V (1989) Accuracy of the AVHRR vegetation index as a predictor of biomass, primary productivity and net CO2 flux. Vegetatio 80: 71–89.
12. Studer S, Stockli R, Appenzeller C, Vidale PL (2007) A comparative study of satellite and ground-based phenology. Int. J Biometeorol 51: 405–414.
13. White MA, Thornton PE, Running SW (1997) A continental phenology model for monitoring vegetation responses to interannual climatic variability. Glob Biogeochem Cycles 11: 217–234.
14. Beck PSA, Atzberger C, Høgda KA, Johansen B, Skidmore AK (2006) Improved monitoring of vegetation dynamics at very high latitudes: A new method using MODIS NDVI. Remote Sens Environ 100: 321–334.
15. Zhang X, Friedl MA, Schaaf CB, Strahler AH, Hodges JCF, et al. (2003) Monitoring vegetation phenology using MODIS. Remote Sens Environ 84: 471–475.

16. Tombre IM, Høgda KA, Madsen J, Griffin LR, Kuijken E, et al. (2008) The onset of spring and timing of migration in two arctic nesting goose populations: the pink-footed goose Anser bachyrhynchus and the barnacle goose Branta leucopsis. J Avian Biol 39: 691–703.
17. Demment MW, Van Soest PJ (1985) A nutritional explanation for body-size patterns of ruminant and nonruminant herbivores. Am Nat 125: 641–672.
18. Fryxell JM (1991) Forage quality and aggregation by large herbivores. Am Nat 138: 478–498.
19. Van der Graaf S, Stahl J, Klimkowska A, Bakker JP, Drent RH (2006) Surfing on a green wave – how plant growth drives spring migration in the Barnacle Goose *Branta leucopsis*. Ardea 94: 565–577.
20. Doiron M, Legagneux P, Gauthier G, Levesque E (2013) Broad-scale satellite Normalized Difference Vegetation Index data predict plant biomass and peak date of nitrogen concentration in Arctic tundra vegetation. Appl Veg Sci 16: 343–351.
21. Mueller T, Olson KA, Fuller TK, Schaller GB, Murray MG, et al. (2008) In search of forage: predicting dynamic habitats of Mongolian gazelles using satellite-based estimates of vegetation productivity. J Appl Ecol 45: 649–658.
22. Hamel S, Garel M, Festa-Bianchet M, Gaillard JM, Cote SD (2009) Spring Normalized Difference Vegetation Index (NDVI) predicts annual variation in timing of peak faecal crude protein in mountain ungulates. J Appl Ecol 46: 582–589.
23. Ryan SJ, Cross PC, Winnie J, Hay C, Bowers J, et al. (2012) The utility of normalized difference vegetation index for predicting African buffalo forage quality. J Wildl Manage 76: 1499–1508.
24. Bartlam-Brooks HLA, Beck PSA, Bohrer G, Harris S (2013) In search of greener pastures: Using satellite images to predict the effects of environmental change on zebra migration. J Geophys Res-Biogeo 118: 1427–1437.
25. Bohrer G, Beck PS, Ngene SM, Skidmore AK, Douglas-Hamilton I (2014) Elephant movement closely tracks precipitation-driven vegetation dynamics in a Kenyan forest-savanna landscape. Mov Ecol 2: 2.
26. Stoner DC, Wolfe ML, Mecham C, Mecham MB, Durham SL, et al. (2013) Dispersal behaviour of a polygynous carnivore: do cougars Puma concolor follow source-sink predictions? Wildlife Biol 19: 289–301.
27. Owen M (1980) Wild geese of the world. London: Batsford. 236.
28. Bischof R, Loe LE, Meisingset EL, Zimmermann B, Van Moorter B, et al. (2012) A Migratory Northern Ungulate in the Pursuit of Spring: Jumping or Surfing the Green Wave? Am Nat 180: 407–424.
29. Wang TJ, Skidmore AK, Zeng ZG, Beck PSA, Si YL, et al. (2010) Migration patterns of two endangered sympatric species from a remote sensing perspective. Photogramm Eng Remote Sens 76: 1343–1352.
30. Prop J, Vulink T (1992) Digestion by barnacle geese in the annual cycle - the interplay between retention time and food quality. Funct Ecol 6: 180–189.
31. Black JM, Prop J, Larsson K (2007) Wild goose dilemmas. Groningen, The Netherlands: Branta Press. 254.
32. Van Wijk RE, Kölzsch A, Kruckenberg H, Ebbinge BS, Muskens G, et al. (2012) Individually tracked geese follow peaks of temperature acceleration during spring migration. Oikos 121: 655–664.
33. Van der Graaf AG (2006) Geese on a green wave: Flexible migrants in a changing world. (PhD thesis), University of Groningen, Groningen.
34. Eichhorn G, Drent RH, Stahl J, Leito A, Alerstam T (2009) Skipping the Baltic: the emergence of a dichotomy of alternative spring migration strategies in Russian barnacle geese. J Anim Ecol 78: 63–72.
35. Eichhorn G, Afanasyev V, Drent RH, van der Jeugd HP (2006) Spring stopover routines in Russian Barnacle Geese *Branta leucopsis* tracked by resightings and geolocation. Ardea 94: 667–678.
36. Hübner CE, Tombre IM, Griffin LR, Loonen M, Shimmings P, et al. (2010) The connectivity of spring stopover sites for geese heading to arctic breeding grounds. Ardea 98: 145–154.
37. Ogilvie MA, Boertmann D, Cabot D, Merne O, Percival SM, et al. (1999) Barnacle Goose *Branta leucopsis*: Greenland. In: Madsen J, Cracknell G, Fox T editors. Goose Populations of the Western Palearctic. A Review of Status and Distribution. Rond, Denmark: National Environmental Research Institute. 344.
38. Beck PSA, Wang TJ, Skidmore AK, Liu XH (2008) Displaying remotely sensed vegetation dynamics along natural gradients for ecological studies. Int J Remote Sens 29: 4277–4283.
39. Jonsson AM, Eklundh L, Hellstrom M, Barring L, Jonsson P (2010) Annual changes in MODIS vegetation indices of Swedish coniferous forests in relation to snow dynamics and tree phenology. Remote Sens Environ 114: 2719–2730.
40. Liston GE, Sturm M (2002) Winter precipitation patterns in arctic Alaska determined from a blowing-snow model and snow-depth observations. J Hydrometeorol 3: 646–659.
41. Beck PSA, Jönsson P, Høgda KA, Karlsen SR, Eklundh L, et al. (2007) A ground-validated NDVI dataset for monitoring vegetation dynamics and mapping phenology in Fennoscandia and the Kola peninsula. Int J Remote Sens 28: 4311–4330.
42. Burgan RE (1996) Use of remotely sensed data for fire danger estimation. Earsel advances in Remote Sens 4: 1–8.
43. Nielsen A, Steinheim G, Mysterud A (2013) Do different sheep breeds show equal responses to climate fluctuations? Basic Appl Ecol 14: 137–145.
44. Griffin LR (2008) Identifying the pre-breeding areas of the Svalbard Barnacle Goose *Branta leucopsis* between mainland Norway and Svalbard: an application of GPS satellite-tracking techniques. Vogelwelt 129: 226–232.
45. Ens BJ, Bairlein F, Camphuysen CJ, de Boer R, Exo KM, et al. (2008) Tracking of individual birds. Report on WP 3230 (bird tracking sensor characterization) and WP 4130 (sensor adaptation and calibration for bird tracking system) of the FlySafe basic activities project: SOVON-onderzoeksrapport 2008/10. SOVON Vogelonderzoek Nederland, Beek-Ubbergen, the Netherlands.
46. ARGOS/CLS. (2011) Argos user's manual, from http://www.grouptechnologies.com.au/downloads/apex/apex-argos-mk2-user-manual.pdf.
47. Drent R, Eichhorn G, Flagstad A, Van der Graaf A, Litvin K, et al. (2007) Migratory connectivity in Arctic geese: spring stopovers are the weak links in meeting targets for breeding. J Ornithol 148: 501–514.
48. Duriez O, Bauer S, Destin A, Madsen J, Nolet BA, et al. (2009) What decision rules might pink-footed geese use to depart on migration? An individual-based model. Behav Ecol 20: 560–569.
49. Pendlebury C, Zisman S, Walls R, Sweeney J, McLoughlin E, et al. (2011) Literature review to assess bird specis connectivity to Special Protection Areas: Scottish Natural Heritage Commissioned Report No. 390, UK.
50. Wilmshurst JF, Fryxell JM, Bergman CM (2000) The allometry of patch selection in ruminants. Proceedings of the Royal Society B-Biological Sciences 267: 345–349.
51. Justice CO, Townshend JRG, Holben BN, Tucker CJ (1985) Analysis of the phenology of global vegetation using meteorological satellite data. Int J Remote Sens 6: 1271–1318.
52. Prop J, Devries J (1993) Impact of snow and food conditions on the reproductive-performance of barnacle geese Branta leucopsis. Ornis Scand 24: 110–121.
53. Bergman G (1978) Effects of wind conditions on the autumn migration of waterfowl between the White Sea area and the Baltic region. Oikos 393–397.
54. Butler PJ, Woakes AJ, Bishop CM (1998) Behaviour and physiology of Svalbard Barnacle Geese Branta leucopsis during their autumn migration. J Avian Biol 29: 536–545.
55. Prop J, Black JM, Shimmings P, Owen M (1998) The spring range of barnacle geese Branta leucopsis in relation to changes in land management and climate. Biological Conservation 86: 339–346.
56. Prop J, Black JM, Shimmings P (2003) Travel schedules to the high arctic: barnacle geese trade-off the timing of migration with accumulation of fat deposits. Oikos 103: 403–414.
57. Kokko H (1999) Competition for early arrival in migratory birds. J Anim Ecol 68: 940–950.
58. Van der Jeugd HP, Eichhorn G, Litvin KE, Stahl J, Larsson K, et al. (2009) Keeping up with early springs: rapid range expansion in an avian herbivore incurs a mismatch between reproductive timing and food supply. Glob Chang Biol 15: 1057–1071.
59. Tullus A, Kupper P, Sellin A, Parts L, Sober J, et al. (2012) Climate change at northern latitudes: rising atmospheric humidity decreases transpiration, N-uptake and growth rate of hybrid aspen. Plos One 7: e42648.
60. Chmielewski FM, Rotzer T (2001) Response of tree phenology to climate change across Europe. Agric For Meteorol 108: 101–112.
61. Kölzsch A, Bauer S, Boer Rd, Griffin L, Cabot D, et al. (2014) Forecasting spring from afar? Timing of migration and predictability of phenology along different migration routes of an avian herbivore. J Anim Ecol doi:10.1111/1365-2656.12281.

6

Impact of Non-Native Terrestrial Mammals on the Structure of the Terrestrial Mammal Food Web of Newfoundland, Canada

Justin S. Strong[1,2], **Shawn J. Leroux**[1]*

1 Department of Biology, Memorial University of Newfoundland, St. John's, NL, Canada, 2 Biology Department, Concordia University, Montreal, QC, Canada

Abstract

The island of Newfoundland is unique because it has as many non-native terrestrial mammals as native ones. The impacts of non-native species on native flora and fauna can be profound and invasive species have been identified as one of the primary drivers of species extinction. Few studies, however, have investigated the effects of a non-native species assemblage on community and ecosystem properties. We reviewed the literature to build the first terrestrial mammal food web for the island of Newfoundland and then used network analyses to investigate how the timing of introductions and trophic position of non-native species has affected the structure of the terrestrial mammal food web in Newfoundland. The first non-native mammals (house mouse and brown rat) became established in Newfoundland with human settlement in the late 15th and early 16th centuries. Coyotes and southern red-backed voles are the most recent mammals to establish themselves on the island in 1985 and 1998, respectively. The fraction of intermediate species increased with the addition of non-native mammals over time whereas the fraction of basal and top species declined over time. This increase in intermediate species mediated by non-native species arrivals led to an overall increase in the terrestrial mammal food web connectance and generality (i.e. mean number of prey per predator). This diverse prey base and sources of carrion may have facilitated the natural establishment of coyotes on the island. Also, there is some evidence that the introduction of non-native prey species such as the southern red-backed vole has contributed to the recovery of the threatened American marten. Long-term monitoring of the food web is required to understand and predict the impacts of the diverse novel interactions that are developing in the terrestrial mammal food web of Newfoundland.

Editor: Stephanie S. Romanach, U.S. Geological Survey, United States of America

Funding: JSS was supported by a Undergraduate Summer Research Award from the Natural Sciences and Engineering Research Council of Canada (NSERC) and SJL was supported by a Discovery Grant from NSERC. The funders had no role in study design, data collection and analysis, decision to publish, or preparation of the manuscript.

Competing Interests: The authors have declared that no competing interests exist.

* Email: sleroux@mun.ca

Introduction

Non-native species invasions are a major threat to the persistence of native biodiversity and ecosystem functioning [1–3]. The problem is pervasive as invasive species from most major taxonomic groups including plants (e.g. [4]), invertebrates (e.g. [5]), amphibians (e.g. [6]) and mammals (e.g. [7]) have established themselves in many of the World's biomes. The rate of biological invasions has increased with our increasing human footprint and in some regions the number of non-native species outnumbers the number of native ones. The majority of studies of biological invasions focus on how invasive species and recipient ecosystem traits influence the dispersal and establishment of one or a few non-native species [8], [9]. Few studies, however have investigated the impacts of a community of non-native species on native populations and even fewer have addressed community and ecosystem properties [10], [11]. Consequently, there is a pressing need to understand and predict the impacts of non-native species assemblages on native community and ecosystem properties.

The island of Newfoundland, Canada (Fig. 1) presents a case where the number of non-native and transient terrestrial mammals outnumbers the number of extant native terrestrial mammals. Specifically, there are 13 extant terrestrial mammal species native to the island compared to 13 non-native and 3 transient mammals to the island (Table 1). Native mammals include American Beaver (*Castor canadensis*), American Black Bear (*Ursus americanus*) and American Marten (*Martes americana*). The Grey Wolf (*Canis lupus*) is native to Newfoundland but was extirpated in the early 1930s [12]. Among the introduced non-native mammals are the particularly abundant Red Squirrel (*Tamiasciurus hudsonicus* – introduced in 1963), Snowshoe Hare (*Lepus americanus* – introduced in 1864), and American Moose (*Alces americanus* – introduced in 1904). The Coyote (*Canis latrans*), arrived in 1985 after crossing on ice from Nova Scotia and has established itself on the island through natural range expansion (Table 1). These invasive species may have a large role in the terrestrial mammal food web, where species such as moose may now be found at considerably high densities [13], [14]. In this study, we investigated the impact of the non-native mammal assemblage on the structural properties (e.g. link density, connectance) of the terrestrial mammal food web in Newfoundland.

Compared with mainland regions, islands can be more susceptible to disturbances from the arrival of exotic species [15]. This is due in part to fewer species, lack of natural parasites

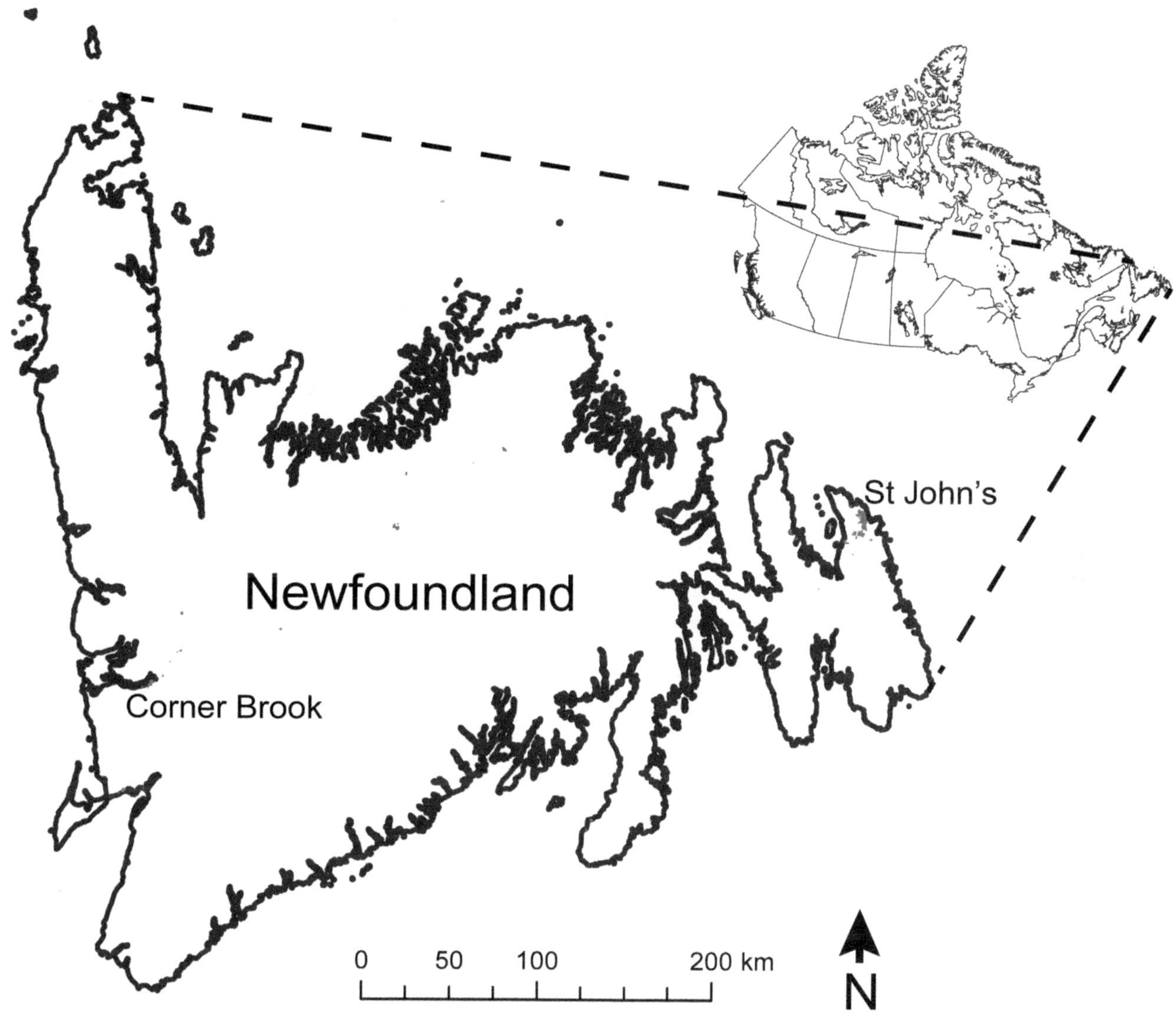

Figure 1. Island of Newfoundland. Map of the island of Newfoundland with a map of Canada inset.

or predators, more endemic species, and simpler food webs on islands [16]. These traits have generally led to higher rates of species extinctions in insular ecosystems [16], [17]. Biological invasion is advanced as one of the leading causes of biodiversity loss and disturbance on island ecosystems [18]. Specifically, the introduction of non-native mammals can have significant impacts on the flora and fauna of insular ecosystems [19], [20]. For instance, the introduction of the Small Asian Mongoose (*Urva auropunctata*) to Mauritius, Fiji, West Indies, Caribbean islands, and Hawai'i has been catastrophic for some species endemic to these islands. Originally introduced to control rat populations, this non-native mammal has preyed heavily on native fauna [21]. In the Caribbean islands alone, the decline or extinction of 14 of 16 skink species can be attributed to the Small Asian Mongoose [22]. Multiple evolutionary explanations exist for why insular species are sensitive to disturbances from alien species. For example, insular species may be maladapted for competing and co-existing with non-native species; evolution in geographic isolation may result in a lack of traits that facilitate persistence when subject to predation and interference pressures from non-native species [23]. Islands may also support an abundance of resources to allow the spread of non-native species. Therefore, non-native species may rapidly become established and proliferate in island ecosystems, often reaching abundances that are unsustainable within the environment [16]. Although Newfoundland is a continental island and not a "true" island (i.e. oceanic), there is evidence that biodiversity on continental islands also is susceptible to human disturbances such as species introductions (e.g. Australia) [24], [25].

Previous studies have proposed that the effects of non-native species on native food webs are mediated by trophic position (reviewed in [10]). Non-native predators may be especially detrimental for native biodiversity in some areas [25]. For example, a recent meta-analysis showed that non-native predators have a much larger negative effect on population size and reproduction of vertebrate prey than native predators [25]. Conversely, the introduction of non-native prey may confer positive effects on native predators as they can act as viable food sources (reviewed by [26]).

The order of establishment of sequential species introductions also may influence how a non-native species assemblage affects a native food web [27], [28]. For example, Pope et al. [27] show that salmonids introduced to the Klamath Mountains region of California as early as the 1800s have facilitated the establishment and expansion of the non-native Pacific Coast Aquatic Garter Snake (*Thamnophis atratus*) with negative impacts on native Cascades Frogs (*Rana cascadae*). The idea of synergistic or facilitative interactions between non-native species is central to the concept of an "invasional meltdown" (sensu [29]) and evidence of an invasional meltdown has been demonstrated in systems such as the Great Lakes [30].

Ecological network analysis has been applied widely to assess the impact of species loss on food web structure, stability and function (e.g. [31–34]). Several recent studies have applied network analysis to investigate the impact of species or groups of species, particularly parasites, on food web properties such as connectance, link density, and prey:predator ratio [9], [35–37]. Woodward and Hildrew [38] applied network analysis to investigate the impacts of an invasive predator on native stream food web properties. However, we are aware of few studies that apply network analysis to investigate the impacts of a community of invasive mammal species on native food web properties (but see examples from insect communities in [39], plant-herbivore communities in [40], and plant-pollinator communities in [41]). Studies of specific introduced species to Newfoundland such as American Moose and Red Squirrel have demonstrated significant impacts of these herbivores on forest regeneration and structure [13], [14]. Our goal is to assess the impact of the full suite of non-native terrestrial mammals on the island's terrestrial mammal food web. Specifically, our research goals are: 1) assemble the first terrestrial mammal food web for the island of Newfoundland, 2) assess impacts of non-native mammals on native mammal food web properties over time, and 3) determine whether changes to food web structure arising from the addition of non-native species can be attributed to specific trophic positions. We hypothesized that the trophic position (e.g. herbivore, predator) of introduced species is the main determinant of its impact on food web structure.

Materials and Methods

Food web construction

We consulted the Government of Newfoundland and Labrador's Department of Conservation and the Environment's list of terrestrial mammals of Newfoundland [42] for our list of terrestrial mammals on the island. To create a food web describing feeding relationships among these species we extracted data from published studies that involved diet analyses for mammal species on the island. We performed an online search in Web of Science and the Memorial University of Newfoundland online library catalogue for peer-reviewed journals, using the keywords "diet" or "food habits" in combination with "Newfoundland" and the species in question (e.g. "American Black Bear", "Red Fox"). For mammal species without diet analyses for Newfoundland, we used diet data from other studies mostly from the Canadian boreal region (Table 1). We included non-mammal prey items in our food web.

The studies we reviewed used various techniques to infer the diet of each species including visual (e.g. [43]) and DNA identification [44] of scat contents, stomach-contents analysis (e.g. [45]) stable-isotope analysis (e.g. [46]), and quantitative fatty-acid-signature analysis (e.g. [47]), as well as various forms of field observation (e.g. evidence of browsing on vegetation [48]). The measures used in those studies include percent frequency of occurrence of various food items within total sample or within a specified subcategory of the total sample (e.g. percent frequency of occurrence within samples containing bird remains, [49]), percent of total volume on a year-round and per-season basis (e.g. total percent volume per season, [45]), mean percent prey composition (e.g. [50]), mean percent relative density (e.g. [51]), percent of dry fecal weight (e.g. [52]), percent contribution to fatty acid signatures (e.g. [47]), as well as measures pertaining to field observations, such as percent of stems utilized in an area and percent composition of browse pile (e.g. [48]). Where available, we collected information on sample size, season, and year. From the resulting database, we created a matrix that identified all feeding

Table 1. Extant native terrestrial mammals on the island of Newfoundland with sources for dietary data used in our study.

Common name	Scientific name	Order	Data origin[α]
American Beaver [F2]	*Castor canadensis*	Rodentia	Newfoundland
American Black Bear [F3]	*Ursus americanus*	Carnivora	Newfoundland
American Marten [F3]	*Martes americana*	Carnivora	Newfoundland
Arctic Hare[F1]	*Lepus arcticus*	Lagomorpha	Newfoundland
Canadian Lynx [F3]	*Lynx canadensis*	Carnivora	Newfoundland
Caribou [F2]	*Rangifer tarandus*	Artiodactyla	Newfoundland
Common Muskrat [F2]	*Ondatra zibethicus*	Rodentia	Maine, New York
Ermine [F3]	*Mustela erminea*	Carnivora	Newfoundland
Little Brown Myotis [F1]	*Myotis lucifugus*	Chiroptera	Illinois
Meadow Vole [F2]	*Microtus pennsylvanicus*	Rodentia	Newfoundland
North American River Otter [F3]	*Lontra canadensis*	Carnivora	Newfoundland
Northern Myotis [F1]	*Myotis septentrionalis*	Chiroptera	Illinois, Central Appalachians
Red Fox [F3]	*Vulpes vulpes*	Carnivora	Newfoundland

[α]see [53] for full details on data.
[F1]Herbivores and insectivores without predators (F_1 functional group).
[F2]Herbivores and insectivores with predators (F_2 functional group).
[F3]Predators (F_3 functional group).
Common and scientific species names follow Wilson and Reeder [83] except for Caribou.

links among the terrestrial mammal assemblage in Newfoundland. The detailed diet and feeding links databases are available on figshare [53].

The studies we collected showed considerable variability in the resolution at which food items and groups of food items were presented. For example, Riewe [54] categorized vegetation in Red Fox (*Vulpes vulpes*) diet as grasses, *Abies*, or berries, whereas Bridger [55] simply reported on the single class "vegetation". We grouped food items into categories deemed appropriate to examine the effects of non-native mammals on native mammal feeding interactions, some groups being at a different resolution than others. Ultimately these groups were defined based on the lowest resolution of each food item for each species. Data on mammals are reported at the species level. Data on bird food items are reported to the level of Order (members of the Orders Anseriformes, Columbiformes, Galliformes, Gruiiformes, Passeriformes), except aquatic birds (a group that contains the Orders Charadriiformes, Procellariiformes, and Pelecaniformes). The category of aquatic birds is the finest resolution to which we could classify these food items.

Plant food items were designated as grasses, herbaceous plants, mosses, lichens, fungi, fruits or deciduous shrubs/trees. Balsam Fir (*Abies balsamea*) was placed in its own category apart from the other evergreen shrubs and trees because it is an important dietary component for many species in Newfoundland [56–59]. Our category "herbaceous plants" includes freshwater aquatic plants, "fruits" includes berries, and "mosses" includes mosses and other bryophytes. The food item "terrestrial invertebrates" includes insects and "marine mammals" includes only cetaceans and pinnipeds.

The remaining food item groups were marine invertebrates and fish. Some diet studies used broad or vague categories (e.g. "achlorophyllous plant matter", "birds", "bone", "eggshells", "feathers", "flesh", "garbage", "green vegetation", "leaves", "miscellaneous vegetation", "nuts/seeds", "plant material", "roots", "seeds", and "woodchips"). Oftentimes detailed explanations of these groups were not included in the studies. We incorporated those data in our detailed database but not to construct the mammal feeding links matrix. In addition, some studies from outside Newfoundland included diet items that are not present on the island, or are domestic animals (e.g. lemmings, deer, chicken, cow, elk). Again, these data were included in the detailed database only.

Food web analysis

We calculated species richness (S), total number of links (L), connectance ($C = L/S^2$), link density (L/S), fraction of basal species (i.e. species that do not consume other species), fraction of intermediate species (species that consume and are consumed by other species), fraction of top species (i.e. species that are not consumed), prey:predator ratio (i.e. the number of basal and intermediate species divided by the number of predatory and intermediate species), vulnerability (i.e. mean number of predators per prey species; sensu [60]), and generality (i.e. mean number of prey per predator species; sensu [60]). We began by computing these food web properties for the native terrestrial mammal food web only. To assess the influence of non-native species arrivals over time, non-native mammals were added to the native mammals food web sequentially in the order in which they arrived on the island, with non-native species being left in the web as more recently introduced species were added. We then quantified each food web metric for the addition of every new non-native species.

There is considerable empirical evidence relating non-native species functional groups to their ecological impacts (reviewed in [10]). Consequently, we assessed the effect of different non-native functional groups on food web properties by separating native and non-native species into three functional groups defined by our feeding links matrix. The first functional group, F_1, was "Herbivores and insectivores without predators", the second functional group, F_2, was "Herbivores and insectivores with predators", and the third functional group, F_3, was "Predators". Tables 1 and 2 outline which native and non-native mammal species belong to each functional group. We quantified the effect of non-native trophic position on food web properties by first removing each native and non-native mammal from the same trophic group and calculating the following effect size ratios for each metric; $(Full\ web - F_{1,i})/Full\ web$,; $(Full\ web - F_{2,j})/Full\ web$, *and*; $(Full\ web - F_{3,k})/Full\ web$, where i, j, k are indices representing species from functional groups F_1, F_2, and F_3, respectively. For example, the effect of removing American Moose (functional group F_2) on link density was calculated as link density of the full food web minus link density of the food web without American Moose divided by the link density of the full food web. We calculated the mean effect of removing non-native mammals from each functional group, x, on our food web metrics as non-native mean $((Full\ web - F_{x,i})/Full\ web)$/native mean$((Full\ web - F_{x,i})/Full\ web)$. This allowed us to tease apart the effect of removing a native vs non-native species from the functional group on food web properties. Wolves were not included in our analyses because most non-native species arrived after the wolf was extirpated from Newfoundland and our focus was to investigate the impacts of non-native species on the native food web. Also, we excluded three transient or vagrant mammals species: Arctic Fox (*Alopex lagopus*), Polar Bear (*Ursus maritimus*), and Hoary Bat (*Lasiurus cinereus*) because they are only seasonal visitors to the island. Finally, we excluded Bank Vole (*Myodes glareolus*) and American Bison (*Bison bison*) from our analyses because they were both introduced to offshore islands with the latter introduction not being successful. All analyses were conducted with the foodweb package [61] in R v 2.15.1 [62]. Gephi [63] was used to construct our food web diagram.

Results

The extant native terrestrial mammals of Newfoundland come from the orders Carnivora (6 species), Rodentia (3 species), Chiroptera (2 species), Artiodactyles (1 species), and Lagomorpha (1 species). We obtained dietary data on 10 of these species from studies conducted on the island of Newfoundland (Table 1). The non-native terrestrial mammals of Newfoundland come from the orders Rodentia (7 species), Carnivora (4 species), Artiodactyles (1 species), Chiroptera (1 species), and Soricomorpha (1 species). We obtained dietary data for 4 of these non-native species from studies conducted on the island of Newfoundland (Table 1). The remaining non-native mammal diet data came from elsewhere in North America (e.g. Ontario, Alaska). The diet data reveals that the 11 non-native terrestrial mammal species considered in our food web analysis (i.e. excluding the 3 transient species) have, on average, 5 different resource types (sd = 2.19) whereas the native terrestrial mammal species have, on average, 7.08 different resources (sd = 5.65).

The earliest non-native mammal species to colonize Newfoundland (Fig. 1) were the House Mouse and Brown Rat which arrived with human settlement in the late 15th and early 16th centuries (Table 2). The Southern Red-Backed Vole is the last non-native mammal species to colonize the island ca. 1998 (Table 2). There

Table 2. Non-native and transient terrestrial mammals on the island of Newfoundland with sources for dietary data used in our study.

Common name	Scientific name	Order	Year of introduction[φ]	Data origin[α]
American Mink [F3]	*Neovison vison*	Carnivora	1935	Newfoundland
American Moose [F2]	*Alces americanus*	Artiodactyla	1904	Newfoundland
Arctic Fox	*Alopex lagopus*	Carnivora	transient	Alaska, Greenland, Sweden
Bank Vole	*Myodes glareolus*	Rodentia	1967	excluded
American Bison	*Bison bison*	Artiodactyla	1964	excluded
Cinereus Shrew [F2]	*Sorex cinereus*	Soricomorpha	1958	Ontario
Coyote* [F3]	*Canis latrans*	Carnivora	–	Newfoundland
North American Deermouse [F1]	*Peromyscus maniculatus*	Rodentia	<1968	Ontario, Virginia, Indiana
Eastern Chipmunk [F1]	*Tamias striatus*	Rodentia	1962	Ohio
Hoary Bat	*Lasiurus cinereus*	Chiroptera	transient	Manitoba
House Mouse [F1]	*Mus musculus*	Rodentia	with human settlement	Indiana
Brown Rat [F2]	*Rattus norvegicus*	Rodentia	with human settlement	Alaska
Polar Bear	*Ursus maritimus*	Carnivora	transient	Canadian Arctic
Red Squirrel [F2]	*Tamiasciurus hudsonicus*	Rodentia	1963	Quebec
Snowshoe Hare [F2]	*Lepus americanus*	Rodentia	1864	Newfoundland
Southern Red-Backed Vole [F2]	*Myodes gapperi*	Rodentia	1998	Ontario

*Coyote arrived on the island of Newfoundland through natural range expansion in 1985.
[φ]From [42], except southern red-backed vole [66].
[α]See [53] for full details on data.
[F1]Herbivores and insectivores without predators (F_1 functional group).
[F2]Herbivores and insectivores with predators (F_2 functional group).
[F3]Predators (F_3 functional group).

are 30 species (including basal resource categories) in the native mammal food web of Newfoundland and 41 species after the establishment of 11 non-native mammals on the island (Fig. 2). As expected, the total number of links and link density in the food web increased steadily with the addition of non-native mammals over time (Fig. 3). The trend in connectance over time also was positive but this trajectory was variable with the addition of some non-native mammals leading to a decline in connectance. The fraction of intermediate species in the food web increased from 0.4 to 0.56 with the addition of non-native mammals over time (Fig. 3). As more intermediate species colonized the island, the fraction of basal and top species in the food web declined from 0.57 to 0.41 and 0.03 to 0.02, respectively. The prey:predator ratio in the food web declined steadily from 2.23 to 1.67 with the addition of non-native mammals to the food web. Finally, food web vulnerability declined from 4.17 to 3.61 and generality increased from 3.1 to 3.61 with the arrival of non-native mammals to the island (Fig. 3).

The removal of non-native herbivores and insectivores without predators (i.e. F_1) led to lower (0.96 x) total number of links, connectance, link density and generality than the removal of native mammals from the same functional group (Fig. 4). There was no difference in the fraction of basal, fraction of intermediate, fraction of top, prey:predator ratio, and vulnerability to the loss of non-native vs native herbivores and insectivores with or without predators (Fig. 4). The removal of non-native herbivores and insectivores with predators led to a small increase (1.01–1.02 x) in the total number of links, connectance, link density, and generality than the removal of native mammals from the same functional group (Fig. 4). The removal of non-native predators led to an increase (1.02–1.05 x) in the total number of links, connectance, link density, fraction of intermediate, prey:predator ratio, and generality, a decrease in the fraction of top (0.33 x) and no change in the fraction of basal and vulnerability compared to the removal of native predators (Fig. 4).

Discussion

The island of Newfoundland has as many non-native terrestrial mammals as native terrestrial mammals (Table 1, 2). A few studies have investigated the impacts of non-native American Moose, Red Squirrel and Snowshoe Hare on the island's flora (e.g. [13], [14]) but no studies have investigated the impact of the full assemblage of non-native terrestrial mammals on Newfoundland's mammal food web structure. Our diet and network analysis of terrestrial mammals in Newfoundland confirmed our hypothesis that the trophic position of non-native species would be the main determinant of non-native species impacts on food web structure. Specifically, our data show that non-native mammals have influenced the island's terrestrial mammal food web primarily by increasing the fraction of intermediate (i.e. herbivores and insectivores) species, which has led to increasing food web connectance and generality (Fig. 3).

The colonization of 11 non-native mammals to Newfoundland has resulted in a terrestrial mammal food web with 1.3x the number of carnivore species, 2x the number of herbivore and insectivore species without predators and 3x the number of herbivore and insectivore species with predators compared to the native mammal food web (Fig. 2, 3). These changes to food web structure has led to greater connectance in the terrestrial mammal food web on the island because, on average, herbivores and insectivores have more food web links than predators. Recent theoretical work based on the analysis of Jacobian matrices suggests that an increasing number of prey species per predator

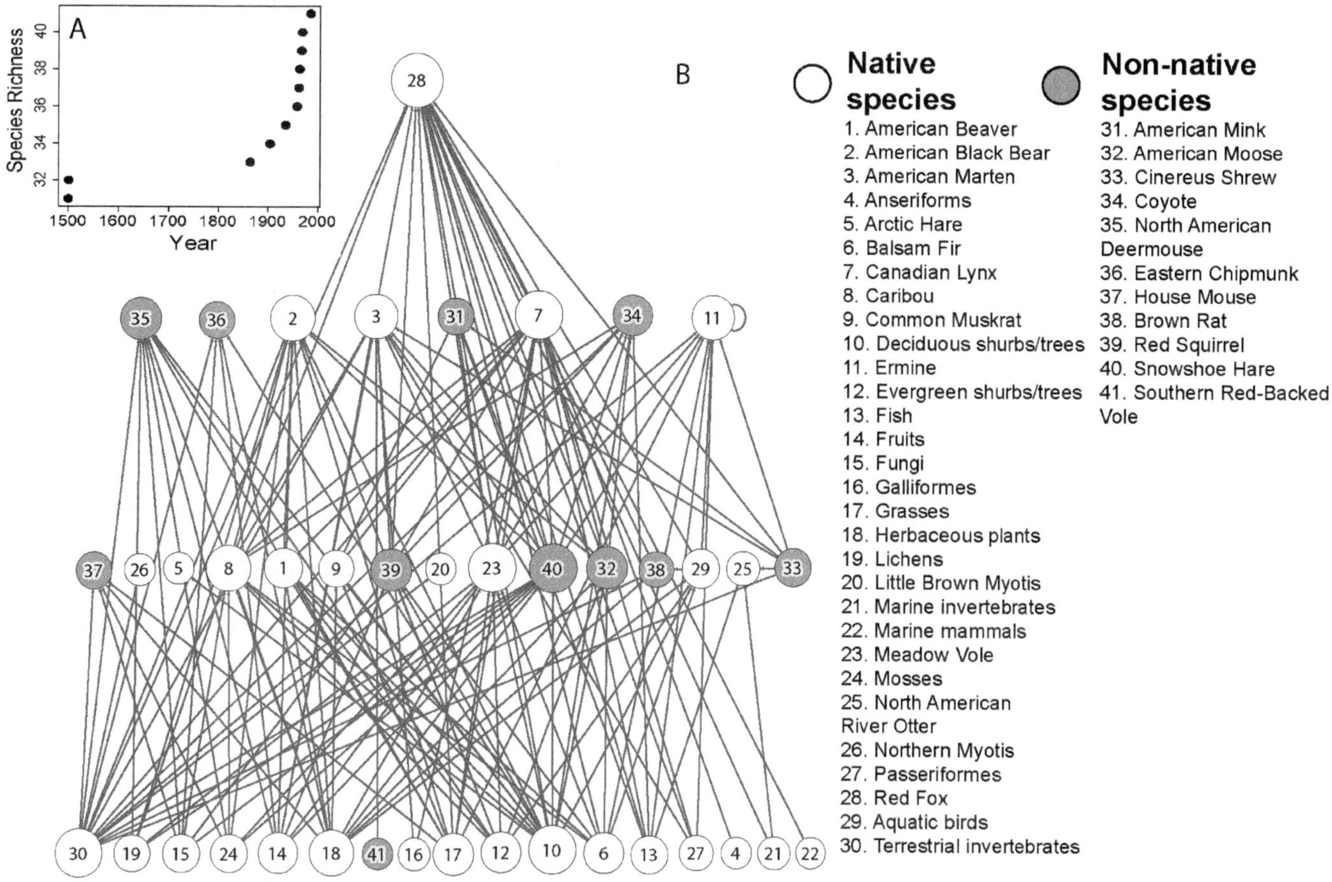

Figure 2. Terrestrial mammal food web of Newfoundland. A, year that each non-native species considered in our study was introduced vs total terrestrial mammal species richness in Newfoundland. B, Newfoundland terrestrial mammal food web with native and non-native terrestrial mammals shaded in different colours. Each node represents a different species or group of species and each link represents a feeding relationship. For example, terrestrial invertebrates (30) are prey for Cinereus Shrew (33).

may confer stability to food webs [64]. In Newfoundland, non-native species introductions have led to a threefold increase in the number of prey species per predator species and there is already evidence that a greater prey base has potentially contributed to the recovery of a native top predator species on the island. Specifically, American Marten were listed as endangered by the Committee on the Status of Endangered Wildlife in Canada (COSEWIC) in 1996 but a re-assessment in 2007 resulted in marten being listed in a lower risk category: threatened [56]. Introduced Snowshoe Hares and Southern Red-Backed Voles may be leading to a more diversified diet for marten and contributing to its recovery [65], [66]. However, other factors such as reduced mortality due to a decline in snaring by-catch and habitat loss also are contributing to marten recovery [56].

While the introduction of prey species may confer some benefits to the terrestrial mammal food web in Newfoundland, it is too early to determine the long-term effects of these introductions on the native mammal assemblage. For example, abundant non-native moose may provide ample carrion to support Coyotes and this has the potential to increase predation pressure on the native Caribou population. This apparent competition between moose and caribou via canid predators has been observed in other boreal ecosystems [67], [68].

The Coyote is one of the most recent species to establish itself in Newfoundland. In most of their range, Coyotes have a diverse diet consisting of a number of small- to medium-sized mammals (e.g. snowshoe hare, voles) and carrion (e.g. [69], [70]). We suggest that the large increase in terrestrial mammal prey and carrion base in Newfoundland may have facilitated the successful establishment of Coyotes to the island. If this interpretation is true, we may have the first evidence of an invasional meltdown (sensu [29]) in Newfoundland where positive interactions among non-native species could lead to cumulative impacts on native ecosystems and the facilitation of additional non-native species colonizations. The indirect effects of Coyote predation on native mammals in Newfoundland should be a focus of future terrestrial mammal studies because this mesopredator has significant impacts on food webs in other ecosystems [71], [72].

It may seem trivial that prey must be present on an island or patch in order for predators to successfully establish themselves (naturally or via human introduction). But ecologists have built this fundamental trophic interaction constraint into theoretical models only recently [73], [74]. For example, in the trophic theory of island biogeography, predators can only colonize and persist on islands that are already occupied by their main prey species [73]. The inclusion of this trophic constraint led to improved predictions of real food web community composition compared to the classic theory of island biogeography [73]. Extensions of this theory to consider multiple preys may help explain the persistence of predator populations over large spatial extents.

The full terrestrial mammal food web of Newfoundland with 11 non-native mammals has 1.37x the number of mammals and

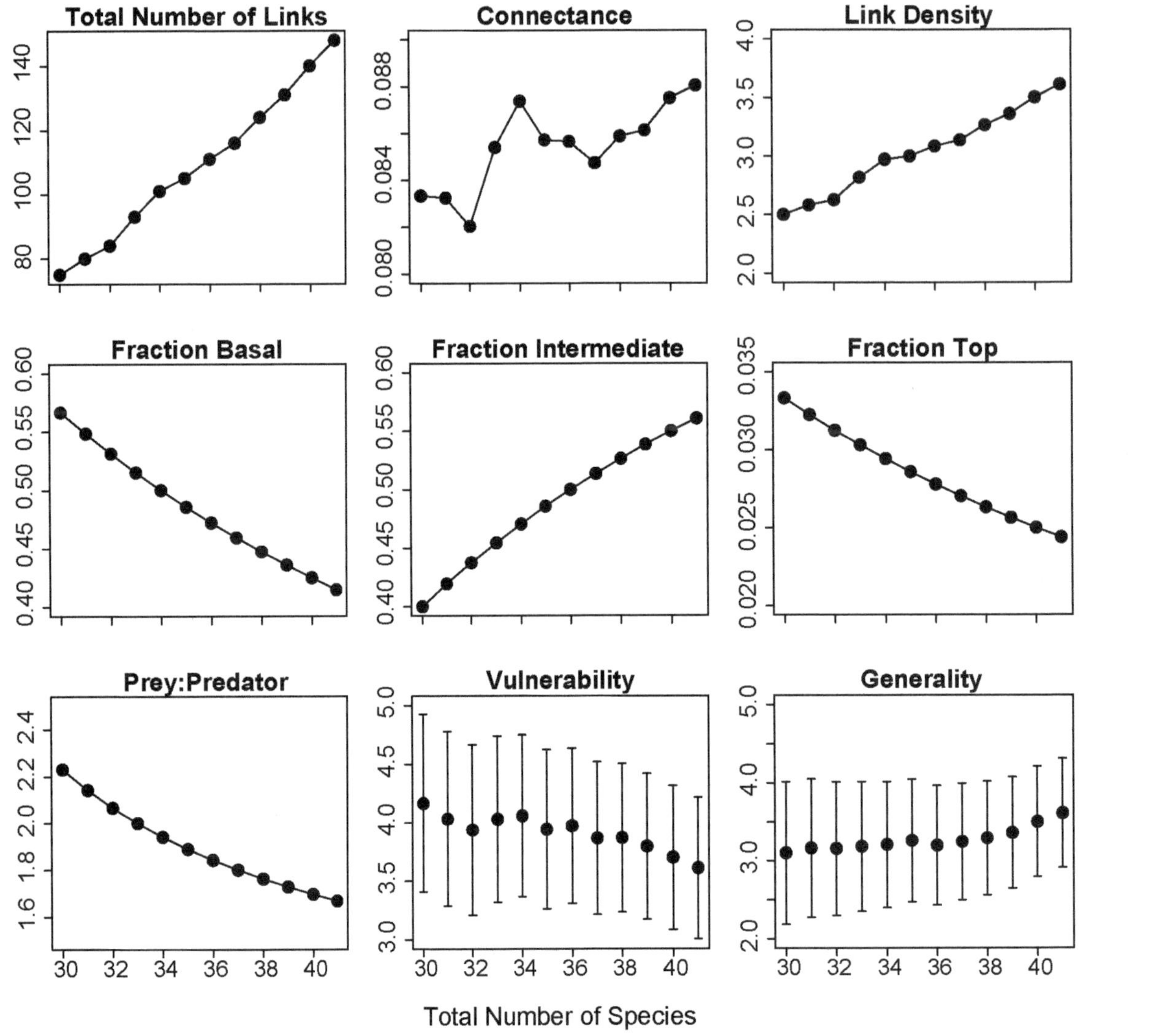

Figure 3. Effects of non-native mammals on food web properties. Change in terrestrial mammal food web properties with the sequential addition of non-native species on the island of Newfoundland. The native food web has 30 species and every point after this represents the addition of one non-native species added in chronological order (see Table 2).

1.97x the number of links of the native terrestrial mammal food web of Newfoundland (Fig. 2, 3). However, the removal of non-native herbivores and insectivores with predators or non-native predators led to an increase in the total number of links, connectance, link density and generality compared to the removal of native mammals from these trophic groups (Fig. 4). This suggests that native mammals from these groups are more connected to other species in the food web than non-native mammals. For example, the mean number of prey per native predator (American Black Bear, American Marten, Canadian Lynx, Ermine, North American River Otter, Red Fox) is 11.2 whereas the mean number of prey per non-native predator (Coyote, American Mink) is 6.5. Conversely, the removal of non-native herbivores and insectivores without predators led to a decline in the total number of links, connectance, link density and generality compared to the removal of native species from this trophic group (Fig. 4). This suggests that non-native herbivores and insectivores without predators were more connected to other species in the food web than native mammals from this trophic group. Indeed, all three native herbivores and insectivores without predators (Arctic Hare, Little Brown Myotis, Northern Myotis) only have one prey item each whereas the three non-native herbivores and insectivores without predators (Eastern Chipmunk, North American Deermouse, House Mouse) have on average 6.33 prey items each.

Connectance and link density of species in food webs are primary determinants of the robustness of food webs to species losses or additions [31], [34], [36], [64], [75]. Species that have many links to other species in a food web have a greater potential to influence community structure [31]. But recent research highlights that the connectivity, trophic level, and traits of species removed (or added) will determine the cascading impacts of removing (or adding) this species [31], [36], [64]. For example, the loss of a prey species in a web with generalist predators and a diverse prey base may lead to few secondary extinctions [76]. Our study differs from most other food web network analyses in that we are concerned with the impacts of species addition instead of species loss on food web structure. The additions of species will

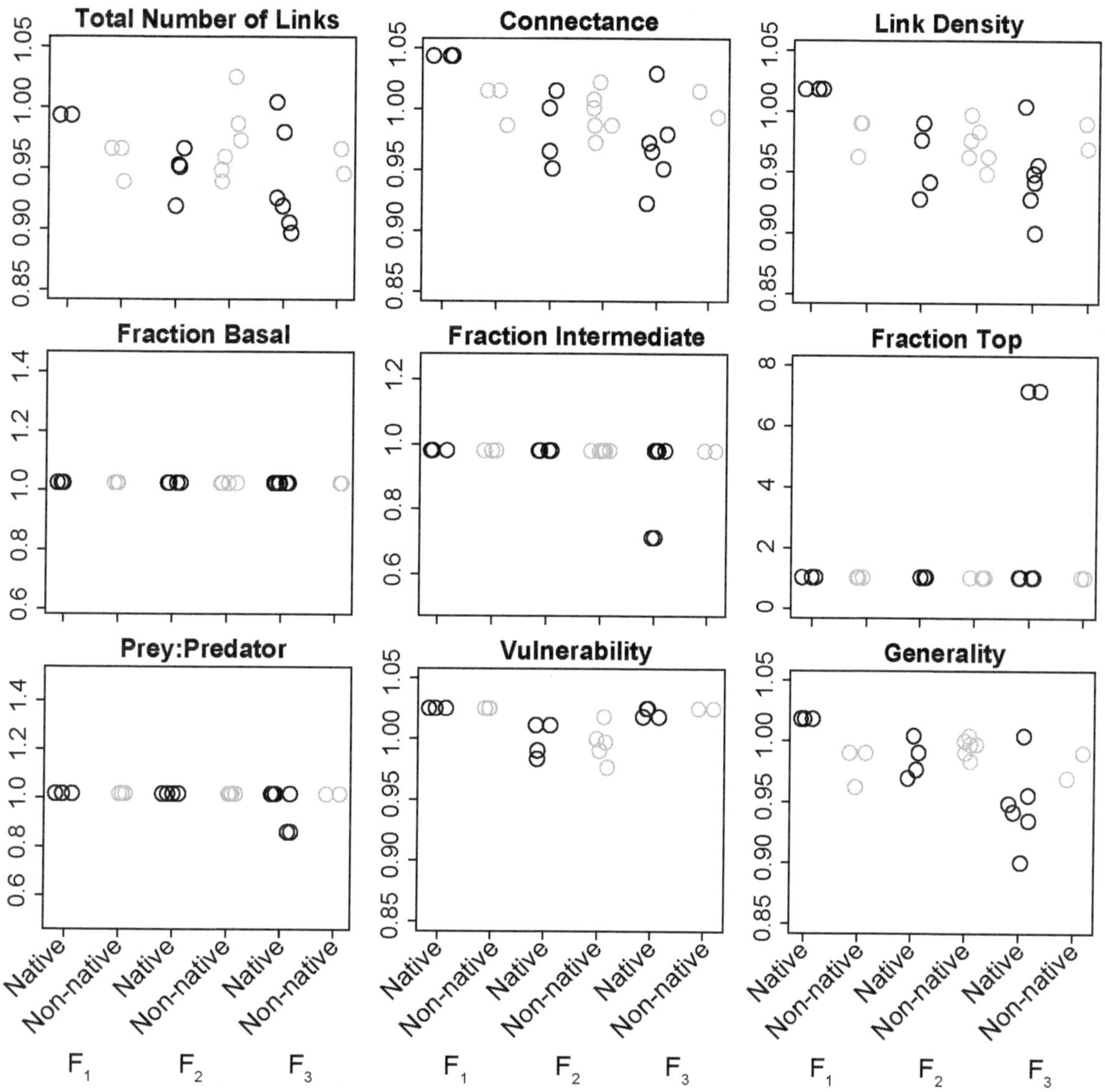

Figure 4. Effects of non-native mammal functional groups on food web properties. Change in Newfoundland terrestrial mammal food web properties for the removal of native vs non-native species of herbivores and insectivores without predators (F_1), herbivores and insectivores with predators (F_2), and predators (F_3). See methods for specific definitions of F_1, F_2, and F_3.

invariably lead to novel trophic interactions [77–79] and novel interactions may lead to complex effects on food web properties. For example, the introduction of the Southern Red-Backed Vole may lead to direct competition for resources and space with the native Meadow Vole but also indirect competition (i.e. apparent competition) among these small mammals (sensu [80]) via shared predators (e.g. American Marten). Teasing apart the relative strength of direct and indirect competition among species may be key for understanding species coexistence [81]. Long-term monitoring of populations is required to understand the impacts of novel interactions in the terrestrial mammal food web of Newfoundland.

This research represents one of the few studies to apply network analysis to investigate the impacts of non-native species on native food web properties. We began by assembling the first terrestrial mammal food web for the island of Newfoundland based on a review of empirical diet analyses studies. Although likely an incomplete representation of the diet of all mammal species in Newfoundland (e.g. we did not include non-mammal predators such as *Falconiformes spp.*), our data represent the best knowledge of the diet of Newfoundland mammals and forms a good starting point for identifying key gaps in diet data. Non-native mammals have resulted in a 2.5x increase in the number of herbivorous and insectivorous terrestrial mammals in Newfoundland. Through network analysis we show that this increase in intermediate species has led to the development of a terrestrial mammal food web with higher connectance and generality. Future work should consider the cumulative effects of global changes such as land-use and climate change and species harvesting along with the arrivals of non-native species on the structure and functioning of the terrestrial mammal food web of Newfoundland [10], [42], [82].

Acknowledgments

We thank L. Hermanutz, J. Gosse, T. Poisot, E. Vander Wal, T. Miller, L.-F. Bersier and anonymous reviewers for feedback that greatly improved our manuscript.

Author Contributions

Conceived and designed the experiments: SJL JSS. Performed the experiments: JSS SJL. Analyzed the data: JSS SJL. Wrote the paper: SJL JSS.

References

1. Lockwood JL, Hoopes MF, Marchetti MP (2013) Invasion Ecology, 2nd edition. Malden, Massachusetts, USA: Wiley-Blackwell.
2. Ricciardi A (2007) Are modern biological invasions an unprecedented form of global change? Conserv Biol 21: 329–336.
3. Vitousek PM, D'Antonio CM, Loope LL, Rejmanek M, Westbrooks R (1997) Introduced species: a significant component of human-caused global change. N Z J Zool 21: 1–16.
4. Liao C, Peng R, Luo Y, Zhou X, Wu X, et al. (2008) Altered ecosystem carbon and nitrogen cycles by plant invasion: a meta-analysis. New Phytol 177: 706–714.
5. Ward JM, Ricciardi A (2007) Impacts of Dreissena invasions on benthic macroinvertebrate communities: a meta-analysis. Divers Distrib 13: 155–165.
6. Beard KH, Vogt KA, Kulmatiski A (2002) Top-down effects of a terrestrial frog on forest nutrient dynamics. Oecologia 133: 583–593.
7. Nentwig W, Kuhnel E, Bacher S (2009) A generic impact-scoring system applied to alien mammals in Europe. Conserv Biol 24: 302–311.
8. Johnson PTJ, Older JD, Solomon CT, Vander Zanden MJ (2009) Interactions among invaders: community and ecosystem effects of multiple invasive species in an experimental aquatic system. Oecologia 159: 161–170.
9. Romanuk TN, Zhou Y, Brose U, Berlow EL, Williams RJ, Martinez ND (2009) Predicting invasion success in complex ecological networks. Phil Trans R Soc B 364: 1743–1754.
10. Ricciardi A, Hoopes MF, Marchetti MP, Lockwood JL (2013) Progress toward understanding the ecological impacts of nonnative species. Ecol Monogr 83: 263–282.
11. Ehrenfeld JG (2010) Ecosystem consequences of biological invasions. Annu Rev Ecol Evol Syst 41: 59–80.
12. Pimlott DH (1959) Reproduction and productivity of Newfoundland moose. J Wildl Manage 23: 381–401.
13. Gosse J, Hermanutz L, McLaren B, Deering P, Knight T (2011) Degradation of boreal forests by nonnative herbivores in Newfoundland's national parks: recommendation for ecosystem restoration. Nat Ar J 31: 331–339.
14. McLaren BE, Roberts BA, Djan-Chekar N, Lewis KP (2004) Effects of overabundant moose on the Newfoundland landscape. Alces 40: 45–59.
15. Pysek P, Jarosik V, Hulme PE, Pergl J, Hejda M, et al. (2012) A global assessment of invasive plant impacts on resident species, communities and ecosystems: the interaction of impact measures, invading species' traits and environment. Glob Chang Biol 18: 1725–1737.
16. Courchamp F, Chapuis J-L, Pascal M (2003) Mammal invaders on islands: impact, control and control impact. Biol Rev 78: 347–383.
17. Baillie JEM, Hilton-Taylor C, Stuart SN (2004) A global species assessment. Gland, Switzerland: IUCN.
18. Millenium Ecosystem Assessment (2005) Ecosystems and Human Well-being: Biodiversity Synthesis. Washington, DC: World Resources Institute.
19. Ebenhard T (1998) Introduced birds and mammals and their ecological effects. Swedish Wildl Res 13: 1–107.
20. Lever C (1994) Naturalized animals: the ecology of successfully introduced species. London: Poyser History.
21. Lowe S, Browne M, Boudjelas S (2001) 100 of the world's worst invasive alien species: a selection from the global invasive species database. Aukland, New Zealand: Global Invasive Species Programme.
22. Hedges SB, Conn CE (2012) A new skink fauna from Caribbean islands (Squamata, Mabuyidae, Mabuyinae). Zootaxa 3288: 1–244.
23. Atkinson IAE (2001) Introduced mammals and models for restoration. Biol Conserv 99: 81–96.
24. Ceballos G, Brown JH (1995) Global patterns of mammalian diversity, endemism, and engangerment. Conserv Biol 9: 559–568.
25. Salo P, Korpimaki E, Banks PB, Nordstrom M, Dickman CR (2007) Alien predators are more dangerous than native predators to prey propulations. Proc R Soc B 274: 1237–1243.
26. Carlsson NOL, Sarnelle O, Strayer DL (2009) Native predators and exotic prey – an acquired taste? Front Ecol Environ 7: 525–532.
27. Pope KL, Garwood JM, Welsh Jr HH, Lawler SP (2008) Evidence of indirect impacts of introduced trout on native amphibians via facilitation of a shared predator. Biol Conserv 141: 1321–1331.
28. Walsworth TE, Budy P, Thiede GP (2013) Longer food chains and crowded niche space: effects of multiple invaders on desert stream food web structure. Ecol Freshw Fish 22: 439–452.
29. Simberloff D, Von Holle B (1999) Positive interactions of nonindigenous species: invasional meltdown? Biol Invasions 1: 21–32.
30. Ricciardi A (2001) Facilitative interactions among aquatic invaders: is an "invasional meltdown" occurring in the Great Lakes? Can J Fish Aquat Sci 58: 2513–2525.
31. Dunne JA, Williams RJ, Martinez ND (2002) Network structure and biodiversity loss in food webs: robustness increases with connectance. Ecol Lett 5: 558–567.
32. Pascual M, Dunne JA (2006) From small to large ecological networks in a dynamic world. In: Pascual M, Dunne JA, editors. Ecological Networks: Linking Structure to Dynamics in Food Webs. Toronto, Canada: Oxford University Press. 3–24.
33. Riede JO, Rall BC, Banasek-Richter C, Navarrete SA, Wieters EA, et al. (2010) Scaling of food-web properties with diversity and complexity across ecosystems. Advances in Ecological Research: Ecological Networks. Burlington: Academic Press. 139–170.
34. Sole RV, Montoya JM (2001) Complexity and fragility in ecological networks. Proc R Soc B 268: 2039–2045.
35. Britton JR (2012) Introduced parasites in food webs: new species, shifting structures? Trends Ecol Evol 28: 93–99.
36. Dunne JA, Lafferty KD, Dobson AP, Hechinger RF, Kuris AM, et al. (2013) Parasites affect food web structure primarily through increased diversity and complexity. PLoS Biol 11: e1001579.
37. Lafferty KD, Allesina S, Arim M, Briggs CJ, De Leo G, et al. (2008) Parasites in food webs: the ultimate missing links. Ecol Lett 11: 533–546.
38. Woodward G, Hildrew AG (2001) Invasion of a stream food web by a new top predator. J Anim Ecol 70: 273–288.
39. Bezemer TM, Harvey JA, Cronin JT (2014) Response of native insect communities to invasive plants. Annu Rev Entomol 59: 119–141.
40. Pearse IS, Harris DJ, Karban R., Sih A (2013) Predicting novel herbivore-plant interactions. Oikos 122: 1554–1564.
41. Lopezaraiza-Mikel ME, Hayes RB, Whalley MR, Memmott J (2007) The impact of an alien plant on a native plant-pollinator network: an experimental approach. Ecol Lett 10: 539–550.
42. Department of Conservation and Environment (2013) Introduced land mammals to Newfoundland. St John's: Provincial Government of Newfoundland and Labrador. Available: http://www.env.gov.nl.ca/env/snp/programs/education/animal_facts/mammals/. Accessed 15 May, 2013.
43. Day SM (1997) Aspects of Newfoundland black bear (Ursus americanus hamiltoni) food habits and habitat use in human-influenced environments [M.Sc.]. Acadia University, Wolfville, Canada. 107 p.
44. Dodd LE, Chapman EG, Harwood JD, Lacki MJ, Rieske LK (2012) Identification of prey of Myotis septentrionalis using DNA-based techniques. J Mammal 93: 1119–1128.
45. Saunders JK (1963) Food habits of the lynx in Newfoundland. J Wildl Manage 27: 384–390.
46. Major HL, Jones IL, Charette MR, Diamond AW (2007) Variations in the diet of introduced Norway rats (Rattus norvegicus) inferred using stable isotope analysis. J Zool 271: 463–468.
47. Thiemann GW, Iverson SJ, Stirling I (2008) Polar bear diets and arctic marine food webs: insights from fatty acid analysis. Ecol Monogr 78: 591–613.
48. Northcott TH (1971) Feeding habits of beaver in Newfoundland. Oikos 22: 407–410.
49. Dalerum F, Angerbjorn A (2000) Arctic fox (Alopex lagopus) diet in Karupelv Valley, East Greenland, during a summer with low density. Arctic 53: 1–8.
50. Milakovic B, Parker KL (2011) Using stable isotopes to define diets of wolves in northern British Columbia, Canada. J Mammal 92: 295–304.
51. Larter NC (1999) Seasonal changes in arctic hare, Lepus arcticus, diet composition and differential digestibility. Can Field-Nat 113: 481–486.
52. Seccombe-Hett P, Turkington R (2008) Summer diet selection of snowshoe hares: a test of nutritional hypotheses. Oikos 117: 1874–1884.
53. Strong JS, Leroux SJ (2013) Terrestrial mammal food web of Newfoundland. figshare http://dx.doi.org/10.6084/m9.figshare.855617.
54. Riewe RR (1971) Ecology of the meadow vole, Microtus pennsylvanicus terraenovae, on the islands of Notre Dame Bay, Newfoundland, Canada. [PhD] University of Manitoba, Winnipeg, Canada. 231 p.
55. Bridger KE (2005) A comparative study of the dietary habits and helminth fauna of Canada lynx (Lynx canadensis), red fox (Vulpes vulpes) and eastern coyote (Canis latrans) on insular Newfoundland, Canada. [MSc] Memorial University of Newfoundland, St John's, Canada. 151 p.
56. Committee on the Status of Endangered Wildlife in Canada (2007) COSEWIC assessment and update status report on the American marten (Newfoundland population) Martes americana atrata in Canada. Ottawa.
57. Dodds DG (1960) Food competition and range relationships of moose and snowshoe hare in Newfoundland. J Wildl Manage 24: 52–60.
58. Ferron J, Ouellet JP, Lemay Y (1986) Spring and summer time budgets and feeding behaviour of the red squirrel (Tamiasciurus hudsonicus). Can J Zool 64: 385–391.
59. Thompson ID, Curran WJ (1995) Habitat suitability for marten of second-growth balsam fir forests in Newfoundland. Can J Zool 73: 2059–2064.
60. Schoener TW (1989) Food webs from the small to the large. Ecology 70: 1559–1589.

61. Perdomo G (2012) foodweb: visualisation and analysis of food web networks. R package version 1.0.
62. R Development Core Team (2011) R. v.3.0.2. R, a language and environment for statistical computing. R Foundation for Statistical Computing, Vienna, Austria.
63. Gephi (2014) The open graph viz platform. Available https://gephi.org/. Accessed on 21 April, 2014.
64. Gross T, Rudolf L, Levin SA, Dieckmann U (2009) Generalized models reveal stabilizing factors in food webs. Science 325: 747–750.
65. Gosse JW, Hearn BJ (2005) Seasonal diets of Newfoundland martens, Martes americana atrata. Can Field-Nat 119: 43–47.
66. Hearn BJ, Neville JT, Curran WJ, Snow DP (2006) First record of the southern red-backed vole, Clethrionomys gapperi, in Newfoundland: implications for the endangered Newfoundland marten, Martes americana atrata. Can Field-Nat 120: 50–56.
67. Seip DR. (1992) Factors limiting woodland caribou populations and their interrelationships with wolves and moose in southeastern British Columbia. Can J Zool 70: 1494–1503.
68. Wittmyer HU, Sinclair ARE, McLellan BN (2005) The role of predation in the decline and extirpation of woodland caribou. Oecologia 144: 257–267.
69. Pekins PJ, Mautz WW (1990) Energy requirements of eastern coyote. Canadian J Zool 68: 656–659.
70. Wells MC, Bekoff M (1982) Predation by wild coyotes: behavioral and ecological analyses. J Mammal 63: 118–127.
71. Prugh LR, Stoner CJ, Epps CW, Bean WT, Ripple WJ, et al. (2009) The rise of the mesopredator. BioScience 59: 779–791.
72. Ripple WJ, Wirsing AJ, Wilmers CC, Letnic M (2013) Widespread mesopredator effects after wolf extirpation. Biol Conserv 160: 70–79.
73. Gravel D, Massol F, Canard E, Mouillot D, Mouquet N (2011) Trophic theory of island biogeography. Ecol Lett 14: 1010–1016.
74. Pillai P, Loreau M, Gonzalez A (2010). A patch-dynamic framework for food web metacommunities. Theor Ecol 3: 223–237.
75. May RM (1972) Will a large complex system be stable? Nature 238: 413–414.
76. Srinivasan UT, Dunne JA, Harte J, Martinez ND (2007) Response of complex food webs to realistic extinction sequences. Ecology 88: 671–682.
77. Hobbs RJ, Higgs E, Harris JA (2009) Novel ecosystems: implications for conservation and restoration. Trends Ecol Evol 24: 599–605.
78. Pearse IS, Altermatt F (2013) Predicting novel trophic interactions in a non-native world. Ecol Lett 16: 1088–1094.
79. Staniczenko PPA, Lewis OT, Jones NS, Reed-Tsochas F (2010) Structural dynamics and robustness of food webs. Ecol Lett 13: 891–899.
80. Holt RD (1977) Predation, apparent competition, and the structure of prey communities. Theor Popul Biol 12: 197–229.
81. Chesson P, Kuang JJ (2008) The interaction between predation and competition. Nature 456: 235–238.
82. Didham RK, Tylianakis JM, Gemmell NJ, Rand TA, Ewers RM (2007) Interactive effects of habitat modification and species invasion on native species decline. Trends Ecol Evol 22: 489–496.
83. Wilson DE, Reeder DM (2005) Mammal species of the world: a taxonomic and geographic reference, 3rd ed. Johns Hopkins University Press, Baltimore, USA. 2142 p.

7

Dual Roles of an Algal Farming Damselfish as a Cultivator and Opportunistic Browser of an Invasive Seaweed

Kimberly A. Peyton[1*¤], Lauren M. Valentino[1], Karen P. Maruska[2]

1 Hawaii Institute of Marine Biology, University of Hawaii – Manoa, Kaneohe, Hawaii, United States of America, 2 Department of Biological Sciences, Louisiana State University, Baton Rouge, Louisiana, United States of America

Abstract

Herbivory is a fundamental process determining reef resilience, and while algal farming damselfishes can help shape benthic assemblages, an understanding of their contribution to areas outside of defended territories is relatively unexplored. Here, we demonstrate how the farming damselfish *Stegastes marginatus* plays a dual role in benthic structuring by 1) contributing to persistence of the invasive macroalga *Acanthophora spicifera* within a Hawaiian marine protected area, where the macroalga occurred exclusively inside *Stegastes* territories, and 2) behaving as an opportunistic browser of the exotic alga outside their territorial borders. Greater than 50% of the biomass of tethered *A. spicifera* was consumed within one-hour when placed outside *Stegastes* territories, compared to <5% lost from tethers within territories or herbivore exclusion cages. *In situ* remote video revealed that tethered *A. spicifera* located outside territories was grazed primarily by the surgeonfish *Acanthurus nigrofuscus* (~68% of total bites) and, surprisingly, by *S. marginatus* (~27% of total bites) that left their territories to feed on this resource on 107 occasions during 400 min of filming. Further, for over half of those occurrences where *S. marginatus* grazed on the tethered macroalga outside of territories, they fed alongside conspecifics and other species, displaying little of the aggressiveness that characterizes this damselfish. These results show that *S. marginatus* plays a wider role in determining benthic assemblages than previously recognized, acting both as cultivators of a canopy-forming invasive macroalga within their territories, and as opportunistic browsers in neighboring sites. Consequently, *S. marginatus* can affect benthic species composition across their territory borders. These results provide a rare example of interspecific facilitation of an exotic alga by an indigenous marine fish. Accounting for fish behaviors more broadly is important to further our understanding of ecological processes that shape reef ecosystems to improve management of MPAs that often support extensive farming damselfish populations.

Editor: Erik Sotka, College of Charleston, United States of America

Funding: K.A.P was supported by Hawaii Coral Reef Initiative Grant Number: NA09NOS4260242 awarded to F. Thomas. K.P.M. was supported by start-up funds from the Department of Biological Sciences in the College of Science at Louisiana State University. The funders had no role in study design, data collection and analysis, decision to publish, or preparation of the manuscript.

Competing Interests: The authors have declared that no competing interests exist.

* Email: peyton@hawaii.edu

¤ Current address: State of Hawaii, Department of Land and Natural Resources -Division of Aquatic Resources, Honolulu, Hawaii, United States of America.

Introduction

Herbivory by reef fishes profoundly influences the biomass, diversity, and canopy height of algal assemblages, typically resulting in coral-epilithic algae dominated communities [1,2]. However, studies that examine how the range of behaviors exhibited by herbivorous fish species contributes to shaping coral reef ecosystems are limited [3–6]. Worldwide declines in coastal fish populations have had cascading effects on benthic reef communities, with the loss of herbivores credited as a key factor driving phase shifts on coral reefs towards assemblages overgrown with fleshy macroalgae [1,2,7–9]. Hawaiian coral reefs have been further exacerbated by herbivore losses because the fleshy macroalgae that typically replace corals are often exotic (= invasive) species [10,11]. Even highly palatable invasive species, such as the red alga *Acanthophora spicifera*, readily proliferate on Hawaiian coral reefs where herbivore abundance is reduced [12]. Thus, knowledge of how herbivore behaviors shape species composition is crucial for a more comprehensive understanding of reef ecosystem dynamics and implementation of management strategies to preserve them.

The identities of herbivores and their relative impacts in the Hawaiian Archipelago, as well as other reef systems, have primarily been inferred from field surveys, feeding assays and gut content analyses [12]. When herbivore behavior is considered, fishes are typically designated as exclusively farmers or foragers [3] with most foragers largely grazing on turf algae while fewer species consume macroalgae (i.e. browser) [13]. Research utilizing *in situ* remote video recordings, however, has captured novel foraging behaviors and revealed more nuanced relationships between coral reef algae and herbivores than previously understood [4,14–16]. In one compelling example, *in situ* remote video clearly showed that a single species, regarded as an uncommon invertebrate consumer (dusky batfish *Platax pinnatus*), was in reality the primary consumer responsible for the reversal of an experimentally induced phase shift from fleshy macroalgae to coral-epilithic algae [4]. How the potential range of fish behaviors, including latent ones, may affect coral reef resilience is poised to receive further consideration.

While herbivorous fishes contribute to coral reef communities by continually cropping back algal canopies as foragers, one prevalent guild of herbivores, the territorial farming damselfishes (Pomacentridae), are an exception to this paradigm because these fish typically increase macroalgal standing biomass, productivity and diversity on coral reefs [17–21] by promoting growth of palatable macroalgae and defending it from other grazers [22–24]. As a result, territorial damselfishes have received considerable attention. Algal farming damselfishes typically maintain territories with distinct benthic assemblages by aggressively defending their cultivated resources from other herbivorous fishes, which has led to suggestions that farming damselfishes may function as keystone species [18,25]. The visible contrast between benthic assemblages inside damselfish territories versus adjacent, undefended areas is often striking [23,26]. However, behavior of farming damselfish in shaping reef assemblages outside their territories has rarely been considered. The two exceptions were observations that *Stegastes planifrons* does not feed outside its territory [19] and *Dischistodus prosopoteania* attempted to defend a palatable macroalga tethered outside of its territory [23]. Examining *in situ* behaviors of farming damselfish using remote video can further our understanding of ecological processes that shape coral reef ecosystems and improve management of coral reefs, including marine protected areas (MPA) that support extensive populations of farming damselfishes.

The goal of this study was to use *in situ* remote video recordings and benthic cover analyses to examine what role the algal farming damselfish, the Hawaiian Gregory *Stegastes marginatus* (= Pacific Gregory *S. fasciolatus*), may have on the invasion success of the exotic macroalga *A. spicifera* in a Hawaiian MPA.

Materials and Methods

Ethics statement

The State of Hawaii Department of Land and Natural Resources (DLNR) issued Special Activity Permit No. 2010-51 for this field study because it took place in Pupukea Marine Life Conservation District (MLCD), property that is owned and administered by the DLNR.

Study site and benthic characterizations

Observations and experiments were conducted in tide pools (<1.5 m depth) of the Pupukea Marine Life Conservation District, a MPA on the North Shore of Oahu, Hawaii (N21°38′5″ W158°3′48″) established in 1983. To characterize and compare benthic species composition inside versus outside *S. marginatus* territories, 15 haphazardly chosen *S. marginatus* territories and 15 nearby non-defended areas were mapped, labeled and photographed. *Stegastes marginatus* territories were identified by noting the extent and spatial orientation of each resident's farmed patch(s), defined as the substrate where the majority of maintenance takes place [27]. Using a camera mounted on a fixed distance photoquad stand [28], standardized 147 cm^2 images (3–4 images per territory) of the substrate were taken haphazardly within farming patches. To characterize benthic cover outside territories, photographs (3–4 images per area) were taken adjacent (ca. 1–5 m away) to *S. marginatus* territories where other fish species were seen grazing without being chased by the damselfish, and were matched in terms of depth, hydrodynamic conditions, and benthic topography to the inside territory areas. Images were taken in June 2010 and then repeated in the same 15 paired sites in September 2010. Photographs were analyzed using Coral Point Count (CPCe V3.6) [29] by projecting 30 stratified random points per image, and identifying the taxa beneath each point as turf algae, *A. spicifera*, other macroalgae, coralline algae, other live (i.e. urchin, coral) or sand. Field notes taken concurrent with photoquad sampling were used to confirm identifications. Percent cover values from the multiple images taken within a single territory or non-territory areas were then averaged to yield a mean percent cover for that location. As the primary goal here was to test for differences in *A. spicifera* abundance between inside and outside the *S. marginatus* territories, and finding that *A. spicifera* was completely absent (percent cover = 0%) outside of all the territories (confirmed by multiple site visits in October 2009 and April to September 2010; searched for at least 30 min per visit), statistical tests on percent cover data were not necessary.

To estimate the biomass of *A. spicifera* (wet weight) that was maintained inside *S. marginatus* territories and to test whether it persisted through time, all of this macroalga within a randomly placed 10 cm^2 quadrat was collected, blot dried and weighed to the nearest 0.01 g. Biomass samples were collected in April–June and again in September 2010 from 3–4 of the mapped territories with one quadrat per territory per sampling period. Also, the areal coverage of *A. spicifera* was measured in the 15 mapped (photoquad) territories using flexible tape measures that could be laid along the reef contours. To account for the irregular shapes of *A. spicifera* patches inside the damselfish territories, patch area was measured to the nearest 1 cm by positioning one weighted tape measure along the patch length as a reference and then used a second tape measure to record patch width at 10 cm intervals. Further, because our scientific collecting permit specified that only *A. spicifera* could be manipulated or sampled in the MPA, canopy height was used rather than biomass to test for relative differences in macroalgae abundance inside and outside territories. Mean canopy height was determined by measuring canopy height of benthic algae at five haphazardly selected points within each of the 15 mapped territories and each of the 15 nearby non-defended areas.

Fish grazing intensity

To determine relative grazing intensity on the invasive macroalga 5–7 branches (7–12 cm tall) were collected inside *S. marginatus* territories, blotted, weighed and then woven into tethers constructed of 20 cm long twisted raffia fibers with a weight (ca. 225 g) tied to each end. Tethered *A. spicifera* was randomly assigned to one of three different treatments: (1) outside *S. marginatus* territories to test the grazing rate of herbivores on the alga without damselfish defenses (outside territory; $n = 7$); (2) inside *S. marginatus* territories (inside territory; $n = 7$); and (3) controls consisting of a 1.3 cm^2 mesh plastic-coated wire cage measuring 15×9×10 cm (cage control; $n = 6$) placed outside territories to account for any biomass loss due to handling because *A. spicifera* fragments easily [30]. A single 20 cm tether was used for each deployment and was placed in unique locations among trials within each respective treatment group. With 15 mapped territories and a total of 14 uncaged replicates (7 inside, 7 outside), each territory was used only once and randomly assigned to a treatment, leaving one unused territory. For inside territory replicates, tethers were placed haphazardly within ca. 20 cm radius of the center of a feeding patch. Before placing tethers in outside territory treatments, nearby resident fish were observed to determine the borders of the territory on the day of the experiment and then a tether was haphazardly placed in a neutral area ca. 1–1.5 m beyond the territory border. Care was exercised to avoid deploying tethers near any *A. spicifera* donor territories. All tethers

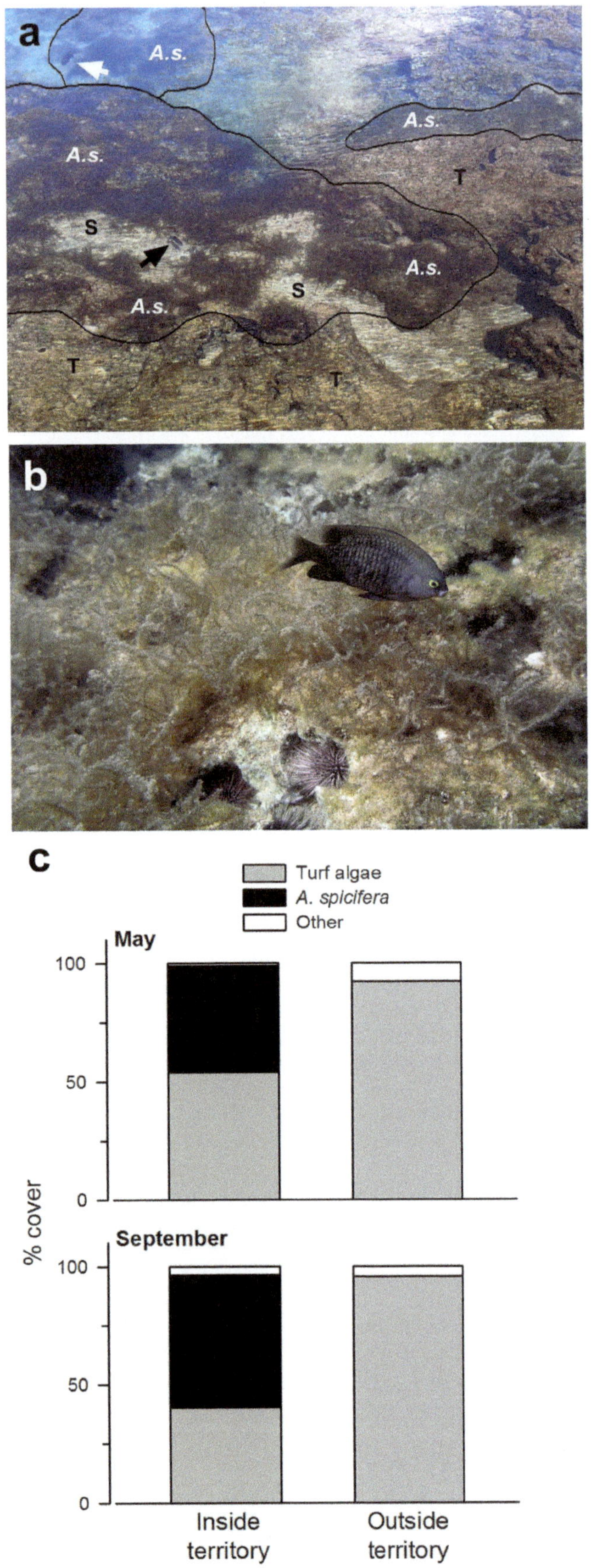

Figure 1. Invasive macroalga *Acanthophora spicifera* occurs exclusively within the territories of the algal farming damselfish *Stegastes marginatus*. A) Oblique view of *Stegastes marginatus* (arrow) feeding patches with dense growths of the invasive red macroalga *Acanthophora spicifera* (A.s.), turf algae (T) and sand (S). B) *S. marginatus* defending a lush canopy of *A. spicifera*. C) Percent cover of benthos inside and outside *S. marginatus* territories in May and September 2010.

were deployed for one-hour and then remaining macroalga was blotted and reweighed to determine percent change in wet weight.

Behavioral observations

To identify and quantify interspecific differences in foraging behavior on the tethered macroalga treatments described above, an underwater camera (MicroVideo, #MVC2120-WP) was used to film browser activities for both inside and outside territory treatments for one-hour. The underwater camera was fixed to a low profile stand and cabled to an external digital camcorder (Canon Optura 20) that was positioned out of the water at least 15 m from a tether. Replicates were conducted singly and in random order over several days between the hours of 0800 and 1200 (i.e., 3–4 sequential deployments per day). This timing and experimental setup was necessary because our permit within this MPA specified that our trials be performed only during morning hours to minimize interaction with the public. From the video footage, the herbivore species, number of individuals grazing and total number of bites on the tethered macroalga were quantified independently by two observers. The number of occasions when one or more browsers were simultaneously feeding on the tethered *A. spicifera* was also quantified. Simultaneous feeding occurrences were defined as occasions when one *S. marginatus* was consuming tethered algae at the same time another fish (*S. marginatus* or other species) was also present at the tether. One of our original predictions was that *S. marginatus* might try to defend tethered macroalga in neutral areas and attempt to expand their territory border; if for example, they perceived *A. spicifera* as a valuable resource. Thus, to examine this possibility and test whether *S. marginatus* behaved aggressively while feeding on the tethered *A. spicifera* placed outside of their territories, the number of times they chased another fish away from the tether was also quantified. Bite rate values and grazing occurrences obtained by the two observers were then averaged for each replicate (differences between the two observers ranged from 0% to 12%). All statistical comparisons were made with SigmaPlot 11.0 (Systat, Inc., San Jose, CA.), and names of statistical tests and corresponding values are reported in the results.

Results

Acanthophora spicifera inside and outside of *Stegastes* territories

Discreet patches of the invasive red alga *A. spicifera* occurred exclusively inside territories of the farming damselfish *S. marginatus* within the Pupukea MLCD (Fig. 1a & b). In May, feeding patches within the *S. marginatus* territories consisted of 45±4.3% (SE) *A. spicifera* and 53.9±4.3% filamentous turf algae (Fig. 1c). In contrast, outside territory areas were dominated by filamentous turf algae (92.1±2.3%), whereas *A. spicifera* (0%) was not distinguishable in photoquadrats or by close examination of the substratum *in situ*. Re-sampled in September, benthic composition remained different between inside (*A. spicifera*: 56.0±3.4%; filamentous turf algae: 40.3±3.3%) and outside (*A. spicifera*:

0%; filamentous turf algae: 92.2±2.1%) territory areas. Further distinguishing the two algal assemblages, the mean canopy height inside *S. marginatus* territories was 6.3±0.7 cm (SE) in April, 5.1±0.6 cm in May and 7.6±0.9 cm in September, whereas canopy height of algal assemblage was <0.5 cm outside territories across all sampling times.

Average standing biomass of *A. spicifera* inside *S. marginatus* territories was 6.1±0.9 g wet weight 10 cm^{-2} (SE) in April; 11.5±2.3 g in May; 17.2±1.2 g in June and 8.2±1.7 g in September with tetrasporophytes (spore producing thalli) present throughout the study period. In the 15 territories measured in May, *A. spicifera* in the farmed patches covered 3,942.4±1800.1 cm^2 $territory^{-1}$ (range = 0–25,607.3 cm^2 $territory^{-1}$) with an average standing biomass of 2,168.5±99.1 g wet weight $territory^{-1}$ (range = 0–14,084.1 g wet weight $territory^{-1}$). Although wave exposure was not directly measured, the only damselfish territory without invasive seaweed was also the only territory located adjacent to the primary tidal channel and therefore subjected to higher water motion than the other 14 territories. Since the brittle thalli of *A. spicifera* easily fragment [30], the proximity of this territory to increased hydrodynamic action could account for the absence of the invasive alga in this area. Inside territories, *A. spicifera* was consumed at a rate of 4.5±2.2% h^{-1} (SE), while inside cage control biomass loss was negligible (1.3±0.2% h¯1; Fig. 2a). When the invasive macroalga was placed outside territories, and removed from the defenses of *S. marginatus*, on average more than half of the biomass (58.7±10.8% h^{-1}) was consumed within one hour and the percent loss was significantly higher than that of inside territories or cage controls (ANOVA, $F_{2,17} = 23.63$, $p<0.001$; Tukey's, $p<0.001$).

Fish grazing inside and outside of *Stegastes* territories

To determine the fish species responsible for consuming the invasive macroalga and to compare their relative contributions, remote video recordings revealed that mean bite rates (bites min^{-1}) were significantly greater when the tethered *A. spicifera* was placed outside (range = 6.1–31.9 bites min^{-1}) compared to inside (range = 0–1.0 bites min^{-1}) *S. marginatus* territories (t-test, $t = -3.088$, df = 12, $p = 0.009$; $n = 7$; Fig. 2b). Tethered *A. spicifera* within caged controls were not accessible to fishes and therefore not consumed. Inside territories, 100% of the few bites taken were by the resident *S. marginatus*, while outside territories, the majority of bites (67.9±7.1% SE) were taken by the surgeonfish *Acanthurus nigrofuscus* (Fig. 2c). Surprisingly, 26.5±7.6% of bites outside *S. marginatus* territories were also taken by *S. marginatus* that left their own nearby territories to feed on the undefended *A. spicifera* in adjacent areas. The remaining bites (5.6±4.3%) were taken by the Hawaiian sergeant fish (*Abudefduf abdominalis*), milletseed butterflyfish (*Chaetodon miliaris*), or convict tang (*Acanthurus triostegus*) (Fig. 2c).

Remarkably, there were a total of 107 separate occasions when *S. marginatus* left their own nearby territories to feed on the tethered *A. spicifera* located in neutral areas (~once every 4 min), a behavior that occurred in all 7 outside territory treatment locations by 2–5 different visiting fish per treatment location (Fig. 3). Total filming time for the outside territory treatments was 400 min, with 37.9±35.4% of that time spent by *S. marginatus* feeding on *A. spicifera* outside of their territories. Following deployment of the tethered *A. spicifera*, the average latency to first bite by a *S. marginatus* was 10.3±6.1 (SE) min. (range = <1–46 min), and by any herbivorous fish was 2.7±1.1 min. (range = <1–9 min). There were also numerous occasions (59 in 400 min) when *S. marginatus* fed on the tethered macroalga in the outside territory treatments along with either a conspecific (indicating that this treatment was consistently located in non-territorial areas) or another species (*A. nigrofuscus*, *A. abdominalis*, or *C. miliaris*) (Movie S1; Fig. 3). These simultaneous feeding

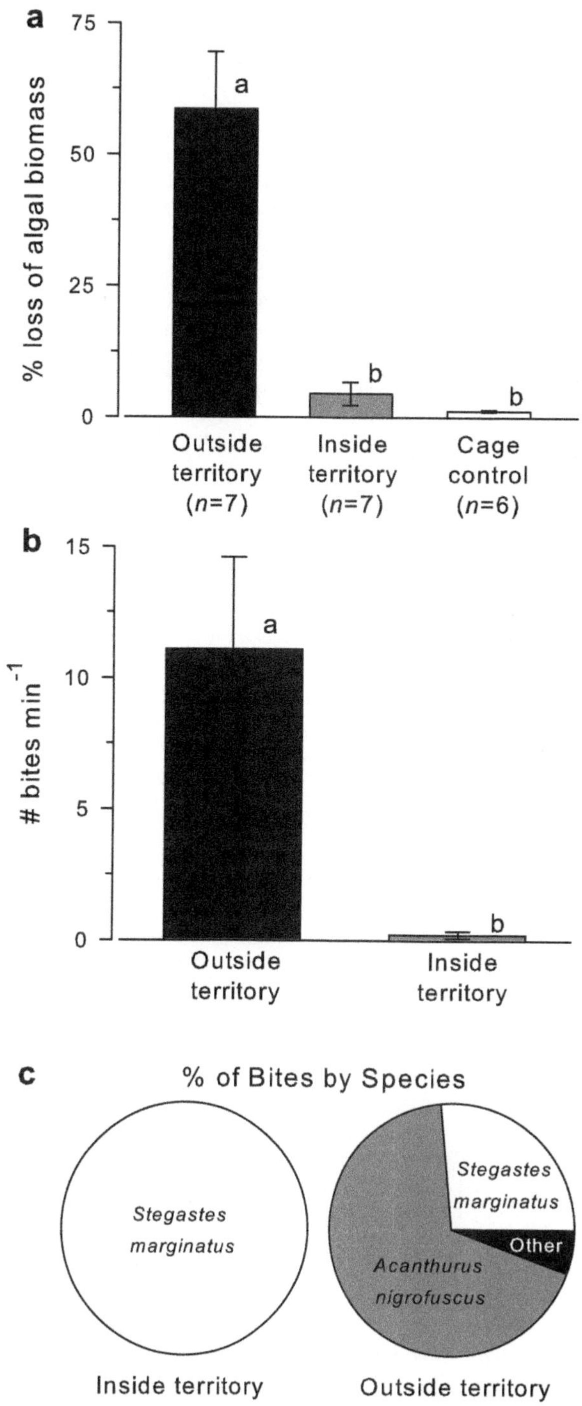

Figure 2. Herbivore grazing intensity on *A. spicifera* was highest outside the territorial defenses of *S. marginatus*. A) Percent loss of *A. spicifera* algal biomass from tethers placed inside *S. marginatus* territories, outside territories, or in caged controls. B) Number of bites min^{-1} on tethered *A. spicifera* placed inside and outside *S. marginatus* territories. C) Percentage of bites taken by different fish species inside and outside *S. marginatus* territories. Data are plotted as mean ± SE and bars with different letters indicate significant differences at $p<0.05$.

events also occurred in all 7 outside territory treatment locations, and represented 23.0±26.7% of total filming minutes in these outside territory replicates. Because *S. marginatus* is compelled to continuously return to defend its own territory, it was also relevant to account for that time by partitioning out the total *S. marginatus* feeding minutes (e.g., all minutes with ≥1 *S. marginatus* feeding on *A. spicifera* in the outside territory treatment) to further discern the behavior of *S. marginatus* grazing in neutral areas. Surprisingly, 58.2±23.4% of total *S. marginatus* feeding minutes had one or more *S. marginatus* feeding with one or more other fish species. During simultaneous feeding events, aggression by *S. marginatus* was surprisingly low, as there were only 4 instances (6.8% of simultaneous feeding occurrences) when they chased another fish away from the tethered *A. spicifera*. This low aggression is in stark contrast to the typical defensive behavior of resident *S. marginatus* that vigorously chase away conspecifics or other herbivorous fishes such as *A. nigrofuscus* when they approach or enter their territory.

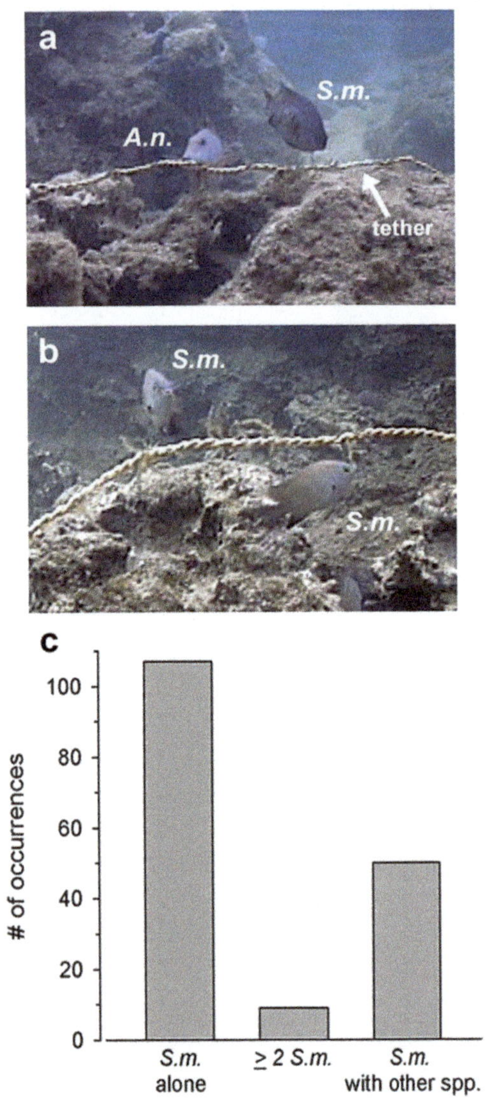

Figure 3. *Stegastes marginatus* acts as an opportunistic browser on *A. spicifera* outside its territorial borders. Examples of *Stegastes marginatus* feeding outside their territories on tethered *Acanthophora spicifera* alongside (A) the surgeonfish *Acanthurus nigrofuscus* (*A.n.*), and (B) another *S. marginatus* (*S.m.*). C) The total number of occurrences from 400 min of filming when *S. marginatus* was observed feeding on tethered *A. spicifera* either alone (*S.m.* alone), along with another *S. marginatus* (≥2 *S.m.*), or with another fish species in the outside territory treatments (*S.m.* with other spp.).

Discussion

Using *in situ* remote video, our results demonstrate that an algal farming damselfish, *S. marginatus*, performs a dual function by shaping macroalgal assemblages both within and outside of its territorial borders, thereby serving as both a cultivator and an opportunistic browser, respectively. In contrast, earlier studies, which relied on other approaches, primarily focused on how algal farming damselfishes influence benthic composition *within* their guarded territories [18,20,22,24,25,27] and rarely considered their ability to structure communities adjacent to their territories [19,23]. Behaviors by these highly site-attached territorial fish, such as chasing intruding browsers, and weeding and fertilization, result in visually distinct and often more diverse algal assemblages within territorial borders compared to adjacent areas [17,20–24,27,31]. *In situ* remote video is an effective approach used in herbivory experiments, revealing new insights into animal behaviors and plant-animal interactions on coral reefs [4,15,16,32]. For example, a sleeping functional group was identified after a batfish displaying opportunistic herbivory was identified as the primary consumer responsible for reversal of a macroalgal dominated phase shift [4]. In our study, video recordings captured behaviors that showed *S. marginatus* can function in dual roles by shaping distinct benthic assemblages both as a farmer-defender of a palatable macroalga within its territory and as an opportunistic browser of the same algal species in neighboring sites. Without these video observations we would have incorrectly concluded that algal farming damselfish function only to harbor an introduced species inside their territories and thereby simply contribute to its invasion success. Consequently, our experiment demonstrates for the first time that a farming damselfish may influence algal assemblages outside its territories to a greater degree than previously recognized.

Our *in situ* remote video experiments also revealed a novel behavior of these territorial aggressive fish, namely, that *S. marginatus* shared available food resources with conspecifics and other herbivores when *A. spicifera* was located outside of their territorial borders. It is unlikely that this behavior is an attempt to expand the size of their territories because there was little evidence of *S. marginatus*-initiated aggression at the tethered *A. spicifera* placed in neutral areas. This is in stark contrast to the aggressive chasing typically performed by resident *S. marginatus* when other herbivores enter or swim nearby their territory borders [18,25] [personal observations]. Although this farming species' aggressive behaviors during territory defense are well-documented, *S. marginatus* was not previously known to feed with other herbivores or exhibit resource sharing [18,25,33]. Despite the large number of studies on territorial farming damselfishes [24], there is only limited information on their behavior outside of their defended areas for comparison. For example, Hoey and Bellwood [23] noted that on a single occasion the farming damselfish *Dischistodus prosopotaenia* chased other herbivores grazing on the macroalga *Sargassum* that was experimentally tethered outside but adjacent to its territory, suggesting that it perceived it as a defendable resource rather than an opportunistic food source. Moreover, an Atlantic species, *Stegastes planifrons*, reportedly does not graze at all beyond its territorial borders [19]. Our experiment demonstrated that because *S. marginatus* grazed repeatedly alongside other fish species, as well as conspecifics, on a valued food resource when it was located outside the territory, that this damselfish species is an opportunistic browser, at least within the study site examined. In other taxa, foraging trips outside of a territory are often triggered by low food availability inside the territory [34,35], however, our study provides a rare example of a territorial species traveling outside of its defended area to feed on an identical resource (*A. spicifera*) that is also abundant within the territory. Future studies are needed, however, to determine what consequences this has for the economics of territory defense in *S. marginatus*, and whether this phenomenon occurs in other farming damselfishes or in other reef locales.

Opportunistic browsing of *S. marginatus* occurs when a nearby resource is perceived as at least equivalent to the defended resources. In this example, the *A. spicifera* as a resource was clearly valued by *S. marginatus* because this macroalga proliferated inside of their feeding territories. Additional evidence of this macroalga's palatability is supported by the 107 occasions during the experiment that multiple *S. marginatus* individuals left their territories to graze on the rhodophyte in neutral areas. The strong feeding preference that *S. marginatus* displays for *A. spicifera* is also reflected by other non-fish herbivores in Hawaii [12,36] including the green sea turtle [37,38]. Damselfish territories function as refugia for macroalgae [17,19–21,23,25] and results presented here are consistent with this conclusion because a conspicuous *A. spicifera* canopy was present only because it was protected from herbivory by *S. marginatus* (e.g., alga was absent outside of damselfish territories). To our knowledge, this is the first example of a farming damselfish specifically cultivating an exotic algal species. Algal farming damselfishes cultivate food species that occur nearby, however, these resources can be rare otherwise [21–23].

As an invasive species in the Hawaiian Islands, *A. spicifera* is a focus for management efforts [39–42] because its rapid growth can result in displacement of native species in coral reef habitats [43–45]. *Acanthophora spicifera* is also the most widely distributed invasive algae in the Hawaiian archipelago [45] even though many herbivorous fishes consume, and even prefer, this exotic over native species, such as *Padina japonica* and *Dictyospaeria cavernosa* [12]. The relationship between *S. marginatus* and *A. spicifera* is an interesting example of interspecific facilitation of an exotic species by a coral reef fish. Specifically, this facilitation is an example of mutualism because the interspecific interactions benefit both species [46]. As a highly palatable species [12], *A. spicifera* can only establish in sites, such as this Hawaiian MPA, where herbivore populations would otherwise exclude it because the macroalga has gained the tenacious defenses of *S. marginatus*. This is further supported by the fact that based on the intensity of grazing on *A. spicifera* tethered outside *S. marginatus* territories in the Pupukea MLCD, we estimate that 100% of the tethered *A. spicifera* would be consumed in approximately 1–6 hrs, a grazing rate that is comparable for this invasive species in other locations (e.g., Kaneohe Bay, Hawaii) [12].

Positive interactions also include reproductive mutualisms, in which one species benefits another, for example, by increasing propagule output [47]. By protecting the exotic rhodophyte from herbivory, *S. marginatus* enables *A. spicifera* to form canopies of adult thalli, which produce propagules, both as spores and vegetative fragments. Spore production in macroalgae contributes to both propagule dispersal as well as persistence in favorable areas [48]. This is important because after wave swells scour the brittle canopies of *A. spicifera* in Pupukea MLCD in winter, the macroalga has repeatedly regrown in the damselfish territories the following spring [personal observations, 2011–2014]. Further, the scoured canopies of *A. spicifera* become vegetative fragments that have the capacity to re-attach, also an important propagule dispersal strategy for this species [30]. Thus, the defenses of *S. marginatus* ensure the persistence of a reproductively-viable *A. spicifera* crop in this reef ecosystem.

As a case of a native species harboring an invasive one, the *S. marginatus* – *A. spicifera* relationship can inform management strategies, including within MPAs. Marine protected areas have been legislated to enhance fish stocks with the expectation that this action will increase browser densities and, in turn, curtail the expansion of fleshy macroalgae on coral reefs [49–51]. Here, we provide clear evidence that foraging fishes in a MPA serve as a biotic control for an invasive macroalga in Hawaii. On the one hand, this result is encouraging because, for example, a MPA on Maui (Kahekili Herbivore Fisheries Management Area (HFMA)) was established as a management tool to increase herbivore densities and thereby reduce the standing biomass and spread of fleshy macroalgae, including exotics such as *A. spicifera* (Department of Land and Natural Resources 2009). On the other hand, managers must contend with reefs that support both invasive macroalgae as well as colonial territories of *S. marginatus* because these fish territories can be a source of propagules of an exotic species within a marine managed area. Thus, the dual role of these farming damselfish as cultivators of invasive macroalgae within their territories and as opportunistic browsers of the same algal species in adjacent areas has broad implications for understanding the dynamics of reef ecosystems, invasion success of exotic macroalgae, and management.

Acknowledgments

Thanks to F. Thomas, R. Toonen and J. Lemus for their gracious support. We also thank the reviewers who provided constructive comments on this manuscript. Contribution no. 1596 from Hawaii Institute of Marine Biology.

Author Contributions

Conceived and designed the experiments: KAP KPM. Performed the experiments: KAP KPM LMV. Analyzed the data: KAP KPM. Contributed reagents/materials/analysis tools: KAP KPM. Wrote the paper: KAP KPM LMV.

References

1. Hughes TP (1994) Catastrophes, phase shifts, and large-scale degradation of a Caribbean coral reef. Science 265: 1547–1551.
2. Bellwood DR, Hughes TP, Folke C, Nystrom M (2004) Confronting the coral reef crisis. Nature 429: 827–833.
3. Ceccarelli DM, Jones GP, McCook LJ (2005) Foragers versus farmers: contrasting effects of two behavioural groups of herbivores on coral reefs. Oecologia 145: 445–453.
4. Bellwood DR, Hughes TP, Hoey AS (2006) Sleeping functional group drives coral-reef recovery. Curr Biol 16: 2434–2439.
5. Burkepile DE, Hay ME (2008) Herbivore species richness and feeding complementarity affect community structure and function on a coral reef. Proc Natl Acad Sci U S A 105: 16201–16206.
6. Hoey AS, Bellwood DR (2011) Suppression of herbivory by macroalgal density: a critical feedback on coral reefs? Ecol Lett 14: 267–273.

7. Mumby PJ, Dahlgren CP, Harborne AR, Kappel CV, Micheli F, et al. (2006) Fishing, trophic cascades, and the process of grazing on coral reefs. Science 311: 98–101.
8. Cheal AJ, Emslie M, Aaron MM, Miller I, Sweatman H (2013) Spatial variation in the functional characteristics of herbivorous fish communities and the resilience of coral reefs. Ecol Appl 23: 174–188.
9. Cheal AJ, MacNeil MA, Cripps E, Emslie MJ, Jonker MR, et al. (2010) Coral-macroalgal phase shifts or reef resilience: links with diversity and functional roles of herbivorous fishes on the Great Barrier Reef. Coral Reefs 29: 1005–1015.
10. Rogers S, Cox E (1999) Rate of spread of introduced rhodophytes *Kappaphycus alvarezii*, *Kappaphycus striatum*, and *Gracilaria salicornia* and their current distributions in Kaneohe Bay, Oahu, Hawaii. Pac Sci 53: 232–241.
11. Conklin EJ, Smith JE (2005) Abundance and spread of the invasive red algae, *Kapppaphycus* spp., in Kaneohe Bay, Hawaii and and experimental assessment of management. Biol Invasions 7: 1029–1039.
12. Stimson J, Larned ST, Conklin EJ (2001) Effects of herbivory, nutrient levels, and introduced algae on the distribution and abundance of the invasive macroalaga *Dictyosphaeria cavernosa* in Kaneohe Bay, Hawaii. Coral Reefs 19: 343–357.
13. Hoey AS, Bellwood DR (2009) Limited functional redundancy in a high diversity system: single species dominates key ecological process on coral reefs. Ecosystems 12: 1316–1328.
14. Mantyka CS, Bellwood DR (2007) Macroalgal grazing selectivity among herbivorous coral reef fishes. Mar Ecol Prog Ser 352: 177–185.
15. Hoey AS (2010) Size matters: macroalgal height influences the feeding response of coral reef herbivores. Mar Ecol Prog Ser 411: 299–302.
16. Burkepile DE, Hay ME (2011) Feeding complementarity versus redundancy among herbivorous fishes on a Caribbean reef. Coral Reefs 30: 351–362.
17. Sammarco PW (1983) Effects of fish grazing and damselfish territoriality on coral reef algae and algal community structure. Mar Ecol Prog Ser 13: 1–14.
18. Hixon MA, Brostoff WN (1983) Damselfish as keystone species in reverse: intermediate disturbance and diversity of reef algae. Science 220: 511–513.
19. Hinds PA, Ballantine DL (1987) Effects of the Caribbean threespot damselfish *Stegastes planifrons* (Cuvier), on algal lawn composition. Aquat Bot 27: 299–308.
20. Gobler CJ, Thibault DB, Davis TW, Curran PB, Peterson B, et al. (2006) Algal assemblages associated with Stegastes sp. territories on Indo-Pacific coral reefs: characterization of diversity and controls on growth. J Exp Mar Ecol 336: 135–145.
21. Hata H, Kato M (2006) A novel obligate cultivation mutualism between damselfish and *Polysiphonia* algae. Biol Lett 2: 593–596.
22. Ceccarelli DM, Jones GP, McCook LJ (2005) Effects of territorial damselfish on an algal-dominated coastal coral reef. Coral Reefs 24: 606–620.
23. Hoey AS, Bellwood DR (2010) Damselfish territories as a refuge for macroalgae on coral reefs. Coral Reefs 29: 107–118.
24. Ceccarelli DM, Jones GP, McCook LJ (2001) Territorial damselfishes as determinants of the structure of benthic communities on coral reefs. Oceanogr Mar Biol Annu Rev 39: 355–389.
25. Hixon MA, Brostoff WN (1996) Succession and herbivory: effects of differential fish grazing on Hawaiian coral-reef algae. Ecol Monogr 66: 67–90.
26. Russ GR (1987) Is removal of algae by grazers reduced inside territories of tropical damselfishes? J Exp Mar Biol Ecol 110: 1–17.
27. Ceccarelli DM (2007) Modification of benthic communities by territorial damselfish: a multi-species comparison. Coral Reefs 26: 853–866.
28. Preskitt LB, Vroom P, Smith C (2004) A rapid ecological assessment (REA) quantitative survey method for benthic algae using photo quadrats with SCUBA. Pac Sci 58: 201–209.
29. Kohler KE, Gill SM (2006) Coral point count with Excel extensions (CPC3): a visual basic program for the determination of coral and substrate coverage using random point count methodology. Computers and Geosciences 32: 1259–1269.
30. Kilar J, McLachlan J (1986) Ecological studies of the alga, *Acanthophora spicifera* (Vahl) Boerg. (Ceramiales: Rhodophyta): vegetative fragmentation. J Exp Mar Biol Ecol 104: 1–21.
31. Schacter CR, Albright LB, Dubofsky EA, Fitzsimmons JN, Focht R, et al. (2014) Risk-sensitive resournce defense in a territorial reef fish. Environ Biol Fish 97: 813–819.
32. Mallet D, Wantiez L, Lemouellic S, Vigliola L, Pelletier D (2014) Complementarity of rotating video and underwater visual census for assessing species richness, frequency and density of reef fish on coral reef slopes. PLoS ONE 9: e84344.
33. Losey GS (1982) Ecological cues and experience modify interspecific aggression by the damselfish, *Stegastes fassciolatus*. Behavior 81: 14–37.
34. Hartwick E (1978) The use of feeding areas outside of the territory of breeding Black Oystercatchers. Wilson Bull 90: 650–652.
35. Tsukada H (1997) A division between foraging range and territory related to food distribution in the red fox. J Ethol 15: 27–37.
36. Stimson J, Cunha T, Philippoff J (2007) Food preferences and related behavior of the browsing sea urchin *Tripneustes gratilla* (Linnaeus) and its potential for use as a biological control agent. Mar Biol 151: 1761–1772.
37. Arthur K, Balazs G (2008) A comparison of immature green turtle (*Chelonia mydas*) diets among seven sites in the Main Hawaiian Islands. Pac Sci 62: 205–217.
38. Russell DJ, Balazs GH (2009) Dietary shifts by green turtles (*Chelonia mydas*) in Kaneohe Bay region of the Hawaiian Islands: a 28-year study. Pac Sci 63: 181–192.
39. O'Doherty DC, Sherwood AR (2007) Genetic population structure of the Hawaiian alien invasive seaweed *Acanthophora spicifera* (Rhodophyta) as revealed by DNA sequencing and ISSR. Pac Sci 61: 223–233.
40. Weijerman M, Most R, Wong K, Beavers S (2008) Attempt to control the invasive red alga *Acanthophora spicifera* (Rhodophyta: Ceramiales) in a Hawaiian fishpond: an assessment of removal techniques and management options. Pac Sci 62: 517–532.
41. Lapointe BE, Bedford BJ (2011) Stormwater nutrient inputs favor growth of non-native macroalgae (Rhodophyta) on Oahu, Hawaiian Islands. Harmful Algae 10: 310–318.
42. Dailer ML, Smith JE, Smith CM (2012) Responses of bloom forming and non-bloom forming macroalgae to nutrient enrichment in Hawaii, USA. Harmful Algae 17: 111–125.
43. Doty M (1961) *Acanthophora*, a possible invader of the marine flora of Hawaii. Pac Sci 15: 547–552.
44. Russell DJ (1992) The ecological invasion of Hawaiian reefs by two marine red algae, *Acanthophora spicifera* (Vahl) Boerg. and *Hypnea musciformis* (Wulfen) J. Ag. and their association with two native species, *Laurencia nidifica* J. Ag. and *Hypnea cervicornis* J.Ag. ICES Marine Science Symposium 194: 110–125.
45. Smith J, Hunter C, Smith C (2002) Distribution and reproductive characteristics of nonindigenous and invasive marine algae in the Hawaiian Islands. Pac Sci 56: 299–316.
46. Stachowicz JJ (2001) Mutualism, facilitation, and the structure of ecological communities. Bioscience 51: 235–246.
47. Richardson DM, Allsopp N, D'Antonio CM, Milton SJ, Rejmanek M (2000) Plant invasions–the role of mutualisms. Biol Rev Camb Philos Soc 75: 65–93.
48. Santelices B (1990) Patterns of reproduction, dispersal and recruitment in seaweeds. Oceanogr and Mar Biol: an annual review 28: 177–276.
49. Friedlander AM, Brown E, Monaco ME (2007) Defining reef fish habitat utilization patterns in Hawaii: comparisons between marine protected areas and areas open to fishing. Mar Ecol Prog Ser 351: 221–233.
50. Mumby PJ, Harborne AR (2010) Marine reserves enhance the recovery of corals on Caribbean reefs. PLoS ONE 5: e8657.
51. Selig ER, Bruno JF (2010) A global analysis of the effectiveness of marine protected areas in preventing coral loss. PLoS ONE 5: e9278.

8

Invasiveness Does Not Predict Impact: Response of Native Land Snail Communities to Plant Invasions in Riparian Habitats

Jitka Horáčková[1,2]*, Lucie Juřičková[2], Arnošt L. Šizling[3], Vojtěch Jarošík[1,4†], Petr Pyšek[4,1]

1 Department of Ecology, Charles University in Prague, Faculty of Science, Prague 2, Czech Republic, **2** Department of Zoology, Charles University in Prague, Faculty of Science, Prague 2, Czech Republic, **3** Center for Theoretical Study, Charles University and the Academy of Sciences of the Czech Republic, Prague 1, Czech Republic, **4** Department of Invasion Ecology, Institute of Botany, Academy of Sciences of the Czech Republic, Průhonice, Czech Republic

Abstract

Studies of plant invasions rarely address impacts on molluscs. By comparing pairs of invaded and corresponding uninvaded plots in 96 sites in floodplain forests, we examined effects of four invasive alien plants (*Impatiens glandulifera, Fallopia japonica, F. sachalinensis*, and *F.×bohemica*) in the Czech Republic on communities of land snails. The richness and abundance of living land snail species were recorded separately for all species, rare species listed on the national Red List, and small species with shell size below 5 mm. The significant impacts ranged from 16–48% reduction in snail species numbers, and 29–90% reduction in abundance. Small species were especially prone to reduction in species richness by all four invasive plant taxa. Rare snails were also negatively impacted by all plant invaders, both in terms of species richness or abundance. Overall, the impacts on snails were invader-specific, differing among plant taxa. The strong effect of *I. glandulifera* could be related to the post-invasion decrease in abundance of tall nitrophilous native plant species that are a nutrient-rich food source for snails in riparian habitats. *Fallopia sachalinensis* had the strongest negative impact of the three knotweeds, which reflects differences in their canopy structure, microhabitat humidity and litter decomposition. The ranking of *Fallopia* taxa according to the strength of impacts on snail communities differs from ranking by their invasiveness, known from previous studies. This indicates that invasiveness does not simply translate to impacts of invasion and needs to be borne in mind by conservation and management authorities.

Editor: Brock Fenton, University of Western Ontario, Canada

Funding: This work was supported by grant of the Grant Agency of Charles University in Prague no. 40007 and institutional resources of Ministry of Education, Youth and Sports of the Czech Republic (http://www.msmt.cz/). PP was supported by long-term research development project no. RVO 67985939 (Academy of Sciences of the Czech Republic) and acknowledges the support by Praemium Academiae award from the Academy of Sciences of the Czech Republic (http://www.cas.cz/). The funders had no role in study design, data collection and analysis, decision to publish, or preparation of the manuscript.

Competing Interests: The authors have declared that no competing interests exist.

* Email: jitka.horackova@gmail.com

† Deceased.

Introduction

Invasive species are one of the major biotic stressors in native ecosystems all over the world [1–3], affecting the diversity of resident biota at various scales [4–10]. Plants are the most frequently studied group of invaders [11,12] and in the last decades, extensive literature has accumulated on how they impact ecosystem structure, functioning and services [13–20].

The majority of studies on impacts of plant invasions focus on the same trophic level, i.e., what effects invasive species have on the performance of populations, species and communities of resident plants. In their global review of available data on impact, Pyšek et al. ([21]; their Table 1) found that effects of plant invasions on plant diversity are addressed about twice as frequently as those on animal diversity e.g., [22–25]; see [9,26] for meta-analyses. However, invasive plants may alter interactions between trophic groups via the co-introduction of alien pollinators, seed dispersers, herbivores and predators, that cause profound disruptions to plant reproductive mutualisms [27], and by changing the biotic environment they may also impact reproductive output and population status of animal species [28,29]. Invasions not only have major implications for biodiversity, but by forging novel functions in resident ecosystems, they also limit the effectiveness of restoration efforts that can be followed by unpredictable responses [18,30].

Studies addressing the impacts of plant invasions on macroinvertebrates mostly reported significant reductions in species abundance, richness and diversity of arthropod communities [31–38] although in some studies this effect was restricted only to some groups [39]. Rarely, the studies reported shifting in food guilds [37]. However, studies exploring the impact of invasive plants on the abundance, species richness and diversity of molluscs, one of the model groups of herbivore generalists, are rather rare [40–46]. It has been shown that, for example, mollusc abundance decreased in areas invaded by *Tamarix ramosissima* in the southwestern United States [42] and in the riverine *Fallopia* stands in western Germany [43] and Switzerland [45]. Additionally, the litter of alien grasses from the genera *Avena* and *Bromus* reduced the number of snails in the Mediterranean biome of Australia [41]. However, the abundance of molluscs was not

Table 1. Quantitative summary of the effects of the four invasive plants studied on species numbers and abundances of land snail communities separated into groups (see text for criteria).

Invading plant	Snail category					
	Total		Small		Rare	
	Species number	Abundance	Species number	Abundance	Species number	Abundance
Fallopia sachalinensis	↓ 41.7	↓ 69.6	↓ 48.0			↓ 89.6
F. japonica			↓ 48.0			↓ 65.3
F.×bohemica			↓ 48.0			↑ 19.5
Impatiens glandulifera	↓ 16.3		↓ 48.0			↓ 28.8

Species numbers indicate percentage reduction in invaded compared to control plots, the arrow a decrease or increase in invaded plots. Empty cells refer to non-significant effects. Abundance is expressed as the number of living snail individuals.

significantly affected in vegetation invaded by *Spartina anglica* in Australia [40] and both gastropod species richness and abundance even increased following invasion by *I. glandulifera* in northern Switzerland [46]. The results are thus rather scarce and contradictory and none of the studies compared the impact of several invasive plants on multiple criteria of mollusc performance. Such impacts are, however, likely to differ; mollusc assemblages were shown to respond strongly to the change in vegetation, with associated changes in calcium content and humidity being the most important factors determining their occurrence [46–48]. Therefore, the close dependence of land-snail assemblages on soil and vegetation, resulting from their food preferences, makes this group of invertebrates a promising model for studying the impact of plant invasion on higher trophic levels. It can be assumed that invasive plants differing in stature, canopy structure, and chemical composition of tissues would exert different impacts on the structure and composition of land-snail communities.

Here we examine the effects of four invasive alien plants on communities of land snails inhabiting invaded stands. The plants studied are all highly invasive in the Czech Republic [49] and include representatives of contrasting life forms: clonal perennials (three taxa of the genus *Fallopia*) versus an annual species (*I. glandulifera*). The impact of these invaders on plant diversity has been thoroughly documented (see below), but there is a lack of information on changes they induce in the species richness of terrestrial snail communities. To get insight into this issue we address the following questions: (1) Do invasive alien plants exert impacts on species richness and abundance of land snail communities? (2) If so, do the impacts differ with respect to particular invasive plant taxa? Finally, using the three *Fallopia* congeners for which there is a thorough knowledge of mechanisms of invasion in central Europe that makes it possible to rank them according to their invasiveness e.g., [50,51], we ask (3) whether their ranking according to invasiveness corresponds to that based on the strength of impact on land snail communities?

Materials and Methods

Ethics statement

No permits and approvals were required for the field work, as sampling sites were under neither nature nor law protection.

Invasive plants studied

Fallopia japonica (Houtt.) Ronse Decr. var. *japonica* and *F. sachalinensis* (F. Schmidt) Ronse Decr. (Polygonaceae) are stout rhizomatous perennials native to East Asia, introduced to Europe (the former as a single female clone that spread across the continent) as garden ornamentals and fodder plants in the 19th century [52,53]. In the Czech Republic, both species are classified as invasive [10] and the genus *Fallopia* is represented also by the invasive hybrid *F.×bohemica* (Chrtek and Chrtková) J.P. Bailey, that is likely to have arisen on this continent several times independently and is also known from the native range of the parental species [53]. The first record of *F. japonica* var. *japonica* in the wild is from 1902, that of *F. sachalinensis* from 1921, and the earliest record of the hybrid *F.×bohemica* is from 1950. The invasion occurred in the second half of the 20th century, the hybrid lagged behind the two parental species but proceeded faster [54] due to its competitive superiority over the parents [50,51]. In the early 2000s, *F. japonica* var. *japonica* was recorded from 1335 localities, *F. sachalinensis* from 261 and the hybrid from 382 [54]. Their dispersal is mainly vegetative through regeneration from rhizome and stem segments transported with contaminated soil and water [51,55]. All three taxa became invasive (sensu [56,57]) in a number of habitats including riparian, where they reach high covers and reduce species richness and diversity of invaded vegetation [49]. The invasion by *Fallopia* taxa exhibits the most severe impact on species richness and diversity among central-European alien plants, reducing the number of species present prior to invasion by 66–86%, depending on the taxon [23]. *Fallopia* taxa affect infrastructure by damaging roads and flood-prevention structures, and increasing the erosion potential of rivers [58,59].

Impatiens glandulifera Royle (Balsaminaceae) is an annual species, up to 2.5 m tall, native to the Himalayas, introduced as a garden ornamental to Europe in 1839 and first recorded as escaped in 1855 [60]. In the Czech Republic, it was first recorded outside cultivation in 1896 [10], but rapid invasion only started in the mid-20th century [61]. *Impatiens glandulifera* is a dominant species of nitrophilous herbaceous fringes of rivers, willow galleries of loamy and sandy riverbanks and of riverine reed vegetation [49]. The species produces higher biomass than its congeners and is plastic in terms of response to nutrient availability and shading, but it also exhibits some genetically based population differentiation [62,63]. Due to its massive spread and extensive populations in riparian habitats, it is considered a conservation problem [64]. However, despite forming populations with a high cover of up to 90%, it does not markedly reduce the numbers of species co-occurring in invaded stands, although invasion does alter species composition in favour of ruderal species [23,65], but see [66]. *Impatiens glandulifera* was also shown to reduce the availability of pollinators for co-occurring native species [67].

The number of plant taxa included in the study was constrained by the fact that they represent a complete set of widespread invasive aliens in riparian habitats in the Czech Republic, with the only additional species being *Helianthus tuberosus*, that, however, invades different vegetation types than natural floodplain forests addressed in our study [49]. The massive invasion of all taxa under study on the rivers in the Czech Republic started at comparable times, around the mid-20th century [61,68], with some local differences [69]; in study sites the invasive species were permanently present for at least 10–15 years therefore it is unlikely that possible differences in residence times among localities affected the results.

Study area and field sampling

Field work was conducted from 2006 to 2011. In total, 96 sites with a maximum distance of 279.5 km were located in floodplain forests, in alluvia of six rivers of the lower Elbe catchment area in the western part of the Czech Republic, Central Europe (see Supporting Information, Table S1). For each of the four invasive plant taxa, one pair of 10×10 m plots was established in each of the sampling sites (*I. glandulifera*, n = 16 paired plots at 32 sites; *F. japonica*, n = 10 paired plots at 20 sites; *F. sachalinensis*, n = 10 paired plots at 20 sites; *F.×bohemica*, n = 12 paired plots at 24 sites). One plot of the pair was located in invaded vegetation where the cover of the invader was 70–100%, the second non-invaded (control) plot was placed in a close vicinity to ensure that the habitat conditions matched as closely as possible to the invaded part [23]. Both plots in the studied pair were sampled once only on the same day. Most of the plots within pairs were placed within the distance of 200 m (median value 159 m) and with four exceptions, all plots were paired within one kilometre. In a few cases, the invader occurred in the non-invaded plot, but its cover range of 1–2% could not have any effect on vegetation or land snail species. That the invasion was the main factor in which plots within the pair differed was confirmed by direct measurements in both plots of the pair of environmental characteristics that might be important predictors of land snail species richness and composition [70,71]. Of these we controlled for elevation (as a surrogate for climate), soil pH, and soil Ca content.

Land snail communities were sampled in the same plot as vegetation using a standard sampling procedure [72]. To document the presence of large and especially dendrophilous species (that rarely occur in litter samples), one person searched by eye for half an hour in all appropriate microhabitats within the whole plot, from which the litter sample could not be taken (e.g., dead wood, stones, tree trunks). Slugs were not included in data analysis because their activity depends mostly on weather conditions [73], and our sampling method was not suitable to record slugs quantitatively. The leaf litter samples with topsoil, twigs and vegetation were taken from four randomly selected quadrats (each measured 25×25 cm^2) at each plot. These subsamples were amalgamated, air-dried and all shells were sorted out using sieves of different mesh size. All empty shells, including their fragments, were excluded from analyses in order to reduce potential bias caused by (1) including species that were not living in the locality but that had their shells redeposited by floods, (2) not including species living in the locality whose accumulated empty shells were removed by accidental flooding [74], and (3) a different length of shell degradation time in various floodplain forest types, which depends mainly on humidity [70,75] and topsoil calcium content [76]. For these reasons in our analyses we only used the total numbers of living land snail species (further referred to as "total species") and the total number of individuals per species. Species included in any of the four threat categories used in the Red List of molluscs of the Czech Republic [77] were labelled as "rare species" and considered as indicators of the state of molluscan assemblages. Species with shells smaller than 5 mm [78] were classified as "small species". We distinguish this category because of biologically important variation of snail size in relationship to ecological [79] and geographical [80] scales. Specifically, large snails are often associated with moist conditions and low latitudes and thus their representation in communities is not uniform. Total number of species, the total number of individuals and their categorization into rare and small species, are shown in Table S2.

Statistical analysis

Because the data were hierarchically structured in the sense that the invaded and non-invaded plots were nested within locations, to account for the spatial dependencies within locations statistical models were constructed by introducing a random effect for the locations. Invasion status of the plots (invaded/non-invaded) and plant taxon (*F. sachalinensis*, *F. japonica*, *F.×bohemica* and *I. glandulifera*; further termed 'plant species') were fixed factors and location a random intercept, implicitly introducing the compound symmetrical correlation structure [81]. To check whether the models adequately accounted for the spatial dependences in the data, meaning that the models did not violate the basic assumption of the independence of errors of the observations due to spatial autocorrelation [82,83], we used a spline correlogram with 1000 resamples for bootstrap [84–86] based on Moran's *I* [87,88], to investigate residuals of the models [89].

Numbers of total and small snail species were square-root or square-root+1 transformed, and numbers of total and small individuals log or log+1 transformed to normalize the data e.g., [90], and analyzed by linear mixed models (LMMs) using the function *lme* [91]. Numbers of rare species and individuals could not be transformed to normal distribution due to a large number of zero counts, and these data were therefore analyzed by generalized linear mixed models (GLMMs) with Poisson errors, using the functions *glmmPQL* [92] and *lmer* [93]. LMMs and the function *glmmPQL* also made it possible to calculate intra-class correlation, i.e., the associations between non-invaded and invaded plots within locations, and distinguish, after explaining the part of variance due to differences between the paired plots, the part of residual variance within paired plots at a particular location from the part of residual variance among the locations. Fitted models were checked by plotting appropriate residuals against fitted values and predictors, and by Q-Q plots e.g., [81]. Calculations were done in R 2.12.1 [94].

Finally, a binomial test across all snail species for all four invasive plant species was performed in order to assess the effect of the invaders on each snail species separately.

Results

The effect of spatial autocorrelations was eliminated. This was so for all linear mixed models, and generalized linear mixed models obtained by application of *glmmPQL* function for rare species and *lmer* function for rare individuals (see Figure S1). This means that the explanatory variables were properly included in the models and their effects on the recorded mollusc species adequately measured, successfully accommodating for the spatial autocorrelation within the invaded and non-invaded plots.

High values of associations between non-invaded and invaded plots within sites, as well as relatively low residual variance within paired plots at each site and, at the same time, relatively high residual variance among sites (Table S3), indicate that the invaded and non-invaded plots within each site were appropriately selected. This is so because these results suggest that there was a relatively high similarity of the environmental factors listed above within the pairs of invaded and non-invaded plots within each site.

A simple pairwise test indicated significant reduction of plant and snail species richness at the invaded sites ($p<0.01$, df = 57, for details see legend to the Fig. 1A), but none or insignificant difference in elevation, soil pH, and soil Ca content between the invaded and non-invaded sites (Fig. S2). Moreover the snail species richness was independent from plant species richness in both the invaded and non-invaded plots as well as when using pooled data, where all the plots were analyzed together (Fig. 1B). Hence we conclude that neither the reduction of plant species richness nor difference in environmental factors between the invaded and non-invaded sites can be a direct driver of snail species richness. We therefore interpret the reduction of snail species richness as a consequence of the focal plant invaders presence/absence at the sites.

Except for small snails, the effect of individual plant taxa on the numbers of all snail species (i.e., including small and rare) and individuals was statistically different, as indicated by the significant plant species×invasion status interactions (Tables 2 and 3). *Fallopia sachalinensis* had the greatest negative effect on snail communities, significantly decreasing the total number of species and individuals, and the number of rare individuals. *Impatiens glandulifera* had a significant negative effect on the total number of species and on rare individuals. *Fallopia japonica* significantly decreased the number of rare individuals. Surprisingly, *F.* ×*bohemica* significantly increased the number of rare individuals. All invasive species had the same, significant negative effect on the number of small snail species (Fig. 2, Tables 1 and 4).

Detailed binomial analyses showed, after the Bonferroni correction applied at the significance level of 0.01, that *I. glandulifera* decreased abundances of 11 species and increased those of eight species of the 51 in total; *F.* ×*bohemica* decreased and increased abundances of 11 and six species, respectively, of 54 in total; *F. japonica* decreased abundances of seven species and increased abundances of two species of 50 in total; and *F. sachalinensis* decreased abundances of 19 of 43 species. The snails whose abundances were significantly higher at sites with the invasive plant present compared to non-invaded ones belonged mostly to small, leaf litter-dwelling species (see Table S4).

Discussion

Impacts on snail communities are invader-specific

Our study shows that invasive plants in temperate riparian habitats significantly affect species composition and structure of land-snail communities, and that these impacts vary with respect to the ecological groups of snails (i.e., they depend on woodland, open-country, mesic and/or aquatic character of particular snail species). Overall, the significant impacts range from 16 to 48% reduction in terms of species numbers, and 29–90% reduction in abundance. However, unlike in previous studies that mostly addressed the impacts of a single invasive plant species on mollusc communities e.g., [40,42,43,45,46], but see [41], our results provide insights into how impacts differ with respect to the identity of the invader.

That *I. glandulifera* was the plant with the second strongest impact on snail communities in the study, the only one besides *F. sachalinensis* that decreased total snail numbers, is rather surprising. This plant forms less homogenous and less dense cover than *Fallopia* taxa, and was reported to exert relatively minor impact on species richness of native plants following invasion in the Czech Republic [23,65]. This plant species has been documented to cause an increase in the richness and abundance of gastropods in deciduous forests in Switzerland, attributed to higher humidity in invaded sites [46]. It needs to be noted, however, a greater impact on native plant species richness than that recorded in the Czech Republic was reported from the UK [22]. The strong effect of this invader could be related to the decreased abundance, following invasion, of tall nitrophilous

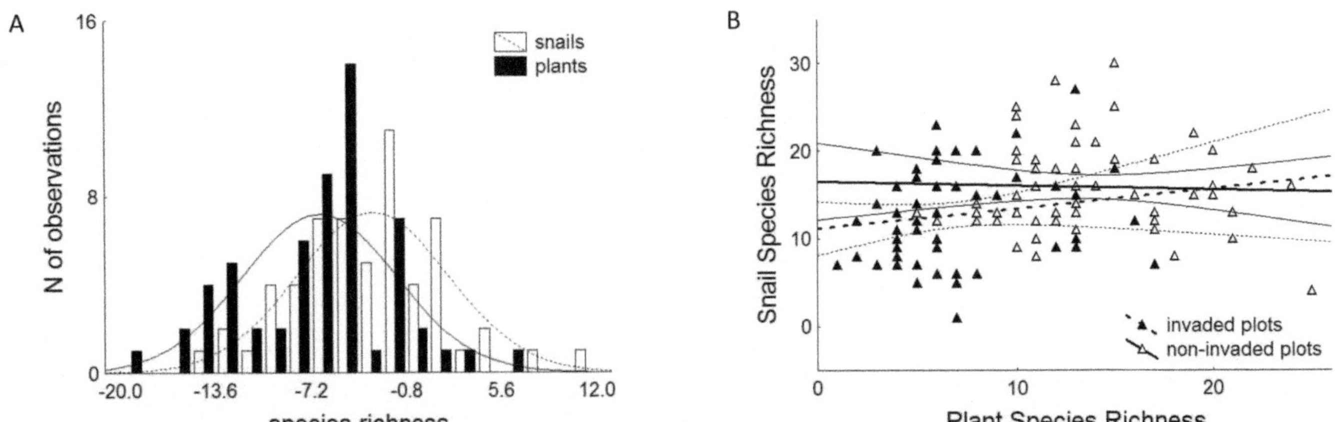

Figure 1. Frequency distribution of the pairwise residuals between species richness of invaded and non-invaded plots across the whole dataset. A) Frequency distribution of the pairwise residuals between species richness of invaded and non-invaded plots (richness of an invaded plot minus richness of the non-invaded plot) across the whole dataset. As the mean values of residuals in both the taxa (−6.7, and −3.1 for plants and snails, respectively) lies below zero value, and their two-sided 99% confidence intervals ([−8.5; −4.8] and [−4.9; −1.3], N = 58, df = 57) does not overlap zero, we conclude that the presence of the focal plant invaders reduce simultaneously plant and snail species richness. **B**) The lack of significant relationship (straight lines – mean trends, curved lines 95% confidence intervals) between the plant and snail species richness (in both the invaded-full lines, open symbols- and non-invaded plots-dashed lines, full symbols- as well as when using pooled data, where all the plots were analyzed together-is not shown) suggests that the reduction of plant species richness at the invaded plots is not a direct driver of the observed snail species richness reduction.

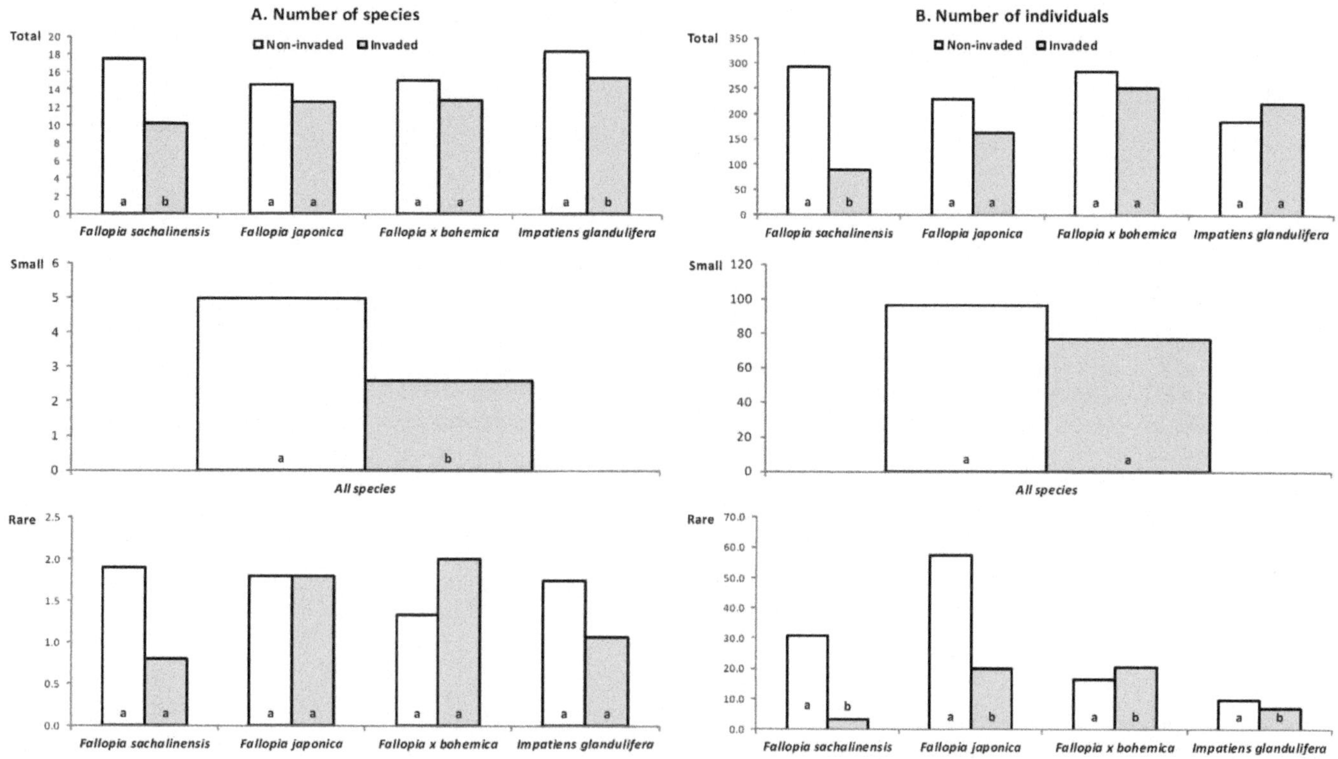

Figure 2. Average numbers of snail species and individuals from 48 paired invaded and non-invaded plots. Average numbers of total, small and rare snail species (A) and individuals (B) from 48 paired plots at individual sites for the species of invasive plants studied (*Fallopia sachalinensis, F. japonica, F.×bohemica* and *Impatiens glandulifera*). Counts for small snail species and individuals are shown together for all invasive plants as these numbers changed consistently for all plant species (non-significant invasion status×plant species interaction in Table 2). Paired columns followed by different letters differ significantly ($P<0.05$). Full statistics are given in Table 4.

native species (e.g., *Urtica dioica* and *Aegopodium podagraria*) that are a nutrient-rich food source for snails in riparian habitats. *Urtica*-dominated stands are characteristic of the understory of native floodplain forests in the study area, and harbour typical woodland snail fauna that includes a number of rare species.

Fallopia sachalinensis had the strongest negative impact on land-snail communities. Invasion by this species markedly decreased total species number and abundance of snails, as well as the number of small species and abundance of rare species. This is in contrast with the recorded impacts of the other two *Fallopia* taxa, which were much less profound and did not affect the snail community as a whole; their effects were only evident with regard to small and/or rare species. Interestingly, Stoll et al. [45] who examined the impact of a single *Fallopia* species, *F. japonica*, on snail communities in northern Switzerland arrived to opposite conclusions. This invasion reduced average snail richness but the impact differed with respect to shell size; the decreases in species richness were even more pronounced in large, long-lived species as compared to slugs and small, short-lived snails. Moreover, in our study there was a positive effect of *F. ×bohemica*, *F. japonica*, and *I. glandulifera* on numbers of individuals of rare snail species. Nevertheless, the overall pattern of impacts markedly differing among the three closely related taxa is surprising if compared to how they affect the plant species richness of invaded communities. The degree to which plant species richness is reduced following invasion is rather high and consistent for all three *Fallopia* taxa. They exhibit one of the most severe impacts on species richness and diversity among central-European alien plants, reducing the number of species present prior to invasion by 66–86%, depending on the taxon [23]; see also [45] for *F. japonica*.

Laboratory experiments may shed a light on the differences among the three *Fallopia* taxa in respect with the impact on snail communities. Laboratory experiments have shown that there exists a pronounced phytotoxic effect of *Fallopia* leaf extracts on seed germination. *Fallopia sachalinensis* exerts the largest negative effect on germination of *Urtica dioica*, the most abundant native species commonly growing in floodplain habitats invaded by *Fallopia* taxa in the studied area, while *F. ×bohemica* consistently has the lowest inhibitory effect [97]. Although these results do not provide direct evidence for differential effect of the individual *Fallopia* taxa on snail communities, they clearly show that litter quality differs among the *Fallopia* species and their hybrid. Importantly, this difference in phytotoxicity of leaf litter for seed germination is consistent with the different impact of the individual *Fallopia* taxa on snail community; *F. sachalinensis* had consistently the strongest negative impact on land snail communities, while for *F. ×bohemica* there was a positive effect. In addition, the high negative effect of *F. sachalinensis* leaf litter on germination of native plant species can further exacerbate the negative effect of this species on snail communities by an indirect way, via the suppression of the important native food plant *U. dioica*.

Body size affects the response of snails to invasion

Only snail community characteristics for which the impact was not invader-specific is the proportion of small species relative to

Table 2. ANOVA tables for the numbers of total and small snail species and individuals.

Source of variation	Species						Individuals					
	Total			Small			Total			Small		
	Df	F	P	Df	F	P	Df	F	P	Df	F	P
Site status	1, 44	25.707	<0.0001	1, 44	6.589	<0.05	1, 44	8.181	<0.01	1, 44	3.159	<0.1
Plant species	3, 44	1.499	NS	3, 44	0.935	NS	3, 44	1.129	NS	3, 44	0.207	NS
Status×Plant	**3, 44**	**3.115**	**<0.05**	3, 44	1.640	NS	**3, 44**	**6.800**	**<0.001**	3, 44	1.844	NS

ANOVA tables for the numbers of total and small snail species and individuals, analyzed by linear mixed models with plot invasion status (invaded/non-invaded) and invading plant species (*Fallopia sachalinensis*, *F. japonica*, *F.×bohemica* and *Impatiens glandilifera*) as fixed effects and sites with paired invaded/non-invaded plots as random intercepts. Significant invasion status×plant species interactions are in bold. Rare species and individuals were analyzed by generalized linear models (GLMMs) for which ANOVAs are not available. Results of t-tests for the fixed effects of these GLMMs are in Table 3.

Table 3. Results of t-tests for the numbers of rare snail species and individuals.

Source of variation	Rare species					Rare individuals			
	Value	Std. Error	Df	t-value	P	Estimate	Std. Error	z-value	P
Intercept	0.144	0.294067	44	0.489	NS	**2.0096**	**0.484**	**4.154**	**<0.001**
Status invasive	0.405	0.242285	44	1.674	NS	**0.2171**	**0.096**	**2.273**	**<0.05**
Fallopia japonica	0.257	0.424453	44	0.607	NS	1.1191	0.713	1.569	NS
Fallopia sachalinensis	0.270	0.425662	44	0.634	NS	−0.1536	0.733	−0.209	NS
Impatiens glandulifera	0.204	0.382983	44	0.534	NS	−0.4884	0.646	−0.757	NS
Invasive×*Fallopia japonica*	−0.405	0.348307	44	−1.164	NS	**−1.2764**	**0.126**	**−10.126**	**<0.001**
Invasive×*Fallopia sachalinensis*	**−1.270**	**0.398501**	**44**	**−3.188**	**<0.01**	**−2.4814**	**0.209**	**−11.875**	**<0.001**
Invasive×*Impatiens glandulifera*	**−0.90446**	**0.33462**	**44**	**−2.70294**	**<0.01**	**−0.5574**	**0.157**	**−3.556**	**<0.001**

Results of t-tests for the numbers of rare snail species and individuals, analyzed by generalized linear mixed models with plot invasion status (invaded/non-invaded) and invading plant species (*Fallopia sachalinensis*, *F. japonica*, *F.×bohemica* and *Impatiens glandulifera*) as fixed effects and sites with paired invaded/non-invaded plots as random intercepts. Rare species were analyzed using the function *glmmPQL* and rare individuals using *lmer* in R [94]. Results were obtained by the function *summary* and show fixed effects based on treatment contrasts where the intercept is for the plot invasion status 'non-invaded' and invading plant species *F.×bohemica*. Significant invasion status×plant species interactions are in bold.

Table 4. Results of full statistical analyses describing numbers of total, small and rare snail species and individuals between non-invaded and invaded plots.

Numbers of species	Total				Small				Rare			
	Difference	F	Df	P	Difference	F	Df	P	Difference	t-value	Df	P
Fallopia sachalinensis	**−1.070 ±0.189 (−7.3)**	**32.519**	**1, 9**	**<0.001**	**−0.229 ±0.091 (−2.4)**	**6.332**	**1, 47**	**< 0.05**	−0.865 ±0.438 (−1.1)	1.473	9	NS
Fallopia japonica	−0.298 ±0.256 (−2.0)	1.350	1, 9	NS	**−0.229 ±0.091 (−2.4)**	**6.332**	**1, 47**	**< 0.05**	0.000 ±0.150 (0.0)	0.000	9	NS
Fallopia ×bohemica	−0.315 ±0.177 (−2.3)	3.176	1, 11	NS	**−0.229 ±0.091 (−2.4)**	**6.332**	**1, 47**	**<0.05**	0.405 ±0.206 (0.7)	1.964	11	NS
Impatiens glandulifera	**−0.390 ±0.169 (−3.0)**	**5.359**	**1, 15**	**<0.05**	**−0.229 ±0.091 (−2.4)**	**6.332**	**1, 47**	**<0.05**	−0.499 ±0.236 (−0.7)	2.117	15	NS

Numbers of individuals	Total				Small				Rare		
	Difference	F	Df	P	Difference	F	Df	P	Difference	z-value	P
Fallopia sachalinensis	**−1.408 ±0.309 (−203.9)**	**20.821**	**1, 9**	**<0.001**	−0.437 ±0.252 (−17.3)	2.998	1, 47	< 0.1	**−2.264 ±0.186 (−27.6)**	**12.185**	**<0.001**
Fallopia japonica	−0.556 ±0.367 (−66.3)	2.291	1, 9	NS	−0.437 ±0.252 (−17.3)	2.998	1, 47	<0.1	**−1.059 ±0.082 (−37.5)**	**12.876**	**<0.001**
Fallopia ×bohemica	−0.188 ±0.250 (−31.5)	0.568	1, 11	NS	−0.437 ±0.252 (−17.3)	2.998	1, 47	<0.1	**0.217 ±0.095 (4)**	**2.273**	**<0.05**
Impatiens glandulifera	0.229 ±0.186 (36.2)	1.526	1, 15	NS	−0.437 ±0.252 (−17.3)	2.998	1, 47	<0.1	**−0.340 ±0.124 (−2.8)**	**2.74**	**<0.01**

Results of full statistical analyses describing numbers of total, small and rare snail species and individuals between non-invaded plots and plots invaded by plant species *Fallopia sachalinensis*, *F. japonica*, *F.×bohemica* and *Impatiens glandulifera*. Difference is a change in counts between non-invaded and invaded plots, with negative values indicating a decrease in snail numbers on invaded sites. Values ± standard errors are on transformed scales, numbers in parentheses are original counts. Significant differences between invaded and non-invaded plots for the individual plant species are in bold.

the total snail community. Our results indicate that small snail species are a group especially prone to reduction in species richness resulting from plant invasions; their numbers in invaded plots consistently decreased regardless of the identity of the invader (but see [45]). This holds also for plots invaded by *F. japonica* and *F. ×bohemica*, where total snail numbers were not affected, but the proportion of small snails decreased by 48% (Table 1). Stoll et al. [45] argued that small snail species in *F. japonica* invaded plots feed on algae, fungi and leaf litter, hence are less impacted by invasion than herbivorous large snails suffering from low palatability of knotweed tissues caused by high concentrations of phenolic compounds and lignin [96,97]. On the other hand, slow decomposition of knotweed litter [97,98] most likely results in a limited availability of food for small snail species, causing their reduction in invaded plots.

The post-invasion shift in snail species size hierarchies can be also linked to large snail species controlling a greater proportion of available resources than the smaller ones [80]. It can be hypothesized that under deteriorated conditions and namely reduced diversity of available food after the invasions, small snails are more affected than large ones that are superior in utilization of the limited resources [99]. An additional explanation could be that large snails inhabiting riparian vegetation are capable of profiting from the presence of tall invasive plants due to their climbing behaviour which is not the case of epigeic small snails.

High invasiveness does not automatically translate into strong impact

The three *Fallopia* taxa addressed in our study represent a thoroughly investigated study system for which there is detailed information on the history of invasion, ecology and traits conferring invasiveness in the invaded range in Europe. Previous research consistently points to an increased invasiveness of the hybrid compared to both parental species. The hybrid was reported to spread faster [54], and its abundance in the landscape can be related to better regeneration capacity from rhizome fragments [51]; it is also more difficult to control [50,55] and was a superior competitor to both parents when grown together in an experimental garden (P. Pyšek et al. unpublished data). In other studies, one of the parents performed poorly, such as with *F. sachalinensis* in a field study addressing the establishment of the three taxa [100] or *F. japonica* in a laboratory study investigating phytotoxic effects on germination of native species [95]; however, in both studies the hybrid was, together with the other parent, superior to the poorly performing one.

That the ranking of *Fallopia* taxa according to the strength of impacts on snail communities markedly differs from that according to their invasiveness as measured in the above studies, points to the fact that invasiveness does not simply translate to impacts. This is in accordance with conclusions of Ricciardi and Cohen [7] who found no correlations between invasiveness of alien plants, mammals, fishes, invertebrates, amphibians and reptiles, and their impact on biodiversity on a broad scale. Although the issue requires further study the possibility that the mechanisms of invasion and impact may not be strongly linked needs to be taken into account by managers. For our study it needs to be borne in mind that the impact on snail communities is only one particular type of impacts of plant invasions. Therefore, our results also emphasize the necessity of employing a variety of response measures when studying impacts of invasive species, as what we measure to a large extent determines whether or not the impact of a particular invasion appears serious [21,20].

The results of our study convey an important message for conservation authorities in the Czech Republic. Riparian habitats serve as refugia for many snail species that lost the majority of their natural habitats in the fragmented, intensively used landscape. Invasions of riparian zones by alien plants are an important factor further contributing to deterioration of snail habitats, and knotweeds are among the major invaders of these habitats. Focusing management effort on the hybrid, as the taxon with the greatest potential to spread [54], and paying the least widespread parent, *F. sachalinensis*, less attention, would be justified if one was primarily concerned with plant diversity. Without knowledge of impacts on snails, as documented in our study, this might seem the best strategy in general. However, based on a more comprehensive picture of taxon-specific impacts that vary with respect to the affected group of biota, and with specific conservation goals in mind, our results may help to inform conservation policy in a given area. For example, in regions with high land snail diversity and conservation value, allocation of resources to *Fallopia* control should reflect the ranking of taxa according to impact on snails.

Supporting Information

Figure S1 Spline autocorrelation statistics for residuals of models describing the numbers of total, small and rare snail species and individuals.

Figure S2 Frequency distribution of residuals between environmental parameters of the invaded and non-invaded plots.

Table S1 Overview of non-invaded and invaded sites used in this study.

Table S2 Overview of all recorded land snail species.

Table S3 Results of full statistical analyses describing intraclass correlations between plots and partition of residual variance for numbers of total, small and rare snail species and individuals between invaded and non-invaded plots.

Table S4 Results of binomial analyses describing differences in total abundances of each snail species in the invaded and non-invaded plots.

Acknowledgments

We thank Christina Alba for improving our English, and commenting on the manuscript.

Author Contributions

Conceived and designed the experiments: LJ JH. Performed the experiments: JH. Analyzed the data: VJ ALŠ. Contributed reagents/materials/analysis tools: JH. Wrote the paper: PP VJ ALŠ JH LJ.

References

1. Sala OE, Chapin FS, Armesto JJ, Berlow E, Bloomfield J, et al. (2000) Global biodiversity scenarios for the year 2100. Sci 287: 1770–1774.
2. Millenium Ecosystem Assessment (2005) Ecosystems and Human Well-being: Synthesis. Washington, DC.: Island Press. 160p.
3. Pimentel D, Zuniga R, Morrison D (2005) Update on the environmental and economic costs associated with alien-invasive species in the United States. Ecol Econ 52: 273–288.
4. Wilcove DS, Rothstein D, Dubow J, Phillips A, Losos E (1998) Quantifying threats to imperiled species in the United States. BioSci 48: 607–615.
5. Mack RN, Simberloff D, Lonsdale WM, Evans H, Clout M, et al. (2000) Biotic invasions: causes, epidemiology, global consequences, and control. Ecol Appl 10: 689–710.
6. Gurevitch J, Padilla D (2004) Are invasive species a major cause of extinctions? Trends Ecol Evol 19: 470–474.
7. Ricciardi A, Cohen J (2007) The invasiveness of an introduced species does not predict its impact. Biol Inv 9: 309–315.
8. Powell KI, Chase JM, Knight TM (2011) A synthesis of plant invasion effects on biodiversity across spatial scales. Am J Bot 98: 539–548.
9. Vilà M, Espinar JL, Hejda M, Hulme PE, Jarošík V, et al. (2011) Ecological impacts of invasive alien plants: a meta-analysis of their effects on species, communities and ecosystems. Ecol Lett 14: 702–708.
10. Pyšek P, Danihelka J, Sádlo J, Chrtek Jr J, Chytrý M, et al. (2012) Catalogue of alien plants of the Czech Republic (2nd edition): checklist update, taxonomic diversity and invasion patterns. Preslia 84: 155–255.
11. Pyšek P, Richardson DM, Pergl J, Jarošík V, Sixtová Z, et al. (2008) Geographical and taxonomic biases in invasion ecology. Trends Ecol Evol 23: 237–244.
12. Vilà M, Basnou C, Pyšek P, Josefsson M, Genovesi P, et al. (2010) How well do we understand the impacts of alien species on ecosystem services? A pan-European, cross-taxa assessment. Front Ecol Environ 8: 135–144.
13. Gordon DR (1998) Effects of invasive, non-indigenous plant species on ecosystem processes: lessons from Florida. Ecol Appl 8: 975–989.
14. Parker IM, Simberloff D, Lonsdale WM, Goodell K, Wonham M, et al. (1999) Impact: toward a framework for understanding the ecological effects on invaders. Biol Invasions 1: 3–19.
15. Tilman D (1999) The ecological consequences of changes in biodiversity: a search for general principles. Ecol 80: 1455–1474.
16. Dangles O, Jonsson M, Malmqvist B (2002) The importance of detritivore species diversity for maintaining stream ecosystem functioning following the invasion of a riparian plant. Biol Invasions 4: 441–446.
17. Levine JM, Vilà M, D'Antonio CM, Dukes JS, Grigulis K, et al. (2003) Mechanisms underlying the impacts of exotic plant invasions. Proc R Soc Lond B 270: 775–781.
18. Pyšek P, Richardson DM (2010) Invasive species, environmental change and management, and health. Annu Rev Environ Res 35: 25–55.
19. Gaertner M, Richardson DM, Privett SDJ (2011) Effects of alien plants on ecosystem structure and functioning and implications for restoration: insights from three degraded sites in South African Fynbos. Environ Manage 48: 57–69.
20. Hulme PE, Pyšek P, Jarošík V, Pergl J, Schaffner U, et al. (2013) Bias and error in current knowledge of plant invasions impacts. Trends Ecol Evol 28: 212–218.
21. Pyšek P, Jarošík V, Hulme PE, Pergl J, Hejda M, et al. (2012) A global assessment of invasive plant impacts on resident species, communities and ecosystems: the interaction of impact measures, invading species' traits and environment. Glob Change Biol 18: 1725–1737.
22. Hulme PE, Bremner ET (2006) Assessing the impact of *Impatiens glandulifera* on riparian habitats: partitioning diversity components following species removal. J Appl Ecol 43: 43–50.
23. Hejda M, Pyšek P, Jarošík V (2009) Impact of invasive plants on the species richness, diversity and composition of invaded communities. J Ecol 97: 393–403.
24. Cushman JH, Gaffney KA (2010) Community-level consequences of invasion: impacts of exotic clonal plants on riparian vegetation. Biol Invasions 12: 2765–2776.
25. Hladyz S, Åbjörnsson K, Giller PS, Woodward G (2011) Impacts of an aggressive riparian invader on community structure and ecosystem functioning in stream food webs. J Appl Ecol 48: 443–452.
26. Gaertner M, Breeyen AD, Hui C, Richardson DM (2009) Impacts of alien plant invasions on species richness in Mediterranean-type ecosystems: a meta-analysis. Progr Phys Geogr 33: 319–338.
27. Traveset A, Richardson DM (2006) Biological invasions as disruptors of plant–animal reproductive mutualisms. Trends Ecol Evol 21: 208–216.
28. Leslie AJ, Spotila JR (2001) Alien plants threaten Nile crocodile (*Crocodylus niloticus*) breeding in Lake St. Lucia, South Africa. Biol Conserv 98: 347–355.
29. Dibble KL, Meyerson LA (2012) Tidal flushing restores the physiological condition of fish residing in degraded salt marshes. PLoS ONE 7: e46161.
30. Milton SJ, Wilson JRU, Richardson DM, Seymour CL, Dean WRJ, et al. (2007) Invasive alien plants infiltrate bird-mediated shrub nucleation processes in arid savanna. J Ecol 95: 648–661.
31. Slobodchikoff CN, Doyen JT (1977) Effects of *Ammophila arenaria* on sand dune arthropod communities. Ecol 58: 1171–1175.

32. Lambrinos JG (2000) The impact of the invasive alien grass *Cortaderia jubata* (Lemoine): stapf on an endangered mediterranean-type shrubland in California. Divers Distrib 6: 217–231.
33. Herrera AM, Dudley TL (2003) Reduction of riparian arthropod abundance and diversity as a consequence of giant reed (*Arundo donax*) invasion. Biol Invasions 5: 167–177.
34. Greenwood H, O'Dowd DJ, Lake PS (2004) Willow (*Salix×rubens*) invasion of the riparian zone in south-eastern Australia: reduced abundance and altered composition of terrestrial arthropods. Divers Distrib 10: 485–492.
35. Ernst CM, Cappuccino N (2005) The effect of an invasive alien vine, *Vincetoxicum rossicum* (Asclepiadaceae), on arthropod populations in Ontario old fields. Biol Invasions 7: 417–425.
36. Gratton C, Denno RF (2006) Arthropod food web restoration following removal of an invasive wetland plant. Ecol Appl 16: 622–631.
37. Topp W, Kappes H, Rogers E (2008) Response of ground-dwelling beetle (Coleoptera) assemblages to giant knotweed (*Reynoutria* spp.) invasion. Biol Invasions 10: 381–390.
38. Wilkie L, Cassis G, Gray M (2007) The effects on terrestrial arthropod communities of invasion of a coastal heath ecosystem by the exotic weed bitou bush (*Chrysanthemoides monilifera* ssp. *rotundata* L.). Biol Invasions 9: 477–498.
39. Durst SL, Theimer TC, Paxton EH, Sogge MK (2008) Temporal variation in the arthropod community of desert riparian habitats with varying amounts of saltcedar (*Tamarix ramosissima*). J Arid Environ 72: 1644–1653.
40. Hedge P, Kriwoken LK (2000) Evidence for effects of *Spartina anglica* invasion on benthic macrofauna in Little Swanport estuary, Tasmania. Austral Ecol 25: 150–159.
41. Lenz TI, Moyle-Croft JL, Facelli JM (2003) Direct and indirect effects of exotic annual grasses on species composition of a South Australian grassland. Austral Ecol 28: 23–32.
42. Kennedy TA, Finlay JC, Hobbie SE (2005) Eradication of invasive *Tamarix ramosissima* along a desert stream increases native fish density. Ecol Appl 15: 2072–2083.
43. Kappes H, Lay R, Topp W (2007) Changes in different trophic levels of litter-dwelling macrofauna associated with giant knotweed invasion. Ecosyst 10: 734–744.
44. Gerber E, Krebs C, Murrell C, Moretti M, Rocklin R, et al. (2008) Exotic invasive knotweeds (*Fallopia* spp.) negatively affect native plant and invertebrate assemblages in European riparian habitats. Biol Conserv 141: 646–654.
45. Stoll P, Gatzsch K, Rusterholz R, Baur B (2012) Response of plant and gastropod species to knotweed invasion. Basic Appl Ecol 13: 232–240.
46. Ruckli R, Rusterholz H, Baur B (2013) Invasion of *Impatiens glandulifera* affects terrestrial gastropods by altering microclimate. Acta Oecol 47: 16–24.
47. Martin K, Sommer M (2004) Relationships between land snail assemblage patterns and soil properties in temperate-humid forest ecosystems. J Biogeogr 31: 531–545.
48. Juřičková L, Horsák M, Cameron RDA, Hylander K, Míkovcová A, et al. (2008) Land snail distribution patterns within a site: the role of different calcium sources. Eur J Soil Biol 44: 172–179.
49. Pyšek P, Chytrý M, Pergl J, Sádlo J, Wild J (2012) Plant invasions in the Czech Republic: current state, introduction dynamics, invasive species and invaded habitats. Preslia 84: 576–630.
50. Bímová K, Mandák B, Pyšek P (2001) Experimental control of *Reynoutria* congeners: a comparative study of a hybrid and its parents. In: Brundu G, Brock J, Camarda I, Child l, Wade M, editors. Plant Invasions: Species Ecology and Ecosystem Management. Leiden: Backhuys Publishers. pp. 283–290.
51. Pyšek P, Brock JH, Bímová K, Mandák B, Jarošík V, et al. (2003) Vegetative regeneration in invasive *Reynoutria* (Polygonaceae) taxa: the determinant of invasibility at the genotype level. Am J Bot 90: 1487–1495.
52. Beerling DJ, Bailey JP, Conolly AP (1994) *Fallopia japonica* (Hout.) Ronse Decraene (*Reynoutria japonica* Houtt., *Polygonum cuspidatum* Sieb. & Zucc.). J Ecol 82: 959–979.
53. Bailey JP, Conolly AP (2000) Prize-winners to pariahs: a history of Japanese knotweed s.l. (Polygonaceae) in the British Isles. Watsonia 23: 93–110.
54. Mandák B, Pyšek P, Bímová K (2004) History of the invasion and distribution of *Reynoutria* taxa in the Czech Republic: a hybrid spreading faster than its parents. Preslia 76: 15–64.
55. Bímová K, Mandák B, Pyšek P (2003) Experimental study of vegetative regeneration in four invasive *Reynoutria* taxa. Plant Ecol 166: 1–11.
56. Richardson DM, Pyšek P, Rejmánek M, Barbour MG, Panetta FD, et al. (2000) Naturalization and invasion of alien plants: concepts and definitions. Divers Distrib 6: 93–107.
57. Blackburn TM, Pyšek P, Bacher S, Carlton JT, Duncan RP, et al. (2011) A proposed unified framework for biological invasions. Trends Ecol Evolut 26: 333–339.
58. Beerling DJ (1991) The effect of riparian land use on the occurrence and abundance of Japanese knotweed (*Reynoutria japonica*) on selected rivers in South Wales. Biol Cons 55: 329–337.
59. Reinhardt F, Herle M, Bastiansen F, Streit B (2003) Economic Impact of the Spread of Alien Species in Germany. Frankfurt am Main: Biological and Computer Sciences Division, Dept. of Ecology and Evolution. 229p.
60. Beerling DJ, Perrins DM (1993) Biological flora of the British Isles: *Impatiens glandulifera* Royle (*Impatiens Roylei* Walp.). J Ecol 81: 367–381.
61. Pyšek P, Prach K (1993) Plant invasions and the role of riparian habitats: a comparison of four species alien to central Europe. J Biogeogr 20: 413–420.
62. Skálová H, Moravcová L, Pyšek P (2011) Germination dynamics and seedling frost resistance of invasive and native *Impatiens* species reflect local climatic conditions. Persp Pl Ecol Evol Syst 13: 173–180.
63. Skálová H, Havlíčková V, Pyšek P (2012) Seedling traits, plasticity and local differentiation as strategies of invasive species of *Impatiens* in central Europe. Ann Bot 110: 1429–1438.
64. DAISIE (eds.) (2009) Handbook of Alien Species in Europe. Berlin: Springer. 381 p.
65. Hejda M, Pyšek P (2006) What is the impact of *Impatiens glandulifera* on species diversity of invaded riparian vegetation? Biol Conserv 132: 143–152.
66. Bremner ET, Hulme PE (2006) Assessing the impact of *Impatiens glandulifera* on riparian habitats: partitioning diversity components following species removal. J Appl Ecol 43: 43–50.
67. Chittka L, Schürkens S (2001) Successful invasion of a floral market. Nature 411: 653–654.
68. Pyšek P, Prach K (1995) Invasion dynamics of *Impatiens glandulifera*: a century of spreading reconstructed. Biol Conserv 74: 41–48.
69. Malíková L, Prach K (2010) Spread of alien *Impatiens glandulifera* along rivers invaded at different times. Ecohydrol Hydrobiol 10: 81–85.
70. Čejka T, Horsák M, Némethová D (2008) The composition and richness of danubian floodplain forest land snail faunas in relation to forest type and flood frequency. J Mollusc Stud 74: 37–45.
71. Dvořáková J, Horsák M (2012) Variation of snail assemblages in hay meadows: disentangling the predictive power of abiotic environment and vegetation. Malacologia 55: 151–162.
72. Cameron RAD, Pokryszko BM (2005) Estimating the species richness and composition of land mollusc communities: problems, consequences and practical advice. J Conchol 38: 529–548.
73. Rollo CD (1991) Endogenous and exogenous regulation of activity in *Deroceras reticulatum*, a weather-sensitive terrestrial slug. Malacol 33: 199–220.
74. Ilg Ch, Foeckler F, Deichner O, Henle K (2009) Extreme flood events favour floodplain mollusc diversity. Hydrobiol 621: 63–73.
75. Horáčková J, Horsák M, Juřičková J (2014) Land snail diversity and composition in relation to ecological variations in Central European floodplain forests and their history. Community Ecol 15: 44–53.
76. Cernohorsky N, Horsák M, Cameron RAD (2010) Land snail species richness and abundance at small scales: the effects of distinguishing between live individuals and empty shells. J Conchol 40: 233–241.
77. Beran L, Juřičková L, Horsák M (2005) Mollusca. In: Farkač J, Král D, Škorpík M, editors. Red List of Threatened Species in the Czech Republic - Invertebrates. Prague: AOPK ČR. pp. 67–69.
78. Kerney MP, Cameron RAD, Jungbluth JH (1983) Die Landschnecken Nord- und Mitteleuropas. Hamburg and Berlin: Verlag Paul Parey. 384 p.
79. Goodfriend GA (1986) Variation in land-snail shell form and size and its causes: a review. Syst Biol 35:204–223.
80. Hausdorf B (2007) The interspecific relationship between abundance and body size in central European land snail assemblages. Basic Appl Ecol 8: 125–134.
81. Zuur AF, Ieno EN, Waker NJ, Saveliev AA, Smith GM (2009) Mixed Effects Models and Extensions in Ecology. New York: Springer. 574p.
82. Legendre P (1993) Spatial autocorrelation: Trouble or new paradigm? Ecol 74: 1659–1673.
83. Lichstein JW, Simons TR, Shriner SA, Franzreb KE (2002) Spatial autocorrelation and autoregressive models in ecology. Ecol Monogr 72: 445–463.
84. Cliff AA, Ord JK (1981) Spatial processes: models and applications. London: Pion.
85. Bjørnstadt ON, Falck W (2001) Nonparametric spatial covariance functions: estimation and testing. Environ Ecol Stat 8: 53–70.
86. Bjørnstadt ON (2008) Ncf: Spatial Nonparametric Covariance Functions. R package version 1.1-1. Available: http://onb.ent.psu.edu/onb1/R. Accessed 29 May 2012.
87. Sokal RR, Oden NL (1978) Spatial autocorrelation in biology: 1. Methodology. Biol J Linn Soc 10: 199–228.
88. Legendre P, Legendre L (1998) Numerical Ecology. Amsterdam: Elsevier. 853 p.
89. Rhodes JR, McAlpine CA, Zuur AF, Smith GM, Ieno EN (2009) Chapter 21. GLMM applied on the spatial distribution of koalas in a fragmented landscape. In: Zuur AF, Ieno EN, Walker NJ, Saveliev AA, Smith GM, editors. Mixed Effects Models and Extensions in Ecology. New York: Springer. pp. 469–492.
90. Sokal RR, Rohlf FJ (1995) Biometry: the principles and practise of statistics in biological research, third ed. New York: Freeman. 887p.
91. Pinheiro J, Bates D, DebRoy S, Sarkar D, the R Core team (2009) Nlme: Linear and Nonlinear Mixed Effects Models. R package version 3: 1–93.
92. Venables WN, Ripley BD (2002) Modern Applied Statistics with S, fourth ed. New York: Springer. 497p.
93. Bates D, Maechler M, Bolker B (2011) lme4: Linear Mixed-Effects Models Using S4 Classes. R package version 0.999375-39. Available: http://CRAN.R-project.org/package=lme4. Accessed 15 April 2013.
94. R Development Core Team (2009) R: A Language and Environment for Statistical Computing. Vienna: R Foundation for Statistical Computing. Available: http://www.R-project.org. Accessed January 2012.
95. Moravcová L, Pyšek P, Jarošík V, Zákravský P (2011) Potential phytotoxic and shading effects of invasive *Fallopia* (Polygonaceae) taxa on the germination of dominant native species. NeoBiota 9: 31–47.

96. Vrchotová N, Šerá B, Tříska J (2007) The stilbene and catechin content of the spring sprouts of *Reynoutria* species. Acta Chromatogr 19: 21–28.
97. Pálková K (2007) Palatabilita druhů, jejich bionomické vlastnosti a rychlost rozkladu detritu. Master's Thesis, České Budějovice: University of South Bohemia in České Budějovice. 68 p.
98. Mincheva T, Barni E, Varese GC, Brusa G, Cerabolini B et al. (2014) Litter quality, decomposition rates and saprotrophic mycoflora in *Fallopia japonica* (Houtt.) Ronse Decraene and in adjacent native grassland vegetation. Acta Oecol 54: 2–35.
99. Baur B, Baur A (1990) Experimental evidence for intra- and interspecific competition in two species of rock-dweling land snails. J Anim Ecol 59: 301–315.
100. Brabec J, Pyšek P (2000) Establishment and survival of three invasive taxa of the genus *Reynoutria* (Polygonaceae) in mesic mown meadows: a field experimental study. Folia Geobot 35: 27–42.

9

Group-Living Herbivores Weigh Up Food Availability and Dominance Status when Making Patch-Joining Decisions

Keenan Stears[1]*, Graham I. H. Kerley[2], Adrian M. Shrader[1]

1 School of Life Sciences, University of KwaZulu-Natal, Pietermaritzburg, South Africa, **2** Centre for African Conservation Ecology, Department of Zoology, Nelson Mandela Metropolitan University, Port Elizabeth, South Africa

Abstract

Two key factors that influence the foraging behaviour of group-living herbivores are food availability and individual dominance status. Yet, how the combination of these factors influences the patch-joining decisions of individuals foraging within groups has scarcely been explored. To address this, we focused on the patch-joining decisions of group-living domestic goats (*Capra hircus*). When individuals were tested against the top four ranked goats of the herd, we found that at patches with low food availability they avoided these dominant patch-holders and only joined subordinates (i.e. costs outweighed benefits). However, as the amount of food increased, the avoidance of the top ranked individuals declined. Specifically, goats shifted and joined the patch of an individual one dominance rank higher than the previous dominant patch holder when the initial quantity of food in the new patch was twice that of the lower ranking individual's patch (i.e. benefits outweighed costs). In contrast, when individuals chose between patches held by dominant goats, other than the top four ranked goats, and subordinate individuals, we found that they equally joined the dominant and subordinate patch-holders. This joining was irrespective of the dominance gap, absolute rank of the dominant patch-holder, sex or food availability (i.e. benefits outweighed costs). Ultimately, our results highlight that herbivores weigh up the costs and benefits of both food availability and patch-holder dominance status when making patch-joining decisions. Furthermore, as the initial quantity of food increases, food availability becomes more important than dominance with regard to influencing patch-joining decisions.

Editor: Alan McElligott, Queen Mary, University of London, United Kingdom

Funding: KS received funding from the University of KwaZulu-Natal, School of Life Sciences (http://lifesciences.ukzn.ac.za/Homepage.aspx). AMS received funding from the South African National Research Foundation (http://www.nrf.ac.za). The funders had no role in study design, data collection and analysis, decision to publish, or preparation of the manuscript.

Competing Interests: The authors have declared that no competing interests exist.

* Email: keenan.stears@gmail.com

Introduction

Living in groups has two broad benefits: reduced predation risk and increased foraging efficiency [1,2]. Reduction in predation risk comes from increased group vigilance [3], smaller domains of danger [4] and the dilution effect [5]. Alternatively, foraging benefits are obtained via social information [6,7]. For example, by watching conspecifics, individuals can increase their ability to find and assess the quality of food patches [8,9]. However, this may lead to individuals joining patch-holders at their patches, resulting in costs such as increased competition and greater social conflicts between group members [10–12].

One factor that can influence patch-joining decisions is the dominance status of the patch-holder [13,14]. Dominant individuals tend to be larger and more aggressive than subordinates and thus can better defend patches [13]. However, social bonds can also influence foraging behaviour because related individuals can form subgroups [15], which can influence food preference [16] and the foraging location of group members [17,18], with forgers spending more time feeding closer to other herd members. In addition, patch quality also influences joining decisions, as foragers prefer to feed from high quality patches [19]. Furthermore, patch quality determines the amount of food that is available to joining individuals once the patch-holder has removed the finder's share (i.e. proportion of food available to the patch-holder before another individual arrives) [20]. In poor quality patches, less food is available to all individuals feeding within a patch, compared to high quality patches. Therefore, to obtain food in poor quality patches, patch-holders tend to defend these patches more aggressively [21].

One of the most important requirements for foragers is to feed in a manner that promotes efficiency and reduces perceived risk. Generally, the value of a patch is enhanced as the abundance of food increases [22] and decreases as the risk of foraging in that patch increases [23]. For example, Shrader *et al.* [24] found that goats reduced their feeding effort from patches that were close to predator cues (dung and urine) compared with patches that did not have predator cues (i.e. high-risk and low-risk patches). Interestingly, gerbils (*Gerbillus allenbyi* and *G. pyramidum*) increased their use of risky patches when the amount of food within these patches was about 4–8 times richer than safer patches [25]. Thus, a forager should feed in patches where it makes a trade-off between high food reward and safety [26].

In group-living animals, dominant individuals tend to aggressively defend resources, so joining a dominant individual will have higher risks compared to joining a subordinate. However, competition may also be a passive process, operating through the avoidance of conflict by subordinates [27]. Various studies have shown that individuals only join patches held by subordinate individuals [12,14]. However, in some instances, dominance can have little effect on tactic use [28,29]. This may be a result of the flexibility and consistency in tactic use across taxa [30]. For example, a study on Japanese macaques (*Macaca fuscata fuscata*) found that when food is abundant, patch choices by subordinates are unaffected by whether the patch is occupied by a dominant or subordinate individual [31]. This suggests that the role of dominance in foraging decisions may be influenced by additional factors such as patch quality [32]. Interestingly, the factors that influence joining decisions in species where dominance is not important are poorly known [30].

Surprisingly, the degree to which herd members weigh up food availability in a patch, and the dominance status of a patch-holder, when making patch-joining decisions, has scarcely been explored in species other than birds [12]. Goats provide a good species with which to test this as they have a linear hierarchy [33,34], use social [9] and personal information [35], and dominant goats join lower ranking individuals at patches [33,36]. As a result, we conducted two experiments to focus on how these two factors influence the patch-joining decisions of a group-living herbivore–the domestic goat (*Capra hircus*). The first experiment focused on how the dominance rank of the top ranked goats in the herd and food availability (i.e. the quantity of food in a patch) influenced patch-joining decisions. To further explore the influence of dominance and food availability, we ran an additional experiment that tested three hypotheses: 1) the dominance gap between a dominant patch-holder and a joining goat is important in determining patch choice, 2) the absolute rank of the dominant patch-holder will influence patch choice, and 3) the sex of the dominant patch-holder as well as the sex of the joining goat will influence patch choice. For the above experiments, if dominance is the driving force behind patch choice, we expect goats to join subordinate patch-holders, irrespective of the amount of food. Alternatively, if food availability is the main factor influencing patch choice, goats should feed from patches with large quantities of food, irrespective of the dominance status of the patch-holder.

Materials and Methods

Ethical statement

We adhered to the protocols of the University of KwaZulu-Natal's Animal Ethics Committee, which cleared all animal husbandry and experimental procedures (permit number: 068/10/Animal). Prior to approval, the application was reviewed by the South African National Council of the Society for the Prevention of Cruelty to Animals (NSPCA). Throughout all the experiments, every effort was made to minimise the harm and stress to the goats, and no goats were injured during the experiments. All data are available in data supplement S1.

Experimental procedure

Experiments ran from April 2010 to May 2011 at the Ukulinga Research Farm, Pietermaritzburg, South Africa. We used a single herd of 45 indigenous veld goats (16♂, 29♀, average age: 26±12 months), thus all the goats were familiar with each other prior to the start of our experiments. For our experiments, we relied on the goats' use of personal information [35,37] and their ability to remember complicated foraging tasks [35] to differentiate and remember the amount of food found in the different patch types (see below).

For eight months prior to running the experiments, the goats were exposed to the testing arenas and all the food patches to allow them to become accustomed to the experimental protocol. This habituated the goats to the different food patches and allowed them to determine the amount of food in each patch (see below). During this time, we also tested the dominance hierarchy and determined whether goats could differentiate between the different amounts of food in each patch.

Goats were housed in a 30×12 m sheltered barn containing water troughs. To ensure that the goats joined patches and fed during the experiments, we did not provide the goats with food overnight. However, they had *ad libitum* access to water. This mimicked common husbandry practices [24]. Each day prior to running the experiments, the goats were allowed to feed for an hour in a 30×15 m rye grass pasture. This limited the possibility that hunger levels would differ substantially between the first and last goats tested each morning. Experiments ran from 7 h00 to 11 h00 each day. After running the experiments, all the goats were released to forage in a natural grassland for the remainder of the day (i.e. ~6 hours).

Dominance relationships

Prior to all experiments, we determined the dominance hierarchy of the herd (Table 1). Initially, we explored the possibility of recording natural agonistic interactions between herd members as a way to estimate the hierarchy because this would limit any added stress or injury to the animals. However, after initial observations, we concluded that it would take an inordinate amount of time to observe interactions between all the different individuals ($N = 45$ goats). This is because each day the goats fed in a large natural grassland where our ability to observe them was limited. Thus, due to the impracticalities of observing interactions in the field, we chose to stage the interactions experimentally.

Each day before we recorded the dominance interactions, the goats were allowed to feed for one hour in a rye grass pasture. This ensured that all goats had similar satiation levels and thus avoided situations where hungrier goats could be more aggressive. To determine the dominance hierarchy, we facilitated interactions between two goats at a feeding patch. For these interactions, a single artificial food patch (L×W×H: 57.0×36.5×23.0 cm) was provided. These patches contained 200 g of commercial sheep pellets (Complete Sheep Finisher, Meadow Feeds, South Africa) poured into one of the corners.

The main risk of injury that we foresaw during these interactions was an individual being cornered against a fence and thus being unable to escape from aggressive contact from the other goat. To prevent this, the interactions were conducted in a large (700×450 cm) pen, which allowed the subordinate goat to move away from the dominant individual after an interaction. Furthermore, we placed the feeding patch in the middle of the pen and released the goats simultaneously from opposite sides of the patch. This allowed the subordinate goat to decide whether it wanted to interact with the dominant goat, a situation that mimics natural conditions (Shrader, pers. obs.). If the subordinate goat did not approach the patch, we did not force an interaction but, rather, classified the feeding goat as dominant.

To initiate the interactions, we released the two goats simultaneously from ~300 cm away on opposite sides of the patch. The goats were allowed to feed from the patch for up to 3 min. Dominance was determined during this time using visual cues [33,34]. In no trial did the two goats interact aggressively for

Table 1. Sex, presence of horns, and dominance rank of the goats calculated using David's Score.

Goat name	Sex	Horned	Dominance rank	David's score
Blue 05	M	Yes	1	977
IG35	M	Yes	2	932
Blue06	M	Yes	3	929
IG41	M	Yes	4	842
Big beige	M	Yes	5	797
Blue04	M	No	6	720
IG1	F	Yes	7	708
Matey	F	Yes	8	619
Mottled	F	Yes	9	618
IG36	M	Yes	10	485
IG32	F	No	11	484
Blue03	F	Yes	12	444
IG22	F	No	13	443
IG42	M	No	14	348
IG46	M	Yes	15	336
Blue01	F	No	16	309
IG51	M	No	17	260
Blue08	M	No	18	233
IG57	M	Yes	19	220
S6K	F	No	20	174
S3K	F	No	21	140
Blue02	F	Yes	22	−4
S2K	F	No	23	−5
S7K	F	No	24	−50
TK1	M	Yes	25	−77
TK2	M	Yes	26	−122
Blue07	F	No	27	−167
IG55	F	Yes	28	−185
IG44	F	No	29	−320
S13K	M	Yes	30	−337
IG7	F	No	31	−382
TK8	M	Yes	32	−451
S12K	F	Yes	33	−461
TK7	F	Yes	34	−493
S10K	F	No	35	−507
S9K	F	No	36	−561
S8K	F	Yes	37	−597
TK5	F	No	38	−638
S14K	F	No	39	−701
TK3	F	No	40	−714
TK6	F	No	41	−769
Blue11	F	Yes	42	−846
TK4	F	No	43	−851
S15K	F	No	44	−889
S16K	F	Yes	45	−964

the entire 3 min period. Rather, a distinction between dominant and subordinate goats was generally made within the first 30 seconds.

Goats that consistently initiated aggressive acts and ultimately prevented or limited the feeding opportunities of the other individual were considered dominant. Dominant goats showed agonistic behaviour through aggression with contact, i.e. butting the flank or head of a feeding goat with their horns or the top of their heads if they did not have horns, and aggression without contact, i.e. threat displays and rush threats, see review by [38]. In contrast, some goats avoided a competitor completely and did not approach or feed from the artificial patch. We considered these individuals subordinate. With or without physical contact, it was easy to identify the dominant and subordinate goats via visual cues [33,34]. Generally, if both goats approached the patch, vocalisations from the dominant goat were enough to cause the subordinate to retreat. Therefore, the agonistic interactions did not always result in physical encounters. In the limited instances (~10% of the interactions) where aggressive physical interactions took place, these interactions were brief (<30 sec) and not severe enough to injure an individual. The maximum number of times that a dominant individual physically made contact with a subordinate before the subordinate retreated was three. Thus, extensive drawn-out aggressive confrontations did not take place during these interactions. In addition, to ensure that individuals were not hurt during the experiments, we observed all interactions between the goats from 500 cm away and thus could intervene and separate the goats if required. Moreover, there were full-time animal-handling staff on site that could assist and veterinarians on call in case of emergencies. However, no interventions or assistance was required due to the brief nature of the interactions.

Interactions were considered to be over once one of the goats withdrew [39]. Once dominance was determined, we removed the subordinate individual to reduce its stress and prevent any further unnecessary interactions. Dominant goats were allowed to continue feeding for the remaining portion of the 3 min to further familiarise them with the experimental setup. After the 3 min, we then released both the subordinate and dominant goats into a grass pasture to feed. This helped to ensure that these goats were not tested more than once a day. In all instances, when the dominant goats entered the pasture, they began feeding immediately and did not show aggravated levels of aggression towards the subordinate goat or the other herd members already within the pasture.

To generate the hierarchy, all individuals in the herd were tested against each other over a period of three months resulting in 1993 interactions. Dyads were chosen randomly because all goats ultimately interacted. Due to the size of the herd and the number of interactions, we only tested each dyad once. The majority of interactions resulted in a clear winner. However, if it was not clear which goat was dominant or subordinate, the two goats were retested on another day. If they still did not interact, we recorded them as being non-interacting [5]. Out of the 1993 interactions, only 13 were non-interacting, and these goats were not tested against each other in the patch-joining experiments (see below).

For each dyad, we identified the dominant and subordinate individuals and constructed a win-loss matrix to determine dominance [40]. The dominant individuals were assigned a value of one and subordinates zero. Dominance was calculated for each individual, using David's score (DS) [41,42]. We used DS because it deals logically with repeated interactions between group members and when there are equal numbers of interactions between dyads, it reduces row-sum scoring [43]. To determine the overall dominance status of each goat within the herd, the goats were ranked according to their DS value. We used the linearity index h' [44], and the win-loss matrix to measure the linearity of the hierarchy based on the dominance ranks from the DS values. When individuals in a dyad did not interact, each goat was given a value of 0.5 in our linearity calculations [44]. To avoid possible dominance-subordinate reversals [33,45], we never tested consecutively ranked individuals against each other in the patch choice experiments.

The top six ranked goats within the herd were male (see Table 1) and patch-holders as well as patch-joining goats were a mixture of male and female goats. As a result, we included sex as a factor in the analysis to determine if it influenced patch-joining decisions. Two factors that were not included in the model were the presence/absence of horns and social bonds. Within the herd, 23 of the 45 goats had horns (51%). Because these individuals were found throughout the hierarchy, horns did not ensure that an individual was highly ranked (Table 1). In addition, the presence of horns did not influence how close individuals fed next to each other [46] and are therefore unlikely to influence patch-joining decisions. Moreover, because we were not testing for the factors that determined dominance, but rather were interested in the hierarchy itself, the presence/absence of horns was not included as a factor.

Due to the lack of information regarding the relatedness of the different goats in the herd, we could not consider social bonds in our analysis. However, due to this limited information, we may have potentially run tests between mothers and their offspring from previous years.

Patch choice

To test for the effect of food availability in a patch, we used four artificial patch types, each with their own distinctive visual cues: black rectangle, white rectangle, black circle and white circle. Rectangular patches (L×W×H: 57.0×36.5×23.0 cm) were similar to the plastic trays used by [24]. We constructed smaller circular patches using a 45 cm diameter round plastic basin that was 10 cm deep. We used patches of different sizes to help the goats differentiate between the patches (i.e. the smaller containers represented smaller amounts of food). We used patch size as an indicator of food availability because herbivores can use sward height [47] and bush size [48] as a measure of food availability under natural conditions (e.g. large bushes likely contain more food than small bushes).

To provide diminishing returns within the different artificial patches and thus make them similar to natural patches [49–52], we attached an equally spaced 3×3 grid of 0.2 cm wire to the top of each patch. In addition, we added 5 ℓ of inedible medium, comprising of equal amounts of dried corncobs and a number of 20 cm lengths of 3 cm plastic tubing to the rectangular patches. The number of plastic tubes per tray was the same in each trial. In the smaller circular patches, we added 2.5 ℓ of the combined inedible medium. The difference in the amount of medium used between the different sized patches was relative to their respective volumes.

To generate patches of different food availability we put 400 g of commercial sheep pellets into the black rectangular patch, 200 g into the white rectangular patch, 100 g into the black circular patch and 40 g into the white circular patch. Finally, we created a 'super patch' by pouring 1000 g of food into a pile on a plastic sheet. This patch did not contain inedible medium because the amount of food completely covered the 5 ℓ of inedible medium used in the rectangular patch. Thus, it would not provide diminishing returns. Our choice of food quantity for the different patches was arbitrary. However, large differences in food availability were used between the patches to make it easier for the goats to identify and remember the different patch types.

During the eight months prior to experimentation, we tested whether the goats were able to visually differentiate between the patches (e.g. [53]). However, goats were not tested against the super patch. Therefore, we used the four different patch qualities, and tested each goat with the different patch combinations (black rectangle vs. white rectangle, white rectangle vs. black circle, etc.). The patch choice trials involved sequences of binary choices between two feeding patches made by individual goats across the different patch quantities. To ensure that the goats could differentiate among the patches, we tested all the goats within the herd twice for each of the six different patch combinations, resulting in 12 days of testing. Despite the majority of the goats being able to identify the best quality patches, we only used individuals as patch-joiners if they consistently differentiated between all the different patch quantity combinations. We did this to ensure that in our experiments, the patch-joiners were able to incorporate patch quality into their patch-joining decisions. The remainder of the goats then acted as patch-holders in the different experiments. However, we continued to test some of these goats throughout the experiment to see if any additional goats would be able to learn to differentiate between the patches. This was useful because some of the original goats needed to be replaced for the second experiment (see below).

We conducted these trials in a 700×450 cm pen with the outer walls covered with black plastic to ensure that the goats focused on the feeding patches (Figure 1). In addition, the end of the pen was divided in half using a non-transparent fence (300×150 cm) that ran from the back wall and placed one patch on either side of the divider. To ensure that the test goat had to enter a side in order to feed, we placed the artificial patches 30 cm inside of each half. This then allowed a subordinate patch-holder ~150 cm of space to back away from an approaching dominant patch-joiner (see Figure 2) and, thus, limit aggressive interactions between these goats.

A patch-joining goat was released from the front of the pen from where it could see both patches. Once released, we recorded which side of the pen the goat entered. The criterion we used to determine whether a goat had entered a side of the pen was when it started to feed from the patch on that side. In each situation, the goat went straight to a side and started feeding. No goat investigated both sides before making a decision or switched between sides during the two minutes we allowed it to feed. To account for an area effect, the positions of the different patches were rotated between successive trials.

Experiment 1: Patch food availability and the top ranked patch-holders

For the following experiment we used a subset of the goats ($N = 25$, see results) as patch-joiners from the initial herd ($N = 45$) because these 25 individuals could successfully differentiate between all the different patch types. The remaining 20 goats, as well as some of the test goats, were used as patch-holders. To test the influence of food availability and dominance status of a patch-holder on patch-joining decisions of individuals, we simultaneously tethered a dominant and a subordinate goat behind separate artificial food patches, of the same quality, on either side of the divider (Figure 2). A subordinate patch-holder was any goat that was lower in rank than the patch-joining goat.

We initially only tested the top ranked goats in the herd because higher ranked goats continuously reinforced their status through aggressive interactions [33]. To determine which of the dominant goats were able to prevent patch-joining, we started with the highest ranked goat and moved down the hierarchy. Overall, we tested the top six ranked goats but found only the top four goats influenced the patch-joining decisions of individuals (see Results). Therefore, the top four dominant individuals (all males) were used in our experiments. In total, 25 different subordinate goats (male and female) were used in this experiment. As with the patch choice experiment, the test goat (i.e. the patch-joiner) was released from the front of the pen and we recorded the dominance status of the patch-holder it chose to join. As above, upon release each goat went straight to a patch and started feeding without hesitation. No individual investigated both sides before making a decision or switched between sides. Goats were allowed to feed for two minutes.

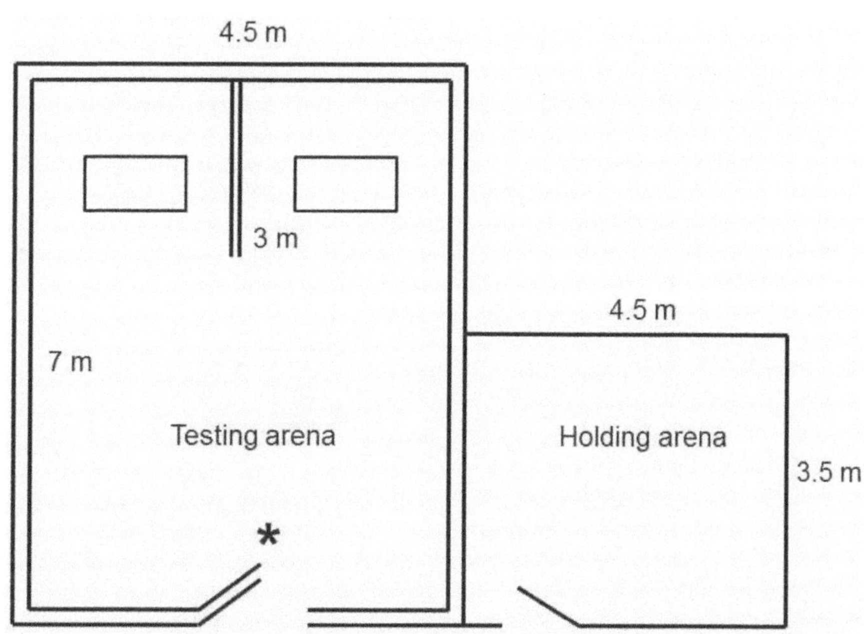

Figure 1. Testing arena used for the different experiments. Double lines denote non-transparent fences. Feeding patches (open rectangle) were placed on each side of the non-transparent divider. The * shows from where we released the patch-joining goat.

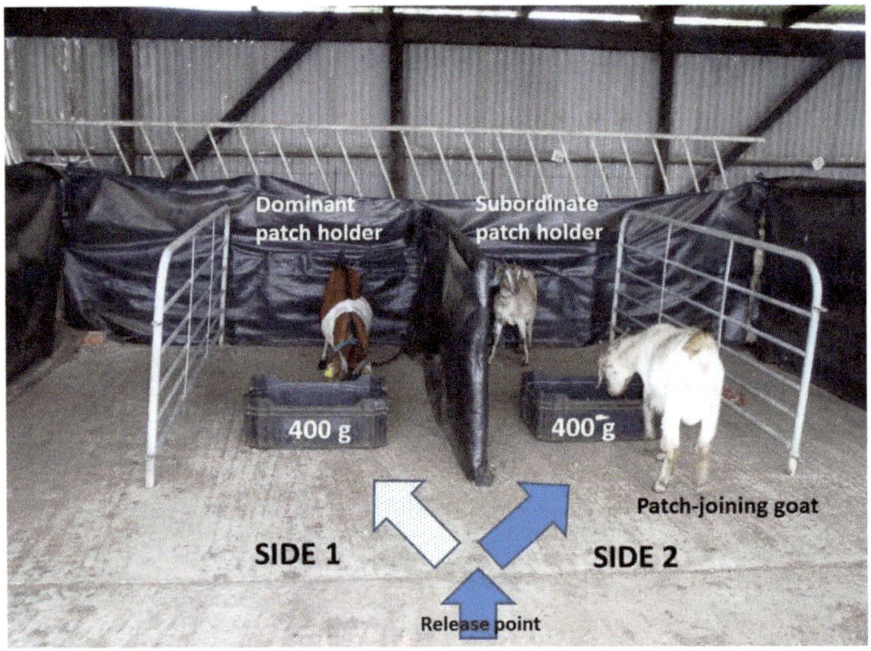

Figure 2. Annotated picture of the experimental design. Each patch-joining goat was released from the release point (~4 m) from each patch. This gave the goats enough time to make a decision on which patch to join based off the dominant status of the patch-holders and the amount of food in the patch. Every goat went straight to a patch and started to feed for two minutes. During this time, no goat switched between trays. If the goat had followed the patterned white arrow, it would have joined the dominant patch-holder. Thus, in the above picture the goat decided to join the subordinate patch-holder.

In contrast to the patch choice trials above, we did not provide the test goats with a choice between different combinations of patch types (e.g. 400 g vs. 100 g). Rather, we placed the same patch type in front of both the dominant and subordinate patch-holders in each trial (e.g. 400 g vs. 400 g). This allowed us to control for food availability and thus determine how dominance status of patch-holders influenced patch-joining decisions at each of the patch food availabilities. Another option could have been to use patches of different quantities in each trial, but this would have made it difficult to tease apart the factors driving patch selection. For example, if a goat joined a patch containing 400 g that was held by a subordinate, as opposed to a dominant individual's 100 g patch, it would be unclear whether it was the amount of food or the subordinate status of the patch-holder that drove the patch-joining decision. Moreover, the use of different combinations of patches would require a greater number of trials than were run in our present experimental design. Thus, our experimental setup allowed us to minimise the number of tests that were required and, therefore, minimise any potential stress to the goats. As above, we only tested each goat once a day.

To limit discomfort of the patch-holders during this experiment, we tethered them to the back of the pen using a rope attached to a dog collar that we placed around their necks. The rope was long enough (200 cm) for the patch-holder to stand next to the patch, but did not allow it to feed. Although patch-holders were not able to feed, they still were, to some degree, able to defend their patch from goats joining them (e.g. rush and horn threats). Patch-holders were only tethered for two minutes per trial.

We tested each patch-joiner against each of the top four ranked goats at the different food quantities resulting in 20 days of testing. For each test, a different subordinate patch-holder was used (i.e. we never paired the same subordinate with the same top ranked goat across the different patch quantities). Furthermore, we switched the position of the dominant and subordinate goats between successive trials to control for a patch-joiner's preference for a specific side of the pen. The position of the dominant and subordinate was randomly chosen for the first trial for each patch-joiner.

In addition, we explored whether the sex of the joining goat influenced patch-joining decisions. However, the small sample size of males ($N = 4$) in the herd of test goats prevented us from making any reliable inferences about the influence of sex. As a result, a more detailed analysis of the influence of sex on patch-joining decisions was conducted in the second experiment.

Experiment 2: Patch quality and dominance status

To further explore the influence of dominance and patch quality, we ran an additional experiment. This allowed us to test the additional hypotheses that 1) dominance gap influences patch choice, 2) absolute rank of the dominant goat influences patch choice, and, finally, 3) the sex of the dominant and joining goat influences patch choice. We were able to test the influence of sex in this experiment due to the inclusion of three additional males (see below).

For this experiment, only 22 (7♂; 15♀) individuals were used that could differentiate between the different patches. These individuals comprised 19 (4♂; 15♀) of the goats used in the first experiment, and three additional males that, with supplementary training, had learned to differentiate between the patch types. We limited the addition of individuals to these three males, because during the initial training they were able to distinguish five out of the six patch quantity combinations. Moreover, the ranks of these males were from throughout the dominance hierarchy. Thus, their inclusion did not increase the number of either high or low ranking individuals. Ultimately, these males were added to increase the sample size because six of the goats from the previous experiment were removed. These removed goats preferred to jump out of the testing arena rather than feed from the patches in this experiment.

Hypothesis 1. To determine the influence of dominance gap and patch quality on patch choice, we used the same experimental design as in the previous experiment (i.e. simultaneously testing a patch-joiner against a dominant and subordinate patch-holder). Each patch-joining goat was tested three times against different dominant individuals that represented a range of different dominance gaps (i.e. close, intermediate and far in rank). The dominance gap was calculated between the patch-joiner and the dominant patch-holder. We considered this gap to be close if the dominant patch-holder was between 3–5 ranks above the patch-joiner, intermediate when these individuals were separated by 6–8 ranks and far if the separation was 9–12 ranks. Subordinate patch-holders comprised any goat that was lower in rank than the patch-joining goat. We subjected all the patch-joiners ($N = 22$) to each combination of three different dominance gaps at each of the five patch qualities, resulting in 15 days of testing. As above, we recorded which goat the patch-joiner chose to join (i.e. dominant or subordinate) at each of the different patch qualities. As with the previous experiment, goats were only tested once a day.

Hypothesis 2. To further explore dominance, we tested how absolute rank of the dominant patch-holder and patch quality influenced patch choice. The absolute rank is the rank of a goat in relation to the rank of all other goats in the herd. Goats were assigned an absolute rank in relation to their position within the dominance hierarchy (i.e. the top ranked goat had an absolute rank of 1 and the lowest ranked goat in the herd had an absolute rank of 45). We used the same experimental design as in the previous experiments. Each patch-joining goat ($N = 22$) was tested against three different dominant individuals that had different absolute ranks. Absolute dominance rank was restricted to a range of 1–36, as opposed to 1–45, because we only used 36 dominant individuals in the experiments. This was because 1) we needed to have goats that were subordinate to the lowest ranking dominant goat, and 2) we needed some goats to act as subordinate patch-holders.

Hypothesis 3. We assessed whether sex influenced patch-joining decisions because both male and female goats were used as dominant patch-holders as well as patch-joiners. Firstly, we separated the data into two categories: 1) when joining goats fed from dominant male patch-holders ($N = 154$), and 2) when joining goats fed from dominant female patch-holders ($N = 176$). We then asked whether the choice (i.e. to feed from a dominant or subordinate patch-holder) of male and female patch-joiners was influenced by the sex of the dominant patch-holder.

Statistical analysis

Dominance calculations were run using DomiCalc [54], while all other statistical analyses were run with PASW (SPSS) v. 19 (IBM Inc.). We used h' to test the linearity of the dominance hierarchy [44]. As in other studies, the dominance hierarchy was considered linear when $h \geq 0.9$ [40,44]. The problem with arranging a win-loss matrix into a specific order is that it may create linear relationships where no such relationships exist. To determine the statistical significance of the linearity of the dominance hierarchy, a sampling process using 10,000 randomisations was used, see [44].

The patch choice experiment involved sequences of binary choices between two feeding patches. Because the same goats were used in all the experiments, individuals were treated as the subjects for repeated measures in generalized estimating equations (GEEs). We used GEEs because of the non-independence of the data [55]. Moreover, a GEE should be used if a subject is tested repeatedly and could potentially remember previous trials. Goats were tested over consecutive days, so not enough time elapsed between the trials for the autocorrelation of their behaviour to be reduced or absent. The model incorporated an exchangeable correlation matrix, binomial error distribution and a logit link function and significance was analysed using score statistics. For graphical representation, we back-transformed data from the logit scale which resulted in asymmetrical confidence intervals (CIs). To assess if goats could visually differentiate between the patches differing in food availability (i.e. food quantity), we analysed the proportion of visits made by goats to the higher food quantity patch offered in the choice experiment. We used means and their 95% CIs to determine whether goat's preference between patches departed from the expected 50% visit to each patch type under random visitation for each food quantity combination.

We used GEEs to determine whether the rank of the four most dominant goats influenced patch-joining decisions. GEEs were used because choices were non-independent because all the 25 joining goats were tested against the same top ranked goats at each patch food quantity. These were choice experiments, so we compared the number of patch-joining goats that avoided the dominant patch-holder at each food quantity. The response variable was choice (i.e. 1 = goat fed from a dominant patch-holder, 0 = goat fed from a subordinate patch-holder) and the factors that were included were rank of the dominant patch-holder (1–4), food availability (40 g, 100 g, 200 g, 400 g, 1000 g) and their interaction. As the amount of food increased, fewer of the top ranked patch-holders were avoided (see Results). By fitting a logarithmic trendline to these data ($y = -1.1611\text{Ln}(x) + 8.3519$), where y is the top ranked goat that is avoided and x is the amount of food, we were able to predict when the quantity of food became more important than patch-holder dominance. Moreover, we used this trendline to generate behavioural titrations to predict how much better a patch needed to be for a patch-joiner to join a higher ranked patch-holder.

To explore the influence of dominance gap, absolute rank and patch quality on patch-joining decisions, we ran two GEEs. We used GEE, because it is a conservative test that takes into account the possible non-independence of the data [55]. For the dominance gap model the response variable was choice (i.e. 1 = goat fed from a dominant patch-holder, 0 = goat fed from a subordinate patch-holder) and the factors were patch food availability (40 g, 100 g, 200 g, 400 g, 1000 g), dominance gap (between 2 and 20 individuals) and their interaction. The absolute rank model had the same response variable but the factors were patch food availability (40 g, 100 g, 200 g, 400 g, 1000 g), absolute rank and their interaction.

To determine whether food availability, the sex of the joining goat and their interaction influenced patch-joining decisions we ran a separate GEE for male and female dominant patch-holders. The sex of the joining goat did not influence how they reacted to a dominant male or female patch-holder (see Results). As a result, we pooled the data and ran a GEE to determine whether the sex of the dominant patch-holder influenced the choice of the joining goat across the different patches. This allowed us to determine whether there was a difference in the proportion of goats joining a dominant or subordinate patch-holder when the dominant goat was male or female.

Results

The interactions from the win-loss matrix confirmed a linear dominance hierarchy ($h' = 0.96$) that was statistically significant (Linearity test using h', based on 10,000 randomisations, $P <$

0.001). In the patch choice experiment, goats showed a clear preference to feed from the patch with the largest food quantity in each food availability combination (Figure 3). The better patch (i.e. more food) received between 71% and 87% of all visits, which differed significantly from the expected 50% under random visitation (Figure 3). From the herd ($N = 45$), 25 individuals (4♂; 21♀) consistently selected the patch with more food across all patch combinations. We used these individuals as patch-joiners in the patch choice experiment against the top ranked goats.

Experiment 1: Patch-joining against top ranked goats

When we tested patch choice of goats against the top four ranked individuals, we found significant effects of rank (GEE: $\chi^2 = 28.118$, $P<0.001$), food availability (GEE: $\chi^2 = 15.419$, $P = 0.004$) and their interaction (GEE: $\chi^2 = 21.019$, $P = 0.05$). Interestingly, individuals avoided high-ranking goats as a function of food availability (Figure 4 a–e). When food availability was low, goats avoided all dominant individuals. However, as the amount of food in the patches increased, the number of dominant individuals that were avoided decreased. This continued up to the highest patch quantity (i.e. 1000 g), where dominance rank did not influence patch-joining decisions, and individuals equally joined top ranked individuals and subordinates at their patches (Figure 5). By fitting a logarithmic trendline to these data ($y = -1.1611\text{Ln}(x) + 8.3519$), where y is the top ranked goat that is avoided and x is food availability, we were able to predict that at ~563 g, the amount of food in a patch becomes more important than patch-holder dominance status with regard to the patch-joining decisions of our goats (Figure 5).

We generated titrations for our study by first determining the patch qualities at which individuals would join the different top ranked goats by using the trendline from our results ($y = -1.1611\text{Ln}(x) + 8.3519$). This resulted in estimates of 563 g for the top ranked patch-holder, 238 g for the second highest, 100 g for the third and 42 g for the fourth. Then by dividing the estimated amount of food in a patch required to join a more dominant patch-holder (e.g. 563 g) by that required to join the next lower ranked patch-holder (e.g. 238 g), we found that a patch needed to be ~2.4 times richer for individuals to consider these two patches to be of equal value.

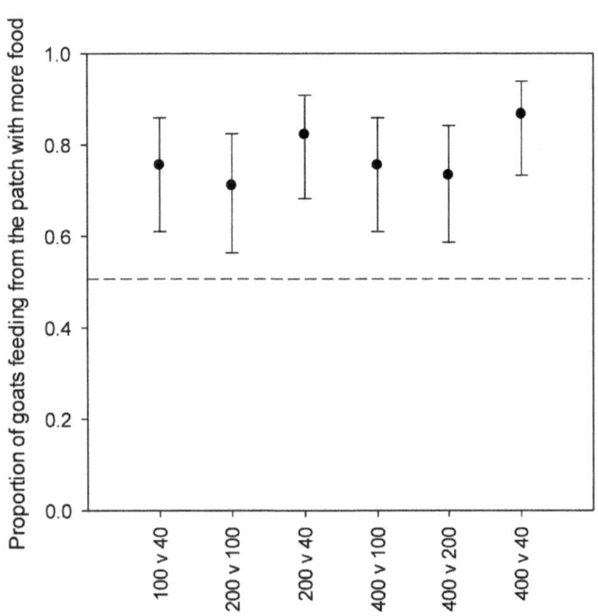

Figure 3. Patch-joining goats prefer to feed from the patch containing more food at each food availability combination. Marginal means (±95% CI) of the proportion of visits to the patch with the highest food availability are plotted. At each food availability combination, there is no overlap with the 0.5 expectation under random visitation, indicating a preference for the patch with the most food.

Experiment 2

Hypothesis 1 and 2: Influence of dominance gap and absolute rank on patch-joining. Contrary to Experiment 1, when we expanded our investigation to include patch-joining decisions across a range of dominance gaps, goats joined patches held by both dominant and subordinate patch-holders equally. As a result, there was no effect of food availability in a patch (GEE: $\chi^2 = 7.366$, $P = 0.118$), dominance gap (GEE: $\chi^2 = 0.759$, $P = 0.385$) or their interaction (GEE: $\chi^2 = 7.390$, $P = 0.117$) on patch-joining decisions. Furthermore, patch-joining decisions were not influenced by the food availability (GEE: $\chi^2 = 7.575$, $P = 0.108$), absolute rank of the dominant patch-holder (GEE: $\chi^2 = 0.389$, $P = 0.533$), or their interaction (GEE: $\chi^2 = 8.657$, $P = 0.070$).

Hypothesis 3: Sex and patch choice. The sex of the joining goat as well as the sex of the dominant goat did not influence patch-joining decisions. When joining goats only joined dominant male patch-holders, there was no significant effect of food availability (GEE: $\chi^2 = 6.214$; $P = 0.184$), sex of the joining goat (GEE: $\chi^2 = 1.888$; $P = 0.184$) or their interaction (GEE: $\chi^2 = 7.734$; $P = 0.102$) on patch-joining decisions (i.e. male and female goats joined both dominant and subordinate patch-holders equally). Moreover, the same result was found when the dominant patch-holder was female (Food availability: GEE: $\chi^2 = 1.742$; $P = 0.783$, Sex of joining goat: GEE: $\chi^2 = 0.739$; $P = 0.390$, Food availability X sex of joining goat: GEE: $\chi^2 = 7.701$; $P = 0.103$). Because male and female patch-joiners did not react differently to male and female dominant patch-holders, we pooled the above data. This then allowed us to determine whether the sex of the dominant patch-holder influenced the proportion of patch-joining goats that joined a dominant or subordinate goat. Once again, food availability (GEE: $\chi^2 = 4.692$; $P = 0.320$), the sex of the dominant goat (GEE: $\chi^2 = 0.520$; $P = 0.471$) and their interaction (GEE: $\chi^2 = 6.074$; $P = 0.194$) did not influence patch-joining decisions.

Discussion

The patch-joining decisions of herbivores can be influenced by both the quantity of food in a patch and the dominance status of patch-holders [12]. However, until now, the combined effect of these two factors has been poorly explored. Generally, individuals prefer to join patches held by subordinate individuals [14,56,57]. In contrast, results from Experiment 1 suggest that goats only joined subordinate patch-holders when the dominant patch-holders were the top ranked goats in the herd. However, as food availability in the patches increased, goats avoided fewer of these top ranked individuals. This suggests that when facing the most dominant goats when the amount of food in a patch is low, patch-holder dominance status dictates patch-joining decisions (i.e. costs of aggression are greater than the potential benefits of food intake). However, as food availability increases, eventually the benefits of feeding in the patch outweigh the costs of aggressive interactions with even the most dominant of patch-holders.

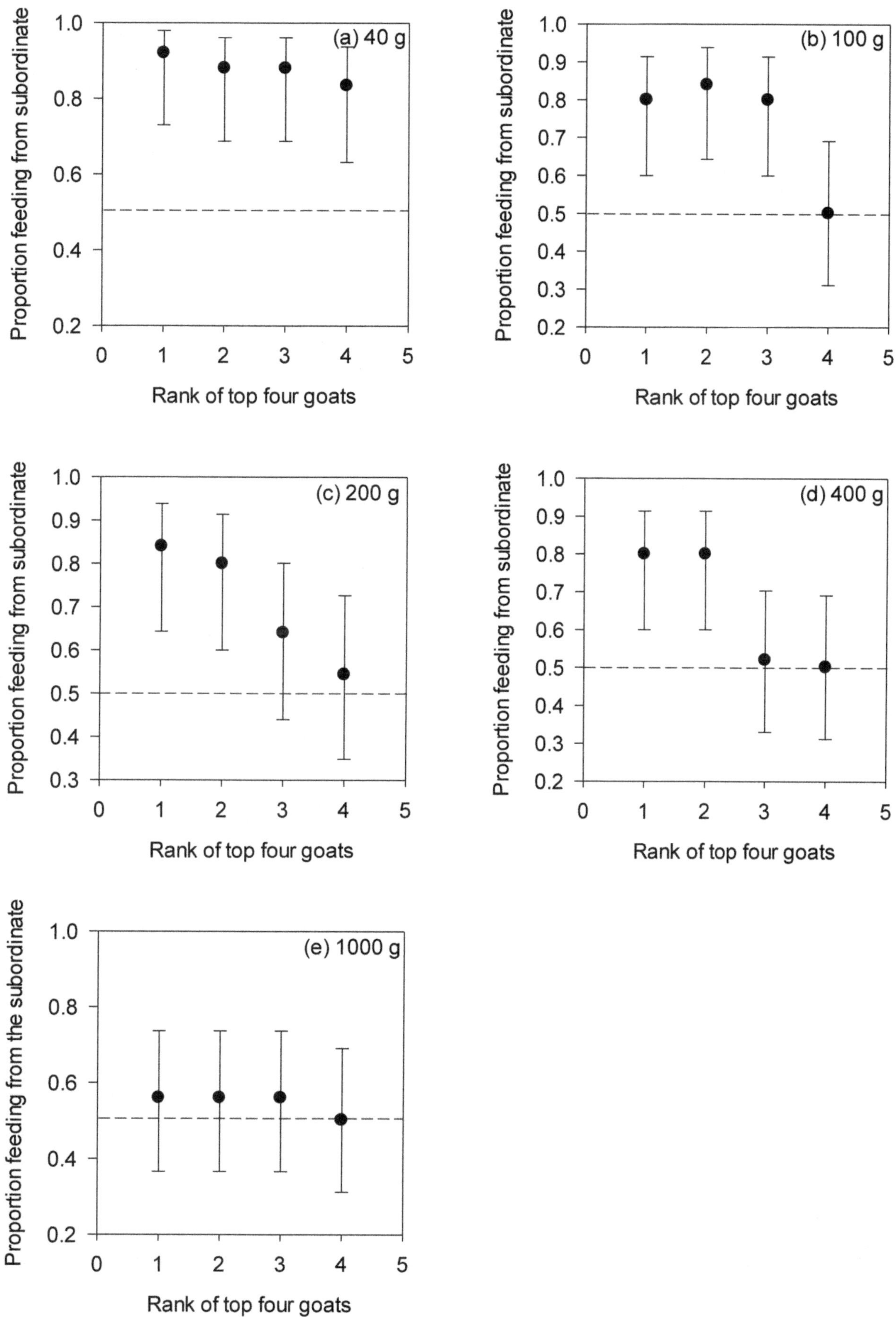
(a) 40 g
(b) 100 g
(c) 200 g
(d) 400 g
(e) 1000 g
Proportion feeding from subordinate
Proportion feeding from the subordinate
Rank of top four goats
0.2
0.3
0.4
0.5
0.6
0.7
0.8
0.9
1.0
0
1
2
3
4
5

Figure 4. Proportion of goats feeding from a subordinate when tested against the top four ranked goats. Back transformed marginal means (±95% CI) of the proportion of goats feeding from a subordinate patch-holder when tested against each of the four most dominant patch-holders. Patches comprise (a) 40 g, (b) 100 g, (c) 200 g, (d) 400 g and (e) 1000 g. No overlap at each rank between the 95% CI and the 0.5 expectation under random visitation indicates an avoidance of that dominant patch-holder. Overlap between the 95% CI and the 0.5 expectation indicates a lack of preference with patch-joining goats joining both dominant and subordinate patch-holders.

When we tested patch-joiners' responses to the patch-holder being subordinate or the four most dominant goats, we found that as predicted in social foraging models [14,56], goats preferred to join patches held by subordinate individuals. However, when the dominant patch-holder was from elsewhere within the hierarchy, individuals seemed to ignore patch-holder rank (i.e. dominance gap and absolute rank) and equally joined both subordinate and dominant patch-holders (Experiment 2). As a result, this greatly increased the number of patches that individuals could join, and thus increased feeding opportunities. Therefore, it seems that for the goats only the top ranked individuals (in our case the top 10%, $N = 4$) were dominant enough to prevent individuals from joining their patches (Experiment 1). However, this ability declined as the amount of food within the patch increased.

One explanation for increasing food availability overriding patch-holder rank could be that at low food availability, the finder's share comprises a large proportion of the available food [20]. This then leaves very little food for the patch-joining goat when it joins the patch. Furthermore, in low food quantity patches, patch-holders are likely to defend the small amount of food within the patch more aggressively [21]. As a result, within these patches, the costs of joining a dominant patch-holder greatly outweigh the potential benefits [11]. However, as the amount food in the patches increases, the amount of food available to a patch-joining goat (i.e. benefit) increases and the intensity of patch defence (i.e. cost) declines.

Behavioural titrations provide a method to quantify dominance with regard to patch use behaviour [25]. In general, individuals should join a patch when the marginal benefits (e.g. the amount of food) exceed the marginal costs (e.g. rank of the patch-holder increases) [25]. We found that the initial amount of food needed to be ~2.4 times richer for individuals to consider these two patches to be of equal value. Thus, at that point, the marginal benefit of joining either patch was the same.

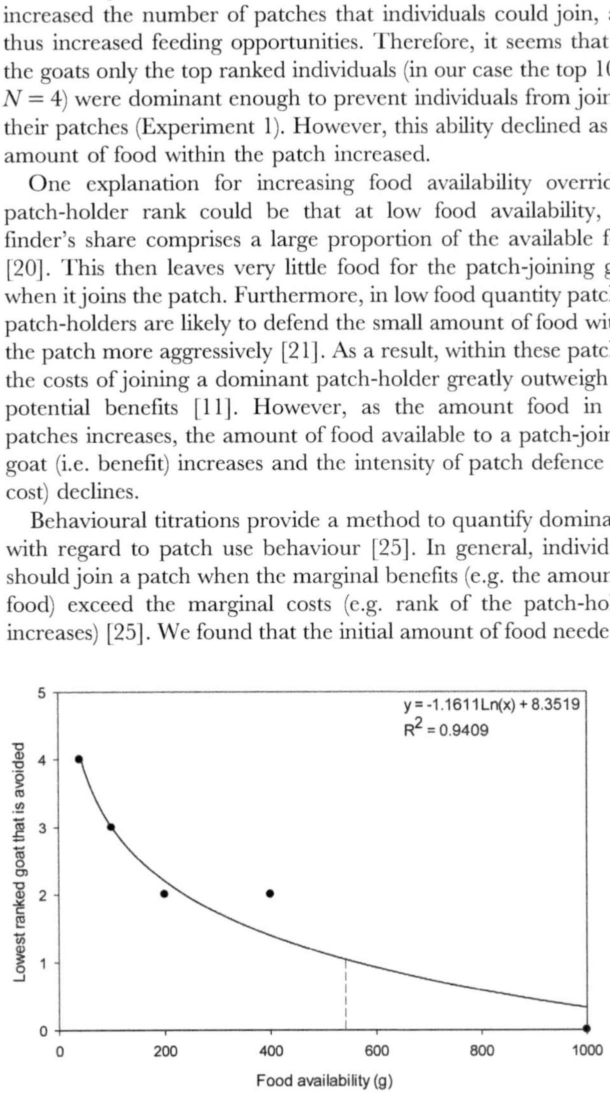

Figure 5. Change in the number of top ranked individuals avoided by patch-joiners as food availability increases. Y-axis represents the lowest ranked goat of the top four ranked individuals that was avoided. Ranking is set so that rank 4 indicates that the top four highest ranked goats were avoided, 3 the top three, 2 the top two and 1 only the top goat was avoided. Dashed line indicates the amount of food (~563 g) where the most dominant individual was not avoided by patch-joining goats, and thus where the quantity of food within a patch overcame dominance as the main factor influencing patch-joining decisions.

In Experiment 2, when we explored the rest of the dominance hierarchy, neither dominance gap nor absolute rank of the dominant patch-holder influenced patch choice. This was surprising because individuals generally tend to avoid feeding from a dominant individual's patch [12,14,34,58]. This could be due to the patch-holders being tethered and thus not posing a threat to the patch joiners. However, we observed patch-holders being aggressive towards patch-joiners (e.g. horn threats and vocalisations) that attempted to approach the feeding tray. Thus, this suggests that the threat of aggression was real.

Another, potentially more likely explanation for patch joiners not avoiding dominant patch-holders could be that the dominance gaps we used were not large enough for dominance to be important. In our experiments, the size of the dominance gaps was limited by herd size as well as by the number of goats that could differentiate between the different patch qualities. Therefore, dominant patch-holders did not necessarily include the top ranked individuals of the herd. Thus, many of the goats would have faced dominant patch-holders that were intermediate in rank overall, and thus not pose enough of a threat to prevent individuals from joining a patch. For example, Barrosso *et al.* [33] found that intermediate ranked goats were less aggressive than top ranked individuals. This may explain why the goats in our study joined both dominant and subordinate patch-holders when the dominant goat was from the middle of the hierarchy. Moreover, it may also explain why the absolute rank of the goats did not influence patch-joining decisions.

Interestingly, our results contradict the findings of a number of studies that suggest that the absolute difference in dominance between individuals is important in patch choice [13,58]. For example, Mexican jays (*Aphelocoma ultramarine*) tend to join patches more frequently if they are held by lower ranking individuals compared to patch-holders that are closer in rank [12]. In addition, other studies have shown that patch-joining individuals only join subordinates at their patch [14,56,57]. However, the majority of this research has focused primarily on birds [57]. Our contrasting results may be due to us focusing on a mammalian herbivore. This suggests that it is difficult to make generalizations about the role of dominance in determining patch choice across taxa, see [59].

For herbivores, food patches (e.g. grass or a shrub) are not very discrete and are thus difficult to defend [40]. When dominant individuals are not able to defend patches, they cannot prevent other individuals from joining the patch [59]. Our results suggest that the only individuals capable of defending a patch were the most dominant individuals. As a result, within a herd there are likely numerous patch-joining opportunities for individuals irrespective of their rank.

The fact that only 25 goats were able to differentiate between all six patch quantity combinations, suggests that they may have had difficulty differentiating between such wide ranges of food quantity combinations. Alternatively, there may have been too many factors (i.e. colour, size and shape) associated with the different food availabilities. If fewer of these variables were used, it is possible that more of the goats could have consistently selected the patch with more food.

Finally, due to the limited information on social bonds, related goats may have been tested against each other. The relatedness between goats can influence foraging decisions, with related individuals feeding closer to each other and thus not interacting aggressively [46]. However, in our dominance relationship experiments, all goats interacted. The 13 non-interacting pairs were never tested against each other in the patch-joining experiments. Therefore, if we tested related goats, then they too displayed dominance behaviours and limited the feeding opportunities of the related subordinates. Thus, if we tested related individuals, it unlikely influenced our results.

Conclusions

An additional outcome of our study is that our results create new opportunities to expand aspects of social foraging theory [7] (e.g. producer-scrounger models). Specifically, as we have focused on a group-living mammalian herbivore, our findings better incorporates this important group into this theory. Moreover, our findings highlight the importance of food availability for these herbivores, and potentially other species, when they make patch-joining decisions. Furthermore, although we did not directly test producer-scrounger interactions, because the goats did not produce, the fact that subordinate individuals joined dominant patch-holders challenges the assumption of these models that individuals can only scrounge from subordinates. By expanding scrounging opportunities for individuals within the models to include scrounging from dominant individuals (except possibly the top 10% of the hierarchy) and incorporating food availability, social foraging theory could be expanded to better explore the foraging decisions made by group-living mammalian herbivores. In addition, our findings on herbivores may help explain why patch-holder dominance sometimes fails to explain patch-joining decisions of other species in producer-scrounger models [28,60], and suggest that certain predictions of the models may not apply to all taxa.

Acknowledgments

We would like to thank Mangesh Depaap, Sam Khumalo and Melissa Schmitt for assistance at Ukulinga Research Farm, Steve Johnson for statistical advice and Han de Vries for providing access to DomiCalc. Burt Kotler, Nicole Hagenah and three anonymous reviewers provided valuable comments on earlier drafts of the manuscript.

Author Contributions

Conceived and designed the experiments: KS GIHK AMS. Performed the experiments: KS. Analyzed the data: KS. Contributed reagents/materials/analysis tools: KS AMS. Contributed to the writing of the manuscript: KS GIHK AMS. Read and approved the manuscript: KS GIHK AMS.

References

1. Clark CW, Mangel M (1986) The evolutionary advantages of group foraging. Theor Popul Biol 3: 45–75.
2. Elgar MA (1989) Predator vigilance and group size in mammals and birds: a critical review of the empirical evidence. Biol Rev 64: 13–33.
3. Pulliam HR (1973) On the advantages of flocking. J Theor Biol 38: 419–422.
4. Hamilton WD (1971) Geometry for the selfish herd. J Theor Biol 31: 295–311.
5. Mooring MS, Hart BL (1992) Animal grouping for protection from parasites: selfish herd and encounter-dilution effects. Behaviour 123: 173–193.
6. Valone TJ (1993) Patch information and estimation: a cost of group foraging. Oikos 68: 258–266.
7. Giraldeau L-A, Caraco T (2000) Social foraging theory. Princeton, NJ: Princeton University Press.
8. Valone TJ, Templeton JJ (2002) Public information for the assessment of quality: a widespread social phenomenon. Philos Trans R Soc Lond B Biol Sci 357: 1549–1557.
9. Shrader AM, Kerley GIH, Kotler BP, Brown JS (2007) Social information, social feeding, and competition in group-living goats (*Capra hircus*). Behav Ecol 18: 103–107.
10. Barnard CJ, Sibley RM (1981) Producers and scroungers: a general model and its application to captive flocks of house sparrows. Anim Behav 29: 543–550.
11. Vickery W, Giraldeau L-A, Templeton JJ, Kramer DL, Chapman CA (1991) Producers, scroungers, and group foraging. Am Nat 137: 847–863.
12. McCormack JE, Jablonski PG, Brown JL (2007) Producer-scrounger roles and joining based on dominance in a free-living group of Mexican Jays (*Aphelocoma ultramarina*). Behaviour 144: 967–982.
13. Barta Z, Giraldeau L-A (1998) The effect of dominance hierarchy on the use of alternative foraging tactics: a phenotype-limited producing-scrounging game. Behav Ecol Sociobiol 42: 217–223.
14. Liker A, Barta Z (2002) The effects of dominance on social foraging tactic use in house sparrows. Behaviour 139: 1061–1076.
15. Boissy A, Dumont B (2002) Interactions between social and feeding motivations on the grazing behaviour of herbivores: sheep more easily split into subgroups with familiar peers. Appl Anim Behav Sci 79: 233–245.
16. Biquand S, Biquand-Guyot V (1992) The influence of peers, lineage and environment on food selection of the criollo goat (*Capra hircus*). Appl Anim Behav Sci 34: 231–245.
17. Scott CB, Provenza FD, Banner RE (1995) Dietary habits and social interactions affect choice of feeding location by sheep. Appl Anim Behav Sci 45: 225–237.
18. Dumont B, Boissy A (2000) Grazing behaviour of sheep in a situation of conflict between feeding and social motivations. Behav Process 49: 131–138.
19. Distell RA, Laca EA, Griggs TC, Demment MW (1995) Patch selection by cattle: maximization of intake rate in horizontally heterogeneous pastures. Appl Anim Behav Sci 45: 11–21.
20. Giraldeau L-A, Dubois F (2008) Social foraging and the study of exploitative behavior. Adv Stud Behav 38: 59–104.
21. Dubois F, Giraldeau L-A (2005) Fighting for resources: the economics of defense and appropriation. Ecology 86: 3–11.
22. MacArthur RH, Pianka ER (1966) On optimal use of a patchy environment. Am Nat 100: 603–609.
23. Lima SL (1985) Maximizing feeding efficiency and minimizing time exposed to predators: a trade-off in the black-capped chickadee. Oecologia 66: 60–67.
24. Shrader AM, Brown JS, Kerley GIH, Kotler BP (2008) Do free-ranging domestic goats show 'landscapes of fear'? Patch use in response to habitat features and predator cues. J Arid Environ 72: 1811–1819.
25. Kotler BP, Blaustein L (1995) Titrating food and safety in a heterogeneous environment: when are the risky and safe patches of equal value? Oikos 74: 251–258.
26. Sih A (1980) Optimal behavior: can foragers balance two conflicting demands? Science 210: 1041–1043.
27. Thouless CR (1990) Feeding competition between grazing red deer hinds. Anim Behav 40: 105–111.
28. Giraldeau L-A, Hogan JA, Clinchy MJ (1990) The payoffs to producing and scrounging: what happens when the patches are divisible? Ethology 85: 132–246.
29. Beauchamp G (2006) Phenotypic correlates of scrounging behavior in zebra finches: role of foraging efficiency and dominance. Ethology 112: 873–878.
30. Beauchamp G (2001) Consistency and flexibility in the scrounging behaviour of zebra finches. Can J Zool 79: 540–544.
31. Ihobe H (1989) How social relationships influence a monkey's choice of feeding sites in the troop of Japanese macaques (*Macaca fuscata fuscata*). Primates 30: 17–25.
32. Wittig RM, Boesch C (2003) Food competition and linear dominance hierarchy among female chimpanzees of the Tai National Park. Int J Primatol 24: 847–867.
33. Barroso FG, Alados CL, Boza J (2000) Social hierarchy in the domestic goat: effect on food habits and production. Appl Anim Behav Sci 69: 35–53.
34. Cté SD (2000) Dominance hierarchies in female mountain goats: stability, aggressiveness and determinants of rank. Behaviour 137: 1541–1566.
35. Briefer EF, Haque S, Baciadonna L, McElligott AG (2014) Goats excel at learning and remembering a highly novel cognitive task. Frontiers in Zoology 11: 20. Available: http://www.frontiersinzoology.com/content/11/1/20. Accessed 23 June 2014.
36. Van DTT, Mui NT, Ledin I (2007) Effect of group size on feed intake, aggressive behaviour and growth rate in goat kids and lambs. Small Ruminant Res 72: 187–196.

37. Baciadonna L, McElligott AG, Briefer EF (2013) Goats favour personal over social information in an experimental foraging task. PeerJ 1: e172. Available: http://dx.doi.org/10.7717/peerj.172. Accessed 23 June 2014.
38. Miranda-de la Lama GC, Mattiello S (2010) The importance of social behaviour for goat welfare in livestock farming. Small Ruminant Res 90: 1–10.
39. Drews C (1993) The concept and definition of dominance in animal behaviour. Behaviour 125: 283–311.
40. Fournier F, Festa-Bianchet M (1995) Social dominance in adult female mountain goats. Anim Behav 49: 1449–1459.
41. Gammell MP, De Vries H, Jennings DJ, Carlin CM, Hayden TJ (2003) David's score: a more appropriate dominance ranking method than Clutton-Brock et al.'s index. Anim Behav 66: 601–605.
42. Bang A, Deshpande S, Sumana A, Gadagkar R (2010) Choosing an appropriate index to construct dominance hierarchies in animal societies: a comparison of three indices. Anim Behav 79: 631–636.
43. David HA (1987) Ranking from unbalanced paired-comparison data. Biometrika 74: 432–436.
44. de Vries H (1995) An improved test of linearity in dominance hierarchies containing unknown or tied relationships. Anim Behav 50: 1375–1389.
45. Reinhardt V, Flood PF (1983) Behavioural assessment in muskox calves. Behaviour 87: 1–21.
46. Aschwanden J, Gygax L, Wechsler B, Keil NM (2008) Social distances of goats at the feeding rack: influences of the quality of social bonds, rank differences, grouping age and presence of horns. Appl Anim Behav Sci 114: 116–131.
47. Illius AW, Clark DA, Hodgson J (1992) Discrimination and patch choice by sheep grazing grass-clover swards. J Anim Ecol 61: 183–194.
48. de Garine-Wichatitsky M, Fritz H, Gordon IJ, Illius AW (2004) Bush selection along foraging pathways by sympatric impala and greater kudu. Oecologia 141: 66–75.
49. Kotler BP, Brown JS (1990) Rates of seed harvest by two species of gerbilline rodents. J Mammal 71: 591–596.
50. Brown JS (1988) Patch use as an indicator of habitat preference, predation risk, and competition. Behav Ecol Sociobiol 22: 37–47.
51. Brown JS (1992) Patch use under predation risk: I. Models and prediction. Ann Zool Fennici 29: 301–309.
52. Kotler BP, Gross JE, Mitchell WA (1994) Applying patch use to assess aspects of foraging behavior in Nubian Ibex. J Wildlife Manage 58: 299–307.
53. Langbein J, Siebert K, Nuernberg G (2008) Concurrent recall of serially learned visual discrimination problems in dwarf goats (*Capra hircus*). Behav Process 79: 156–164.
54. Schmid VS, de Vries H (2013) Finding a dominance order most consistent with a linear hierarchy: an improved algorithm for the I&SI method. Anim Behav 86: 1097–1105.
55. Hardin JW (2005) Generalised Estimating Equations (GEE). In: Everitt B, Howell D, editors. Encyclopedia of Statistics in Behavioural Science. Wiley & Son, Ltd.
56. Lendvai AZ, Liker A, Barta Z (2006) The effects of energy reserves and dominance on the use of social-foraging strategies in the house sparrow. Anim Behav 72: 747–752.
57. King AJ, Isaac NJB, Cowlishaw L (2009) Ecological, social, and reproductive factors shape producer-scrounger dynamics in baboons. Behav Ecol 20: 1039–1049.
58. Stillman RA, Goss-Custard JD, Caldow RWG (1997) Modelling interference from basic foraging behaviour. J Anim Ecol 66: 692–703.
59. Theimer TC (1987) The effect of seed dispersion on the foraging success of dominant and subordinate dark eyed juncos: *Junco hyemalis*. Anim Behav 35: 1883–1890.
60. Giraldeau L-A, Lefebvre L (1987) Scrounging prevents cultural transmission of food-finding behaviour in pigeons. Anim Behav 35: 387–394.

10

No Evidence of Geographical Structure of Salicinoid Chemotypes within *Populus Tremula*

Ken Keefover-Ring[1,2], Maria Ahnlund[3], Ilka Nacif Abreu[1,3], Stefan Jansson[1], Thomas Moritz[3], Benedicte Riber Albrectsen[1,4]*

1 Department of Plant Physiology, Umeå Plant Science Centre, Umeå University, Umeå, Sweden, **2** Department of Entomology, University of Wisconsin-Madison, Madison, Wisconsin, United States of America, **3** Department of Forest Genetics and Plant Physiology, Swedish University of Agricultural Sciences, Umeå Plant Science Centre, Umeå, Sweden, **4** Department of Plant and Environmental Sciences, University of Copenhagen, Frederiksberg, Denmark

Abstract

Salicinoids are well-known defense compounds in salicaceous trees and careful screening at the population level is warranted to fully understand their diversity and function. European aspen, *Populus tremula*, is a foundation species in Eurasia and highly polymorphic in Sweden. We exhaustively surveyed 102 replicated genotypes from the Swedish Aspen collection (SwAsp) for foliar salicinoids using UHPLC-ESI-TOF/MS and identified nine novel compounds, bringing the total to 19 for this species. Salicinoid structure followed a modular architecture of a salicin skeleton with added side groups, alone or in combination. Two main moieties, 2′-cinnamoyl and 2′-acetyl, grouped the SwAsp population into four distinct chemotypes, and the relative allocation of salicinoids was remarkably constant between different environments, implying a highly channeled biosynthesis of these compounds. Slightly more than half of the SwAsp genotypes belonged to the cinnamoyl chemotype. A fraction synthesized the acetyl moiety alone (~7%) or in combination with cinnamoyl (~2%), and close to forty percent lacked either of the two characteristic moieties, and thus resemble *P. tremuloides* in their salicinoid profile. The two most abundant chemotypes were evenly distributed throughout Sweden, unlike geographical patterns reported for SwAsp phenology traits, plant defense genes, and herbivore community associations. Here we present the salicinoid characterization of the SwAsp collection as a resource for future studies of aspen chemical ecology, salicinoid biosynthesis, and genetics.

Editor: Tongming Yin, Nanjing Forestry University, China

Funding: The authors thank the Carl Trygger Foundation for Scientific Research, the Swedish Foundation for Strategic Research, and the Swedish Research Foundation for funding. The staff of the Swedish Metabolomics Center (www.swedishmetabolomicscentre.se) was instrumental in the completion of chemical analyses for this project. The funders had no role in study design, data collection and analysis, decision to publish, or preparation of the manuscript.

* Email: benedicte.albrectsen@umu.se

Introduction

Salicinoids, also known as phenolic glycosides [1] or salicylates [2], are dominant bioactive natural products in the Salicaceae (*Populus* and *Salix*) [3]. Salicinoid diversity ranges from simple structures like salicin to higher order compounds, such as cinnamoylsalicortin (Fig. 1). The relationship between herbivores and salicinoids has mostly involved studies of a few dominant compounds that repel generalist insect defoliators [3–6] and mammalian browsers [7,8], attract specialist herbivores [9,10], or have ambiguous effects on herbivore presence and abundance [11–13]. Salicinoids vary in their toxicity, alone and in combination [14], and it has been suggested that they increase in toxicity with greater molecular complexity [6].

Salicinoids derive from the shikimate-phenylpropanoid pathway [3], which produces other phenolic compounds, such as flavonoids, lignins, tannins, and anthocyanins [15,16]. Several of the genes responsible for specific phenylpropanoid classes have already been described, including those involved in the biosynthesis of monolignols [17,18], flavonoids [15,16], and condensed tannins [19,20]. Salicinoid biosynthesis, however, remains poorly understood [15,16,21,22]. Several evolutionary theories argue that selection should favor a diverse, random, or unpredictable mix of defense compounds in plants [23,24]. For compounds, such as terpenoids, it has been demonstrated that individual profiles and high diversity at the population level results from only a few key genes, that are expressed consistently within a genotype but vary greatly among genotypes [25,26]. Although *Populus* and *Salix* species share many salicinoid compounds, their profiles can also separate salicaceous trees both inter- [3,27] and intraspecifically [28,29]. Furthermore, studies of hybridizing species suggest that inheritance may often be additive with intermediate levels of phenolic compounds in hybrids [30–32], although examples of transgressive inheritance also exist in which hybrids express extreme levels of specific phenolics, including elevated levels of salicinoid compounds [33]. In a recent study, Abreu et al. [29] described an unexpected diversity of salicinoids from five genotypes of European aspen (*Populus tremula* L.). Besides salicortin, tremulacin, salicin, and tremuloidin, the signature salicinoids of the North American sister species *P. tremuloides*

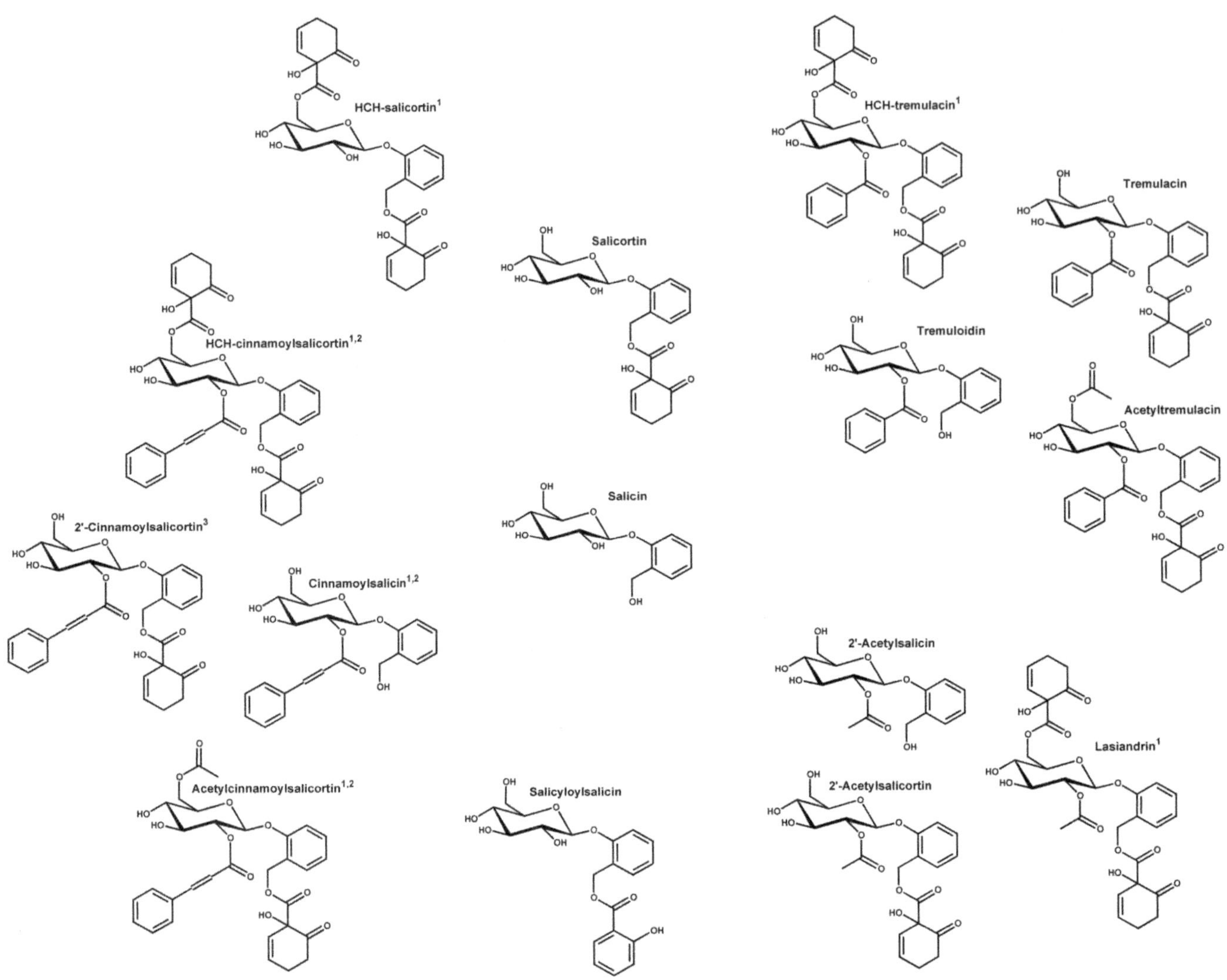

Figure 1. Structural relationship of 19 salicinoids found in the foliage of *Populus tremula* from the SwAsp collection grouped similar to the loading plot (Fig. 2b). 1= new compounds for *P. tremula*, 2= compounds with two isomers present but the conformation of the cinnamoyl group double bond is ambiguous. 3= 2′-(*E*)- and 2′-(*Z*)-cinnamoylsalicortin.

[1,28], *P. tremula* also contained complex salicinoids like 2′-cinnamoylsalicortin, previously only reported for *Salix sericea* [34], and 2′-acetylsalicin and 2′-acetylsalicortin, also found in *S. pentandra* [35]. Abreu et al. [29] showed that concentrations of some of these novel salicinoids in *P. tremula* matched levels in other salicaceous systems that share the same compounds and predicted a minimum of three chemical phenotypes (hereafter chemotypes) of Swedish aspen [29,36].

Historically, screening work of salicinoids used only qualitative methods, such as thin-layer chromatography [27,37,38], as opposed to liquid chromatography-mass spectrometry (LC-MS) techniques presently used for identification and quantification [2,29,33,39]. In addition, species specific profiles have often been based on analyses of a limited number of individuals (e.g., [27,37,38]). At the population level, surveying only a few individuals increases the risk of underestimating natural product diversity and abundance. In salicaceous species, hidden salicinoids could thus potentially be a source of reported ambiguous associations found between phytochemistry and biotic stress agents [11–13].

With aspen's highly diverse salicinoid assemblage [29], the large genetic diversity in Sweden [40], and its complex associated arthropod communities [41], a careful chemical mapping of a larger population of *P. tremula* is warranted. This study presents an exhaustive identification of salicinoid diversity in 319 individual *Populus tremula* trees replicating 102 genotypes from the Swedish Aspen (SwAsp) collection and evaluates the frequency and distribution of the chemotypes across the landscape. To assess the expression in individual trees, salicinoid profiles of the same genotype were also compared across extreme environments.

Materials and Methods

Salicinoid standards

Salicortin, tremuloidin, and HCH-salicortin standards were supplied by Prof. R. L. Lindroth and salicyloylsalicin by Prof. S. D. Mansfield. Salicin and tremulacin standards were purchased from Sigma-Aldrich (St. Louis, MO, USA), and 2′-(*E*)-, 2′-(*Z*)-cinnamoylsalicortin, and tremulacin were isolated from *P. tremula* [39].

Plant material

Two sets of samples were used for this study, consisting of about ten haphazardly chosen, fully expanded leaves from each of 319 individuals of 102 *P. tremula* genotypes from two different environments, belonging to the Swedish Aspen (SwAsp) collection [29,36,41]. SwAsp consists of aspen trees from a range of latitudes (56–67°N) throughout Sweden, collected as roots in 2003 and propagated as genotype replicates. The first greenhouse sample set included leaves from 98 genotypes with 1–2 individuals per genotype that had been growing in standard greenhouse conditions on the Umeå University campus in Umeå, Sweden, since 2005. In 2007, leaves were picked from the greenhouse trees, flash-frozen in liquid nitrogen, and stored at −80°C until chemical analysis (see [29] for details). The second field set, collected in the summer of 2010, consisted of foliage from 41 genotypes from the outdoor SwAsp garden at Sävar (~20 km NW of Umeå), 37 of which were also represented in the greenhouse population. These trees had been planted in 2004 and at the time of sampling had reached an average height of 220 cm (ranging from 65–405 cm) and every genotype was present in the field in replicates of 2–8. Leaves were harvested and immediately placed in separate glassine envelopes, frozen in liquid nitrogen while in the field, brought to a −80°C freezer, lyophilized in a pre-chilled chamber, and then stored at −20°C until chemical analysis. Replicate individuals of all genotypes currently grow in two common gardens (Ekebo 55.9°N, 13.1°E and Sävar 63.4°N, 20.6°E; see also [36] or [41] for location details of original populations). In addition, most of the genotypes are kept in tissue culture at Umeå Plant Science Centre in Umeå, Sweden, and can be propagated upon request.

Chemical analysis

Samples were analyzed using ultra high performance liquid chromatography (UHPLC) with UV and electro-spray ionization time-of-flight mass spectrometry (ESI-TOF/MS) detectors, using the same instrumental conditions as Abreu et al. [29]. Sample preparation differed slightly for the two sample sets, using fresh weight material for the greenhouse and lyophilized foliage for the field samples. Frozen greenhouse leaf material was ground in liquid nitrogen using a mortar and pestle. For the 2010 field samples, we ground the lyophilized leaves on a Retsch (Verder Group, Haan, Germany) ball mill, placing crushed foliage into 20 ml plastic vials along with two 12.5 mm carbide balls and shaking them at 60 Hz for 1 minute. For both sample sets, 10.00±1.00 mg powder was extracted in 1 ml of cold (4°C) methanol: chloroform: water (v:v:v), containing deuterated salicylic acid as an internal standard [29]. After centrifugation in a chilled centrifuge, 200 µl of the extract supernatant from greenhouse samples and 100 µl from the field samples was dried in a speedvac. Just before analysis, the dried greenhouse samples were reconstituted with 20 µl of methanol and 20 µl of a 0.1% v/v aqueous formic acid solution and 25 µl of each for the field samples. Differences in the initial amounts of sample dried and final reconstitution volumes served to equalize the final salicinoid concentrations between fresh-frozen and lyophilized leaf tissue, assuming a ~60 percent difference in water content. Compounds in the reconstituted plant extracts were separated on a C18 UPLC™ column (2.1×100 mm, 1.7 µm) and analyzed by an Acquity photodiode array detector coupled in line with a LCT Premier TOF/MS (all from Waters, Milford, MA, USA) as described in Abreu et al. [29].

MassLynx 4.1 software package (Waters Corp.) allows extraction of single ion chromatograms (±0.15 exact mass unit) from the total ion chromatogram using the QuanLynx module, and was used to search for known and theoretical salicinoids using both the deprotonated ($[M\text{-}H]^-$) and formate adduct ($[M\text{-}H\text{+}FA]^-$) ions (Table S1 in Material S1). QuanLynx software was used to integrate the single ion chromatograms and obtain peak areas, which were normalized with internal standard peak area and individual sample weight. All peak areas represented the formate adduct ion, except for salicyloylsalicin, which included both the deprotonated and formate adduct ion peaks summed, as in Abreu et al. [29].

Salicinoid identification

Salicortin, tremulacin, salicin, tremuloidin, salicyloylsalicin, HCH-salicortin, 2′-(*E*)-, and 2′-(*Z*)-cinnamoylsalicortin were determined using retention times and molecular weight information of purified standards injected on the UHPLC-ESI-TOF/MS [29,39]. The compounds 2′-acetylsalicin, 2′-acetylsalicortin, and acetyltremulacin were identified with LC-MS molecular weights and retention times from previous work [29]. The retention times of lasiandrin (HCH-2′-acetylsalicortin) and HCH-tremulacin were confirmed with molecular mass and with *Salix* species known to have these compounds that were included in the present LC runs [35,42].

UHPLC with high-resolution tandem mass spectrometry (MS/MS) was used to determine the structure of novel salicinoids, using the same chromatographic conditions as Abreu et al. [29]. Peaks were separated on a Hypersil C18 GOLD column (2.1×50 mm, 1.9 µm) using a Thermo Accela LC system coupled to a LTQ Orbitrap MS (all from Thermo Fisher Scientific, Bremen, Germany) and centroid mass spectra of negative ions were collected after collision-induced dissociation (CID) in the LTQ cell at 35 eV. The deprotonated ion was used for all compounds, except for cinnamoylsalicin, which was fragmented using the formate adduct ion.

Statistical analyses

Salicinoid chemotypes were identified using principal component analysis (PCA; SIMCA-P+ v. 12.0 [43]). Due to differences in sampling environment, we used percentages derived from the normalized peak areas of the 19 salicinoids. To further statistically examine the differences between compound profiles, SAS software version 9.1 [44] was used to perform a two-factor multivariate analysis of variance (MANOVA; PROC GLM function with the MANOVA statement) using the same data as above, with sample set (greenhouse or field grown trees) and chemotype (four chemotypes) as factors. Significant MANOVA tests were followed up with ANOVAs for individual compounds.

Amounts of the most common and abundant salicinoids (salicortin and tremulacin) from the field samples were correlated (PROC CORR) separately for trees either low (TL chemotypes) or high in 2′-cinnamoylsalicortin (CN chemotypes), followed by Fisher's Z to determine if the two correlation coefficients differed.

To compute salicinoid clonal repeatability (H^2, broad-sense heritability), we used the R statistical package as described by Robinson et al. [44,45]. Clonal repeatability was calculated the greenhouse and the field samples separately, and for the combined population, when they occurred in replicate of two or more. For all statistical analyses we insured that variables met assumptions of normality, applying transformations where necessary.

Results

New salicinoids from *P. tremula*

In addition to the ten salicinoids described from *P. tremula* by Abreu et al. [29], we found nine new compounds after searching the TOF/MS chromatograms of greenhouse and field foliage samples for 55 known and theoretical ions (Fig. 1, Material S1). The new salicinoids included five molecules similar to existing *P. tremula* compounds but with an additional HCH (hydroxycyclohexen-on-oyl) moiety, including HCH-salicortin, HCH-tremulacin, lasiandrin (HCH-2'-acetylsalicortin), and two isomers tentatively identified as HCH-cinnamoylsalicortin. In addition, we found the newly described 2′-(*Z*)- and 2′-(*E*)-cinnamoylsalicortin isomers [29,39] and two salicinoid isomer pairs tentatively identified as cinnamoylsalicin and acetylcinnamoylsalicortin. Considering the structure of many other salicinoids [3,39], the additional acetyl and HCH groups are most likely attached to C-6′ of glucose and the new isomer pairs probably contain 2′-(*Z*)- and 2′-(*E*)-cinnamoyl groups, respectively. The UV profiles of the new compounds showed typical salicinoid spectra (Table 1, Fig S1a page 4–8 in Material S1). As with these other studies, all new salicinoids with a cinnamoyl moiety had higher second maxima (274–279 nm) compared to those without this functional group (270–274 nm).

The MS/MS spectra of the new *P. tremula* salicinoids produced predictable fragments due to similar disassociation mechanisms (Figure S1b page 9–11 in Material S1). The spectra of all compounds with an additional HCH group [HCH-salicortin, and HCH-tremulacin, lasiandrin (HCH-2′-acetylsalicortin), and HCH-cinnamoylsalicortin] had a single dominant fragment due to the loss of the HCH group (neutral loss of 138) from either the core salicyl group or from the glucose. For all compounds, the *m/z* of this remaining fragment corresponded to their respective deprotonated salicinoid precursors (salicortin, tremulacin, 2′-acetylsalicortin, and 2′-cinnamoylsalicortin, respectively). The next most common fragment for all HCH-containing salicinoids, except for HCH-salicortin, also involved loss of the HCH group. In this case however, the C-O bond of the ether linkage broke proximal to the HCH moiety. For all compounds, these fragments were *m/z* 18 less than breakage of the distal C-O ether bond, indicating dehydration. While this fragmentation mechanism also produced a significant ion for HCH-salicortin (*m/z* 405), its second most abundant MS/MS fragment arose from cleavage and loss of most of the cylcohexeneone ring of either of the compound's HCH groups. The other HCH salicinoids also had relatively abundant fragments due to this breakage pattern. In addition, the spectra of all compounds contained a *m/z* 405 ion; a secondary fragment resulting from a combination of a loss of an HCH group combined with the cleavage of any moiety present at C-2′ of the sugar. A single ion dominated the MS/MS spectra of acetylcinnamoylsalicortin, resulting from the loss of the cinnamoyl group. Subsequent loss of the acetyl moiety led to the appearance of a deprotonated salicortin ion as the second most abundant ion (*m/z* 423). Lastly, fragmentation of cinnamoylsalicin with a formate adduct yielded the deprotonated compound as the primary fragment. Further dissociation of this structure created ions corresponding to a cinnamoyl group and to cinnamoyl-β-D-glucopyranose with subsequent loss of the salicyl moiety.

Chemotypes and compound relationships

All samples contained the signature set of salicinoids in aspen (salicortin, tremulacin, salicin, and tremuloidin) [1,3,27], and in agreement with Abreu et al. [29] a subset of the samples also included novel salicinoids. The salicinoid profiles of SwAsp are presented by genotype and environment in Material S2. PCA analysis separated SwAsp trees into four distinct chemotype groups on the basis of 19 salicinoids with 31.2% of the variation explained by PC1 and 18.4% by PC2 (49.6% cumulative; Fig. 2). These chemotypes were mainly defined by the presence of high amounts of salicinoids with either cinnamoyl moieties (CN, 53% of all genotypes), 2′-acetyl moieties (AC, 8%), both of these moieties (CN-AC, 2%), or very low amounts of either (37%; Fig. 2a). The

Table 1. UV maxima, theoretical and experimental exact masses, molecular formulas, and main high-resolution MS/MS fragments of the new salicinoids from *Populus tremula*.

Compound	λ_{max}	*m/z* [M-H]⁻ Theoreticalmass	LTQ Orbitrap Mass	LTQ Orbitrap Formula	MS/MS fragments (relative intensity)[a]
HCH-salicortin	218, 271	561.1614	561.1611	$C_{27}H_{29}O_{13}$	423 (100), 477 (58.3), 405 (40.0), 299 (13.9), 437 (9.7), 339 (7.3), 293 (6.2), 231 (5.5)
HCH-tremulacin	221, 272	665.1876	665.1859	$C_{34}H_{33}O_{14}$	527 (100), 509 (33.6), 543 (28.6), 405 (14.4), 581 (8.4), 403 (2.8), 389 (1.5), 553 (1.0)
Cinnamoylsalicin	218, 279[b], 219, 279[c]	415.1398	415.1385	$C_{22}H_{23}O_8$	415 (100), 147 (50.9), 309 (39.9), 414 (10.2), 285 (6.6), 509 (6.5), 252 (6.4), 515 (6.4)
Acetylcinnamoylsalicortin	220, 274[b], 218, 278[c]	595.1821	595.1793	$C_{31}H_{31}O_{12}$	447 (100), 423 (15.9), 567 (5.2), 213 (4.3), 285 (4.1), 267 (4.0), 471 (3.5), 341 (2.9)
HCH-cinnamoylsalicortin	222, 277[b] 219, 277[c]	691.2032	691.2023	$C_{36}H_{35}O_{14}$	553 (100), 535 (28.7), 543 (19.5), 405 (10.2), 484 (6.6), 509 (6.5), 252 (6.4), 515 (6.4)
Lasiandrin	219, 270	603.1719	603.1722	$C_{29}H_{31}O_{14}$	465 (100), 447 (22.4), 519 (8.5), 561 (5.8), 543 (4.5), 405 (3.6), 341 (3.0), 423 (2.1)

See Fig. S1a in Material S1 for MS/MS spectra.
a = MS/MS performed on deprotonated isomer 2 of cinnamoyl compounds, except for cinnamoylsalicin, which used the formate adduct [M+FA-H]⁻ (experimental mass *m/z* 461.1433) of isomer 2; b = isomer 1, c = isomer 2 by UHPLC retention times.

salicinoid profile of the last "*tremuloides*-like" (TL) chemotype resembles that of *P. tremuloides* [46]. Most of the AC chemotype trees had very low levels of cinnamoyl-containing salicinoids, however, five individuals from two genotypes had relatively high levels of compounds with both of these moieties, resulting in the separate chemotype designated CN-AC (Fig. 2a). While trees from particular chemotypes generally grouped together in greenhouse and field grown trees, the PCA showed some divergence between the two environments. This was especially evident for individuals from the CN and CN-AC chemotypes and less so for AC and TL trees (Fig. 2a).

The loading scatter plot of the 19 salicinoids from *P. tremula* showed that most compounds grouped according to specific chemical moieties (Fig. 2b). The first component (PC1) separated compounds with a cinnamoyl side chain [2′-(*E*)- and 2′-(*Z*)-cinnamoylsalicortin and the cinnamoylsalicin, acetylcinnamoylsalicortin, and HCH-cinnamoylsalicortin isomers] from compounds with a benzoyl group (tremulacin, tremuloidin, acetyltremulacin, and HCH-tremulacin). The second component (PC2) separated compounds with an 2′-acetyl, (2′-acetylsalicin, 2′-acetylsalicortin, and lasiandrin) from those with an additional HCH group (HCH-salicortin, HCH-tremulacin, and HCH-cinnamoylsalicortin).

Chemotype and environment differences, main salicinoid correlations, and tree origin

The two-factor MANOVA test showed that salicinoid profile differed for both environment (*Wilks'* $\lambda = 0.32$, $F_{19,\ 292} = 32.3$, $P < 0.001$), chemotype (*Wilks'* $\lambda = 0.001$, $F_{57,\ 871} = 154.5$, $P<0.001$), and their interaction (*Wilks'* $\lambda = 0.09$, $F_{57,\ 871} = 18.5$, $P<0.001$; Fig. 3). Individual ANOVA results showed that all salicinoids differed with chemotype, and all compounds, except salicin, 2′-(*E*)-cinnamoylsalicortin, isomer 2 of HCH-cinnamoylsalicortin, and 2′-acetylsalicin, also differed between environment (Fig. 3; Table 2). In general, greenhouse grown trees contained more salicortin, tremuloidin, acetyltremulacin, and 2′-(*Z*)-cinnamoylsalicortin, and field trees contained more tremulacin. For isomer 2 of HCH-cinnamoylsalicortin the difference between environments was only apparent due to the interaction of environment with chemotype. Many other salicinoids interacted between these two factors, usually due to differing ratios of particular salicinoids in samples of the different chemotypes, as in the case of both salicortin and tremulacin. In some instances, the compound patterns were reversed between chemotypes. For instance, AC and CN-AC chemotypes had much more 2'-acetylsalicortin in field grown trees.

Correlations between the amounts of tremulacin and salicortin in field grown individuals with either low (TL chemotypes; $r = 0.87$, $P<0.001$, $N = 49$) or high (CN chemotypes; $r = 0.93$, $P< 0.001$, $N = 83$) levels of 2′-cinnamoylsalicortin showed strong positive relationships in both cases (Fig. 4). The patterns were, however, notably different with CN trees containing considerably less tremulacin with increasing amounts of salicortin (Fisher's $Z = 1.71$, $P<0.05$). The relative amounts of these two salicinoids varied with chemotype from an approximately even relationship of 1 tremulacin: 1 salicortin (in mg g^{-1}) in TL chemotypes to a 1:4 relationship in the CN chemotype.

Trees belonging to the CN and TL chemotypes occurred at sites throughout Sweden (Fig. 5) and their distribution did not follow a simple geographic or clinal pattern. The genotypes containing 2′-acetyl compounds (AC and CN-AC) originated from central and southern Sweden.

Clonal repeatability of salicinoids

Overall, clonal repeatabilities (H^2; broad-sense heritability) for salicinoids in this study were high with the greenhouse population showing slightly higher values compared to field grown trees (Table 3). The most notable examples that contributed to this trend were salicortin, tremuloidin, and the isomer pairs of cinnamoylsalicin and acetylcinnamoylsalicortin. Both acetyltremulacin and lasiandrin had substantially lower clonal repeatabilities in the greenhouse environment than in the field. In addition, all salicinoids with higher clonal repeatabilities in the field contained two HCH moieties.

Discussion

Salicinoid survey of *P. tremula*

On the basis of literature studies (i.e., [3]), previous salicinoid analyses [29,39], and theoretical structures, we identified a total of 19 potentially bioactive salicinoid compounds from *P. tremula*, adding nine new structures to those already described by Abreu et al. [29].

Many of these salicinoids are dominant or characteristic for other species in the Salicaceae [3]. The acetylated compounds 2′-acetylsalicin, 2′-acetylsalicortin, lasiandrin, and acetyltremulacin co-occur in *Salix pentandra* and *S. lasiandra* [35,47,48]. HCH-salicortin (salicortin derivative or disalicortin [49]) and HCH-tremulacin both occur in *S. myrsinifolia* [2,49]. HCH-salicortin was also isolated from *P. fremontii* and its F1 hybrids with *P. angustifolia* [30], and studies with *S. sericea* found 2′-(*E*)-cinnamoylsalicortin [31].

Frequencies and distribution of key compounds in SwAsp

Salicinoid profiles readily divided SwAsp genotypes into four chemotypes, based upon the presence or absence of specific moieties. The 2′-cinnamoyl moiety defined the most abundant chemotype with 53 percent representation in SwAsp. In general, the capability to synthesize cinnamoyl salicinoids appears to be mostly either present or absent in a genotype and thus accounts for the strongest division of the population. The ability to add acetyl moieties appeared less channeled compared to the cinnamoyl addition, but curiously acetylcinnamoylsalicortin had somewhat higher clonal repeatabilities in the less stable field environment (Table 3). Across SwAsp chemotypes we found salicortin, tremulacin, 2′-acetylsalicortin, and the 2′-cinnamoylsalicortins as the dominant salicinoids and when present they usually occurred in relatively high amounts.

The distribution of SwAsp chemotypes throughout Sweden did not resemble the clinal structure reported for phenological traits [36], or the north-south clustering that characterizes inducible defense genes [50]. The two most abundant chemotype groups (CN and TL) were evenly distributed among collection sites, whereas the AC and CN-AC chemotypes mainly originated from the central and southern part of the country. After the last ice age, many plants and animals invaded Sweden from both the north east and the south west and later united in central Sweden (along Limes Norrlandicus) where they either formed hybrid zones or distinct subpopulations [51]. Although the postglacial invasion of aspen in Sweden had weak effects on the genetic differentiation of neutral markers [52], it could have had local effects on adaptive traits [40]. Introgression of defense associated genes naturally occurs in hybrid zones [53,54], and hybrid zones and admixture populations may also promote novel genotypes [30,32,33,53,54]. Evidence of salicinoid inheritance from hybrid zones includes additive inheritance of HCH-salicortin by F_1 hybrids of *P.*

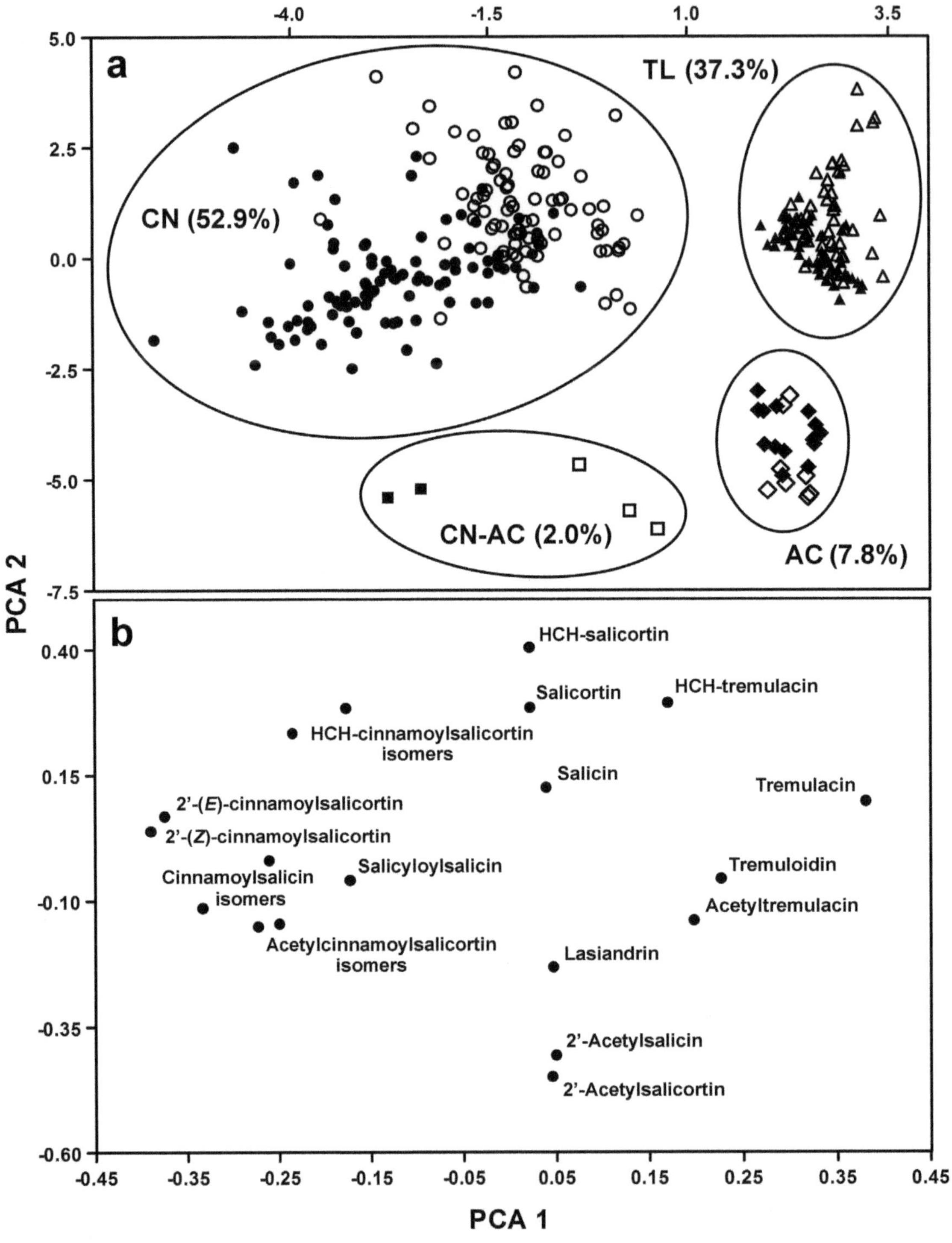

Figure 2. PCA results for chemotypes and compound relationship of 19 salicinoids found in the foliage of *Populus tremula* from the SwAsp collection. a. Score scatter plot of the first two principle components for 319 individuals from 102 genotypes. Solid symbols = greenhouse grown trees, open symbols = field grown trees; circles = CN (2′-cinnamoyl), diamonds = AC (2′-acetyl), squares = CN-AC (2′-cinnamoyl/2′-acetyl), and triangles = TL (*tremuloides*-like) chemotypes. Percentages of clones of the different salicinoid chemotypes in parentheses. **b.** Loading scatter plot of the first two principle components for 19 salicinoids.

fremontii and *P. angustifolia* from the *P. fremontii* parent [30]. Similarly, additive inheritance was found in a *P. tremula-alba* hybrid zone for HCH-salicortin and 2′-acetylsalicortin, whereas HCH-tremulacin was trangressive at higher concentrations [33]. The geographical distribution of the AC and CN-AC SwAsp chemotypes in central and southern Sweden could consequently mirror fitness properties, recent evolution, or introgression.

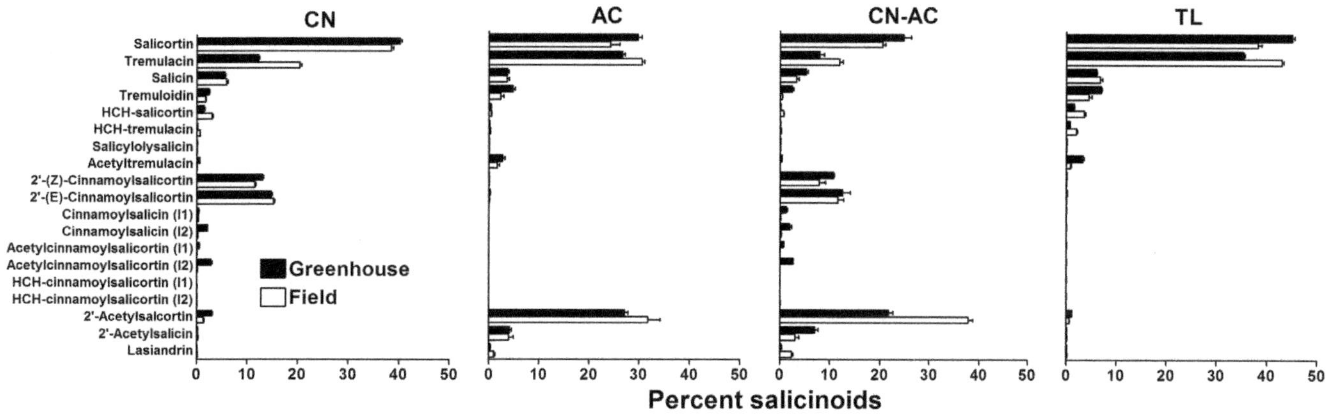

Figure 3. Percentages (± SE) of 19 salicinoids from greenhouse (solid bars) and field (open bars) grown *Populus tremula* trees from the SwAsp collection of the four identified chemotypes, CN (2′-cinnamoyl), AC (2′-acetyl), CN-AC (2′-cinnamoyl/2′-acetyl), and TL (*tremuloides*-like)]. I1 and I2 = isomer 1 and 2, respectively, based upon UHPLC retention times.

Elusive salicinoid biosynthesis

Tsai et al. [16] proposed salicin as the substrate for salicinoid biosynthesis, but studies using labeled compounds could not confirm this suggestion [21] and the biosynthetic route of salicinoids remains poorly understood [15,16,21,22]. Confirming the work by Abreu et al. [29], we found strong correlations between amounts of 2′-acetylsalicin and 2′-acetylsalicortin, as well as associations between cinnamoyl-containing salicinoids. The dynamics of salicinoid pools thus appear to depend on the presence or absence of characteristic chemotype moieties and the relative abundance of common salicinoids vary between chemotypes. For example, the concentration of tremulacin is generally lower in CN chemotype individuals compared to the TL chemotype (Fig. 4). This may reflect competition for salicortin as a substrate for addition of either a cinnamoyl or benzoyl group for synthesis of 2′-cinnamoylsalicortin or tremulacin, respectively.

Table 2. *F* and *P* values from the two-factor ANOVA comparing percentages of individual leaf extract salicinoids from *Populus tremula* trees of four chemotypes (CN, AC, CN-AC, and TL), grown in two environments (greenhouse and field), and their interaction (E*Ct).

Compound	Environment		Chemotype		E*Ct	
	df = 1		*df* = 3		*df* = 3	
	F	*P*	*F*	*P*	*F*	*P*
Salicortin	17.0	<0.001	90.1	<0.001	8.2	<0.001
Tremulacin	60.5	<0.001	436.8	<0.001	3.8	0.011
Salicin	0.4	0.508	10.0	<0.001	0.8	0.497
Tremuloidin	22.6	<0.001	71.8	<0.001	3.8	0.010
HCH-salicortin	10.9	0.001	22.6	<0.001	2.3	0.081
HCH-tremulacin	25.8	<0.001	67.2	<0.001	1.1	0.335
Salicyloylsalicin	26.9	<0.001	14.8	<0.001	2.4	0.066
Acetyltremulacin	22.0	<0.001	35.2	<0.001	1.5	0.225
2′-(*Z*)-Cinnamoylsalicortin	4.9	0.028	28.5	<0.001	4.5	0.004
2′-(*E*)-Cinnamoylsalicortin	0.2	0.645	456.4	<0.001	0.9	0.458
Cinnamoylsalicortin (*I*1)	17.4	<0.001	282.1	<0.001	6.0	0.001
Cinnamoylsalicortin (*I*2)	178.2	<0.001	317.2	<0.001	288.7	<0.001
Acetylcinnamoylsalicortin (*I*1)	14.9	0.000	21.3	<0.001	16.7	<0.001
Acetylcinnamoylsalicortin (*I*2)	17.4	<0.001	38.7	<0.001	30.5	<0.001
HCH-cinnamoylsalicortin (*I*1)	7.6	0.006	607.4	<0.001	21.9	<0.001
HCH-cinnamoylsalicortin (*I*2)	3.2	0.073	855.5	<0.001	2.7	0.048
2′-Acetylsalcortin	17.0	<0.001	90.1	<0.001	8.2	<0.001
2′-Acetylsalicin	2.4	0.126	123.5	<0.001	1.5	0.223
Lasiandrin	18.8	<0.001	27.2	<0.001	2.7	0.048

*I*1 and *I*2 indicate isomers 1 and 2, respectively, designated by UHPLC retention times. See Figure 3 for corresponding data.

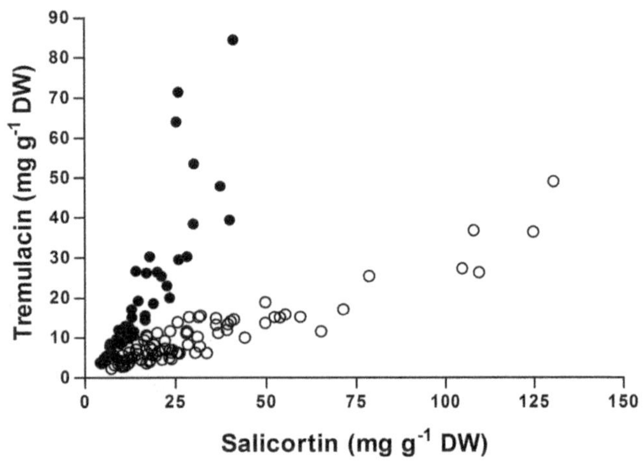

Figure 4. Relationship between the two most abundant salicinoids, tremulacin and salicortin, in field grown *Populus tremula* trees for chemotypes low (TL; solid circles) or high (CN; open circles) in 2′-cinnamoylsalicortin.

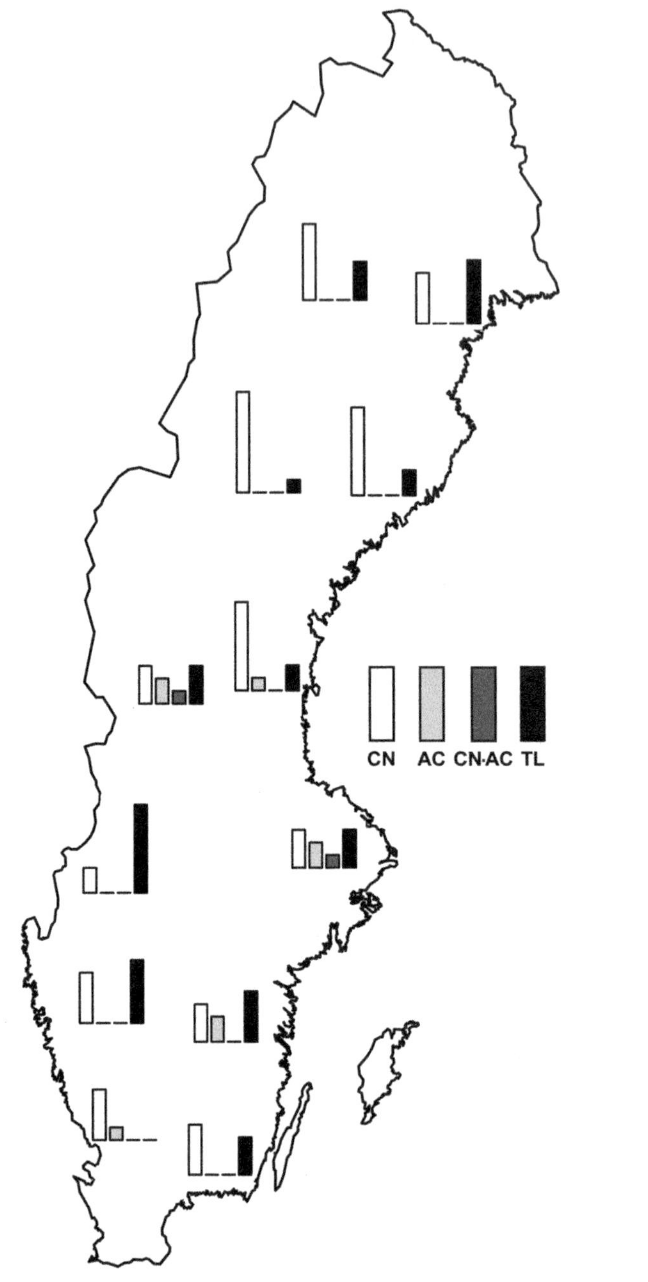

Figure 5. Salicinoid chemotypes of *Populus tremula* trees (the SwAsp collection) collected from ten locations throughout Sweden. White = CN (2′-cinnamoyl), light gray = AC (2′-acetyl), dark gray = CN-AC (2′-cinnamoyl/2′-acetyl), and black = TL (*tremuloides*-like). Bar height corresponds to the number of individuals of each chemotype.

The composite structure of the salicinoids could further suggest that relatively few enzymes are involved in their biosynthesis. Keeling and Bohlmann [26] and Degenhardt et al. [25] have demonstrated how a few key genes that are differently but consistently expressed result in unique terpenoid profiles in individual conifers, creating high terpenoid diversity at the population level. Although at a lower diversity, the salicinoids of SwAsp may be biosynthesized according to a similar strategy. Association studies relate specific traits to genetic patterns [55], and could potentially be a promising way to get insight into the salicinoid biosynthesis. Thus, rather than relating single compounds to gene sequences, association studies may benefit from grouping the compounds on the basis of presence and absence of specific salicinoid moieties (see also [54]).

Salicinoid profile stability and implications for chemical ecology

Salicinoid composition of individual trees across environments with very different growth histories showed overall high heritabilities, confirming that both salicinoid quality (composition) and quantity (abundance) is likely to be highly channeled in aspen, and thus relatively stable in different environments [29,56]. Interestingly, we found that the most represented group in SwAsp, the CN chemotype, also showed the largest plasticity of salicinoids in response to environmental differences. This suggests an elevated level of plasticity in cinnamoyl-containing salicinoid expression and a potential fitness advantage. Salicinoid toxicity to herbivores has been attributed to the HCH moiety, even at low concentrations [3,14,57]. Most of the newly described salicinoids in this study contain one or two HCH groups, and the high heritability of the HCH containing compounds in field samples supports that they may be emphasized in more challenging environments (Table 3). Lindroth et al. [14] found that tremulacin greatly reduced herbivore survival and performance and suggested that the benzoyl group synergizes the toxic effect of the HCH group. Similarly, the cinnamoyl and acetyl groups may also synergize the effects of the HCH moiety on relevant molecules (2′-cinnamoylsalicortin and 2′-acetylsalicortin).

Given the apparent stability of salicinoid profiles in *P. tremula* and the clearly defined chemotypes, the SwAsp collection represents an ideal system for the study of chemical-ecological interactions. With almost the entire collection in tissue culture, and with the present robust salicinoid profiling (Material S2), we can propagate our chemotypes to specifically test properties of resistance and tolerance to various kinds of associated herbivores and fungi [13,45]. In conclusion, we observed a striking division of aspen into four chemotypes, some of which co-occur across latitudes. These chemotypes differ in the dominant moieties and future studies are needed to explore their relative bioactivity, especially including the new dominant compounds (2′-cinnamoylsalicortins and 2′-acetylsalicortin) that are poorly studied. We

Table 3. Clonal repeatabilities (H^2) of 19 salicinoids from 85 clones of *Populus tremula* from greenhouse (GH), field, and combined (All) populations. *I1* and *I2* indicate isomers 1 and 2, designated by UHPLC retention times.

Compound	H^2		
	GH	Field	All
Salicortin	0.87	0.70	0.67
Tremulacin	0.98	0.98	0.92
Salicin	0.46	0.45	0.29
Tremuloidin	0.81	0.59	0.60
HCH-salicortin	0.86	0.91	0.76
HCH-tremulacin	0.85	0.96	0.80
Salicyloylsalicin	0.90	0.84	0.69
Acetyltremulacin	0.95	0.90	0.83
2′-(*Z*)-Cinnamoylsalicortin	0.98	0.96	0.96
2′-(*E*)-Cinnamoylsalicortin	0.99	0.98	0.98
Cinnamoylsalicin (*I1*)	0.79	0.41	0.54
Cinnamoylsalicin (*I2*)	0.94	0.74	0.57
Acetylcinnamoylsalicortin (*I1*)	0.88	0.77	0.51
Acetylcinnamoylsalicortin (*I2*)	0.92	0.75	0.54
HCH-cinnamoylsalicortin (*I1*)	0.88	0.87	0.70
HCH-cinnamoylsalicortin (*I2*)	0.86	0.91	0.83
2′-Acetylsalicortin	0.99	0.99	0.98
2′-Acetylsalicin	0.96	0.95	0.92
Lasiandrin	0.76	0.89	0.73

further suggest that ecological studies, both locally and across latitudes, must take *P. tremula's* salicinoid diversity into account.

Supporting Information

Material S1 Salicinoid identification: Exact masses of 55 salicinoid compounds (**Table S1**), Literature references (**List S1**), UV spectra of nine new salicinoids found in the *P. tremula* foliage (**Fig. S1*a***), and high-resolution MS/MS (**Fig. S1*b***).

Material S2 Average percentages of 19 salicinoids from the foliage of different *Populus tremula* clones (Clone), grown in two different environments (Evir: GH = greenhouse, Sävar = field) and mg g^{-1} for field trees. Chemo = chemotype: CN = 2′-cinnamoyl, AC = 2′- acetyl, CN-AC = 2′-cinnamoyl/2′-acetyl, and TL = *tremuloides*-like. Salicinoids: 1 = salicortin, 2 = tremulacin, 3 = salicin, 4 = tremuloidin, 5 = HCH-salicortin, 6 = HCH-tremulacin, 7 = salicyloylsalicin, 8 = 6′-acetyl-tremulacin, 9 = 2′-(*Z*)-cinnamoylsalicortin, 10 = 2′-(*E*)- cinnamoylsalicortin, 11 = cinnamoylsalicin *I1*, 12 = cinnamoylsalicin I2, 13 = acetylcinnamoylsalicortin *I1*, 14 = acetylcinnamoylsalicortin I2, 15 = HCH-cinnamoylsalicortin *I1*, 16 = HCH-cinnamoylsalicortin I2, 17 = 2′-acetylsalicortin, 18 = 2′-acetylsalicin, 19 = lasiandrin (HCH-2'-acetylsalicortin). *I1* and I2 = isomers 1 and 2, respectively.

Acknowledgments

We thank the Carl Trygger Foundation for Scientific Research, the Swedish Foundation for Strategic Research, and the Swedish Research Foundation for funding. The staff of the Swedish Metabolomics Center (www.swedishmetabolomicscentre.se) was instrumental in the completion of chemical analyses for this project.

Author Contributions

Conceived and designed the experiments: BA IA KKR. Performed the experiments: IA KKR MA BA. Analyzed the data: KKR MA IA. Contributed reagents/materials/analysis tools: BA TM SJ. Wrote the paper: BA KKR SJ MA IA TM.

References

1. Lindroth RL, Hsia MTS, Scriber JM (1987) Characterization of phenolic glycosides from quaking aspen. Biochem Syst Ecol 15: 677–680.
2. Haikio E, Makkonen M, Julkunen-Tiitto R, Sitte J, Freiwald V, et al. (2009) Performance and secondary chemistry of two hybrid aspen (*Populus tremula* L. × *Populus tremuloides* Michx.) clones in long-term elevated ozone exposure. J Chem Ecol 35: 664–678.
3. Boeckler GA, Gershenzon J, Unsicker SB (2011) Phenolic glycosides of the Salicaceae and their role as anti-herbivore defenses. Phytochemistry 72: 1497–1509.
4. Fritz RS, Hochwender CG, Lewkiewicz DA, Bothwell S, Orians CM (2001) Seedling herbivory by slugs in a willow hybrid system: developmental changes in damage, chemical defense, and plant performance. Oecologia 129: 87–97.
5. Albrectsen BR, Gutierrez L, Fritz RS, Fritz RD, Orians CM (2007) Does the differential seedling mortality caused by slugs alter the foliar traits and subsequent susceptibility of hybrid willows to a generalist herbivore? Ecol Entomol 32: 211–220.
6. Constabel CP, Lindroth RL (2010) The impact of genomics on advances in herbivore defene and secondary metabolim in *Populus*. In: S J, R B, A G, editors. Genetics and Genomics of *Populus*: Springer Verlag. pp.279–305.
7. Bailey JK, Schweitzer JA, Rehill BJ, Irschick DJ, Whitham TG, et al. (2007) Rapid shifts in the chemical composition of aspen forests: An introduced herbivore as an agent of natural selection. Biol Invasions 9: 715–722.

8. Wooley SC, Walker S, Vernon J, Lindroth RL (2008) Aspen decline, aspen chemistry, and elk herbivory: Are they linked? Society for Range Management 30: 17–21.
9. Lindroth RL, Scriber JM, Hsia MTS (1986) Differential responses of tiger swallowtail subspecies to secondary metabolites from tulip tree and quaking aspen. Oecologia 70: 13–19.
10. Rank NE (1994) Host-plant effects on larval survival of a salicin-using leaf beetle *Chrysomela aeneicollis* Schaeffer (Coleoptera, Chrysomelidae). Oecologia 97: 342–353.
11. Orians CM, Huang CH, Wild A, Dorfman KA, Zee P, et al. (1997) Willow hybridization differentially affects preference and performance of herbivorous beetles. Entomol Exp Appl 83: 285–294.
12. Hjältén J, Niemi L, Wennstrom A, Ericson L, Roininen H, et al. (2007) Variable responses of natural enemies to *Salix triandra* phenotypes with different secondary chemistry. Oikos 116: 751–758.
13. Albrectsen BR, Witzell J, Robinson KM, Wulff S, Luquez VMC, et al. (2010) Large scale geographic clines of parasite damage to *Populus tremula* L. Ecography 33: 483–493.
14. Lindroth RL, Scriber JM, Hsia MTS (1988) Chemical ecology of the tiger swallowtail: mediation of host use by phenolic glycosides. Ecology 69: 814–822.
15. Chen F, Liu CJ, Tschaplinski TJ, Zhao N (2009) Genomics of secondary metabolism in *Populus*: interactions with biotic and abiotic environments. CRC Crit Rev Plant Sci 28: 375–392.
16. Tsai CJ, Harding SA, Tschaplinski TJ, Lindroth RL, Yuan YN (2006) Genome-wide analysis of the structural genes regulating defense phenylpropanoid metabolism in *Populus*. New Phytol 172: 47–62.
17. Boerjan W, Ralph J, Baucher M (2003) Lignin biosynthesis. Annu Rev Plant Biol 54: 519–546.
18. Sterky F, Regan S, Karlsson J, Hertzberg M, Rohde A, et al. (1998) Gene discovery in the wood-forming tissues of poplar: Analysis of 5,692 expressed sequence tags. Proc Natl Acad Sci U S A 95: 13330–13335.
19. Tanner GJ, Francki KT, Abrahams S, Watson JM, Larkin PJ, et al. (2003) Proanthocyanidin biosynthesis in plants - Purification of legume leucoanthocyanidin reductase and molecular cloning of its cDNA. J Biol Chem 278: 31647–31656.
20. Dixon RA, Xie DY, Sharma SB (2005) Proanthocyanidins - a final frontier in flavonoid research? New Phytol 165: 9–28.
21. Babst BA, Harding SA, Tsai CJ (2010) Biosynthesis of phenolic glycosides from phenylpropanoid and benzenoid precursors in *Populus*. J Chem Ecol 36: 286–297.
22. Morse AM, Tschaplinski TJ, DerviniS C, Pijut PM, Schmelz EA, et al. (2007) Salicylate and catechol levels are maintained in nahG transgenic poplar. Phytochemistry 68: 2043–2052.
23. Shelton AL (2000) Variable chemical defences in plants and their effects on herbivore behaviour. Evol Ecol Res 2: 231–249.
24. Shelton AL (2004) Variation in chemical defences of plants may improve the effectiveness of defence. Evol Ecol Res 6: 709–726.
25. Degenhardt J, Kollner TG, Gershenzon J (2009) Monoterpene and sesquiterpene synthases and the origin of terpene skeletal diversity in plants. Phytochemistry 70: 1621–1637.
26. Keeling CI, Bohlmann J (2006) Diterpene resin acids in conifers. Phytochemistry 67: 2415–2423.
27. Palo RT (1984) Distribution of birch (*Betula* spp.), willow (*Salix* spp.), and poplar (*Populus* spp.) secondary metabolites and their potential role as chemical defense against herbivores. J Chem Ecol 10: 499–520.
28. Lindroth RL, Hwang SY (1996) Clonal variation in foliar chemistry of quaking aspen (*Populus tremuloides* Michx). Biochem Syst Ecol 24: 357–364.
29. Abreu IN, Ahnlund M, Moritz T, Albrectsen BR (2011) UHPLC-ESI/TOFMS determination of salicylate-like phenolic glycosides in *Populus tremula* leaves. J Chem Ecol 37: 857–870.
30. Rehill B, Clauss A, Wieczorek L, Whitham T, Lindroth R (2005) Foliar phenolic glycosides from *Populus fremontii*, *Populus angustifolia*, and their hybrids. Biochem Syst Ecol 33: 125–131.
31. Orians CM, Fritz RS (1995) Secondary chemistry of hybrid and parental willows: phenolic glycosides and condensed tannins in *Salix sericea*, *S. eriocephala*, and their hybrids. J Chem Ecol 21: 1245–1253.
32. Orians CM, Griffiths ME, Roche BM, Fritz RS (2000) Phenolic glycosides and condensed tannins in *Salix sericea*, *S. eriocephala* and their F1 hybrids: not all hybrids are created equal. Biochem Syst Ecol 28: 619–632.
33. Caseys C, Glauser G, Stölting KN, Christe C, Albrectsen BR, et al. (2012) Effects of interspecific recombination on functional traits in trees revealed by metabolomics and genotyping-by-resequencing. Plant Ecol Divers 5: 457–471.
34. Nichols-Orians CM, Clausen TP, Fritz RS, Reichardt PB, Wu JJ (1992) 2'-Cinnamoylsalicortin, a phenolic glycoside from *Salix sericea*. Phytochemistry 31: 2180–2181.
35. Ruuhola T, Julkunen-Tiitto R (2003) Trade-off between synthesis of salicylates and growth of micropropagated *Salix pentandra*. J Chem Ecol 29: 1565–1588.
36. Luquez V, Hall D, Albrectsen BR, Karlsson J, Ingvarsson P, et al. (2008) Natural phenological variation in aspen (*Populus tremula*): the SwAsp collection. Tree Genet. Genomes 4: 279–292.
37. Audette RCS, Blunden G, Steele JW, Wong CSC (1966) Thin-layer chromatography of phenolic glycosides and its use as a screening procedure for genus *Salix*. J Chrom 25: 367–&.
38. Binns WW, Blunden G, Woods DL (1968) Distribution of leucoanthocyanidins phenolic glycosides and imino-acids in leaves of *Salix* species. Phytochemistry 7: 1577–1581.
39. Keefover-Ring K, Carlsson M, Albrectsen BR (2014) 2'-(*Z*)-Cinnamoylsalicortin: A novel salicinoid isolated from *Populus tremula*. Phytochem Lett 7: 212–216.
40. de Carvalho D, Ingvarsson PK, Joseph J, Suter L, Sedivy C, et al. (2010) Admixture facilitates adaptation from standing variation in the European aspen (*Populus tremula* L.), a widespread forest tree. Mol Ecol 19: 1638–1650.
41. Robinson KM, Ingvarsson PK, Jansson S, Albrectsen BR (2012) Genetic variation in functional traits influences arthropod community composition in aspen (*Populus tremula* L.). PloS One 7: e37679.
42. Paunonen R, Julkunen-Tiitto R, Tegelberg R, Rousi M, Heiska S (2009) Salicylate and biomass yield, and leaf phenolics of dark-leaved willow (*Salix myrsinifolia* Salisb.) clones under different cultivation methods after the second cultivation cycle. Ind Crops Prod 29: 261–268.
43. Umetrics (2012) SIMCA-P+ v. 12.0. Umeå, Sweden.
44. SAS Institute (2003) SAS version 9.1. Cary, N.C.: SAS Institute.
45. R Development Core Team (2013) R: A language and environment for statistical computing. Vienna, Austria: R Foundation for Statistical Computing.
46. Lindroth RL, Hsia MTS, Scriber JM (1987) Seasonal patterns in the phytochemistry of three *Populus* species. Biochem Syst Ecol 15: 681–686.
47. Reichardt PB, Merken HM, Clausen TP, Wu JJ (1992) Phenolic glycosides from *Salix lasiandra*. J Nat Prod 55: 970–973.
48. Ruuhola T, Julkunen-Tiitto R, Vainiotalo P (2003) In vitro degradation of willow salicylates. J Chem Ecol 29: 1083–1097.
49. Tegelberg R, Julkunen-Tiitto R (2001) Quantitative changes in secondary metabolites of dark-leaved willow (*Salix myrsinifolia*) exposed to enhanced ultraviolet-B radiation. Physiol Plant 113: 541–547.
50. Bernhardsson C, Ingvarsson PK (2012) Geographical structure and adaptive population differentiation in herbivore defence genes in European aspen (*Populus tremula* L., Salicaceae). Mol Ecol 21: 2197–2207.
51. Hewitt GM (1999) Post-glacial re-colonization of European biota. Biol J Linnean Soc 68: 87–112.
52. Hall D, Luquez V, Garcia VM, St Onge KR, Jansson S, et al. (2007) Adaptive population differentiation in phenology across a latitudinal gradient in European Aspen (*Populus tremula*, L.): A comparison of neutral markers, candidate genes and phenotypic traits. Evolution 61: 2849–2860.
53. Arnold ML (1992) Natural hybridization as an evolutionary process. Annu Rev Ecol Syst 23: 237–261.
54. Rieseberg LH, Sinervo B, Linder CR, Ungerer MC, Arias DM (1996) Role of gene interactions in hybrid speciation: Evidence from ancient and experimental hybrids. Science 272: 741–745.
55. Ingvarsson PK, Street NR (2011) Association genetics of complex traits in plants. New Phytol 189: 909–922.
56. Lindroth RL, Hwang SY, Osier TL (1999) Phytochemical variation in quaking aspen: Effects on gypsy moth susceptibility to nuclear polyhedrosis virus. J Chem Ecol 25: 1331–1341.
57. Reichardt PB, Bryant JP, Mattes BR, Clausen TP, Chapin FS, et al. (1990) Winter chemical defense of Alaskan balsam poplar against snowshoe hares. J Chem Ecol 16: 1941–1959.

11

Coral Reef Disturbance and Recovery Dynamics Differ across Gradients of Localized Stressors in the Mariana Islands

Peter Houk[1,2]*, David Benavente[3], John Iguel[3], Steven Johnson[3], Ryan Okano[3]

1 University of Guam Marine Laboratory, UOG Station, Mangilao, Guam, **2** Pacific Marine Resources Institute, Saipan, Northern Mariana Islands, **3** CNMI Bureau of Environmental and Coastal Quality, Saipan, Northern Mariana Islands

Abstract

The individual contribution of natural disturbances, localized stressors, and environmental regimes upon longer-term reef dynamics remains poorly resolved for many locales despite its significance for management. This study examined coral reefs in the Commonwealth of the Northern Mariana Islands across a 12-year period that included elevated Crown-of-Thorns Starfish densities (COTS) and tropical storms that were drivers of spatially-inconsistent disturbance and recovery patterns. At the island scale, disturbance impacts were highest on Saipan with reduced fish sizes, grazing urchins, and water quality, despite having a more favorable geological foundation for coral growth compared with Rota. However, individual drivers of reef dynamics were better quantified through site-level investigations that built upon island generalizations. While COTS densities were the strongest predictors of coral decline as expected, interactive terms that included wave exposure and size of the overall fish assemblages improved models (R^2 and AIC values). Both wave exposure and fish size diminished disturbance impacts and had negative associations with COTS. However, contrasting findings emerged when examining net ecological change across the 12-year period. Wave exposure had a ubiquitous, positive influence upon the net change in favorable benthic substrates (i.e. corals and other heavily calcifying substrates, $R^2 = 0.17$ for all reeftypes grouped), yet including interactive terms for herbivore size and grazing urchin densities, as well as stratifying by major reeftypes, substantially improved models ($R^2 = 0.21$ to 0.89, lower AIC scores). Net changes in coral assemblages (i.e., coral ordination scores) were more sensitive to herbivore size or the water quality proxy acting independently ($R^2 = 0.28$ to 0.44). We conclude that COTS densities were the strongest drivers of coral decline, however, net ecological change was most influenced by localized stressors, especially herbivore sizes and grazing urchin densities. Interestingly, fish size, rather than biomass, was consistently a better predictor, supporting allometric, size-and-function relationships of fish assemblages. Management implications are discussed.

Editor: Benjamin Ruttenberg, California Polytechnic State University, United States of America

Funding: This study was funded by the National Oceanic and Atmospheric Administration (NOAA) coral-reef monitoring grants administered to the CNMI Coastal Resources Management Office, and USEPA Region IX funding administered to the CNMI Division of Environmental Quality. Financial support for PH was provided through a NOAA domestic coral reef grant award (NA11-NOS4820015). The funders had no role in study design, data collection and analysis, decision to publish, or preparation of the manuscript.

Competing Interests: The authors have declared that no competing interests exist.

* Email: houkp@uguam.uog.edu

Introduction

A vast array of acute disturbances and chronic stressors threaten coral reefs [1–3]. Yet, our knowledge of the role that individual disturbances and stressors play in determining reef dynamics through time remains limited for most locales [4]. One cause of this uncertainty stems from the rarity of ecological datasets that span across sufficient timeframes to capture disturbance and recovery periods, which can help partition the variance associated with population dynamics, and attribute cause, proportionally, to individual stressors. For coral reefs, two putative, localized stressors of primary concern are unsustainable fishing and pollution that may act independently or in combination with disturbance cycles to diminish the growth of reef accreting organisms such as corals [5,6]. Many studies have compared reefs where high and low human influences existed to generalize how water quality and herbivory have impacted coral growth and resulted in macroalgal replacement [7,8]. However, less attention has been given to understanding their individual contributions towards growth dynamics outside of laboratory or manipulative settings [9–11], and thus, our ability to upscale evidence from manipulative studies to entire reefscapes remains limited. In turn, context-dependent roles of localized stressors continue to be the focus of much research, and varying findings provide support for many ideologies [12–14]. When used out of context, or when limited information exists to develop an appropriate context, uncertainty and improperly informed decision making can result.

Acute disturbances such as typhoons, *Acanthaster planci* outbreaks, and climate-induced bleaching are well appreciated for their role in driving coral population dynamics. Disturbance and recovery cycles have traditionally been investigated using coral cover trends, integrated across both local and regional scales, and through time [15,16]. Yet, coral cover can be an inconsistent metric of recovery due to varying natural environmental regimes, such as wave exposure, that dicate coral growth capacity [17–19], and thus the time needed for recovery. Improved assessments of reef condition and calcification potential have emerged from benthic-substrate datasets by simultaneously considering the abundances of macroalgae and other less-calcifying organisms in comparison to corals and other heavily-calcifying organisms [9,20,21]. Such integrated metrics can better account for the inherent environmental variation that drives coral cover trajectories, and represents one useful metric of overall recovery potential when comparing across reefs that is furthered within the present study.

In addition to coverage estimates, shifting species abundances [22] and colony-size distributions [23,24] have also proven to be sensitive indicators of disturbance-and-recovery cycles for coral populations. Studies show that faster growing corals (*Acropora*, *Pocillopora*, *Stylophora*) have lower tolerances to disturbance events, while others (*Porites* and numerous faviids) have a prolonged ability to deal with both acute disturbance and chronic stress [5,25–27]. However, many of the same faster-growing corals may be more resilient (or adaptable) to repeated climate-induced disturbance in the absence of chronic stress [28,29]. Thus, size distributions and abundance patterns of corals that have low thresholds, fast recovery, and the potential for adaptability provides an additional means of evaluating coral population dynamics [20,30].

The premise for the present study is that disturbance and recovery are not necessarily co-dependent processes on coral reefs, and by examining them independently with the aid of refined metrics described above, an improved understanding of causation and predicted resilience can emerge [3,31]. While many studies have described the nature of acute disturbances to reefs from a variety of agents, relatively few have followed recovery dynamics with respect to individual factors. In the Caribbean, Mumby and Harborne [10] reported significantly higher recovery of coral coverage and colony sizes in response to a no-take fishery closure over a 2.5 year period. Similarly, the recovery of coral colony sizes [22], but not recruitment [32], were found to be most heavily dependent upon fishing pressure in Kenya. In American Samoa, Houk et al. [9] found differential recovery of calcifying benthic substrates and coral species evenness to be interactively driven by herbivore biomass and water quality, with their hierarchical influence shifting based upon geological reef settings (i.e., a potential proxy for connectivity with groundwater discharge). In contrast, a meta-analysis of coral reef recovery across all major oceanic basins provided some evidence for unintuitive, reduced recovery rates within fisheries closures following disturbance events [3]. These findings were a perceived artifact of higher pre-disturbance coral coverage within no-take closures, and not attributed to management status. Interestingly, the causes behind recovery trajectories were not consistent across major geographic regions. Clearly recovery is context-dependent with respect to physical settings as well as management regimes [33–37]. Considering that predicting reef futures is becoming more of a priority for both local and global management efforts, further study is needed to interpret when and if generalizations may exist, and at what spatial scales they might operate at.

We examined 12-year trends in coral-reef assemblages across fore-reefs in the Commonwealth of the Northern Mariana Islands (CNMI) during a time period when significant disturbance impacts and differential recovery occurred. We first described the nature of *Acanthaster planci* (i.e., Crown-of-Thorn Starfish, COTS) population dynamics during the study period. COTS population trends were augmented with typhoon records to define pre-, during, and post-disturbance time periods across our 12-year study. Coral cover and colony sizes were then evaluated for two study islands that differed in human population, development, and geological setting. Island-scale coral decline and recovery trajectories provided an initial framework for interpreting the nature of coral growth cycles across the study period, and along with previous studies, highlighted primary factors influencing site-specific reef dynamics. Ensuing site-level regressions were performed between coral decline and COTS densities (i.e., the hypothesized, primary disturbance agent), while including key factors that quantified localized stressors and environmental regimes as interactive covariates. Similar site-level regressions were also performed to examine net ecological change across the study period. Combined, our study provided a descriptive framework for disentangling the differences between 'disturbance impacts' and 'net ecological change', and attributed cause, proportionally, to a suite of factors that help to inform management.

Methods

Study location

The Commonwealth of the Northern Mariana Islands (CNMI) represents a series of active volcanic and inactive raised limestone islands located in the Western Pacific Ocean (Figure 1). The present study focused upon the southernmost, inactive limestone islands where the majority of the human population resides. From north to south, the study islands were Saipan (48,220 people, capital island), Tinian (3,136 people), Aguijan (no inhabitants but the nearshore resources are accessed from Saipan and Tinian), and Rota (2,527 people) (Census 2010 statistics, http://commerce.gov.mp/divisions/central-statistics/) (Figure 1).

CNMI's local monitoring program has been collecting standardized benthic, coral, macroinvertebrate, and fish assemblage data since 2000 [38,39]. Monitoring designs were stratified based upon geological reeftypes, management status, and watershed development. Geological reeftypes have previously been described with respect to wave exposure and submarine groundwater discharge through karst watersheds that were attributed to specific coral assemblages and reef growth through time [18]. Cumulatively, Houk and van Woesik [18] described four distinct geological reeftypes in the CNMI: 1) optimal spur-and-groove structures, 2) constructional, high-relief interstitial framework, 3) low-relief framework with limited Holocene deposition found only on Rota, and 4) incipient coral assemblages with little to no deposition. Coral coverage, diversity, and evenness peak on the first two, with spur-and-groove structures being the most optimal settings for modern coral assemblage growth.

The most significant acute disturbance since the inception of monitoring efforts has been high *Acanthaster planci* populations (COTS herein) evident between mid-2003 and 2006. In complement to COTS disturbances, several tropical storms passed nearby the study islands during a similar timeframe, with the strongest being Pongsona (passing 40 km south of Rota, 95 knots maximum sustained windspeed) and Chaba (passing over Rota, 118 knot sustained windspeed) in December 2002 and August 2004, respectively. While large wave events were recorded, observations by the authors before and after these storms suggested negligible direct impacts compared with the onset of COTS, examined further within. COTS events became widespread throughout the North Pacific during this timeframe [40]. However, no information exists to understand any potential linkages between the storm events and the subsequent emergence of COTS. Initial examinations were undertaken by coupling COTS abundance and tropical storm histories to define three time periods within the present study *a priori*: pre-, during, and post-disturbance. Subsequently,

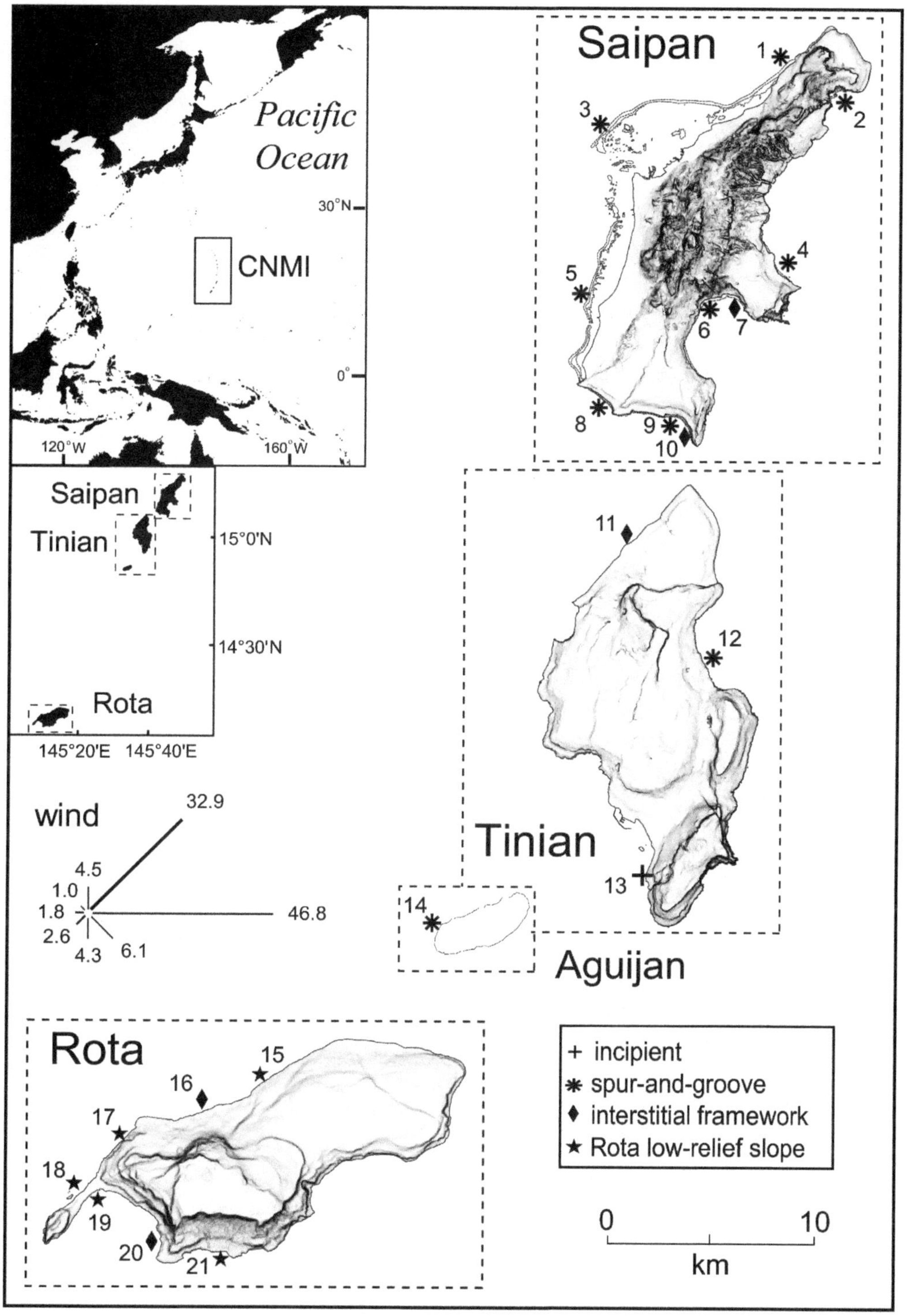

Figure 1. A map of the Western Pacific Ocean and the Commonwealth of the Northern Mariana Islands study islands. Distance between islands is not drawn to scale as dashed boxes indicate individual island entities. Wind vectors show the percent of time that winds originated from each of 8 quadrants (length and corresponding number) as well as the mean annual intensity (thickness). Reeftypes are indicated by symbols referred to in the legend. Topographic lines infer the steepness and size of watersheds.

we quantified how much of the site-level coral decline during the 'disturbance' period could be accounted for by COTS.

Ecological data collection

Data were collected as part of the CNMI Division of Environmental Quality and the CNMI Coastal Resources Management Office coral-reef monitoring program. These programs have the legislative authority to conduct monitoring activities, and given the non-invasive nature of this research, no further permits were required.

Long-term monitoring data have been collected at 1 to 3 year intervals for 25 monitoring stations across the CNMI in most cases, however, longer intervals between repeated site-visits existed in a few instances (Appendix 1). The prerequisite for inclusion in the present study was that coral, benthic, and/or macroinvertebrate data were available for all three timeframes: before, during, and after the disturbance period. Among the 25 monitoring sites, data from 21 met these criteria (Table S1, Figure 1).

Sites were identified by global positioning system coordinates coupled with directional bearings to indicate transect placement. During each survey event, five, 50 m transects were placed at the 8 m depth contour to guide fieldwork. In three instances, homogeneous substrates were not consistently available, and three, 50 m transects were used instead (sites 2, 16, and 18, Figure 1). Benthic substrate abundances were estimated from photographs of 0.5×0.5 m quadrats. In 2000 and 2001, 25 photographs were taken along each transect line, and the substrate under each of 16 data points was identified and recorded. Since 2002, 50 photographs were taken from each transect line, and the substrate under each of 5 data points was recorded. Methods were shifted in order to improve detection limits for temporal change [38]. With respect to the present study, the changes in benthic substrate abundances associated with COTS disturbance were a magnitude of order higher than the expected detection limits based upon 2002 datasets [38]. In all instances, benthic categories chosen for analysis were corals (to genus level), turf algae (non-identifiable turfs typically less than 2 cm), macroalgae (readily identifiable alga typically greater than 2 cm, to genus level if abundant), calcareous encrusting algae known to actively shed epithelial layers or inhibit the survival of juvenile corals (*Peyssonnelia, Pneophyllum*) [41–44], crustose coralline algae (CCA) known to promote reef accretion and juvenile coral settlement, sand, and other invertebrates (genus level if abundant). Using these substrate categories, we defined a benthic substrate ratio by the percent cover of heavily calcifying (corals and CCA) versus non-or-less calcifying (turf, shedding-calcareous, macroalgae) substrates.

Coral assemblage data have been collected since 2003 by the same observer using a standard quadrat-based technique. During each survey, 16 replicate 0.5×0.5 m quadrats were haphazardly tossed at equal intervals along the transect lines. All corals whose centerpoint resided within the quadrat boundary were identified, and the maximum diameter, and the diameter perpendicular to the maximum, were recorded. Surface area was calculated from these measurements assuming colonies were elliptical in nature. For three-dimensional corals, measurements were extended along the colony surface area. Coral taxonomy followed Veron [45]. Species-level data were used for analyses of richness and evenness, while species were grouped by 'sub-genus' (i.e., genus and growth form, digitate *Acropora*, massive *Porites*) to examine multivariate trends and subsequent species weightings described below.

Macroinvertebrate densities were estimated along individual transect lines (noted above) using 50×4m observation belts. All sea cucumbers, sea urchins, shellfish, and other conspicuous macroinvertebrates were identified to species level and recorded. Count data derived from macroinvertebrate belt transects were used to examine COTS and grazing urchin populations through time.

Fish census protocols were recently added into CNMI long-term monitoring program. The present study used data collected between 2011 and 2012 by a single observer. Fish assemblages were estimated from 12 stationary-point-counts (SPC's) conducted at equal intervals along the transect lines, following a modified version of [46]. During each SPC, the observer recorded the name (species) and size of all food-fish within a 5 m circular radius for a period of 3-minutes. Food fish were defined by acanthurids, scarids, serranids, carangids, labrids, lethrinids, lutjanids, balistids, kyphosids, mullids, and holocentrids that are a known to be harvested. Sharks were also included. Size data were binned into 5 cm categories (i.e., 12.5–17.5 cm = 15 cm, inferring the size estimates of 13, 14, 15, 16, an 17 were considered as 15 cm), and converted to biomass using coefficients reported in fishbase (www.fishbase.org). Species data were binned into functional categories based upon maximum adult sizes and family level taxonomy for some graphical interpretations and analyses, such as large/small bodied parrotfishes and surgeonfishes, and large/small bodied snappers and groupers. We defined large-and-small-bodied species based upon estimated mean reproductive sizes greater or less than 30 cm, respectively, based upon fishbase records, or the life-history wizard. Planktivores were excluded from the present analyses due to their low abundance in comparison to other trophic groups, and low sample sizes for defining confidence intervals. Last, fish recruits less than 7.5 cm (i.e., 8–12 cm size class bin) were omitted from all analyses to avoid potential bias from differential recruitment with unknown post-settlement mortality dynamics. Recruits comprised less than 3% of the surveyed population.

Environmental data collection

A proxy for water quality was developed from geographic information system (GIS) layers pertaining to topography, landuse, and human population. Digital elevation models (i.e., topographic data) were used to define watershed boundaries. Landuse data were then overlaid upon the watersheds, and a measure of disturbed land was calculated by combining the coverage of barren land, urbanized vegetation, and developed infrastructure within each watershed (United States Forest Service, http://www.fs.usda.gov/r5). Equal weighting was given to each category because variation in pollution contribution was expected within each. For example, urbanized lands were associated with both septic systems and sewer collection systems that differ with respect to their pollution contribution. Human population density in the watershed adjacent to each site was derived from the 2010 CNMI census. Landuse and human population data were also standardized to weight them equally, and averaged to establish an overall proxy to watershed pollution that represents the simplest assumptions of equal contributions from two known sources of watershed-based pollution. In order to match the low-to-high scale of other localized stressors (i.e., herbivore size), the inverse of the proxy was used in regression modeling.

Wave energy was derived from long-term wind datasets and estimates of fetch [47]. For each site, fetch (i.e., distance of unobstructed open water) was first calculated for 16 radiating lines equally distributed between 20 to 360 degrees. Fully develop sea conditions were considered if unobstructed exposure existed for 20 km or greater. Ten-year windspeed averages were calculated from Saipan airport data (http://www7.ncdc.noaa.gov/), and used as inputs to calculate wave height following Ekebom et al. [47]. Mean height was calculated by:

$$Hm = 0.018 \cdot U^{1.1} \cdot F^{0.45} \quad (1)$$

Where Hm is the wave height (m) for each quadrant, U is the windspeed at an elevation of 10 m, and F is the fetch (km). Windspeed corrections for varying elevations were made following Ekebom et al. [47]. Last, wave height was converted to energy following:

$$E=(1/8)\cdot\rho\cdot g\cdot H^2 \quad (2)$$

Where ρ is the water density (kg/m^3), g is the acceleration due to gravity (9.81 m/s^2), and H is the wave height (m).

Data Analyses

Island comparisons

We first generalized disturbance and recovery trajectories for two islands that differed with respect to human population, development, and geological setting, Rota and Saipan. These islands were selected because they both had sufficient sampling effort and spatial coverage over the study period to provide a generalization of temporal reef dynamics (Figure 1, Table S1). Initial, island-scale examinations were conducted for coral cover and colony sizes. Due to varying geological settings, coral growth capacity is inherently lower on Rota compared with Saipan [18], and thus we focused investigations upon the nature and relative rate of decline and recovery, and not absolute values of change. Coral cover data were site-averaged within each of the study time periods, and repeat measures ANOVA tests with post-hoc Tukey's pairwise tests were conducted between the study periods for each island. Repeat measures ANOVA tests were similarly used to examine coral population density across the disturbance time-frames. Last, differences in coral colony-size distributions were examined using Kolmogorov-Smirnov (K-S) tests that provided a P-value that is generated from the distances between cumulative frequency plots [48].

We next examined sea cucumber and grazing urchin densities across the study periods. Macroinvertebrate count data had zero-inflated, negative binomial distributions that best conformed to zero-inflated, repeat-measures ANOVA models [49]. Zero-inflated models have two distinct parts, one that describes the probability of obtaining a zero count, and one that describes the expected density given a non-zero count. Tests of significance took both simultaneously into account (i.e., hurdle models).

Fish biomass and size were compared between Rota and Saipan within each trophic category noted above: herbivores/detritivores, secondary consumers, and tertiary consumers. Standard comparative tests were used to examine differences in fish size and biomass between the two islands (t-tests, Mann-Whitnet U-test if normality assumptions were not met).

Site-level analyses

Deeper investigations of disturbance and recovery trajectories were conducted at the site-level, in order to match the spatial scale that localized stressors and key environmental regimes such as wave exposure operate at. A stepwise regression modeling process was performed to determine the likelihoods and magnitudes of influence for COTS, pertinent localized stressors, and wave exposure. These factors were considered based upon both initial island-scale analyses and previous studies describing their driving influence upon CNMI's reefs [18], with more details provided below to describe localized stressors. Stepwise modeling to understand the impacts during the disturbance time frame utilized coral cover as the primary, sensitive dependent variable. Modeling to understand net ecological change across the 12-year period used two dependent variables that better described coral reef 'condition' and relied less upon coral cover alone, which is known to differ across species and environments [18]. These were: 1) the net change in the benthic substrate ratio, and 2) net change in multivariate coral ordination scores (Figure 2). Benthic and coral ordination metrics were weakly correlated ($r<0.4$, all comparisons). Coral ordination metrics were calculated based upon two-dimensional principle component ordination plots. Coral species abundance data were log-transformed to produce a Bray-Curtis similarity matrix that quantified multivariate similarities between all sites and years. Similarity matrices were projected in two-dimensional space using principle component ordination (PCO) plots that depicted multivariate differences using axes (i.e., eigenvectors) that hierarchically account for the variation in multi-species datasets (PCO) [50]. Corals that were the strongest drivers of multivariate dissimilarities were overlaid on the plot to indicate the nature of shifting assemblages over the years. The percent net change for coral assemblages represented the two-dimensional PCO-movement (i.e., vector magnitude) away from *Montipora*, *Acropora*, *Stylophora*, and *Pocillopora* assemblages to faviid and *Porites* assemblages, and the magnitude of return proportional to its starting position (Figure 2). These corals cumulatively accounted for the top 30% of the variation in the PCO plot. Percent change values were calculated because they are comparable across sites, whereas absolute values would not be. Second, we used mean values of the benthic substrate ratio defined above. Percent net change was again defined by differences between the post- and pre-disturbance period, divided by the initial pre-disturbance period value (Figure 2). Cumulatively, larger reductions in percent change values indicated a shift to reefs with non-calcifying substrates and smaller, tolerant coral assemblages.

A forward, stepwise regression modeling process was conducted to evaluate the likelihood and magnitude of independent factors in determining the observed biological changes. This process began by searching for factors with a ubiquitous influence upon biological change across all reeftypes. These included COTS in the disturbance timeframe models and wave exposure in the benthic net-change models. No ubiquitous factor was revealed for models describing the net change in coral ordinations, so only individual terms were used. Forward steps were only taken when/if interactive models improved the fit (R^2-values) and likelihood (AIC-values, described below) of predictions. Interactive combinations were considered because independent variables were all scaled in a consistent, low-to-high manner describing a gradient of weak to strong influences, and are known to have contextual effects that are not independent of one another [9,33]. If stepwise terms did not improve the model fit and likelihood they were dropped, not presented in the results, and not further considered. This process continued until all interaction terms were examined. We first examined regression models including sites from all reeftypes grouped together, and subsequently examined subsets of the most favorable reef settings for coral growth to determine if herbivory, water quality, or wave exposure might have context-dependent roles that differed in accordance with reeftype. Comparisons of the explanatory power and likelihood of independent variables were only made across models examining the same reeftypes. Our goal was not to suggest that a single "best-fit" model existed, but rather to highlight that several plausible models existed, and provide the details of each. The influence of each factor was assessed based upon: 1) the (added) model fit, 2) overall presence across the suite of models, and 3) the likelihood scores and influence of outliers.

Regression modeling was performed using R [51]. Independent variables were standardized to provide equal scaling, and a constant value was added to make all numbers positive. Positive numbers were required when log-transformations were needed to ensure residual normality. Models were examined for normality using Shapiro-Wilk tests, and ranked based upon their explanatory

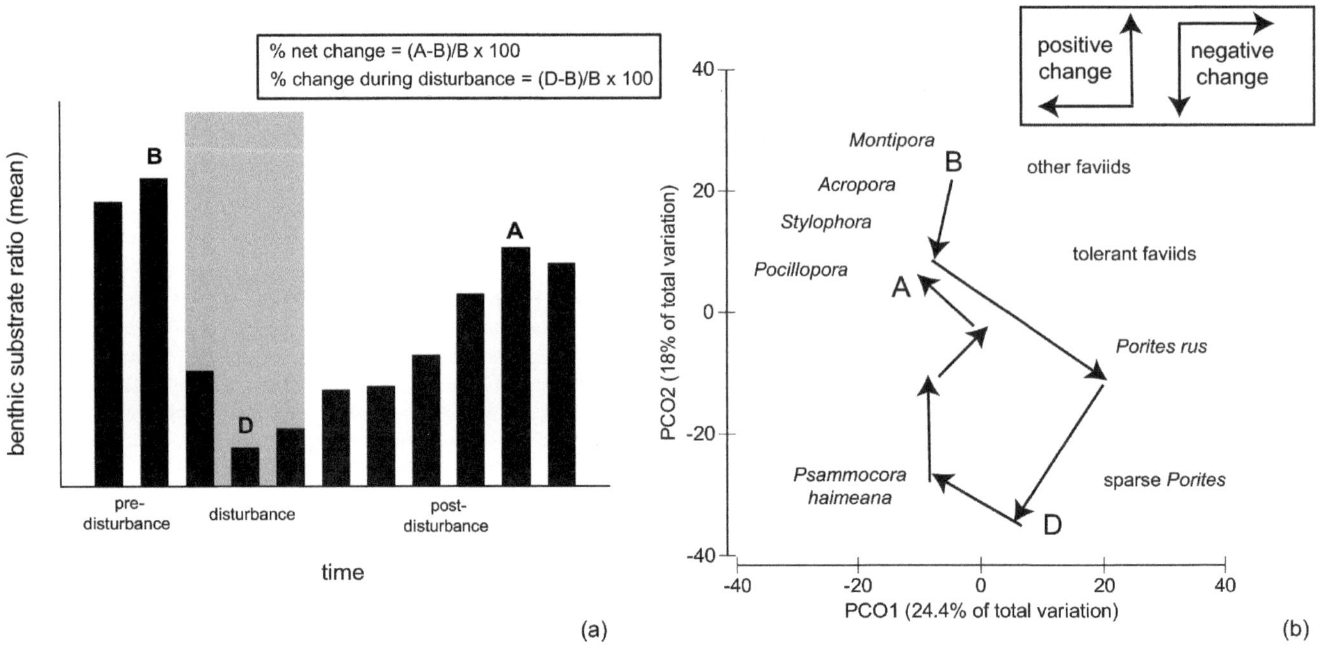

Figure 2. Methods used to calculate the percent decline and net change in (a) the benthic substrate ratio and (b) multivariate measures of the coral assemblages. B, D, A – before, during, and after the disturbance period, respectively. Percent declines were calculated by taking the difference between the minimum and maximum values during and before the disturbance period, and dividing by the pre-disturbance values. Net change values were calculated by taking the difference between maximum values before and after the disturbance period. Values for coral assemblages were calculated based upon their vector magnitude from the origin (0,0), with positive values given for PCO movement towards *Acropora*, *Montipora*, and *Pocillopora* assemblages, and negative values for PCO movement towards tolerant faviids and sparse *Porites*. See methods and Figure 6 for a better description of the corals depicted on the PCO plot.

power (R2-value) and likelihood as measured by Akaike's Information Criterion (AIC). Lower AIC scores indicated a better fit, based upon the least number of parameters and greatest residual normality that together maximize the probability of given outcomes based upon independent predictors. In all instances, the independent predictors used in interactive regression models were not highly correlated ($r<0.30$, $P>0.05$).

Independent variables were selected to represent factors that have a strong influence on CNMI's coral-reef assemblages and dictate disturbance and recovery dynamics in general. These were COTS densities (i.e., as a disturbance agent), wave exposure, the water quality proxy, mean fish size and biomass, and grazing urchin density (defined by the sum of *Echinothrix*, *Echinometra*, and *Diadema* urchins). When examining models dealing with net ecological change, we used mean herbivore/detritivore fish size and biomass. When examining models dealing with coral cover decline during the disturbance period, we used overall fish size and biomass. This was done in accordance with: 1) the literature describing herbivores as key ecosystem engineers facilitating net recovery dynamics [6,22], and 2) existing/emerging relationships between enhanced overall fish size/biomass and reduced COTS impact [12,30,52].

Last, the present study placed a reliance upon 2010/2011 fish data to predict disturbance and recovery trajectories over the past decade. We do not assume that fish populations were static across the disturbance period, as long-term studies across disturbance events elsewhere clearly define fish assemblage dynamics [53,54,55]. However, this approach does assume that site-level relationships were preserved, and relative differences with respect to environmental regimes and/or human exploitation maintained. This was supported by several lines of evidence. Fishery-dependent studies conducted in the CNMI over the past 20 years that showed consistent trends in fish sizes across numerous geographic sectors in the CNMI based upon wave exposure and proximity to human population centers [56,57,58], despite evidence for an overall shift (i.e., decline) in several measures of the fishery resource. These studies also highlighted a shift towards increased herbivore dominance and smaller herbivores over the years, suggesting the relevance of herbivore assemblage metrics. In addition, a recent coral reef resilience assessment conducted for 35 sites around Saipan found that wave exposure and MPA-status together were highly correlated with a fishing pressure metric, and fish abundance trends in the present dataset, which were derived from an independent opinion survey taken by CNMI's resource managers [59]. Beyond CNMI, long-term studies show responses of fish assemblages to disturbances whereby a decline in overall fish diversity, and a decline in the abundance of coral-associated species are most often noted with coral loss [53,54,60]. In response to algal growth following disturbances, these studies also show increases in some acanthurids and/or scarids that can respond rapidly to algal substrate availability, with their abundances gradually decreasing to pre-disturbance levels after several, typically 4 to 5, years. Thus, the present regression models were based upon: 1) environmental gradients and management factors known to persist over longer time periods and more extensive spatial gradients than the present study, and 2) fish abundance datasets that were collected 4 to 5 years following the disturbance period when any pulses of herbivores were expected to diminish, and 3) utilized metrics of food-fish size and biomass that were not influenced by smaller, coral-associated species.

Results

High COTS densities were evident across the CNMI between 2003 and 2006 (Figure 3a, Table S2), concomitant with two tropical storms that passed by the study islands. Together, these findings formed the basis for determining 'before', 'during', and 'after' study periods. Transect-based densities of COTS were largest during the disturbance years on Saipan, however, increased densities were noted across all islands. Following disturbance years, COTS densities declined to pre-disturbance levels for Saipan and Rota, while one unique site on Tinian had an anomalous high density become emergent during 2011 (i.e., site 11 was driving the trend in Figure 3a for Tinian in 2011, Table S2). Cumulatively, COTS densities agreed with 2003 to 2006 as peak disturbance years, but also suggest that Rota and Saipan had similar temporal dynamics, with higher absolute abundances being found on Saipan.

Island comparisons

The disturbance period had a negative impact on coral cover throughout the CNMI, however rates of change were not uniform across the study islands. The largest decline was evident on Saipan, where high human populations and development existed, and smallest on Rota, where human presence and geological foundations for optimal reef growth were lower (Figure 3b). Coral cover declined from 34% to 23% on Saipan (32% decline, F-Statistic = 3.7, P = 0.05, repeat measures ANOVA, pairwise Tukey's post-hoc q-statistic = 3.6, P = 0.05), with no significant recovery since the disturbance years. Conversely, on Rota where coral cover is naturally lower than Saipan as a consequence of island geology, cover had a non-significant decrease from 11.5% to 9.7%, with recovery back to 12%. Yet, Rota coral cover trends were heavily influenced by a single locality where the coral *Porites rus* is abundant and dominant (Site 20 coral cover 30% whereas all other sites have 10%, Figure 1, Table S2). Re-analysis of the coral cover trends with this site omitted indicated that coral cover had a more substantial decline (8.5% to 4.6%) and recovery back to 7.7% (F-Statistic = 4.0, P = 0.06, repeat measures ANOVA). The impact of disturbances to coral-colony sizes and population densities were also markedly different (Figure 4). Mean coral colony-size on Saipan declined from 7.5 to 5.1cm (P<0.001, K-S test), with no change in the years after disturbance (Figure 3). Demographic changes were complemented by significant increases in coral population density during the disturbance years (F-Statistic = 5.96, P = 0.01, repeat measures ANOVA, pairwise Tukey's post-hoc q-statistic = 4.8, P<0.05), mainly due to small faviids and *Porites*. In contrast, mean colony size on Rota had a non-significant decline and recovery (4.3 to 3.7cm, followed by a recovery to 4.1cm). Population density comparisons were also non-significant.

Grazing urchin densities declined across the disturbance period on both islands, yet the magnitude of decline was much larger for Rota, where *Echinothrix* densities were reduced from over 6 individuals per 100 m^2 to less than 2 during the onset of disturbances (Figure 3c, z-statistics = 4.6 and1.8, respectively, for the increased probability of obtaining zero counts and reduction in densities where non-zero counts existed, P<0.01 for both, repeat-measure, zero-inflated hurdle comparisons before and during). Urchin densities were initially lower on Saipan and had a gradual decrease across the study period that was most pronounced in the years following disturbance (z-statistics = 2.8 and1.4, P<0.05, for greater zero counts and reduced densities, respectively, zero-inflated comparisons during and after), representing a timeframe when more abundant COTS existed compared to Rota (Figure 3a). Models indicated that declines were attributed to both higher probabilities of sites with no urchins being present (30% and 38% decline in significant modeled estimates, respectively for Rota and Saipan), and sites with urchins present at lower densities (48% and 58% decline, respectively). No significant differences in sea cucumber densities existed across the study periods.

Fish biomass and density were consistently greater for large-bodied primary and secondary consumers on Rota compared with Saipan (i.e., species that attain larger reproductive sizes, Figure 5). In contrast, smaller-bodied counterparts were higher in biomass and density on Saipan. These trends were most pronounced for large-bodied groupers, snappers, and parrotfish on Rota, and small-bodied acanthurids and parrotfish on Saipan (P<0.05, comparative t-tests and Mann-Whitney U-tests). The findings support that fewer, large-bodied fish with varying functional roles existed on Saipan, with more numerous, small-bodied species comprising a majority of the biomass.

Stepwise regression modeling at the site level

Focusing investigations at the site level revealed that disturbance and recovery cycles varied markedly, with differences attributable, in part, to varying reeftypes and islands that dictated site geomorphology (Figures 6 and 7, Table S3). Percent decline in coral cover during the disturbance period was primarily driven by the observed, transect-based COTS densities regardless of reeftype or island (Table 1). When considering all reeftypes grouped together, only 18% of the variance in coral decline was attributable to COTS (P = 0.06). Yet, the amount of variance explained increased to over 50% when examining the subset of reeftypes with the highest coral growth capacity (52% for interstitial and spur-and-groove reefs combined, and 67% for spur-and-groove reefs alone, P<0.01 for both). The addition of an interaction term describing the mean size of the fish assemblages enhanced the amount of variance accounted for (R^2 = 0.60 and 0.86, P = 0.009 and 0.005, for interstitial and spur-and-groove reefs combined, and spur-and-groove reefs alone, respectively), while wave exposure improve the fit for spur-and-groove reefs alone (R^2 = 0.83, P = 0.003, Table 1). Fish size had a negative association with COTS densities for coral-dominated reefs (r = 0.28, Pearson correlation coefficient, interstitial and spur-and-groove reefs combined), while an interactive term of *fish size×wave exposure* had a stronger association with COTS (r = 0.48), highlighting the inter-dependence among these three factors. In sum, COTS were primary predictors of coral decline (i.e., the disturbance period), while COTS densities and coral impacts were diminished with higher wave exposure and larger fish assemblages, especially on reefs with a high capacity for coral growth.

The general response of coral assemblages to disturbance was a decline in *Acropora*, *Montipora*, *Stylophora*, and *Pocillopora* corals, with faviid and *Porites* assemblages becoming emergent, and differential rates of return in the post-disturbance years. Two-dimensional PCO-plots accounted for 40 to 50% of the variance in coral assemblage similarities, and provided a quantitative basis to assess disturbance and net change across individual sites (Figures 2 and 6). Similarly, photo-quadrat data revealed that benthic substrate ratios were sensitive to disturbance (i.e., the ratio of heavily calcifying corals and crustose coralline algae divided by less-calcifying turf, fleshy-coralline, and macroalgae, Figure 7), with differential levels of decline and net change.

In contrast to regression models describing spatial patterns in coral cover declines, COTS densities did not emerge as a significant predictor of net ecological change. This was the case when considering COTS densities during the disturbance time-

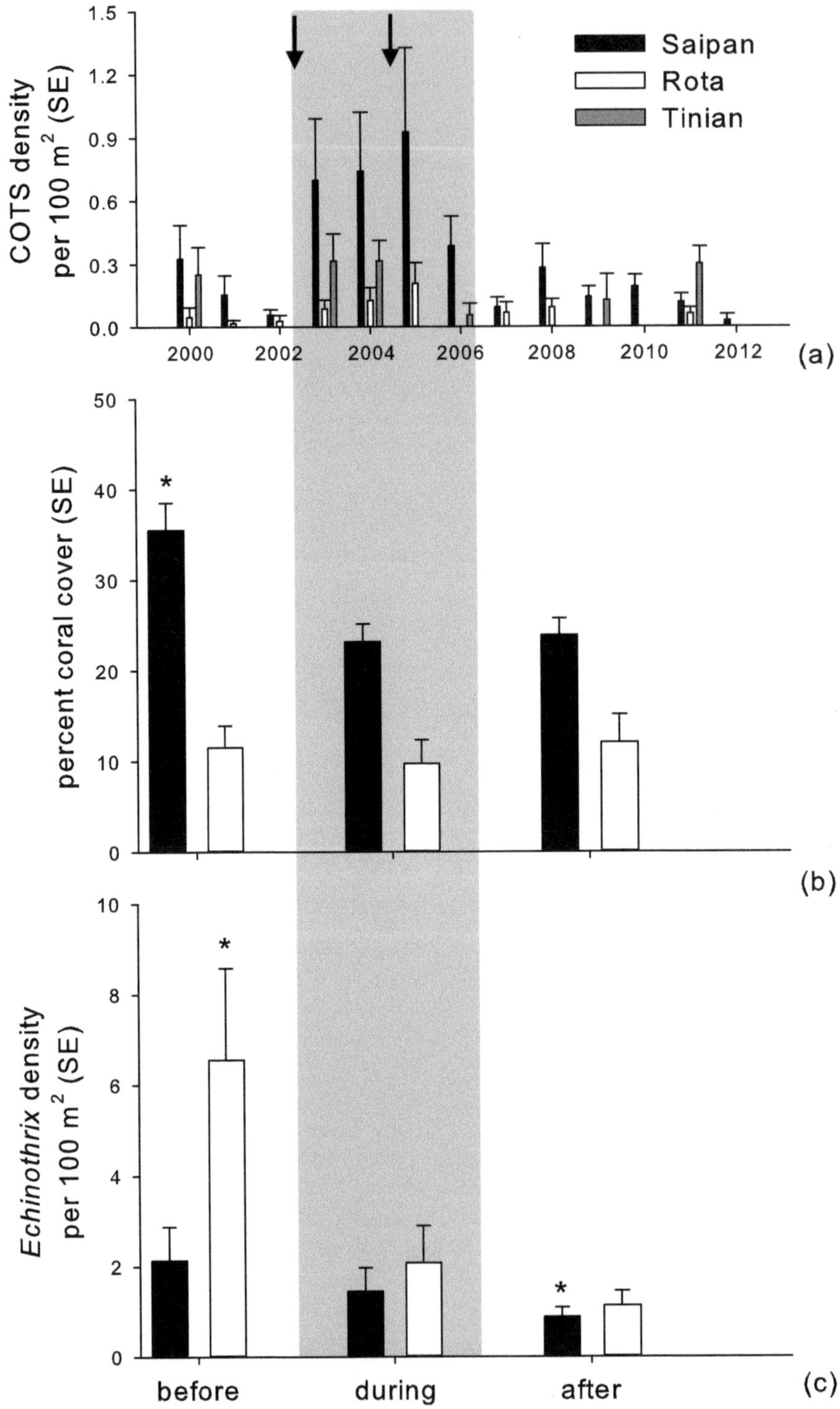

Figure 3. Density of Crown-of-Thorn starfish (COTS) based upon belt-transect data across Saipan, Tinian, and Rota from 2000 to 2012. (a). Densities represent island-based averages that diminish the highest and lowest observations in order to establish patterns across study years. Disturbance reduced coral cover on all islands, but recover trajectories, or the net rate of change, differed by island (b). Coral cover declined on Saipan with no significant recovery (*indicates $P<0.05$, repeat measures ANOVA and post-hoc tests), while a non-significant decline and recovery was noted on Rota. *Echinothrix* urchins also declined (c) in density during the disturbance period, with a further decline in the years after disturbance (*indicates $P<0.05$, zero-inflated hurdle models). Black arrows indicate tropical storms that passed by the study islands during the disturbance timeframe (grey rectangle box indicates the disturbance timeframe).

frame only, or integrated across the entire study period. Stepwise regression modeling for the net change in favorable benthic substrates reported that wave exposure was the strongest individual factor across all reeftypes, accounting for 20% of the

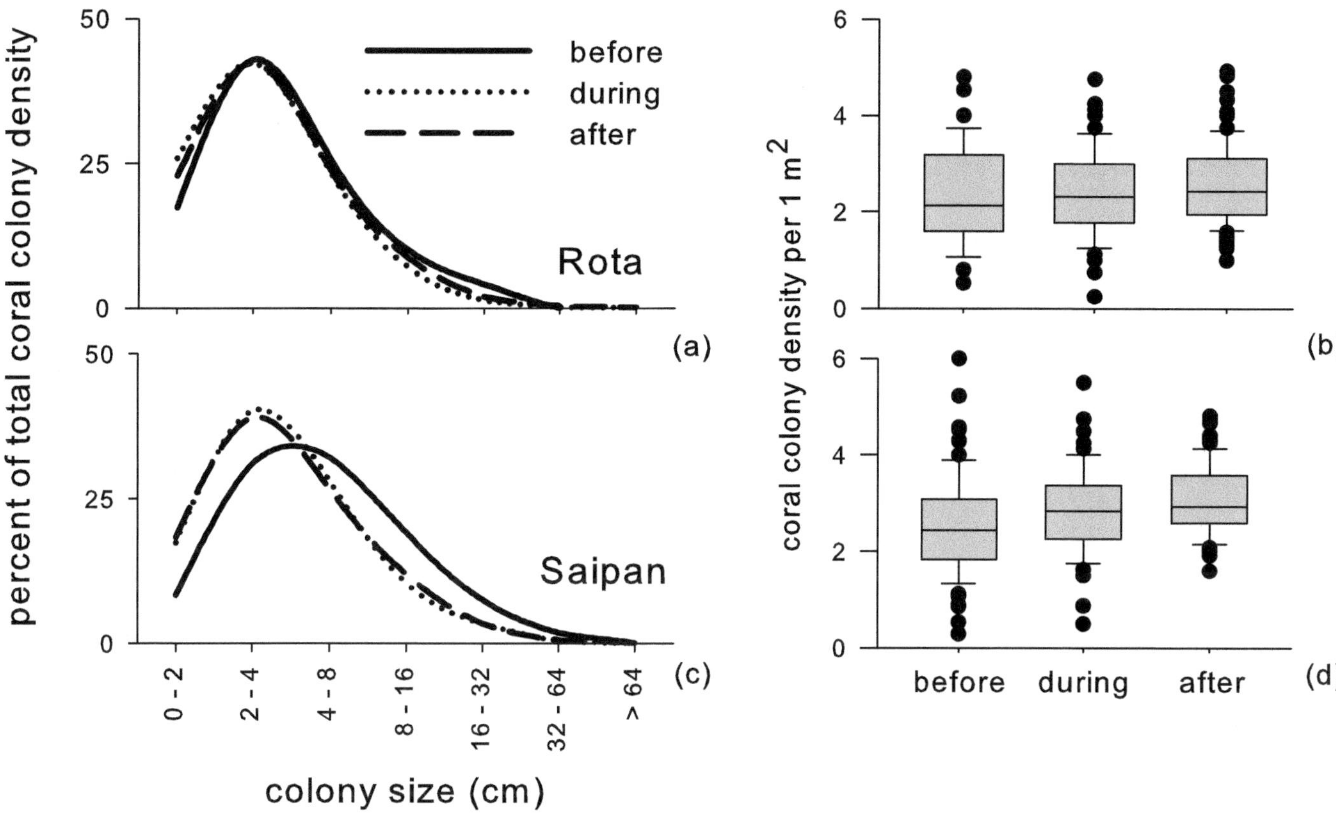

Figure 4. Dynamics of coral colony-size distributions and population densities across the disturbance periods. Rota had a non-significant decline and recovery in colony-size across the study periods (a), as well as a non-significant, sequential increase in population density (b). Reductions in colony-size were evident for Saipan (c) during the COTS period, accompanied by increases in population densities (d), attributed to the emergence of numerous small faviid and *Porites* corals (*indicates $P<0.05$, repeat measures ANOVA with post-hoc tests, see also Fig. 6).

variance ($P = 0.03$, Table 2). The influence of wave exposure grew when considering the subset of reefs with highest coral growth capacity ($R^2 = 0.22$, interstitial and spur-and-groove combined, $R^2 = 0.5$, spur-and-groove only). Yet, improved model fitting consistently required the inclusion of mean herbivore size and/or grazing urchin density that both increased the explanatory power and likelihood of resultant models. Explanatory power increased by 3%, 16%, and 14% when including herbivore size for all reeftypes, interstitial and spur-and-groove reefs combined, and spur-and-groove reefs alone, respectively (Table 2). Explanatory power increased by 0%, 20%, and 33% when including grazing urchin densities for these same reeftypes, respectively. Last, for coral-dominated reeftypes, including the interaction between herbivore size and grazing urchin density increased model fit by 33%. In all instances, interaction models lowered AIC scores suggesting their greater likelihoods.

No single factor consistently emerged as the primary driver of net change in the coral assemblages. Net change was predicted individually and not interactively by both herbivore size and the water quality proxy. When examining sites across all reeftypes, the water quality proxy had a slightly greater explanatory power ($R^2 = 0.28$ versus 0.38, herbivore size and water quality proxy, respectively), however the AIC-based likelihood was lower for water quality due to the relatively strong influence of 1 to 2 sites that diminished the normality of residuals (i.e., residuals still met the requirements of normality, but less so as compared with the herbivore size model, reducing the AIC score). The only other notable models emerged when examining spur-and-groove reefs, whereby herbivore size ($R^2 = 0.44$, $AIC = 18.9$) was a slightly better predictor of net coral assemblage change as compared to the water quality proxy ($R^2 = 0.39$, $AIC = 27.1$, Table 1).

Discussion

Significant coral loss occurred in the CNMI between 2003 and 2006 concurrent with high COTS densities and several typhoons that passed through CNMI. Given that tropical storm paths during these years were consistently in closer proximity to Rota, passing between 60 and 100 km from Saipan, we purport that COTS activity was the primary driver of disturbance that simultaneously impacted the study islands during the mid-2000's. In support, COTS densities were ubiquitous in models explaining the spatial patterns in coral decline during the disturbance years. However, potential synergies and/or linkages between storm activity and COTS were not approached, and deeper mechanisms may have existed.

While starfish densities were similarly elevated during the disturbance years for two islands that differed in human presence and geology, Rota and Saipan, abundances were highest and most persistent on Saipan. These findings help to explain the greater impacts to coral decline on Saipan. Yet, COTS abundances provided no significant explanation of net recovery patterns to the favorable benthic substrates or coral assemblage dynamics across the entire 12 year study period. We hypothesized the island-scale differences in resistance and recovery were attributable to a suite of factors. First, due to varying island geomorphology, Rota naturally had less coral to begin with (i.e., less prey), including a reduction in preferential prey, *Acropora* and *Montipora* [18]. This

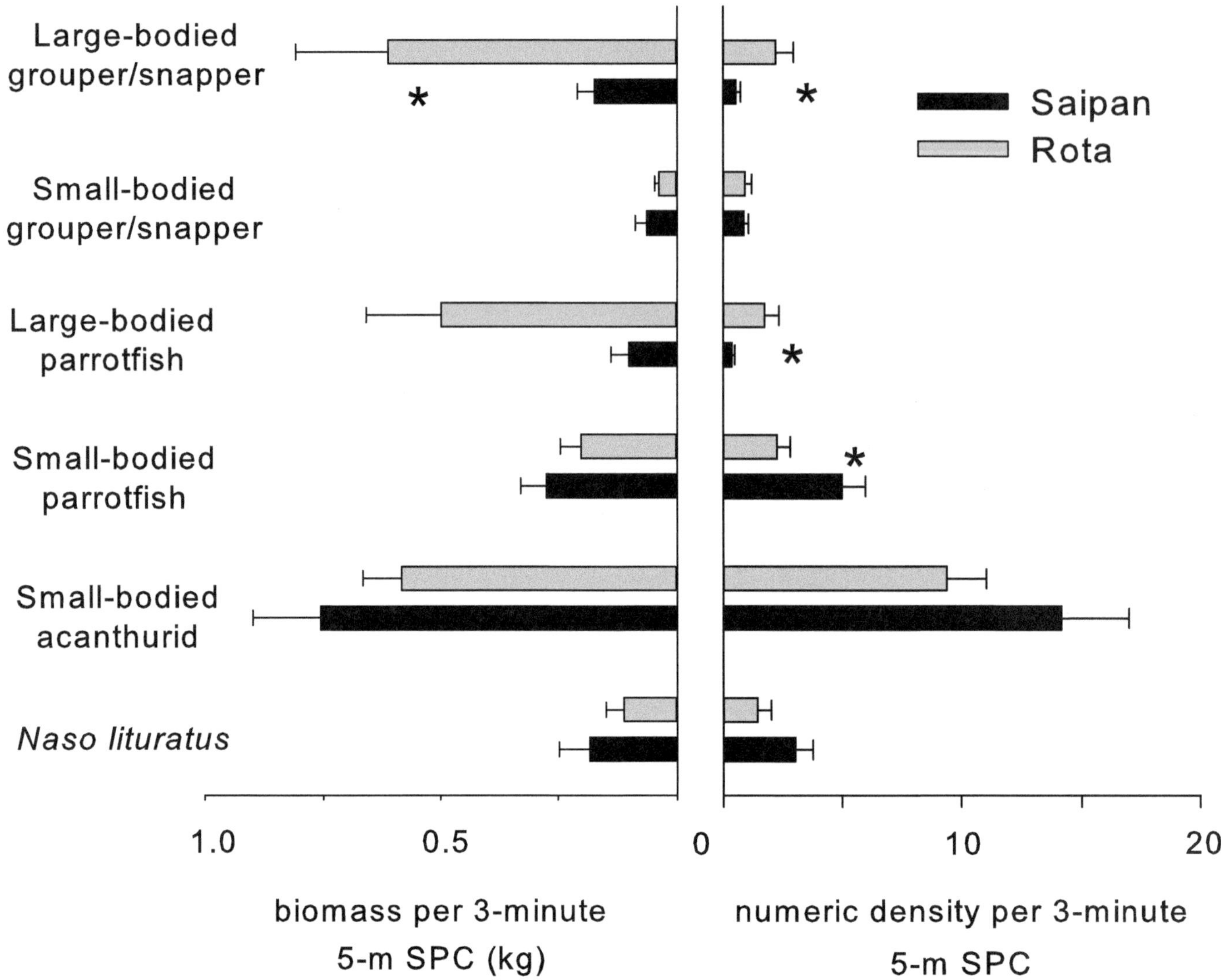

Figure 5. Comparisons of fish biomass and numeric density for several functional fish groups on Saipan and Rota. (*indicates $P<0.05$, comparative tests described in methods).

situation seems most relevant when interpreting why COTS persistence and coral impacts were diminished on Rota. Second, numerous studies continue to support that fish assemblages comprised of larger individuals across all trophic levels are associated with reduced impacts from COTS disturbances [12,30,51], and may help to explain COTS persistence and coral recovery dynamics [10,22]. Larger biomass and body-size of fish assemblages on Rota compared with Saipan supported this notion. Third, nutrient enrichment from watershed runoff is known to contribute to persistent, localized COTS populations [61,62]. CNMI water quality reports have consistently found better water quality on Rota over the past decade, with a 50% reduction in bacteria violations on Rota compared with Saipan in 2012 [63]. When considering where persistent populations existed (site 16 on Rota, sites 5, 6, 7, 8, and 10 on Saipan all had densities of 0.2 individuals per 100 m^2 across several post-disturbance years, Table S2), support for these combined hypotheses grows. Sites 7, 16, and 10 were associated with interstitial framework reefs, posited to have high connections with the karst aquifers compared to others [18] (Figure 1). These sites, in addition to others (5, 6, and 8), all had the smallest overall fish sizes. We synthesize that inherent geological difference as well as localized stressors were influential in describing disturbance dynamics across Rota and Saipan, and utilized site-based analyses to better approach the individual and interactive roles of a suite of factors.

Site-level drivers of change on CNMI's reefs

Wave exposure has long been considered to shape modern coral assemblages and reef growth through time in CNMI [64], whereby full exposure to prevailing northeast trade winds has selected against geological reef development through time (i.e., reeftype 4 noted in the methods). Yet, beside the incipient reef development that exists along much of CNMI's eastern shoreline, significant variation in wave exposure remained among the subset of reeftypes where high coral growth capacity existed, despite having a lower overall magnitude. In fact, this secondary gradient in wave exposure (i.e, low to moderate levels) was the most influential, positive determinant of net benthic substrate change in all reef settings. This may be a result of greater flushing, nutrient transfer rates, and/or the removal of detrital build-up with wave energy [65]. After accounting for wave exposure, the process of grazing, as represented by herbivore/detritivore size and grazing

Table 1. Stepwise regression models predicting coral decline.

Independent variables	Slope	SE	Intercept	R^2	P-Value	AIC
All reeftypes (n = 16)						
$COTS^{-1}$	**6.57**	**3.20**	**0.68**	**0.18**	**0.06**	**46.2**
Interstitial and spur-and-groove reefs (n = 12)						
$COTS^{-1}$	7.84	2.18	0.47	0.52	0.005	25.7
$COTS^{-1}$ × fish_size	1.55	0.43	1.62	0.60	0.009	18.4
Spur-and-groove reefs (n = 7)						
$COTS^{-1}$	8.88	2.43	0.23	0.67	0.01	16.8
$COTS^{-1}$ × log(exposure)	8.21	1.50	0.44	0.83	0.003	12.2
$COTS^{-1}$ × fish_size	2.17	0.38	1.30	0.86	0.005	8.6

Summary of forward, stepwise regression models that examined the drivers of coral decline during disturbance years. Methods describe the suite of independent variables examined and the basis for their selection. Significant independent variables presented below include mean *Acanthaster planci* densities during disturbance years (COTS), wave exposure, and mean fish size for all trophic groups combined. COTS densities were inversely scaled for consistency with other localized stressors (i.e., low-bad/high-good). AIC-scores were used to indicate the relative likelihood of models being able to predict outcomes, and are only comparable within each reeftype grouping.

urchin densities, was the strongest and most reliable predictor of favorable benthic substrates based upon their presence across the suite of models, the added variance accounted for, and the improved model likelihood scores (AIC values). These findings were amplified when stratifying by reeftype, and including only reefs with the most favorable geological foundations for coral growth. Water quality emerged as a significant predictor of change for coral assemblages ordinations along with mean herbivore size, however, we purport contextual roles of water quality that are furthered below.

Synthesizing the findings reveals that localized stressors were most influential to reefs with low to moderate wave exposure, and high inherent capacity for coral growth. Given the distribution of geological reef settings in the CNMI (Figure 1), this means that

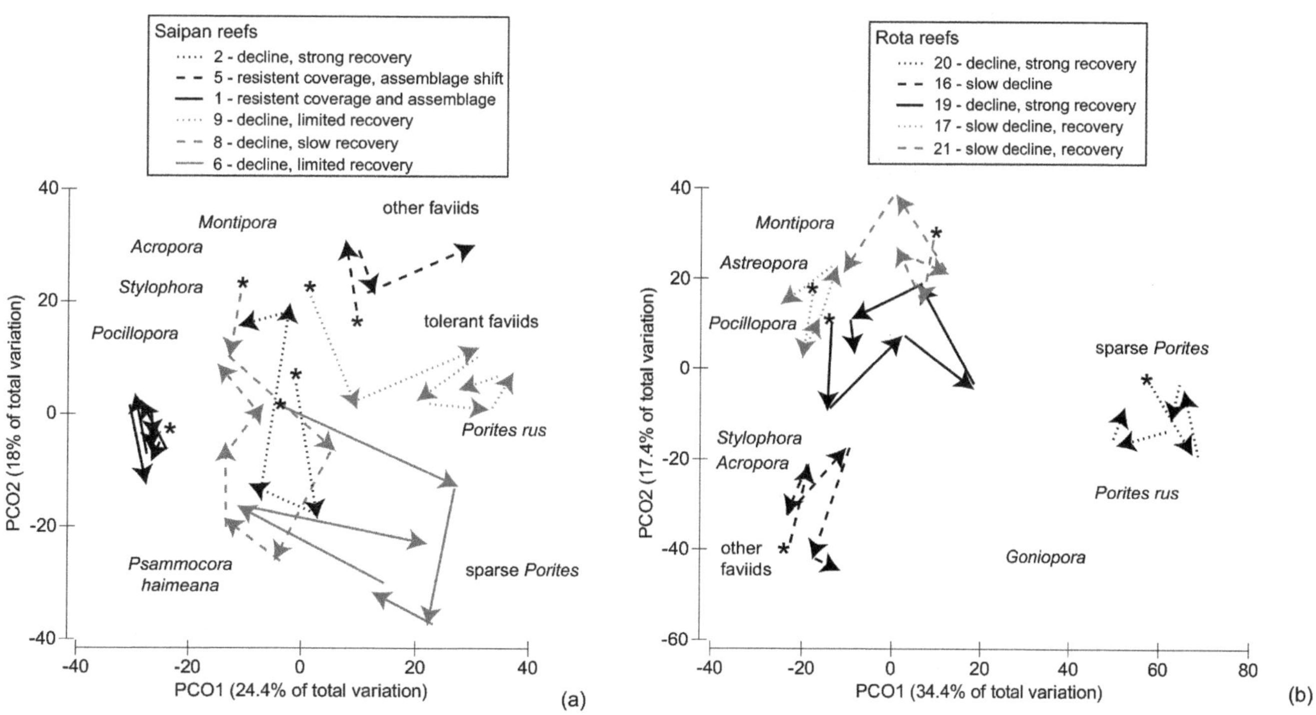

Figure 6. Principle components ordination of coral assemblages for six representative monitoring sites around Saipan (a) and Rota (b). See Figure 1 for site identification and Table S3 for summary statistics. Pre-disturbance assemblages are indicated with an asterisk (*), while vectors depict directional change through time. Sparse *Porites* refers to a dominance of *P. lichen*, *P. vaughani*, and small colonies of other massive species. Tolerant faviids consisted of *Leptastrea purpurea*, *Goniastrea retiformis*, *G. edwardsi*, *Favia matthaii*, *F. pallida*, and *F. favus*. Other faviids consisted of *Favia stelligera*, *Platygyra* spp., *Cyphastrea* spp., and *Favites abdita*.

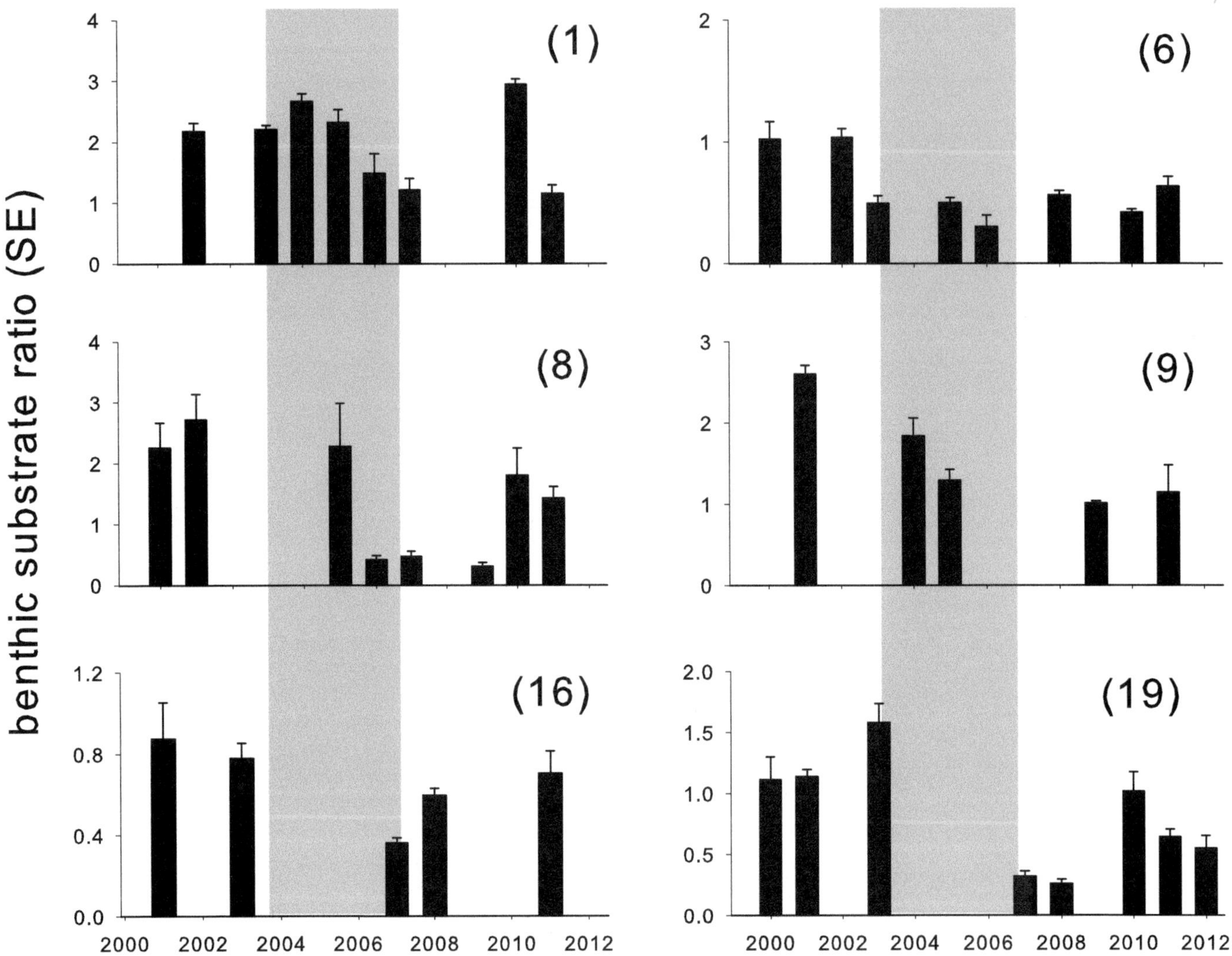

Figure 7. Benthic substrate ratio dynamics for representative monitoring sites around Saipan (sites 1, 9, 8, and 6) and Rota (sites 19 and 16). See Figure 1 for site identification and Table S3 for trends from all sites. Grey bars indicate the disturbance period. Benthic substrate ratios indicate the proportion of heavily-calcifying versus less-or-non-calcifying substrates (*see methods*).

Saipan reefs were weighted disproportionally within the subset of regression models associated with localized stressors, representing 63% of the sites with favorable geological settings (i.e., interstitial or spur-and-groove reefs). Hence, localized stressors were most influential for Saipan, the most populated island where reef-based tourism is centralized, and constitutes a key component of the economy [66].

The present results further an interesting and emerging association between high COTS impacts, smaller-bodied fish assemblages, and low wave exposure. While mechanisms remain unclear and of interest, similar patterns describing diminished COTS impacts with higher fish abundances in successful, no-take marine protected areas have been observed elsewhere [12,30,51]. Fish sizes (i.e., herbivores) were also influential to net change metrics, and the collective findings pertaining to the mean sizes of the fish assemblages rather than their biomass resonated well with power laws that describe relationships between body-size, physiology, and function in ecology [67]. Specific to the present study, power laws have been shown to govern numerous physiological traits such as grazing efficiency [68] and reproductive potential [69], whereby a doubling in fish size equates to an exponential increase in function. Thus, even if similar biomass exists, fish assemblages comprised of mainly small-bodied species are expected to have a reduced ecological function within coral-reef food webs [70,71]. Rasher et al. [72] described that a subset of larger, functionally-dissimilar herbivores play a disproportional role in macroalgal grazing (key constituents include *Chlorurus* spp., *Siganus* spp., and two *Naso* spp., *N. lituratus* and *N. unicornis*), while a suite of other species were more reliant upon generalized detritus and turf grazing from the reef substrate. Species within these functional groups represent highly desirable food fish in CNMI. Market studies have revealed their declining sizes and abundance over the past two decades [55,56,57], and also reported a disproportionally small reef-area-per-person, and mean fish size-at-capture, compared to other Micronesia jurisdictions [57]. We reconcile that both size and functional diversity appear to be key attributes of CNMI herbivore assemblages that are sensitive to harvesting pressure, and influential to coral reef recovery patterns, making their improved management desirable for reef futures.

The reduction in *Echinothrix* urchin densities concomitant with COTS disturbances was a novel association to our knowledge.

Table 2. Stepwise regression models predicting net ecological change.

Benthic substrate ratio net change						
All reeftypes (n = 18)						
Independent variables	*Slope*	*SE*	*Intercept*	R^2	*P-Value*	*AIC*
log(exposure)	3.81	1.64	−0.76	0.20	0.03	50.8
log(exposure) × herb_size	0.62	0.27	1.33	0.23	0.05	44.6
Interstitial and spur-and-groove reefs (n = 12)						
log(exposure)	5.46	2.71	−2.26	0.22	0.07	35.3
log(exposure) × urchin	0.84	0.28	0.74	0.42	0.01	31.7
log(exposure) × herb_size	0.87	0.36	0.85	0.38	0.04	27.4
log(exposure) × herb_size × urchin	0.19	0.06	1.58	0.57	0.01	24.2
Spur-and-groove reefs (n = 7)						
log(exposure)	9.23	3.50	−6.06	0.50	0.05	21.8
log(exposure) × herb_size	1.13	0.35	0.40	0.64	0.03	17.6
log(exposure) × urchin	1.11	0.20	0.08	0.83	0.003	14.4
Coral assemblage net change						
All reeftypes (n = 12)						
herb_size	0.66	0.32	1.37	0.28	0.08	25.0
$poll\_proxy^{-1}$	3.89	1.4	7.14	0.38	0.02	32.2
Spur-and-groove reefs (n = 7)						
$poll\_proxy^{-1}$	4.29	1.71	7.5	0.39	0.04	27.2
herb_size	0.93	0.41	0.64	0.44	0.08	18.9

Summary of forward, stepwise regression models that examined the drivers of net change in the benthic substrate ratio and coral ordination scores across the study period. Methods describe the suite of predictor variables examined and the basis for their selection. Significant independent variables presented below include wave exposure, mean herbivore/detritivore size, mean grazing urchin density, and the water quality proxy. The water quality proxy was inversely scaled (i.e., low-bad/high-good) for consistence with other localized stressors. AIC-scores were used to indicate the relative likelihood of models being able to predict outcomes, and are only comparable within each reeftype grouping.

Field observations and photographs provided anecdotal evidence of competition for refuge within the reef matrix, but clearly these relationships remain speculative. In the event that disturbances not only diminished coral cover but structural complexity as well, (i.e., fewer large *Acropora*, *Pocillopora*, *Stylophora* colonies), a decrease in urchin densities was the anticipated ecological response following disturbance [73]. Yet, urchin declines were concomitant with the onset of disturbance, and continued to decline throughout the study period. Given these trends, deeper investigations into the cause(s) of urchin declines seem warranted.

Last, the water quality proxy was an influential driver of net coral assemblage dynamics across the 12 year study period as well, despite its diminished presence across the suite of regression models (i.e., diminished presence in Tables 1 and 2). These findings support previous relationships between water quality proxies and coral species richness in CNMI [18,20], and reinforce that coral species composition may be a sensitive metric of water quality. It is beyond the purview of this study to formally discuss the linkages between diversity and ecosystem function, however, diversity is well known to facilitate functional redundancy in ecological systems, thereby providing for enhanced resistance and recovery to disturbance. Pollution contribution appeared to be influential to reefs where high human development existed, however, extensive human development was not common within the majority of CNMI's coastal watersheds. In addition, anchor points in the water quality regression models were driven by the presence of interstitial reefs associated with higher groundwater connectivity (Figure 1, Table 2, and Table S3). We conclude that watershed restoration strategies aimed at improving reef condition might obviously focus upon the largest urban centers, but less obviously, focus upon karst watersheds adjacent to high-value reef assemblages. We note that a major watershed restoration project addressing Laolao Bay (eastern Saipan) remains ongoing, and if successful would serve to address one of the key anchor points.

Conclusions

Over the past 12 years in CNMI, a period of high COTS densities led to significant coral declines. Yet, the failure of some reefs to recovery was best attributed to localized stressors, which transformed the substrates opened up by coral loss into persistent stands of turf and macroalgae, less conducive for coral replenishment and recovery. Declining trends were strongest for reefs that have favorable geomorphology (i.e., a gently sloping reef foundation), which disproportionally occur on Saipan alongside lower wave exposure. These same reefs represent centers for reef-based tourism that constitutes a major part of CNMI's economy, highlighting a need to improve upon compromised fish assemblages, grazing urchin populations, and specific localities where water quality concerns exist.

Supporting Information

Table S1 Monitoring site frequencies. Monitoring frequency for each of the long-term sites incorporated into the present study (*see Fig. 1*). Lowercase letters indicate the type of survey conducted in each year: (b) benthic substrate, (i) macroinvertebrate, (c) coral, and (f) fish.

Table S2 Site-based coral coverage and *Acanthaster* density data. Coral coverage and *Acanthaster planci* density summary statistics for each of the long-term monitoring sites incorporated into the present study. Site-based data formed the basis for regression modeling (Table 1). Reeftypes follow: "sg" - optimal spur-and-groove structures, "int" - high-relief, interstitial framework, "rot" - low relief Holocene framework found on Rota only, and "pl" - incipient coral assemblages residing upon a Pleistocene basement (*see methods*).

Table S3 Site-based summary statistics for regression models. Summary statistics for each of the long-term monitoring sites incorporated into the present study. Site-based data formed the basis for regression modeling. Dependent variables included the net change in the benthic substrate ratio and coral assemblages, noted as the sum of the percent decline (−) and subsequent recovery (+) of these ecological metrics (*see methods*). Reeftypes follow: "sg" - optimal spur-and-groove structures, "int" - high-relief, interstitial framework, "rot" - low relief Holocene framework found on Rota only, and "pl" - incipient coral assemblages residing upon a Pleistocene basement (*see methods*).

Acknowledgments

We are grateful for continuous discussions and support from several local and federal resource management agencies, including the United States Environmental Protection Agency Region IX, National Oceanic and Atmospheric Administration Coral Reef Conservation Program, and the Commonwealth of the Northern Mariana Islands (CNMI) Bureau of Environmental and Coastal Quality. We especially thank CNMI agency directors Frank Rabauliman and Frances Castro for logistical support throughout our study. Finally, several reviewers and the topic editor provided constructive criticism that greatly benefitted our manuscript.

Author Contributions

Conceived and designed the experiments: PH. Performed the experiments: PH DB JI SJ RO. Analyzed the data: PH DB JI SJ RO. Contributed reagents/materials/analysis tools: PH. Wrote the paper: PH DB JI SJ RO.

References

1. Hughes TP, Rodrigues MJ, Bellwood DR, Ceccarelli D, Hoegh-Guldberg O, et al. (2007) Phase shifts, herbivory, and the resilience of coral reefs to climate change. Curr Biol 17: 360–365.
2. Baker AC, Glynn PW, Riegl B (2008) Climate change and coral reef bleaching: An ecological assessment of long-term impacts, recovery trends and future outlook. Estuar Coastal Shelf S 80: 435–471.
3. Graham NAJ, Nash KL, Kool JT (2011) Coral reef recovery dynamics in a changing world. Coral Reefs 30: 283–294.
4. Hughes TP, Graham NAJ, Jackson JBC, Mumby PJ, Steneck RS (2010) Rising to the challenge of sustaining coral reef resilience. Trends Ecol Evol 25: 633–642.
5. Fabricius K, De'ath G, McCook L, Turak E, Williams DM (2005) Changes in algal, coral and fish assemblages along water quality gradients on the inshore Great Barrier Reef. Mar Pollut Bull 51: 384–398.
6. Mumby PJ, Dahlgren CP, Harborne AR, Kappel CV, Micheli F, et al. (2006) Fishing, trophic cascades, and the process of grazing on coral reefs. Science 311: 98–101.
7. Costa OS, Leao Z, Nimmo M, Attrill MJ (2000) Nutrification impacts on coral reefs from northern Bahia, Brazil. Hydrobiologia 440: 307–315.
8. Lapointe BE, Barile PJ, Matzie WR (2004) Anthropogenic nutrient enrichment of seagrass and coral reef communities in the Lower Florida Keys: discrimination of local versus regional nitrogen sources. J Exp Mar Biol Ecol 308: 23–58.
9. Houk P, Musburger C, Wiles P (2010) Water Quality and Herbivory Interactively Drive Coral-Reef Recovery Patterns in American Samoa. Plos One 5: e13913.
10. Mumby PJ, Harborne AR (2010) Marine Reserves Enhance the Recovery of Corals on Caribbean Reefs. Plos One 5: e8657.
11. Walsh SM (2010) Ecosystem-scale effects of nutrients and fishing on coral reefs. J Mar Biol doi:10.1155/2011/187248.
12. Dulvy NK, Freckleton RP, Polunin NV (2004) Coral reef cascades and the indirect effects of predator removal by exploitation. Ecol Lett 7: 410–416.
13. Aronson RB, Precht WF (2006) Conservation, precaution, and Caribbean reefs. Coral Reefs 25: 441–450.
14. De'ath G, Fabricius K (2010) Water quality as a regional driver of coral biodiversity and macroalgae on the Great Barrier Reef. Ecol App 20: 840–850.
15. Connell JH, Hughes TP, Wallace CC (1997) A 30-year study of coral abundance, recruitment, and disturbance at several scales in space and time. Ecol Monogr 67: 461–488.
16. Gardner TA, Côté IM, Gill JA, Grant A, Watkinson AR (2005) Hurricanes and Caribbean coral reefs: impacts, recovery patterns, and role in long-term decline. Ecology 86: 174–184.
17. Done TJ (1999) Coral community adaptability to environmental change at the scales of regions, reefs and reef zones. Am Zool 39: 66–79.
18. Houk P, van Woesik R (2010) Coral assemblages and reef growth in the Commonwealth of the Northern Mariana Islands (Western Pacific Ocean). Mar Ecol 31: 318–329.
19. Chollett I, Mumby PJ, Müller-Karger FE, Hu C (2012) Physical environments of the Caribbean Sea. Lim Oceanogr 57: 1233.
20. Cooper TF, Gilmour JP, Fabricius KE (2009) Bioindicators of changes in water quality on coral reefs: review and recommendations for monitoring programmes. Coral Reefs 28: 589–606.
21. Perry CT, Murphy GN, Kench PS, Smithers SG, Edinger EN, et al. (2013) Caribbean-wide decline in carbonate production threatens coral reef growth. Nat Commun 4: 1402.
22. McClanahan TR (2008) Response of the coral reef benthos and herbivory to fishery closure management and the 1998 ENSO disturbance. Oecologia 155: 169–177.
23. Bak RPM, Meesters EH (1998) Coral population structure: the hidden information of colony size-frequency distributions. Mar Ecol Prog Ser 162: 301–306.
24. Bak RPM, Meesters EH (1999) Population structure as a response of coral communities to global change. Am Zool 39: 56–65.
25. Loya Y, Sakai K, Yamazato K, Nakano Y, Sambali H, et al. (2001) Coral bleaching: the winners and the losers. Ecol Lett 4: 122–131.
26. Brown BE, Clarke KR, Warwick RM (2002) Serial patterns of biodiversity change in corals across shallow reef flats in Ko Phuket, Thailand, due to the effects of local (sedimentation) and regional (climatic) perturbations. Mar Biol 141: 21–29.
27. Pratchett MS, Trapon M, Berumen ML, Chong-Seng K (2011) Recent disturbances augment community shifts in coral assemblages in Moorea, French Polynesia. Coral Reefs 30: 183–193.
28. van Woesik R, Sakai K, Ganase A, Loya Y (2011) Revisiting the winners and the losers a decade after coral bleaching. Mar Ecol Prog Ser 434: 67–76.
29. Guest JR, Baird AH, Maynard JA, Muttaqin E, Edwards AJ, et al. (2012) Contrasting Patterns of Coral Bleaching Susceptibility in 2010 Suggest an Adaptive Response to Thermal Stress. Plos One 7: e33353.
30. Houk P, Benavente D, Fread V (2012a) Characterization and evaluation of coral reefs around Yap Proper, Federated States of Micronesia. Biodivers Conserv 21: 2045–2059.
31. Blackwood JC, Hastings A, Mumby PJ (2011) A model-based approach to determine the long-term effects of multiple interacting stressors on coral reefs. Ecol App 21: 2722–2733.
32. McClanahan TR, Maina J, Starger CJ, Herron-Perez P, Dusek E (2005) Detriments to post-bleaching recovery of corals. Coral Reefs 24: 230–246.
33. Burkepile DE, Hay ME (2006) Herbivore vs. nutrient control of marine primary producers: Context-dependent effects. Ecology 87: 3128–3139.
34. Banse K (2007) Do we live in a largely top-down regulated world? J Bioscience 32: 791–796.

35. Mork E, Sjoo GL, Kautsky N, McClanahan TR (2009) Top-down and bottom-up regulation of macroalgal community structure on a Kenyan reef. Estuar Coastal Shelf S 84: 331–336.
36. Sjoo GL, Mork E, Andersson S, Melander I (2011) Differences in top-down and bottom-up regulation of macroalgal communities between a reef crest and back reef habitat in Zanzibar. Estuar Coastal Shelf S 91: 511–518.
37. Wilson SK, Graham NAJ, Fisher R, Robinson J, Nash K, et al. (2012) Effect of Macroalgal Expansion and Marine Protected Areas on Coral Recovery Following a Climatic Disturbance. Conserv Biol 26: 995–1004.
38. Houk P, Van Woesik R (2006) Coral reef benthic video surveys facilitate long-term monitoring in the commonwealth of the Northern Mariana Islands: Toward an optimal sampling strategy. Pac Sci 60: 177–189.
39. Starmer J, Houk P (2008) Marine and Water Quality Monitoring Plan for the Commonwealth of the Northern Mariana Islands. CNMI Division of Environmental Quality, Saipan, MP.
40. Houk P, Bograd S, van Woesik R (2007) The transition zone chlorophyll front can trigger Acanthaster planci outbreaks in the Pacific Ocean: Historical confirmation. J Oceanogr 63: 149–154.
41. Keats DW, Chamberlain YM, Baba M (1997) Pneophyllum conicum (Dawson) comb nov (Rhodophyta, Corallinaceae), a widespread Indo-Pacific non-geniculate coralline alga that overgrows and kills live coral. Bot Mar 40: 263–279.
42. Antonius A (1999) Metapeyssonnelia corallepida, a new coral-killing red alga on Caribbean Reefs. Coral Reefs 18: 301–301.
43. Antonius A (2001) Pneophyllum conicum, a coralline red alga causing coral reef-death in Mauritius. Coral Reefs 19: 418–418.
44. O'Leary JK, Potts DC, Braga JC, McClanahan TR (2012) Indirect consequences of fishing: reduction of coralline algae suppresses juvenile coral abundance. Coral Reefs 31: 547–559.
45. Veron JEN (2000) Corals of the World. Stafford-Smith, Townsville.
46. Bohnsack JA, Bannerot SP (1986) A stationary visual census technique for quantatively assessing community structure of coral reef fishes. National Oceanic and Atmospheric Administration Technical Report NMFS 41.
47. Ekebom J, Laihonen P, Suominen T (2003) A GIS-based step-wise procedure for assessing physical exposure in fragmented archipelagos. Estuar Coastal Shelf S 57: 887–898.
48. Zar J (1999) Biostatistics. Prentice Hall, New Jersey.
49. Zuur AF, Ieno EN, Walker NJ, Saveliev AA, Smith GM (2009) Mixed effects models and extensions in ecology with R. Springer, New York.
50. Anderson M, Gorley R, Clarke K (2008) PERMANOVA+ for PRIMER: Guide to software and statistical methods. PRIMER-E, Plymouth, UK.
51. R Development Core Team RDC (2008) R: A language and environment for statistical computing. R foundation for statistical computing, Vienna, Austria. Available: http://wwwR-projectorg.
52. McCook LJ, Ayling T, Cappo M, Choat JH, Evans RD, et al. (2010) Adaptive management of the Great Barrier Reef: A globally significant demonstration of the benefits of networks of marine reserves. Proc Nat Acad Sci USA 107: 18278–18285.
53. Gilmour JP, Smith LD, Heyward AJ, Baird AH, Pratchett MS (2013) Recovery of an isolated coral reef system following severe disturbance. Science 340: 69–71.
54. Adams TC, Schmitt RJ, Holbrook SJ, Brooks AJ, Edmunds PJ, et al. (2011) Herbivory, connectivity, and ecosystem resilience: response of a coral reef to a large-scale perturbation. PLoS ONE 5: e23717.
55. Glynn PW, Enochs IC, Afflerbach JA, Brandtneris VW, Serafy JE (2014) Eastern Pacific reef fish responses to coral recovery following El Nino disturbances. Mar Ecol Prog Ser 495: 233–247.
56. Graham T (1994) Biological analysis of the nearshore reef fish fishery of Saipan and Tinian. CNMI Division of Fish and Wildlife Technical Report 94–02, Saipan, MP.
57. Trianni MS (1998) Summary and Further Analysis of the Nearshore Reef Fishery of the Northern Mariana Islands. Technical report submitted to the CNMI Division of Fish and Wildlife, Saipan, MP.
58. Houk P, Rhodes K, Cuetos-Bueno J, Lindfield S, Fread V, et al. (2012b) Commercial coral-reef fisheries across Micronesia: A need for improving management. Coral Reefs 31: 13–26.
59. Maynard J, McKagan S, Johnson S, Houk P, Ahmadia G, et al. (2012) Coral reef resilience to climate change in Saipan, CNMI; field-based assessments and implications for vulnerability and future management. Technical report submitted to the CNMI Division of Environmental Quality, Saipan, MP.
60. Halford A, Cheal AJ, Ryan D, Williams DM (2004) Resilience to large-scale disturbance in coral and fish assemblages on the Great Barrier Reef. 85: 1892–1905.
61. Brodie J, Fabricius K, De'ath G, Okaji K (2005) Are increased nutrient inputs responsible for more outbreaks of crown-of-thorns starfish? An appraisal of the evidence. Mar Pollut Bull 51: 266–278.
62. Fabricius K, Okaji K, De'ath G (2010) Three lines of evidence to link outbreaks of the crown-of-thorns seastar Acanthaster planci to the release of larval food limitation. Coral Reefs 29: 593–605.
63. Bearden C, Chambers D, Okano R (2012) Commonwealth of the Northern Mariana Islands Integrated 305(b) and 303(d) Water Quality Assessment Report. CNMI Division of Environmental Quality Saipan, MP.
64. Cloud PE (1959) Geology of Saipan, Mariana Islands, Part 4. Submarine topography and shoal-water ecology. Geological Survey Professional Paper 280-K, Washington, DC.
65. Crossman DJ, Choat JH, Clements KD, Hardy T, McConochie J (2001) Detritus as food for grazing fishes on coral reefs. Limnol Oceanogr 46: 1596–1605.
66. van Beukering P, Haider W, Wolfs E, Liu Y, van der Leeuw K, et al. (2006) The economic value of the coral reefs of Saipan, Commonwealth of the Northern Mariana Islands. Report prepared by Cesar Environmental Consulting for the National Oceanic and Atmospheric Administration, Silver Springs, MD.
67. Peters RH (1983) The ecological implications of body size. Cambridge University Press.
68. Lokrantz J, Nyström M, Thyresson M, Johansson C (2008) The non-linear relationship between body size and function in parrotfishes. Coral Reefs 27: 967–974.
69. Birkeland C, Dayton PK (2005) The importance in fishery management of leaving the big ones. Trends Ecol Evol 20: 356–358.
70. Houk P, Musburger C (2013) Trophic interactions and ecological stability across coral reefs in the Marshall Islands. Mar Ecol Prog Ser 488: 23–34.
71. Bellwood DR, Hoey AS, Hughes TP (2012) Human activity selectively impacts the ecosystem roles of parrotfishes on coral reefs. Proc R Soc B 279: 1621–1629.
72. Rasher DB, Hoey AS, Hay ME (2013) Consumer diversity interacts with prey defenses to drive ecosystem function. Ecology 94: 1347–1358.
73. Graham NAJ, Nash KL (2013) The importance of structural complexity in coral reef ecosystems. Coral Reefs 32: 315–326.

12

Landscape Management of Fire and Grazing Regimes Alters the Fine-Scale Habitat Utilisation by Feral Cats

Hugh W. McGregor[1,2]*, Sarah Legge[3], Menna E. Jones[1], Christopher N. Johnson[1]

1 School of Biological Sciences, University of Tasmania, Hobart, Tasmania, Australia, **2** Australian Wildlife Conservancy, Derby, Western Australia, Australia, **3** Australian Wildlife Conservancy, Mornington Wildlife Sanctuary, Derby, Western Australia, Australia

Abstract

Intensification of fires and grazing by large herbivores has caused population declines in small vertebrates in many ecosystems worldwide. Impacts are rarely direct, and usually appear driven via indirect pathways, such as changes to predator-prey dynamics. Fire events and grazing may improve habitat and/or hunting success for the predators of small mammals, however, such impacts have not been documented. To test for such an interaction, we investigated fine-scale habitat selection by feral cats in relation to fire, grazing and small-mammal abundance. Our study was conducted in north-western Australia, where small mammal populations are sensitive to changes in fire and grazing management. We deployed GPS collars on 32 cats in landscapes with contrasting fire and grazing treatments. Fine-scale habitat selection was determined using discrete choice modelling of cat movements. We found that cats selected areas with open grass cover, including heavily-grazed areas. They strongly selected for areas recently burnt by intense fires, but only in habitats that typically support high abundance of small mammals. Intense fires and grazing by introduced herbivores created conditions that are favoured by cats, probably because their hunting success is improved. This mechanism could explain why, in northern Australia, impacts of feral cats on small mammals might have increased. Our results suggest the impact of feral cats could be reduced in most ecosystems by maximising grass cover, minimising the incidence of intense fires, and reducing grazing by large herbivores.

Editor: Paul Adam, University of New South Wales, Australia

Funding: The research was funded by the Australian Research Council, the Western Australian Department of Parks and Wildlife (Kimberley Science and Conservation Strategy), and supporters of the Australian Wildlife Conservancy. The funders had no role in study design, data collection and analysis, decision to publish, or preparation of the manuscript.

Competing Interests: The authors have declared that no competing interests exist.

* Email: hugh.mcgregor@australianwildlife.org

Introduction

Predator-prey relationships are strongly influenced by the structure and quality of habitat, principally its vegetation [1,2,3]. Variability in vegetation structure may be used by predators to increase hunting success. For example, lions use dense vegetation to hide their approach from prey [4]. Also, it may be used by prey to help them evade predators, such as elk using woodlands as a refuge from wolves [5]. Changes in habitat structure may therefore shift the relationships between predators and prey [6,7]. Such changes can determine the extent to which some prey are threatened with extinction by heavy predation [2,8].

One of most pervasive impacts on vegetation structure arises from changes to fire and grazing regimes. While drivers of such changes vary immensely, the impacts on fauna communities display some general trends. For example, small mammal populations are especially sensitive, and the vast majority of studies detecting declines in populations in response to either intense fire events or intense grazing [9,10,11,12,13], unless they occur in ecosystems with few predators [14,15]. The underlying mechanisms of these declines remain elusive, but are likely to be indirect rather than through direct effects such as being burnt by the fires or trampled by cattle [16]. Instead, such disturbances may improve habitat for predators in ways that increases their impacts on prey [9,17,18], although no field data are available to confirm this.

Many small mammal species are declining in the savannas of northern Australia, and several are threatened with extinction [18,19]. Declines have been greatest in areas subject to intense fires [20,21] and recent experimental evidence also supports an association of grazing by introduced herbivores (cattle, horses, donkeys, buffalo) with the magnitude of small-mammal decline [22]. Both fire and grazing regimes in northern Australia have intensified substantially over recent decades in ways that could contribute to the contemporary native mammal decline. These changes to fire and grazing have generally made grass communities less complex and more open [23].

Predation by feral cats *Felis catus* may also be contributing to the declines. This is suggested by three lines of evidence. First, the declining species fall within the preferred prey-size range of cats [18,24]. Second, mammal populations in complex rocky habitats have been less affected than those in more productive woodlands and savanna, suggesting a predation effect [25,26]. Finally, populations of declining mammal species are more stable in the absence of cats, on island or in large enclosures [27,28]. However, there is a temporal mismatch between the arrival of cats in northern Australia up to 170 years ago [29], and mammal declines observed in the last 20 years [19].

The apparent mismatch in timing of the early arrival of cats and recent mammal declines could be explained by the hypothesis that cat predation has its largest impact when it interacts with fire and grazing regimes established in the more recent past. Reduction of structural complexity of vegetation and increased openness due to fire and grazing might increase the exposure of small mammals to predators, making prey easier to detect and capture [15]. Small mammals are cats' preferred prey [24,30]. If cats preferentially use the open and relatively simple habitats created by fire and grazing, the result could be a higher predation impacts on small mammals. This has been suggested as a possible mechanism for these declines [19,31], however, until now there has been no evidence.

If cats do favour the conditions created by fire and grazing, this should be revealed by the patterns of movement of individual cats in heterogeneous landscapes with variable effects of fire and grazing. We tested this hypothesis using intensive GPS tracking of a large sample of individual cats; both within and outside of a large 40 300 hectare area that has been destocked of all introduced herbivores (cattle, horses, and donkeys) [22], and spanning contrasting fire patterns (mild control fires or intensive wildfires). A dynamic habitat map was created so that fire and vegetation attributes at any location or point in time could be determined. This was used to generate a parsimonious model of fine-scale habitat selection by cats. We predicted that feral cats would select for open grass cover to improve hunting success. If so, we hypothesise that cats would increase their use of habitats that have been recently burnt or intensely grazed, and that this relationship would be stronger in areas of high small-mammal abundance.

Methods

Study area

Our study area encompassed three large properties in the central Kimberley of north-western Australia (**17°01′S, 126°01′E**, see Fig. 1). One property is managed for commercial cattle production (Glenroy, 1455 km^2) and two are ex-pastoral leases managed for conservation by the Australian Wildlife Conservatory (Mornington and Marion Downs Sanctuaries, 3225 km^2 and 2676 km^2 respectively). Habitats are mostly savanna woodlands with a perennial grass layer, dissected by riparian vegetation along the edges of creeks. The region has a tropical monsoon climate with three broad seasons: the wet (December – March), early dry (April – July) and late dry (August – November). Fire is managed on all three properties to promote biodiversity values. This fire management aims to reduce the incidence of extensive, high-intensity, uncontrolled fires in the late dry season using strategic prescribed burning in the early dry season when fires tend to be small and of low intensity because of weather and condition of the grass layer. In addition, when uncontrolled late dry season fires occur, they are suppressed where possible. All large introduced herbivores (cattle, horses and donkeys) have been removed from a 40,300 ha fenced section of Mornington since 2005 [22], - hereafter referred to as the 'destocked' zone. The only other large mammalian predator in the study area is the dingo *Canis dingo*. Whilst they are controlled elsewhere in Australia, they are not persecuted in the study area, and occurred at a density of ~0.2 per km^2 [32].

Cat capture and tracking

Feral cats were captured between September 2010 and June 2013, using either large wire cage traps, leg-hold traps (soft-jaw, size #1.5) or by spotlighting and netting with the assistance of dogs trained to locate and bail cats up trees. If a cat was either bailed up a tree or required examination of possible injury, it was sedated with Zolotil at a rate of 0.5 cc/kg via intramuscular injection. Cats were fitted with GPS collars (Telemetry Solutions Quantum 4000 enhanced). Cats weighing between 2 and 3.3 kg were fitted with a 70 g collar, and those weighing more than 3.3 kg were fitted with a 100 g collar (<3% of body-weight). Sedated cats were released after full muscle control was regained (4–6 hours later), and non-sedated cats were released as soon as possible (2–5 minutes later). When it was necessary to replace GPS collars, the cats were recaptured using the dogs.

GPS collars were deployed on equal numbers of cats in the stocked and destocked zone, and between burnt and unburnt areas. Within burnt areas, the cats were split evenly between areas with low and high intensity fires (Table 1). The GPS units were programmed to record fixes every 15 minutes for two-day bouts, starting and finishing at 12 pm. These bouts were separated by intervals of one, two or fourteen days. All bouts were timed to commence at least 24 hrs after the cat was handled. Units were programmed to search for a satellite for 60 seconds, and to remain on for at least 5 seconds to refine the fix if there was memory from the last fix, or 15 seconds if not.

Habitat variables

Across the study area, we developed habitat maps relating to fire and grazing, along with any other variable likely to influence cat habitat selection or movement. Where necessary, the maps were updated to make them temporally dynamic, so that attributes at any given time and location could be determined. Most descriptors of habitat related to the ground layer, rather than the tree layer. Nine distinct grass communities common in the region (see Table 2) were mapped by examining colour disjunctions on aerial photos while altering light levels in Photoshop Elements v. 8, tracing the boundaries of discrete polygons onto the aerial photos, then geo-rectifying these boundaries in ArcGIS v.10. For example, spinifex *Triodia* spp. grasslands are uniquely green in dry-season aerial photos, while communities dominated by bluegrass *Dichanthium fecundum* are white. The digital map was ground-truthed at 768 plots (described later); the attribution of grass community was correct at 96% of sites.

Fire scars were initially mapped using monthly Landsat 7 remote-sensing imagery available from the US Geological Survey (2011–2013), and fire boundaries were then refined using aerial photography taken from a helicopter flying approximately 300 m above ground. For each burnt area, we assigned the date of burn and intensity (intense = 100% tree scorch and no ground cover remaining unburnt, or mild = all other fires). Relative to the date of each GPS fix, fire was considered in multiple binary variables at 30, 60, 90, 180, 360 or 600 days since fire. The Australian Wildlife Conservancy's stock-proof fence [22] separated the stocked and destocked areas (see Fig. 1).

A dynamic map was created that estimated grass cover at any given location and time since fire, based on a series of models of the response of the grass layer to fire (given grass community, stocking status). They were created from field data, where vegetation attributes were measured at 768 plots (each 10 m^2) across the study area and duration, spread equally across grass communities (see Table 1) and combinations of mild/intense fire, time since fire, and grazed and destocked areas (total of 96 plots per community). At each plot, we estimated the extent of grass

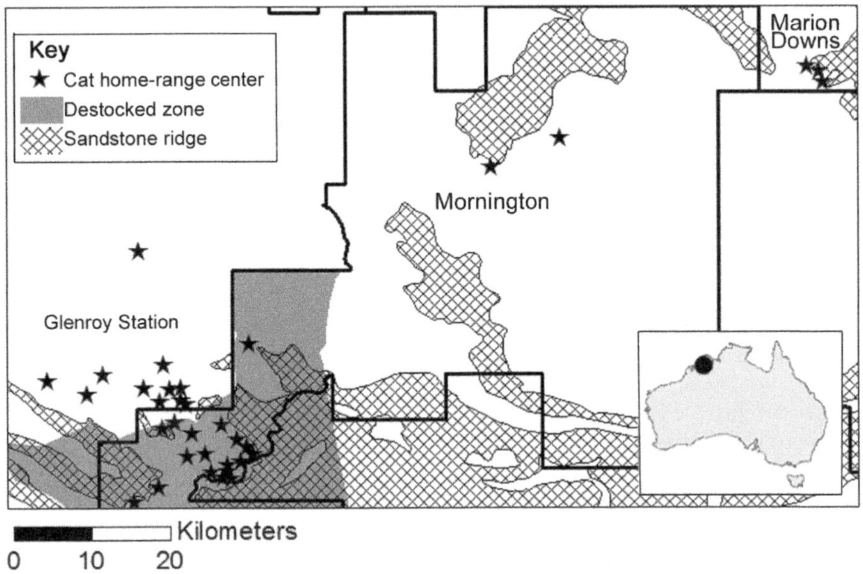

Figure 1. Map of study area in north-west Australia (see inset), including home-range centroids of feral cats used in this study. The dark grey represents the destocked zone.

cover at different heights by adapting a line-intercept method. We inserted a 100 cm pole (diameter of 1.5 cm) vertically through the grass to the ground at 50 points in a systematic grid over the plot. The number of grass intercepts was recorded in height intervals of 0–10 cm, 11–30 cm, and 31–100 cm. This was used to derive grass cover, cover of dense tussocks, and relative biomass at each plot (see Material S1 for more detail on methods). These variables were modelled against grass community, time since last fire, intensity of fire, and stocking status. Models were then used to derive values for all GPS fixes and random points used in discrete choice models (see Material S1).

The influence of grazing on grass biomass was measured as the difference in the average total number of grass intercepts per plot between grass communities in the stocked and destocked sites. We used only unburnt sites for this comparison. These averages were converted into a grazing impact rank. However, these variables would be confounded by correlation as the habitats favourable to cattle would also be favoured by feral cats (e.g. riparian areas), irrespective of impacts of cattle. Therefore, this score or rank was applied to all fixes, grazed or not, and the actual impacts of cattle grazing considered as the difference in this variable's strength between the destocked zone and outside.

Relative small-mammal abundances in different habitats were estimated from the Australian Wildlife Conservancy's annual fauna monitoring data, which is carried out across all three properties [32]. The sample at each site is based on 60 trap-nights using small mammal box traps (Elliott traps) and 12 nights of possum-sized wire traps, with 55 to 64 sites sampled each year - totalling 176 site-years between 2011–3. Small mammals were defined as those weighing between 30 g and 2000 g [18,24], and were predominantly *Rattus tunneyi* and *Pseudomys nanus* (these species comprised 87% of captures). For each year, the average rates of capture of individual small mammals were calculated for each grass community in the grazed and destocked areas (see Table 2). However, as sites were typically larger than the mapped patterns of these grass communities, results from some communities were combined (see Table 2). In the event cats altered their response to a threshold of small mammal abundance, we also included three binary variables of abundance (whether there was more than 1, 2 or 5 small mammal captures per plot) for each habitat.

We measured other spatial and temporal features likely to be important to cats to provide context for their movements. As cats may prefer hunting on edges between open and dense vegetation, we delineated a 25 m buffer either side of any linear boundary where grass cover was <25% on one side and >50% on the other (these were most often fire edges). Water features were mapped in the field, and each assigned a descriptor for the seasons that they contained water (wet season only, wet and early dry, or year-round). This was used to derive distance to known standing water at any given time throughout the study. Elevation and slope were derived from a digital elevation model of 15 m resolution. We

Table 1. The number of cats fitted with GPS collars in each of the different grazing and fire treatments. Destocked means that all introduced herbivores are excluded.

Management	No fire	Mild fire	Intense fire	Total
Stocked	8 (5♂ 3♀)	4 (3♂ 1♀)	4 (3♂ 1♀)	**16**
Destocked	8 (6♂ 2♀)	4 (4♂)	4 (4♂)	**16**
Total	**16**	**8**	**8**	**32**

Table 2. Vegetation communities mapped in the study region.

			Small mammal abundance					
		Grazing rank	Stocked			Destocked		
Grass community	**Dominant grasses**		**2011**	**2012**	**2013**	**2011**	**2012**	**2013**
Riparian forest[A]	*Chrysopogon fallax*	1	0.2	0.3	0.8	1.2	3.4	9.4
	Mnesithea rottboellioides							
	Various introduced grasses							
Alluvial grasslands[A]	*Crysopogon fallax*	2	0.2	0.3	0.8	1.2	3.4	9.4
	Dichanthium fecundum							
Bluegrass plains	*Aristida* spp.	3	1	1.3	4.6	0.3	2.6	2.9
	Dichanthium fecundum							
Canegrass[A]	*Mnesithea rottboellioides*	4	0.2	0.3	0.8	1.2	3.4	9.4
Mixed woodlands	*Aristida* spp.	5	0	1.3	0.6	0.2	3.8	6
	Dichanthium fecundum							
	Heteropogon contortus							
Sandseep	*Triodia* spp.	6	1.7	1.2	1.2	1.8	7.4	16.8
Hillside woodlands[B]	*Sehima nervosum*	7	0	0.6	1	0.4	3.2	4
Spinifex woodlands[B]	*Triodia* spp.	8	0	0.6	1	0.4	3.2	4
Bare ground	No or little	9	0	0	0	0.2	0.2	0.5

[A] and [B] denote grass communities where the small mammal abundance was grouped, as sites were typically larger than the mapped distributions of these communities.
Vegetation classes are ranked by their preference for grazing by domestic stock, with 9 the most impacted and 1 the least (Grazing rank). Small mammal abundance is number of small mammals captured per 100 trap nights in the stocked and destocked zone, from 2011 to 2013.

created a variable representing the spatial home-range context for each cat, by making a kernel density estimate (smooth cross-validation) for all moving fixes of each cat, and delineating kernel isopleths at 50%, 90%, 95% and 99% contours. Temporal variables included sun time [33], season (wet, early dry or late dry), number of months since the end of the wet season, and minimum nightly temperature (HOBO temperature data logger, Micro-DAQ). Finally, a cat's choice of which habitat to select in successive GPS fixes may be dependent on the type of habitat it was last in. This spatial correlation might arise when an animal selects for certain landscape features and tends to remain within them for some time, rather than making a *de novo* selection at every 15 min interval of their walk. Therefore, we fitted a binary 'carry-over' variable, which described whether the vegetation, grazing and fire values were equal to the previous fix.

Organisation of data

All GPS fixes that were likely to be erroneous, biased or to represent a stationary cat were removed from the analysis. Erroneous fixes were those representing implausible 'spikes' in movement, presumably caused by GPS error [34]. We deemed it unlikely that a cat would suddenly change course and speed, then return to the same area it was in 15 minutes ago, so spikes were defined where fixes met all the following criteria: distances from the last fix were >50 m, difference in distance from the preceding and succeeding fix was <10%, and turning angle >170° [34,35]. The HDOP values were not used to filter fixes, as a test of six GPS collars found no relationship between HDOP and distance from the GPS fix to the known location. Fixes that were potentially biased by human disruption of the behaviour of the cat were removed, being those within three hours of field VHF tracking (this was occasionally carried out in order to download data remotely from the cat GPS collar). Fast-moving fixes were also removed, as it was likely cats were moving away from something, rather than choosing habitat. For this purpose a filtering speed of greater than 2 km per hour was used, as this was where the histogram of speeds between fixes reached an asymptote [35], representing a shift in behaviour mode. A test of GPS error within the open savanna found that 95% of fixes had <5 m error (from 634 fixes on six different collars), so fixes less than 10 m from the preceding fix were classed as stationary. We considered only moving fixes, as cats may have different habitat requirements for resting versus hunting.

Data analysis

Habitat selection by cats was analysed using discrete choice modelling [36]. The range of 'available' habitats was calculated for each fix, and then we compared the option selected by the cat to the available habitats. To find the available habitats, we first constructed probability distributions of a cat's step-length and bearing over 15-minute intervals, then used these probability distributions to select five random points to sample the cat's options [36]. Each GPS fix and associated random points were attributed with the habitat variables of interest using the dynamic vegetation map.

We determined resource selection by cats by creating models with all combinations of variables, including different biologically relevant interaction terms, and comparing them within an information theory framework [37]. This produced a total of 916 models. For each interaction term a model was included with all combinations, or with only significant combination terms retained. No pairs of variables with Pearson's correlation greater than 0.5 were included in the same model. Models for habitat selection were created using standard case-control logistic regression models, and were implemented in R [38] using the 'clogit' command from the 'survival' library. Each individual cat was considered as a mixed effect in the models, using Gaussian frailty [39]. The most parsimonious models of cat habitat selection were chosen as those with an AIC score within two points of the highest-ranked model [37], and only these are presented in the results. The cat's selection is measured as an odds ratio representing the magnitude of change in the odds of selection for each unit of the predictor variable. Differences in the odds ratio are relative only to the other habitat choices immediately available to a cat.

Ethics statement

All data collection fulfilled all legal requirements in Australia, and has been approved by University of Tasmania Animal Ethics Committee (A0011661) and Western Australian Department of Parks and Wildlife Animal Ethics Committee (2010/35), with a Regulation 17 licence to research animals (SF009379). All research was conducted with permission on three pastoral leases; Mornington Wildlife Sanctuary, Marion Downs (both managed by the Australian Wildlife Conservancy, ph; +61 8 9191 7014), and Glenroy Station (ph: +61 8 9191 4703). All three leases are located around 17°01′S, 126°01′E. Field studies did not involve protected or endangered species.

Results

In total, 60 cats were captured between September 2010 and June 2013. Three cats were caught in wire cage traps (265 trap nights), 19 in leg-hold traps (940 trap nights) and 38 by spotlighting/netting with trained cat-dogs (221 hours). GPS collars were placed on 37 cats, and at least one month of GPS data was obtained from 32 cats whose locality was spread equally across stocking and fire management treatments (Table 1). There was a strong male bias in the sample of cats: males comprised 78% of all captured cats (47/60) and 78% of cats from which GPS data were obtained (25/32). Of the GPS-collared cats, four had disappeared and their fates were unknown at the end of the study, nine had died naturally, and the rest were euthanased.

From the 32 cats that provided useable GPS data, we obtained a total of 133 047 GPS fixes. Cats were moving 56% of time. Of these moving fixes, 62% were at night (between sunset and sunrise). Removal of erroneous or biased fixes and those representing high-speed movement left 38 472 choices for habitat between successive 15-minute moving fixes.

From the 916 models generated to describe habitat selection by cats, three were ranked within the candidate model set (the model with the lowest AIC score and two other models within two AIC scores of the top model). The top model carried 41% of the weight, compared to 24% and 20% for the second and third ranked models. Of these, the second and third ranked models were almost identical to the top model, but contained interaction terms that did not decrease the models AIC value. As these terms did not improve the maximum likelihood for these models, only the top model was considered further (Table 3).

The top model included negative selection for grass cover (equivalently, positive selection for open areas) and positive selection for edges (see Table 3). Both variables had a significant interaction with small-mammal density in a binary format (>2 captures per 100 trap nights), showing that cats selected particularly strongly for open areas in habitats with higher density of small mammals.

Fire was represented in the top model with fire scars up to 360 days old, fire scars <90 days old, whether the fire was intense, small-mammal abundance, and interaction terms between these

Table 3. Statistics of the top ranked model of cat habitat selection based on GPS data at 15 minute intervals from 32 individuals.

Variable	Odds ratio	robust SE	Z	Pr(>|z|)	
Grass cover with scarce small mammals[a]	−1.26	0.08	−2.88	0.004	**
Grass cover with abundant small mammals[a]	−1.85	0.1	−6.44	0.0001	***
Bare/grass edge with scarce small mammals[a]	1.2	0.07	2.81	0.005	**
Bare/grass edge with abundant small mammals[a]	1.41	0.06	6.26	<0.0001	***
Fire scar <360 days	−1.32	0.07	−4.09	<0.0001	***
Intense fire scar <360 days old	1.54	0.13	3.29	0.001	**
Intense fire scar <90 days old	−2.11	0.13	−5.67	<0.0001	***
Small mammal abundance (square-root)	3.52	0.21	6.03	<0.0001	***
Small mammal abundance, intense fire scar <360 days	−4.93	0.66	−2.4	0.0163	*
Small mammal abundance, intense fire scar <90 days	10.71	0.79	2.99	0.0028	**
Grass communities ranked on grazing impacts, in:					
- stocked areas during day	5.96	0.14	12.56	<0.0001	***
- destocked areas during day	3.52	0.25	5.07	<0.0001	***
- stocked areas over night	2.64	0.17	5.65	<0.0001	***
- destocked areas over night	2.44	0.16	5.67	<0.0001	***
Water proximity (km) by months into dry season	1.56	0.06	−7.77	<0.0001	***
- as above, by minimum nightly temperature (10°C)	1.13	0.03	3.94	<0.0001	***
Elevation (100m)	−1.93	0.28	−2.32	0.0202	*
Home range isopleth[b], adult female	3.53	0.05	27.17	<0.0001	***
Home range isopleth[b], adult male	2.28	0.1	−8.41	<0.0001	***
Home range isopleth[b], sub-adult	1.06	0.49	−0.12	0.9077	
Same habitat as last fix, if within 95% isopleth	2.06	0.02	41.4	<0.0001	***

[a]Scarce and abundant small mammals are defined as less than or greater than two individuals captured per 100 trap nights at Australian Wildlife Conservancy monitoring sites.
[b]Home range isopleth derived at 50, 90, 95 and 99% contours from kernel density estimator.
The odds ratio is the change in selection likelihood per unit of the variable.

variables. Once the odds ratios of these variables were combined, cats showed strong positive selection for areas within 90 days after an intense fire and where density of small-mammal prey was high (Fig. 2). However, in all other circumstances selection for areas burnt by intense or mild fires was negative.

Vegetation types with greater susceptibility to grazing impacts (e.g. riparian areas, Table 2) were strongly selected for in both the grazed and destocked zones. However, this relationship was significantly stronger in the stocked zone than in the destocked zone, especially during the day (see Fig. 3).

Several other variables influenced selection by cats. Cats chose to move towards water. The odds ratio of moving towards water became progressively stronger throughout the dry season (by 0.56 every month into the dry season) and on days with higher minimum temperature (by 0.13 for each 10°C increase). Cats selected against changes in elevation, with odds declining by 0.93 every 100 m. Cats tended to move into higher-use areas within their home range (determined by kernel density estimates), with an interaction with age and sex. Adult females demonstrated the strongest fidelity to home-range isopleths, while for sub-adults of either sex this was not significant ($P = 0.9$). Cats were twice as likely to select for the same habitats as the previous fix ($P<0.001$), assuming the fix was inside a cats' home range (95% isopleth).

Discussion

Our study provides a detailed analysis of the preferences that underlie movement decisions by feral cats in a tropical savanna environment in northern Australia. We show that modifications of habitat produced by grazing and by certain types of fire have strong effects on cat movement behaviour, with the general result that fire and grazing can create habitat conditions which are

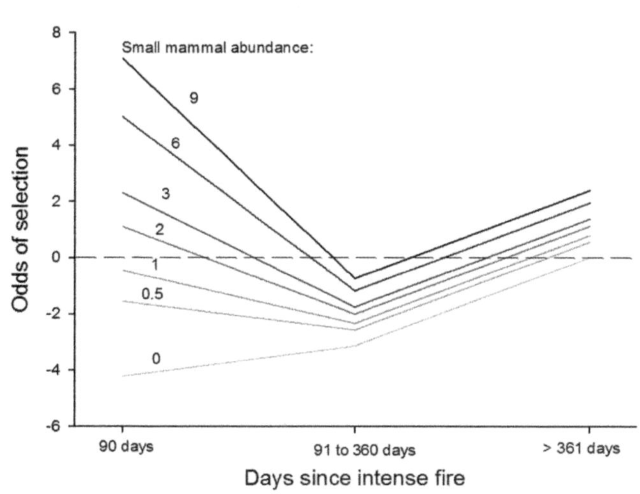

Figure 2. Change in the odds of selection ratios for different days since intense fires, at different average small mammal abundances based on capture rates per 100 trap nights (ranging from 0 to 9, lighter to darker respectively). All other variables in the model are assumed to be constant (see Table 3).

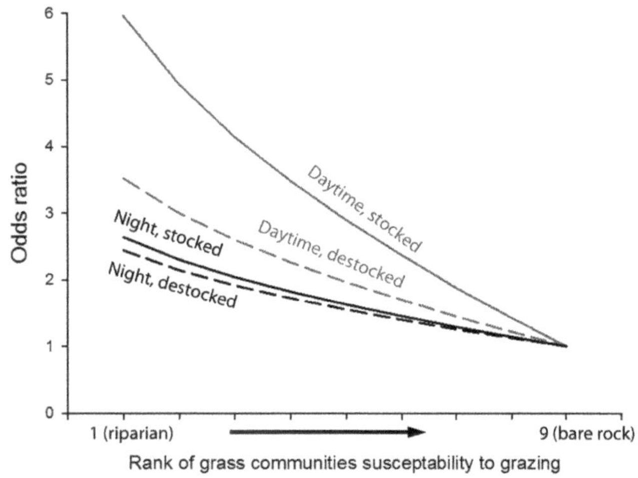

Figure 3. Odds ratios for selection of cats at night (black) and day (grey) in stocked (solid) and destocked (dashed) areas against grass communities ranked by grazing susceptibility. All other variables in model assumed to be constant (see Table 3).

strongly favoured by cats. When faced with choices about where to move, cats consistently selected sites with a more open grass layer, which had recently been subject to intense fires, and which were heavily grazed. Further, the influence of fire intensity on selection of habitat by cats was strongly affected by whether the habitats in question supported high densities of small mammals: sites in mammal-rich habitats that had recently been burnt at high intensity were especially strongly favoured.

This interaction between cat movements, fire and grazing regimes may help explain the recent declines in small mammals in northern Australia. There has been doubt that predation by cats might be driving these declines because of the mismatch in timing between the introduction of cats and small mammal declines, together with the fact that we have no evidence that small-mammal declines have coincided with an increased populations of cats [19]. As our results demonstrate, impacts of cats could have become more severe with the changes in fire and grazing regimes that began around the 1960s [40,41], even if cat density remained constant.

Selection by cats for areas with an open grass layer is a consistent finding from other studies on habitat selection by small felids [42,43], except in situations where moving into open areas leaves them exposed to larger predators [44] or where prey density is low [45]. Our cats' selection for open areas is almost certainly an expression of hunting preferences, as their selection became stronger in areas with higher small-mammal abundances. Although grass cover in itself had a large effect size (odds ratio maximum of 1.8), the effects of grazing and fire patterns on cat movements were much stronger.

Selection for fire scars was strongly dependent on fire intensity and time since fire. In general, cats avoided fire scars up to 360 days after fire. The exception was for recent scars resulting from intense fires in areas where small mammal densities were high, as this created the greatest increase in the odds ratio of any variable in this study. Fire opens up the grass layer thereby leaving prey more exposed and accessible to predators [15]. Intense fires would create pulses where prey would be easy for cats to catch - provided local abundance of prey was high. Cats did not select for recent mild fire scars, even in habitats with high small-mammal abundance. Mild fires typically leave pockets of unburnt vegetation [46], which provide protection for prey [10]. Also, mild fires are often stopped by riparian and alluvial areas [40], and such areas act as refugia for small mammals post-fire [10]. Our results suggest that cats are not able to capitalise on vulnerable small mammals after mild fires, but can after intense fires. This can explain why declines of some threatened native mammals have been so dramatic under regimes of consecutive high intensity fires [20,21].

Grazing by introduced herbivores also affected habitat choice by cats. Certain vegetation types (with more palatable grasses) were more preferred by introduced herbivores than others - typically riparian areas and bluegrass plains [47]. Cats selected these habitats even when they had been destocked, but grazing intensified this preference. Cats may have a stronger selection for these areas in the grazed zone as lower prey densities [22] require them to hunt in areas of relatively higher mammal density in order to obtain enough food. Daytime movements of cats were especially affected by grazing, suggesting that grazing creates favourable conditions for hunting of diurnal prey, such as quail [48].

Adult cat movements at small-scale 15 minute segments were dictated by their overall home range. Females had a stronger home-range fidelity than males, probably due to the importance of staying near dens that are used for rearing young [49]. Sub-adults displayed no such home range fidelity as they were probably in the process of creating and defining their home-ranges.

The top-order predator, the dingo, would also have influenced the movements of cats. Over the area and duration of the study, the dingo populations was abundant (0.2 individuals per km^2) and stable [32], and dingoes would have been a constant threat to cats [50]. A concurrent study in the same area that compared GPS movements of cats and dingoes found that although high use areas of dingoes and cats overlapped, cats were constantly avoiding the locations of individual dingoes [51]. This suggests avoidance of dingoes would have affected the timing of behavioural decisions of cats [52], but not necessarily by excluding them from certain areas [53]. Considering dingoes potential role in trophic regulation [25], further research into the relationship between dingoes, cats, and fire and grazing is warranted.

Conclusions

We demonstrate increased predator activity after intense fires and with grazing by large herbivores, which is likely to increase predation rates on small mammals. Small mammals are the preferred prey of cats, and form a substantial part of the diet of cats in the study region (55% of prey volume, from 33 stomach contents; unpublished data). Furthermore, the preference of cats for open and intensely burned areas depended on small-mammal abundance, and was reduced in habitats where abundance of small mammals was low. These patterns of habitat selection by cats correspond with declines of small mammal populations with intensifying fire and/or grazing regimes in mainland northern Australia, outside complex rocky outcrops [20,21,22,54,55,56,57]. Our work supports the hypothesis that the declines in small mammals across northern Australia are driven by cat predation facilitated by simplification of ground layer structure. However, this evidence is not direct, and other possible mechanisms including trophic alterations [25] and disease [19] may still have a role. Further research needs to measure whether mortality of prey is greater in more open areas. Regardless, the magnitude of the impacts of cats globally [58] suggests that our findings provide a general mechanism for prey decline in ecosystems with grass-dominated understoreys.

The results presented here suggest that manipulation of habitat through careful management of fire and grazing could be used to reduce impacts of cats over large landscapes. Reducing the frequency of intense fires and removing introduced herbivores is likely to be beneficial for small mammals - especially if this management is focussed on naturally mammal-rich habitats, and if it increases ground cover. Vegetation structure is pivotal in creating ideal landscapes for predators to hunt, and/or refuges for prey to hide.

Supporting Information

Material S1 Details on the creation of the dynamic grass cover map.

Acknowledgments

Staff and volunteers from the Australian Wildlife Conservancy assisted with field work, in particular Katherine Tuft, Monica Griffith, Danni Lisle, Isabel Connell, and Lily Leahy. Anja Skroblin provided comments.

Author Contributions

Conceived and designed the experiments: HWM SL MJ CJ. Performed the experiments: HWM SL. Analyzed the data: HWM. Contributed reagents/materials/analysis tools: HWM SL CJ. Wrote the paper: HWM SL MJ CJ.

References

1. Lima SL, Dill LM (1990) Behavioural decisions made under the risk of predation: a review and prospectus. Canadian Journal of Zoology 68: 619–640.
2. Didham RK, Tylianakis JM, Gemmell NJ, Rand TA, Ewers RM (2007) Interactive effects of habitat modification and species invasion on native species decline. TRENDS in Ecology & Evolution 22: 490–496.
3. Griffiths D (1975) Prey availability and the food of predators. Ecology 56: 1209–1214.
4. Hopcraft JCC, Sinclair ARE, Packer C (2005) Planning for success: Serengeti lions seek prey accessibility rather than abundance. Journal of Animal Ecology 74: 559–566.
5. Hebblewhite M, Merrill EH, McDonald TL (2005) Spatial decomposition of predation risk using resource selection functions: an example in a wolf-elk predator-prey system. Oikos 111: 101–111.
6. Koenen MT, Leslie D, Gregory MS (1996) Habitat changes and success of artificial nests on an alkaline flat. Wilson Bulletin 108: 292–301.
7. Dickson BG, Beier P (2002) Home-range and habitat selection by adult Cougars in Southern California. Journal of Wildlife Management 66: 1235–1245.
8. Gilliam JF, Fraser DF (1987) Habitat selection under predation hazard: test of a model with foraging minnows. Ecology 68: 1856–1862.
9. Yarnell RW, Scott DM, Chimimba CT, Metcalfe DJ (2007) Untangling the roles of fire, grazing and rainfall on small mammal communities in grassland ecosystems. Oecologia 154: 387–402.
10. Banks SC, Dujardin M, McBurney L, Blair D, Barker M, et al. (2011) Starting points for small mammal population recovery after wildfire: recolonisation or residual populations? Oikos 120: 26–37.
11. Tabenia S, Ojedaa RA (2003) Assessing mammal responses to perturbations in temperate aridlands of Argentina. Journal of Arid Environments 55: 715–726.
12. Jones A (2000) Effects of cattle grazing on North American arid ecosystems: a quantitative review. Western North American Naturalist 60: 155–164.
13. Horn KJ, McMillan BR, St Clair SB (2012) Expansive fire in Mojave Desert shrubland reduces abundance and species diversity of small mammals. Journal of Arid Environments 77: 54–58.
14. Southgate R, Palmer C, Adams MA, Masters P, Triggs B, et al. (1996) Population and habitat characteristics of the Golden Bandicoot (*Isoodon auratus*) on Marchinbar Island, Northern Territory. Wildlife Research 23: 647–664.
15. Conner LM, Castleberry SB, Derrick AM (2011) Effects of mesopredators and prescribed fire on hispid cotton rat survival and cause-specific mortality. Journal of Wildlife Management 75: 938–944.
16. Driscoll DA, Lindenmayer DB, Bennett AF, Bode M, Bradstock RA, et al. (2010) Fire management for biodiversity conservation: key research questions and our capacity to answer them. Biological Conservation 143: 1928–1939.
17. Sutherland EF, Dickman CR (1999) Mechanisms of recovery after fire by rodents in the Australian environment: a review. Wildlife Research 26: 405–419.
18. Fisher DO, Johnson CN, Lawes MJ, Fritz SA, McCallum HI, et al. (2013) The current decline of tropical marsupials in Australia: is history repeating? Global Ecology and Biogeography 23: 181–190.
19. Woinarski JCZ, Legge S, Fitzsimons JA, Traill BJ, Burbidge AA, et al. (2011) The disappearing mammal fauna of northern Australia: context, cause, and response. Conservation Letters 4: 1–10.
20. Firth RSC, Brook BW, Woinarski JCZ, Fordham DA (2010) Decline and likely extinction of a northern Australian native rodent, the Brush-tailed Rabbit-rat *Conilurus penicillatus*. Biological Conservation 143: 1193–1201.
21. Pardon G, Brook BW, Griffiths AD, Braithwaite RW (2003) Determinants of survival for the northern brown bandicoot under a landscape-scale fire experiment. Journal of Animal Ecology 72: 106–115.
22. Legge S, Kennedy MS, Lloyd R, Murphy SA, Fisher A (2011) Rapid recovery of mammal fauna in the central Kimberley, northern Australia, following the removal of introduced herbivores. Austral Ecology 36: 791–799.
23. Liedloff AC, Coughenour MB, Ludwig JA, Dyer R (2001) Modelling the trade-off between fire and grazing in a tropical savanna landscape, northern Australia. Environment International 27: 173–180.
24. Dickman CR (1996) Overview of the impacts of feral cats on Australian native fauna. Sydney: University of Sydney.
25. Radford IJ, Dickman CR, Start AN, Palmer C, Carnes K, et al. (2014) Mammals of Australia's tropical savannas: A conceptual model of assemblage structure and regulatory factors in the Kimberley region. PLos ONE 9: e92341.
26. Oakwood M (2000) Reproduction and demography of the northern quoll, *Dasyurus hallucatus*, in the lowland savanna of northern Australia. Australian Journal of Zoology 48: 519–539.
27. Woinarski JCZ, Ward S, Mahney T, Bradley J, Brennan K, et al. (2011) The mammal fauna of the Sir Edward Pellew island group, Northern Territory, Australia: refuge and death-trap. Wildlife Research 38: 307–322.
28. Frank ASK, Johnson CN, Potts J, Alaric F, Lawes MJ, et al. (2014) Experimental evidence that feral cats cause local extirpation of small mammals in Australia's tropical savanna. Journal of Applied Ecology - online early.
29. Abbott I (2002) Origin and spread of the cat, *Felis catus*, on mainland Australia, with a discussion of the magnitude of its early impact on native fauna. Wildlife Research 29: 51–74.
30. Kutt AS (2012) Feral cat (Felis catus) prey size and selectivity in north-eastern Australia: implications for mammal conservation. Journal of Zoology 287: 292–300.
31. Ziembicki MR, Woinarski JCZ, Mackey B (2013) Evaluating the status of species using Indigenous knowledge: Novel evidence for major native mammal declines in northern Australia. Biological Conservation 157: 78–92.
32. Legge S, Webb T, Smith J, Tuft K, James A (2014) The ecological health of Mornington and Marion Downs Wildlife Sanctuaries, 2013 Report. Perth, WA: Australian Wildlife Conservancy.
33. Torrea I, Díazb M, Martínez-Padillac J, Bonalb RD, Viñuelae J, et al. (2007) Cattle grazing, raptor abundance and small mammal communities in Mediterranean grasslands. Basic and Applied Ecology 8: 565–575.
34. Bjørneraas K, van Moorter B, Rolandsen CM, Herfindal I (2010) Screening global positioning system location data for errors using animal movement characteristics. Journal of Wildlife Management 74: 1361–1366
35. Recio MR, Mathieu R, Virgos E, Seddon PJ (2014) Quantifying fine-scale resource selection by introduced feral cats to complement management decision-making in ecologically sensitive areas. Biological Invasions 1: 1–13.
36. Cooper AB, Millspaugh JJ (1999) The application of discrete choice models to wildlife resource selection studies. Ecology 80: 566–575.
37. Burnham KP, Anderson DR (1998) Model selection and multimodel inference: A practical information-theoretic approach: Springer Science & Business Media, Inc.
38. R Development Core Team (2008) R: a language and environment for statistical computing. Vienna: R Foundation for Statistical Computing.
39. Therneau TM, Grambsch PM, Pankratz S (2003) Penalized survival models and frailty. Journal of Computational and Graphical Statistics 12: 156–175.
40. Russell-Smith J, Yates CJ, Edwards A, Allan GE, Cook GD, et al. (2003) Contemporary fire regimes of northern Australia, 1997–2001: change since Aboriginal occupancy, challenges for sustainable management. International Journal of Wildland Fire 12: 283–297.
41. Woinarski JCZ, Mackey B, Nix HA, Traill BJ (2007) The nature of northern Australia: Natural values, ecological processes and future prospects. Canberra, ACT: ANU Press.
42. Moseby KE, Stott J, Crisp H (2009) Movement patterns of feral predators in an arid environment – implications for control through poison baiting. Wildlife Research 36: 422–435.
43. Rajaratnam R, Sunquist M, Rajaratnam L, Ambu L (2007) Diet and habitat selection of the leopard cat (*Prionailurus bengalensis borneoensis*) in an agricultural landscape in Sabah, Malaysian Borneo. Journal of Tropical Ecology 23: 209–217.
44. Genovesi P, Besa M, Toso S (1995) Ecology of a feral cat *Felis catus* population in an agricultural area of northern Italy. Wildlife Biology 1: 233–237.
45. Hall LS, Kasparian MA, Van Vuren D, Kelt DA (2009) Spatial organization and habitat use of feral cats (*Felis catus L.*) in Mediterranean California. Mammalia 64: 19–28.
46. Russell-Smith J, Ryan PG, Cheal DC (2002) Fire regimes and the conservation of sandstone heath in monsoonal northern Australia: frequency, interval, patchiness. Biological Conservation 104: 91–106.

47. Tomkins N, O'Reagain P (2007) Global positioning systems indicate landscape preferences of cattle in the subtropical savannas. The Rangeland Journal 29: 217–222.
48. Brown RL (1982) Effects of livestock grazing on Mearns Quail in southeastern Arizona. Journal of Range Management 35: 727–732.
49. Krofel M, Skrbinšek T, Kos I (2013) Use of GPS location clusters analysis to study predation, feeding, and maternal behavior of the Eurasian lynx. Ecological Research 28: 103–116.
50. Moseby KE, Neilly H, Read JL, Crisp HA (2012) Interactions between a top order predator and exotic mesopredators in the Australian rangelands. International Journal of Ecology 2012.
51. Brook LA (2013) Predator guild interactions in northern Australia: behaviour and ecology of an apex predator, the dingo *Canis lupus dingo*, and an introduced mesopredator, the feral cat *Felis catus*. Townsville, QLD: James Cook University.
52. Brook LA, Johnson CN, Ritchie EG (2012) Effects of predator control on behaviour of an apex predator and indirect consequences for mesopredator suppression. Journal of Applied Ecology 49: 1278–1286.
53. Wang Y, Fisher DO (2012) Dingoes affect activity of feral cats, but do not exclude them from the habitat of an endangered macropod. Wildlife Research 39: 611–620.
54. Woinarski JCZ, Armstrong M, Brennan K, Fisher A, Griffiths AD, et al. (2010) Monitoring indicates rapid and severe decline of native small mammals in Kakadu National Park, northern Australia. Wildlife Research 37: 116–126.
55. Kutt AS, Woinarski JCZ (2007) The effects of grazing and fire on vegetation and the vertebrate assemblage in a tropical savanna woodland in north-eastern Australia. Journal of Tropical Ecology 23: 95–106.
56. Andersen AN, Cook GD, Corbett LK, Douglas MM, Eager RW, et al. (2005) Fire frequency and biodiversity conservation in Australian tropical savannas: implications from the Kapalga fire experiment. Austral Ecology 30: 155–167.
57. Legge S, Murphy S, Heathcote J, Flaxman E, Augusteyn J, et al. (2008) The short-term effects of an extensive and high-intensity fire on vertebrates in the tropical savannas of the central Kimberley, northern Australia. Wildlife Research 35: 33–43.
58. Loss SR, Will T, Marra PP (2012) The impact of free-ranging domestic cats on wildlife of the United States. Nature Communications 4: 1396.

13

Seed Dispersers, Seed Predators, and Browsers Act Synergistically as Biotic Filters in a Mosaic Landscape

Regino Zamora[1]*, Luis Matías[1,2]

1 Grupo de Ecología Terrestre, Departamento de Ecología, Facultad de Ciencias, Universidad de Granada, Granada, Spain, 2 Biological and Environmental Sciences, School of Natural Sciences, University of Stirling, Stirling, United Kingdom

Abstract

In this study, we analize the functional influence of animals on the plants they interact with in a mediterranean mountain. We hypothesise that seed dispersers, seed predators, and browsers can act as biotic filters for plant communities. We analyse the combined effects of mutualistic (seed dispersal) and antagonistic (seed predation, herbivory) animal interactions in a mosaic landscape of Mediterranean mountains, basing our results on observational and experimental field. Most of the dispersed seeds came from tree species, whereas the population of saplings was composed predominantly of zoochorous shrub species. Seed predators preferentially consumed seeds from tree species, whereas seeds from the dominant fleshy-fruited shrubs had a higher probability of escaping these predators. The same pattern was repeated among the different landscape units by browsers, since they browsed selectively and far more intensely on tree-species saplings than on the surrounding shrubs. In synthesis, our work identifies the major biotic processes that appear to be favoring a community dominated by shrubs versus trees because seed dispersers, predators, and herbivores together favored shrub dispersal and establishment versus trees.

Editor: Jin Chen, Key Laboratory of Tropical Forest Ecology, Xishuangbanna Tropical Botanical Garden, Chinese Academy of Sciences, China

Funding: This work was supported by Ministerio de Ciencia e Innovación (Spanish Government) Project CGL2011-29910 to R.Z. and by EU Marie Curie Fellowship to L.M (FP7-2011-IEF-300825). The funders had no role in study design, data collection and analysis, decision to publish, or preparation of the manuscript.

Competing Interests: The authors have declared that no competing interests exist.

* Email: rzamora@ugr.es

Introduction

The demographic consequences of plant-animal interactions have constituted a key topic for ecological studies in recent decades [1]. It has been commonly accepted that mutualistic and antagonistic interactions have opposite consequences and therefore induce different responses in plant demography [2–4]. By selective dispersal of the seeds contained in the fruits consumed, mutualistic animal species are able to alter the relative composition and abundance of plant communities, favouring species with which they interact most frequently (*sensu* [5]). Likewise, antagonistic animal species (seed predators and browsers) can have similar effects on plant populations, selectively consuming more palatable species or those with lower defences while avoiding less palatable species, thereby filtering community composition and granting a competitive edge to species with physical or chemical defences [6–9]. In this context, the relative abundance of the different species in a given plant community would be viewed as the result of the selection by mutualistic and/or antagonistic biotic filters over the regional species pool [10–12].

In Mediterranean ecosystems, contrary to temperate and tropical ones, summer drought has been traditionally considered the major limiting factor for plant recruitment among a wide diversity of habitats such as lowland forests [13,14], mountain forests [15,16], shrublands [17,18], and semiarid ecosystems [19]. However, recent evidence such as the high proportion of zoochorous species or the role of seed predators and browsers point to the under-appreciated importance of biotic factors on the dynamics of woody-species communities [4,15,20,21]. Thus, the identification of selection patterns, and their spatio-temporal consistency, would provide a fuller understanding of the consequences of plant-animal interactions as biotic filters for the regeneration of the woody community across different landscape units, representing a gradient between "natural" or "wild" ecosystems on one hand to intensively managed systems on the other [22]. In fact, a growing body of evidence supports the rearrangement of native biotic assemblages subjected to human-induced disturbances where some species may come to dominate ecological communities while others are detrimentally impacted [23]. This community reorganization processes are in fact occurring in the tropics [24], but the extent to which this phenomenon applies to other terrestrial biomes remains largely unknown. More importantly, we hardly know the relative importance and spatio-temporal consistency of biotic and abiotic mechanisms causing species replacements and community reorganization.

In this study, we analyse the combined effects of seed dispersers, seed predators, and browsers in a mosaic landscape composed of patches of native forest and degraded habitats representative of Mediterranean mountains. We sampled landscape mosaics shaped by human disturbance, all chosen to represent gradients of habitat and fruit resource availability. For this we conducted simultaneous field experiments on three plant-animal interactions gathered in the same area and at the same time. In previous studies [20,25,26,27], we analysed the magnitude and relative importance of abiotic and biotic factors on recruitment limitation (by comparing seed vs. establishment limitation) for the woody community in three Mediterranean sites differing in type of degradation: native forest (used as control), reforestation stands, and post-fire shrubland. We found that, overall, recruitment in the woody community was severely limited by both seed dispersal and establishment. However, tree species were more establishment-limited than shrubs. Trees and shrubs represent the two dominant woody vegetation types in terrestrial ecosystems, and our analyses show that their formation will depend on the impact of the abiotic and biotic filters on vegetation dynamics on a specific community.

In this context, taking into account that recruitment in this Mediterranean woody community is both severely seed and establishment limited due to both abiotic and biotic factors [20,25,27], we questioned the role that seed dispersers, seed predators, and browsers play as ecological filters for the woody-plant community across different landscape units. More specifically, we asked: 1) Do mutualistic and antagonistic plant-animal interactions have similar effects on the different functional groups of a woody community? 2) What are the demographic consequences of the mutualistic and antagonistic selective filters in the recruitment bank across a degraded mosaic landscape? 3) To what extent mutualistic and antagonistic animals does create a favourable selective environment for the same functional types of plants, irrespective of habitat? With this approach we seek to determine the combined role of seed dispersers, seed predators and browsers as biotic filters for the regeneration of the woody community across different landscape units, representing a gradient of human management.

Materials and Methods

The present study was conducted at Sierra Nevada National Park, within an area located between 1600 and 1900 m a.s.l. (37°05′N, 3°28′W, SE Spain). All necessary permits for the field studies described herein (which did not involve endangered or protected species) were obtained thanks to Javier Sanchez, Director of Sierra Nevada Nacional Park. The climate in this area is continental Mediterranean, with annual rainfall of 811 mm (mean values for the 1990–2010 period) concentrated in autumn and spring. Winters are cold whereas summers are hot and dry, mean temperatures of the coldest (January) and hottest months (July) are 3.6 and 21.5°C, respectively. The frugivorous guild in this area is composed of small birds (Robin, *Erithacus rubecula*, and Blackcap, *Sylvia atricapilla*; 12–20 g) and medium-sized birds (thrushes, *Turdus* spp.; 60–120 g). Most of these species are sedentary in the study area, except the Redwing (*T. iliacus*) and the Ring-ouzel (*T. torquatus*), which are long-distance migrant birds. All of these species are omnivorous, have a frugivorous diet during autumn–early winter, and are legitimate seed dispersers of many fruit-bearing plants in Mediterranean mountains [20,28]. Previous studies in the same study area [29,30] have reported by means of field observations and trapping that the main post-dispersal seed predators are woodmouse (*Apodemus sylvaticus*) and wild boar (*Sus scrofa*), while the browser guild is composed of ungulates, wild (Spanish ibex, *Capra pyrenaica*) as well as domestic (sheep, goats, and cattle). This assemblage have a full representation of the different types of animals interacting with plants (dispersers, seed predators, browsers), allowing us to examine the relative importance of the biotic filters on the configuration of the plant community from seed dispersal to late recruitment.

As a result of a long history of land use, the study area is a mosaic landscape composed of patches of three main habitats: native forest, dominated by *Pinus sylvestris* mixed with other trees such as *Acer opalus*, *Sorbus aria* or *Taxus baccata*, as well as a dense shrubby understorey formed by fleshy-fruited shrubs such as *Crataegus monogyna*, *Berberis vulgaris*, *Juniperus communis* or *Prunus ramburii*, and dry-fruited shrubs such as *Salvia lavandulifolia*, *Cytisus scoparius* or *Ononis aragonensis*, among others. The next two habitats were reforestation stands, planted in the 1950s and dominated by *P. sylvestris*, where we distinguished two different management levels: (2) dense stands (unmanaged, ca. 1040 trees ha^{-1}) and (3) cleared stands (with a pine density reduced by about 50%) and (4) post-fire shrublands, dominated by the aforementioned fleshy- and dry-fruited shrubs. The first habitat represents a low degradation habitat, whereas the reforestation stands and the shrubland constitute two contrasting types of, novel, human-created habitats [22].

Each landscape unit was represented by three plots of similar size (mean = 0.53 ha±0.06, average distance between the three plots of the same landscape unit = 688 m), giving a total of 12 study plots (a map showing the spatial distribution of the study plots can be found in [25]). In each landscape unit, we determined the diversity and abundance of tree and shrub adults (i.e. reproductive individuals), by counting all the trees and shrubs in each plot. We also collected the following data from each plot:

1) *Seed rain* was quantified by the use of seed traps from October 2003 to April 2005 (two complete dispersal seasons). Two paired seed traps (0.04 m^2 surface) were located at 15 fixed points in each plot. Traps were protected against post-dispersal seed predation by a wire mesh of 1-cm grid size. Contents were collected every two months, and all seeds were counted and identified to the species level. From fleshy-fruited species, only bare seeds (those with fruits digested by animals and thus effectively dispersed) were taken into account for analyses, while all seeds were counted for anemochorous and autochorous species (see [25] for a full description).
2) *Post-dispersal seed predation* was tested by recording the removal of seeds of three tree species (*P. sylvestris*, *A. opalus*, and *S. aria*) as well as two fleshy-fruited shrub species (*C. monogyna* and *B. vulgaris*) exposed simultaneously to predators in the field. At the same points where seed traps were located, two squares (20×20 cm) of plastic mesh were pegged to the soil and two seeds of each species were glued in a random position with a low-odour thermoplastic adhesive to them; the squares were 2 m apart. The seeds were exposed to predators in March 2004 and 2005 (after the natural seed-dispersal period, but before seedling emergence) and monitored after 30 days, recording the proportion of seeds consumed by predators (see [26] for a full description).
3) *Sapling abundance and diversity* was monitored using two 1-m^2 quadrats located at the same points where seed dispersal and predation were estimated. The identity and number of non-reproductive saplings older than two years was recorded in late spring 2004 and 2005 (May-June; see [25] for a full description).
4) *Damage by browsers* was quantified in established saplings of tree species (up to 2 m high) and in shrubs by the determination of browsing intensity in September 2008. This was calculated as the accumulated proportion of apical shoots consumed by ungulates (mainly domestic and wild goats, *Capra aegagrus* and *C. hispanica*, respectively) during the last four years with respect to the total number of buds by the identification of the scars by shoot consumption [31,32]. All

buds were examined for herbivory in small saplings or shrubs (<1,5 m in heigh) and, due to their large size, 100 randomly selected buds in large ones (>1,5 m in height) noting the proportion of damaged buds. In each plot, we measured 20 randomly selected individuals from each of the most abundant tree and shrubby species (*P. sylvestris*, *A. opalus*, *S. aria*, *Q. ilex*, *C. monogyna*, *B. vulgaris*, *P. ramburii*, *Rosa canina*, *Juniperus communis*, *Adenocarpus decorticans*, *S. lavandulifolia*, *C. scoparius* and *O. aragonensis*).

Since we focus on the effect of mutualistic and antagonistic interactions at the community level, and given that the woody community includes a great number of species with different architecture and dispersal type (Table 1), all species were functionally grouped based on growth habit and dispersal type. As zoochorous trees were absent from the degraded habitats (Table 2), we included all tree species within the same group, thus clasifying the community into *trees*, *fleshy-fruited shrubs*, and *dry-fruited shrubs*.

Data analysis

Differences in seed rain, predation rate and in browsing damage among habitats and functional groups were tested by analyses of variance, and differences between functional groups within habitats by means of Bonferroni *post hoc* tests. In addition, overall dispersal probabilities were calculated as the proportion of sampling stations receiving at least one seed from each of the different functional groups, with all habitats pooled. Similarly, for predation we used the probability of a single seed being consumed by rodents, and for browsing damage, the probability of an

Table 1. Species included within the different functional groups according to their growth architecture and dispersal type and their density, relative abundance and number of dispersed seeds in the study area.

Species	Dispersal type	Density (ind. ha^{-1})	Relative abundance (%)	Seed dispersal (seeds ha^{-1})
Trees				
Acer opalus Boiss.	Anemochorous	3.0	0.12	75.7
Pinus nigra Arnold	Anemochorous	42.6	1.70	33.9
Pinus sylvestris L.	Anemochorous	320.8	12.83	589.4
Sorbus aria (L.) Crantz	Zoochorous	0.4	0.02	0.4
Taxus baccata L.	Zoochorous	1.2	0.19	21.4
Total trees		**368**	**14.85**	**720.8**
Fleshy-fruited shrubs				
Amelanchier ovalis Medik.	Zoochorous	1.2	0.05	1.3
Berberis vulgaris L.	Zoochorous	46.7	1.87	79.8
Cotoneaster granatensis Boiss.	Zoochorous	1.7	0.07	0.4
Crataegus monogyna Jacq.	Zoochorous	103.3	4.13	90.6
Juniperus communis L.	Zoochorous	20.0	0.80	36.3
Juniperus sabina L.	Zoochorous	5.0	0.20	0.6
Lonicera arborea Boiss.	Zoochorous	5.9	0.24	13.1
Lonicera xylosteum L.	Zoochorous	58.3	2.33	6.25
Prunus ramburii Boiss.	Zoochorous	216.7	8.67	12.6
Rosa canina L.	Zoochorous	21.7	0.87	29.3*
Rosa pimpinellifolia L.	Zoochorous	58.3	2.33	-
Rosa sicula Tratt.	Zoochorous	41.7	1.67	-
Rosa stylosa Desv.	Zoochorous	13.3	0.53	-
Rubus ulmifolius Schott.	Zoochorous	3.3	0.13	8.4
Total FFS		**597.1**	**23.88**	**278.6**
Dry-fruited shrubs				
Adenocarpus decorticans Boiss.	Autochorous	3.3	0.13	0.5
Astragalus granatensis Lam.	Autochorous	26.7	1.07	-
Cytisus scoparius (L.) Link	Autochorous	6.7	0.27	0.4
Erinacea anthyllis Link	Autochorous	145.0	5.80	-
Genista cinerea (Vill.) DC.	Autochorous	8.3	0.33	-
Ononis aragonensis Asso	Autochorous	286.7	11.47	4.4
Salvia lavandulifolia Vahl.	Autochorous	1055.0	42.20	42.4
Total DFS		**1531.7**	**61.26**	**47.7**

All plots are pooled together.
*All seeds from the different species of the Genus *Rosa* are included together due to the difficulties to distinguish them.

Table 2. Density fo individuals (in individuals ha^{-1}) by functional groups (AT: anemochorous trees; ZT: zoochorous tress; FFS: fleshy-fruited shrubs; DFS: dry-fruited shrubs) on the different study plots.

Habitat	Plot	AT	ZT	FFS	DFS
Native	1	104	97	617	32
	2	91	12	310	6
	3	87	4	123	19
Cleared	1	675	0	105	0
	2	454	0	63	25
	3	432	0	171	77
Dense	1	858	0	10	0
	2	1128	0	20	31
	3	1138	0	73	68
Shrubland	1	21	0	1172	2370
	2	21	0	1607	1293
	3	66	0	647	3117

established sapling to be browsed. All analyses were performed with JMP 7.0 (SAS Inc.).

Results and Discussion

Our results suggest that plant-animal interactions have a consistent influence on the functional components of the vegetation. By integrating field data on seed dispersal, seed predation, and browsing, we were able to infer how birds and mammals influence the recruitment and abundance of three functional groups in a Mediterranean landscape. Notably, seed-dispersing birds, as well as seed predators and browsers appear to favour the same type of vegetation across a degraded mosaic landscape.

By dispersing shrub species with higher recruitment probabilities than trees, birds become the first architects of these Mediterranean landscapes, which have a higher establishment probability than do tree species [25,27]. Only 6.8% of the total seeds dispersed by birds were from zoochorus tree species, i. e. *T. baccata* and *S. aria*, the 98.8% of them were dispersed into native habitat (mean values for seeds of anemochorous trees dispersed into native habitat = 1023±280 seeds/ha; mean values for zoochorous trees = 85±84, F = 10.3, P = 0.03, $df = 1$) In fact, although most of the dispersed seeds came from tree species, mainly the anemochorous *P. sylvestris* (Fig. 1a, b; Table 1), the understory was dominated by fleshy-fruited shrubs (F = 101.5, $P<0.0001$; $df = 2$) irrespective of the habitat type (F = 0.7, $P = 0.5$; $df = 3$). Birds are important long-distance dispersers [33] and their movements may reduce differences in seed availability among landscape units [17,20,34,35]. Thus, we found seeds of a wide diversity of zoochorous species in all habitats, even when parents were absent or scarce.

Although there were differences in the predation intensity among habitats (F = 50.6, $P<0.0001$; $df = 3$), seed predators preferentially consumed seeds from tree species (F = 493.2, $P<0.0001$; $df = 1$; Fig. 1c), whereas seeds from the dominant fleshy-fruited shrubs had a higher probability of escaping predation (Fig. 2). In the nearby oak forests of Sierra Nevada, acorn predation is even higher than for pine [30,36], indicating that the two most abundant tree genera in Mediterranean mountains (*Pinus* and *Quercus*) undergo a high post-dispersal predation rate. Consequently, post-dispersal seed predation has the potential to filter the species pool available for recruitment of the woody community in a similar way in all landscape units, irrespective of the degree of degradation, reducing the number of propagules of dominant tree species and favouring further a shrub-like landscape, which is the type of landscape unit where rodents can find both food and refuge [26,37]. Although we do not have data on seed predation in dry-fruited shrubs (mainly ballistic dispersal) in our study area, other studies indicate that their predation is much lower than for tree species, and are especially depredated by weevils before dispersal and by ants afterwards [38–41].

We found a strong browsing pressure (62% of individuals browsed and 21% damage on average for all functional groups and habitats) in the landscape mosaic of Sierra Nevada, similar to values found in other Mediterranean mountains or temperate woodlands [32,42,43]. However, this pressure did not affect all species in the sapling bank equally, since domestic and wild ungulates selectively browsed tree-species saplings far more intensely than the surrounding shrubs (F = 199.4, $P<0.0001$; $df = 2$; Fig. 1d, e). Besides the differences in herbivory pressure and intensity among habitats (F = 24.6, $P<0.0001$; $df = 3$), the same pattern of higer browsing over tree species was maintained. Previous studies in the same area [44,45] have shown the same gradient of herbivore preferences for the same species, indicating the consistency of the reported pattern. This preference involves higher browsing damage for large numbers of tree saplings, which consequently underwent a reduction in annual height growth [32]. Furthermore, browsing in Mediterranean areas is more severe in summer when other food resources (herbaceous plants) for ungulates are less abundant, leaving less time for saplings to recover from the herbivory before winter [46]. All these factors exacerbate browsing damage, extending the time needed for saplings to reach maturity, and thereby lengthening the exposure time to browsers [20,42].

Browsers therefore are able to constrain the successional trajectory in a native forest, changing the probabilities of transition from the recruitment pattern of the sapling bank to the adults of the canopy [31,47,48]. Furthermore, irrespective of the habitat type considered (whether native forest, pine plantations or shrubland), seed dispersal, seed predation, and browsing act synergistically to favour shrub recruitment and disfavour tree recruitment. In addition, we could expect the selective filtering

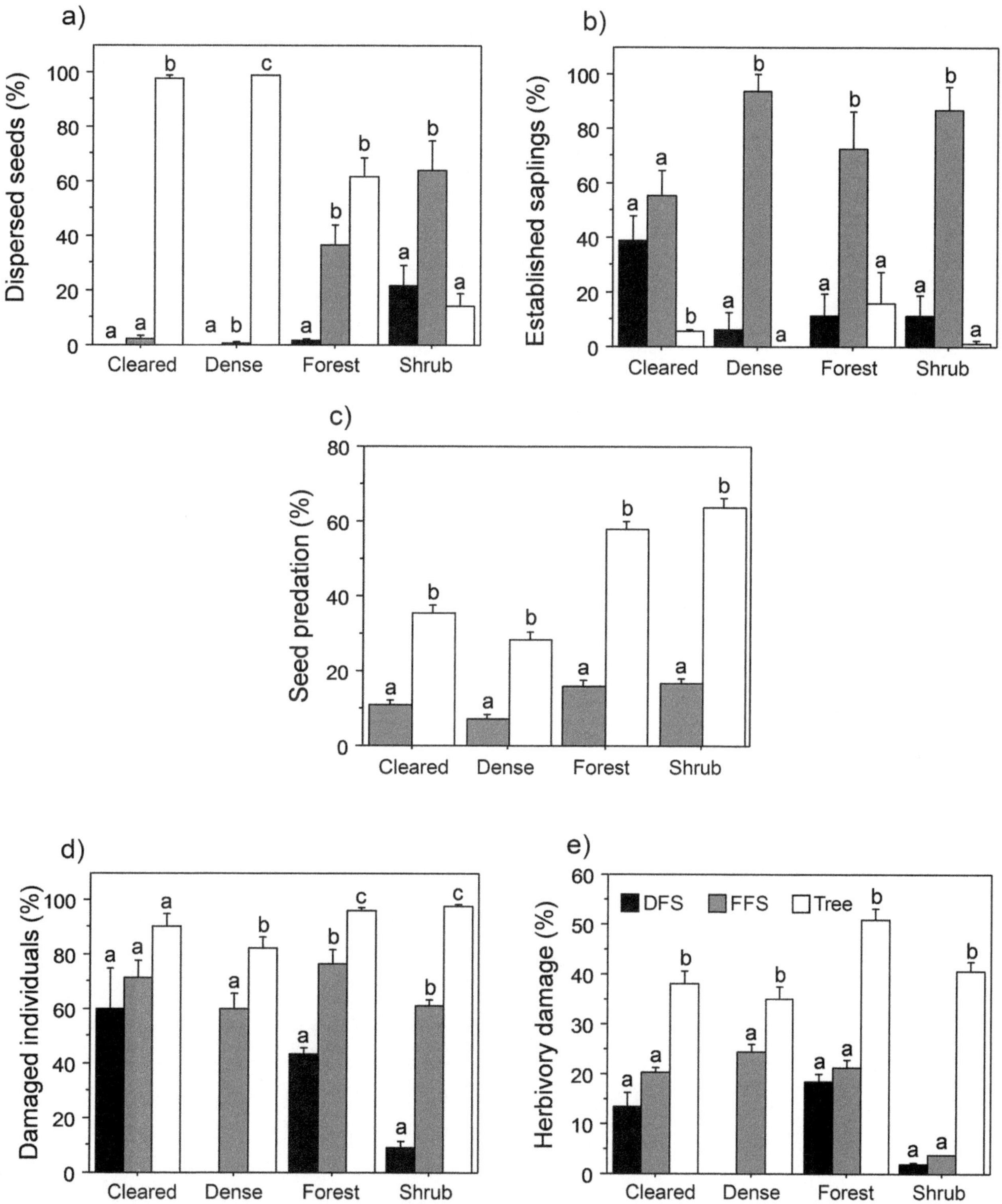

Figure 1. Intensity of the different biotic interactions among the different habitats (cleared and dense reforestation stands, native forest, and shrubland). A) Percentage of dispersed seeds from dry-fruited shrubs (DFS, black bars), fleshy-fruited shrubs (FFS, grey bars) and tree species (Tree, open bars). A total of 14,300 seeds were collected during the two study years: 5700 seeds of trees, 6000 of fleshy-fruited shrubs, and 2600 of dry-fruited shrubs (Mendoza et al. 2009). B). Percent of the established sapling bank belonging to the different functional groups among the different habitats. C) Percentage of depredated seeds in fleshy-fruited shrubs (FFS: *B. vulgaris* and *C. monogyna*; grey bars) and in tree species (*P. sylvestris*, *A. opalus*, and *S. aria*; open bars). D) Proportion of individuals damaged by browsers from the different functional groups among habitats. E) Percentage of buds damaged by browsers on attacked individuals of the different functional groups among habitats. Different letters denote differences at $P<0.05$ within habitats. Error bars represents +1SE.

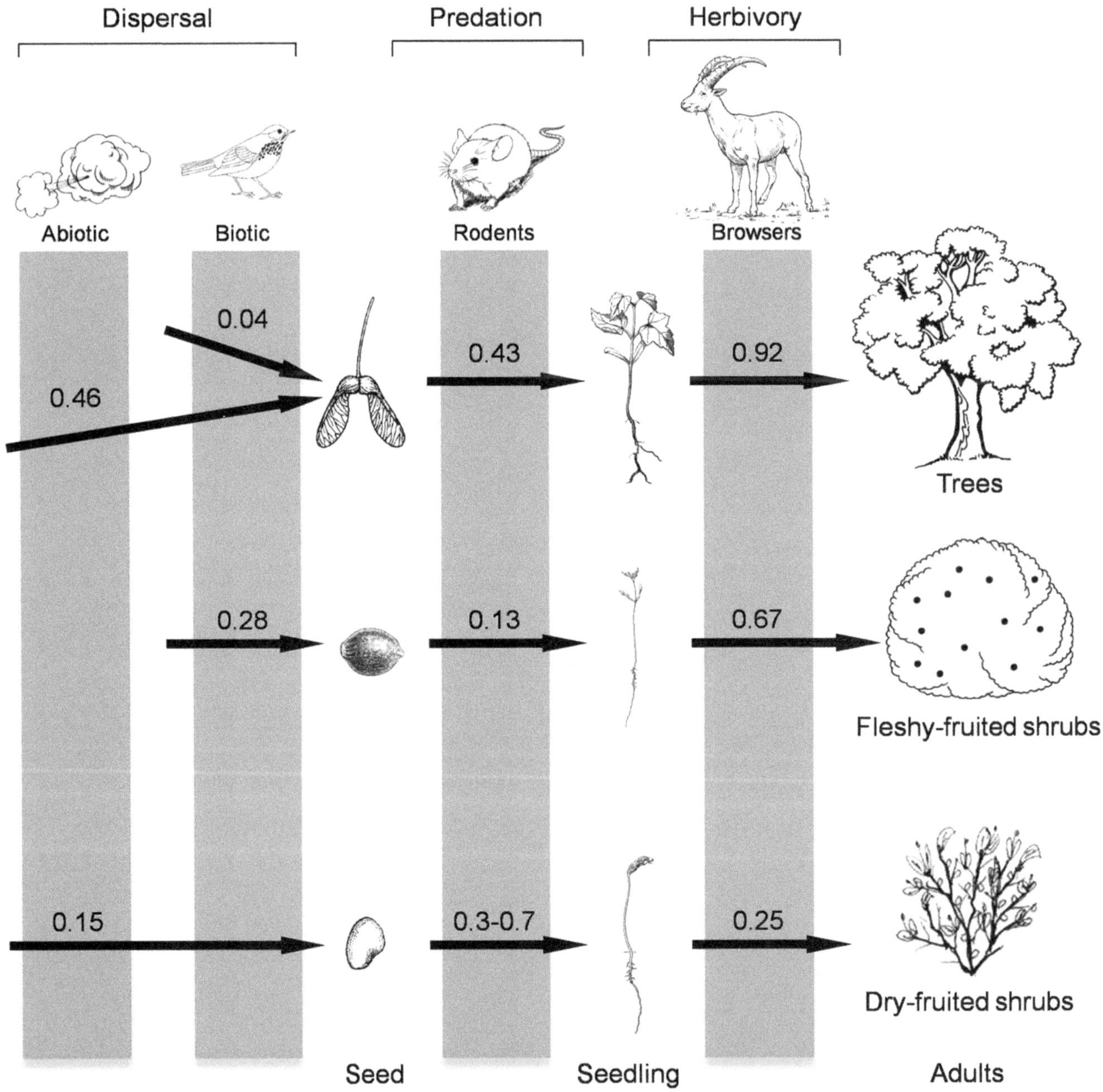

Figure 2. Probabilities (from 0 to 1) for the different plant functional groups (trees, fleshy-fruited shrubs and dry-fruited shrubs) of being affected by the different mutualistic and antagonistic ecological filters (dispersal, seed predation, and herbivory) among the different development phases. Dispersal data are the mean probability of a sampling station receiving at least one seed of the different groups; predation data are the probability of a single seed being consumed by rodents (dry-fruited shrubs range data based on the literature [36,38]); and herbivory data represents the probability of an established individual (saplings in the case of trees) to be browsed. Data are pooled for the different habitats.

process on woody vegetation imposed by mutualistic and antagonic animals to be enhanced by climatic conditions. As fleshy-fruited shrub species were the only ones that recruited in very dry years [25], our results imply that all landscape units, regardless of the degree of degradation, have a potential successional trend towards shrub dominance. This overall surge of shrubby species agrees with the shrub-encroachment trends described for many Mediterranean ecosystems [27,49].

Our empirical results indicate that seed dispersers, seed predators, and browsers may act as a niche-based filter that contributes to a deterministic species assembly in this forest community, a process that may parallel effects of drought in aquatic mesocosm communities [50], or the effect of resource availability on forest recruitment [11]. These findings indicate for the first time that, although seed predation and browsing pressure differs quantitatively between adjacent landscape units, the selective filtering of the seed bank and saplings of woody species tends to be parallel among them. Furthermore, seedlings of fleshy-fruited shrubs are especially dominant in the understory of the two degraded landscape units: reforestation stands and shrubland [25], indicating a trend towards more abundance of fleshy-fruited shrubs in the overall mountain landscape. On the other hand, we should take into account that the dominance in shrub recruitment could also benefit, via plant-plant facilitation, the establishment of many other shrub and tree species [51]. Paradoxically, the expansion of fleshy-fruited shrubs does not lead inexorably to a total exclusion of tree species but also generates good niches for tree regeneration. As we have previously demonstrated, the presence of habitat-modifying pioneer shrubs may enhance species diversity by providing structural refuge to a broad array of woody shrubs and tree species which in turn would promote overall diversity of the woody community, especially in the most degraded landscape units [51].

In synthesis, our work identifies the major biotic processes that appear to be favoring a community dominated by shrubs versus trees because seed dispersers, predators, and herbivores together favored shrub dispersal and establishment versus trees among the different landscape units, irrespective of the degradation phase. The final outcome of this "animal-driven process" could be an overall increase in the functional similarity of the woody plant communities across the mosaic landscape. The concomitant proliferation of shrubs instead trees echoes a process already described in several neotropical fragmented forest landscapes (including Amazonia and the Atlantic Forest) comprising a case of widespread substitution of a wide range of native old-growth tree species ('losers'), by native pioneer ('winners') [24]. In addition, this type of "novel" shrub-dominated community is better adapted to cope with the expected changes in climate for the coming decades in this area under a climate-change scenario [27,52], because shrubs are less vulnerable to drought and browsing than are tree species. Over the long term, biotic as well as abiotic factors may converge to promote increased dominance of the most drought-tolerant species, creating a favourable selective environment for woody plants with a similar combination of functional traits (zoochorous, unpalatable, drought resistant) and thus enhancing shrubland expansion in the overall mountain landscape, irrespective of the degree of past human management.

Acknowledgments

We would like to thank Drs. C.M. Herrera, D. García, L. Gómez-Aparicio, E. Schupp and T. Carlo for their helpful comments in an early version of the manuscript and D. Nesbitt for language advise.

Author Contributions

Conceived and designed the experiments: RZ. Performed the experiments: RZ LM. Analyzed the data: LM. Wrote the paper: RZ LM.

References

1. Herrera CM, Pellmyr O (2002) Plant-Animal Interactions: An Evolutionary Approach. Oxford: Blackwell Science.
2. Thompson JN (1998) Variation in interspecific interactions. Ann Rev Ecol Syst 19: 65–87.
3. Strauss SY, Irvin RE (2004) Ecological and Evolutionary consequences of multispecies plant-animal interactions. Ann Rev Ecol Syst 35: 435–466.
4. García D, Zamora R, Amico G (2011) The spatial scale of plant-animal interactions: effects of resource availability and habitat structure. Ecol Monogr 81: 103–121.
5. Herrera CM (1985) Determinants of plant-animal coevolution: the case of mutualistic dispersal of seeds by vertebrates. Oikos 44: 132–141.
6. Hulme PE (1997) Post-dispersal seed predation and the establishment of vertebrate dispersed plants in Mediterranean scrublands. Oecologia 111: 91–98.
7. Weber P, Rigling A, Bugmann H (2008) Sensitivity of stand dynamics to grazing in mixed *Pinus sylvestris* and *Quercus pubescens* forests: A modelling study. Ecol Model 210: 301–311.
8. Baraza E, Hódar JA, Zamora R (2009) Consequences of plant-chemical diversity for domestic goat food preference in mediterranean forests. Acta Oecol 35: 117–127.
9. Post E, Pedersen C (2008) Opposing plant community community responses to warming with an without herbivores. Proc Nat Ac Sci 105: 12353–12358.
10. Keddy PA (1992) Assembly and response rules: two goals for predictive community ecology. J Veg Sci 3: 157–164.
11. Myers JA, Harms KE (2011) Seed arrival and ecological filters interact to assemble high-diversity plant communities. Ecology 92: 676–686.
12. Wisz MS, Pottier J, Kissling WD, Pellissier L, Lenoir J, et al. (2013) The role of biotic interactions in shaping distributions and realised assemblages of species: implications for species distribution modelling. Biol Rev 88: 15–30.
13. Traveset A, Gulias J, Riera N, Mus M (2003) Transition probabilities from pollination to establishment in a rare dioecious shrub species (*Rhamnus ludovici-salvatoris*) in two habitats. J Ecol 91: 427–437.
14. Pulido FJ, Díaz M (2005) Regeneration of a Mediterranean oak: A whole-cycle approach. Ecoscience 12: 92–102.
15. Herrera CM, Jordano P, López-Soria L, Amat JA (1994) Recruitment of a mast-fruiting, bird-dispersed tree: bridging frugivore activity and seedling establishment. Ecol Monogr 64: 315–344.
16. Castro J, Zamora R, Hódar JA, Gómez JM (2005) Alleviation of summer drought boosts establishment success of *Pinus sylvestris* in a Mediterranean mountain: An experimental approach. Plant Ecol 181: 191–202.
17. García D, Zamora R, Hódar JA, Gómez JM (1999) Age structure of *Juniperus communis* L. in the Iberian Peninsula: Conservation of remnant populations in Mediterranean mountains. Biol Cons 87: 215–220.
18. Rey PJ, Alcántara JM (2000) Recruitment dynamics of a fleshy-fruited plant (*Olea europaea*): Connecting patterns of seed dispersal to seedling establishment. J Ecol 88: 622–633.
19. García-Fayos P, Gasque M (2006) Seed vs. microsite limitation for seedling emergence in the perennial grass *Stipa tenacissima* L. (Poaceae). Acta Oecol 30: 276–282.
20. Zamora R, Hodar JA, Matías L, Mendoza I (2010) Positive adjacency effects mediated by seed disperser birds in pine plantations. Ecol Appl 20: 1053–1060.
21. Carlo TA, Tewksbury JJ (2014) Directness and tempo of avian seed deposition increases emergence of wild chiltepins in desert grasslands. J. Ecol 102: 248–255.
22. Hobbs RJ, Arico S, Aronson J, Baron JS, Bridgewater P, et al. (2006) Emerging Ecosystems: theoretical and management aspects of the new ecological order. Glob Ecol Biogeogr 15: 1–7.
23. Tabarelli M, Peres CA, Melo FPL (2012) The 'few winners and many losers' paradigm revisited: Emerging prospects for tropical forest biodiversity. Biol Cons 155: 136–140.
24. Lewis SL, Lloyd J, Sitch S, Mitchard ETA, Laurance WF (2009) Changing ecology of tropical forests: Evidence and drivers. Annu Rev Ecol Evol Syst 40: 529–49.
25. Mendoza I, ZamoraR, Gómez-Aparicio L, Matías L (2009) Recruitment limitation of forest communities in a degraded Mediterranean landscape. J Veg Sci 20: 367–376.
26. Matías L, Mendoza I, Zamora R (2009) Consistent pattern of habitat and species selection by post-dispersal seed predation in a Mediterranean mosaic landscape. Plant Ecol 203: 137–147.
27. Matías L, Zamora R, Castro J (2012) Rare rainy events are more critical than drought intensification for woody recruitment in Mediterranean mountains: a field experiment simulating climate change. Oecologia 169: 833–844.
28. Zamora R (1990) The fruit diet of ring-ouzels (*Turdus torquatus*) wintering in the Sierra Nevada (south-east Spain). Alauda 58: 67–70.
29. Castro J, Gómez JM, García D, Zamora R, Hódar JA (1999) Seed predation and dispersal in relict scots pine forests in southern spain. Plant Ecol 145: 115–123.
30. Gómez JM (2004) Importance of microhabitat and acorn burial on *Quercus ilex* early recruitment: non-additive effects on múltiple demographic processes. Plant Ecol 172: 287–297.
31. Zamora R, Gómez JM, Hódar JA, Castro J, García D (2001) Effect of browsing by ungulates on sapling growth of Scots pine in a Mediterranean environment: consequences for forest regeneration. For Ecol Manage 144: 33–42.
32. Herrero A, Zamora R, Castro J, Hódar JA (2012) Limits of pine forest distribution at the treeline: herbivory matters. Plant Ecol 213: 459–469.
33. Carlo TA, García D, Martínez D, Gleditsch JM, Morales JM (2013) Where do seeds go when they go far? Integrating distance and directionality of avian seed dispersal in heterogeneous landscapes. Ecology 92: 301–307.
34. Jordano P, Garcia C, Godoy JA, García-Castaño JL (2007) Differential contribution of frugivores to complex seed dispersal patterns. Proc Nat Ac Sci 104: 3278–3282.
35. García D, Martínez D (2012) Species richness matters for the quality of ecosystem services: a test using seed dispersal by frugivorous birds. Proc Royal Soc B 279: 3106–3113.
36. Puerta-Piñero C (2010) Intermediate spatial variations on acorn predation shapes Holm oak establishment within a Mediterranean landscape context. Plant Ecol 210: 213–224.
37. García-Castaño JL, Kollmann J, Jordano P (2006) Spatial variation of post-dispersal seed removal by rodents in highland microhabitats of Spain and Switzerland. Seed Sci Res 16: 213–222.
38. Espadaler X, Gómez C (1996) Seed production, predation and dispersal in the mediterranean myrmecochore *Euphorbia characias* (Euphorbiaceae). Ecography 19: 7–15.

39. Sheppard AW, Hodge P, Paynter Q, Rees M (2002) Factors affecting invasion and persistence of broom *Cytisus scoparius*. J Appl Ecol 39: 721–734.
40. Prider JN, Facelli JM, Watling JR (2011) Multispecies interactions among a plant parasite, a pollinator and a seed predator affect the reproductive output of an invasive plant, *Cytisus scoparius*. Austral Ecol 36: 167–175.
41. Herrera-Reddy AM, Carruthers RI, Mills NJ (2012) Integrated management of scotch broom (*Cytisus scoparius*) using biological control. Inv Plant Sci Manage 5: 69–82.
42. Danell K, Niemela P, Varvikko T, Vuorisalo T (1991) Moose browsing on Scots Pine along a gradient of plant productivity. Ecology 72: 1624–1633.
43. Gill RMA (1992) A review of damage by mammals in north temperate forests. 3. Impact on trees and forests. Forestry 65: 363–388.
44. Baraza E, Zamora R, Hódar JA (2006) Conditional outcomes in plant-herbivore interactions: Neighbours matter. Oikos 113: 148–156.
45. Gómez-Aparicio L, Gómez JM, Zamora R (2007) Spatiotemporal patterns of seed dispersal in a wind-dispersed mediterranean tree (*Acer opalus* subsp. *granatense*): Implications for regeneration. Ecography 30: 13–22.
46. Hester AJ, Millard P, Baillie GJ, Wendler R (2004) How does timing of browsing affect above- and below-ground growth of *Betula pendula*, *Pinus sylvestris* and *Sorbus aucuparia*? Oikos 3: 536–550.
47. Gómez JM, Hódar JA, Zamora R, Castro J, García D (2001) Ungulate damage on Scots pines in Mediterranean environments. effects of association with shrubs. Can J Bot 79: 739–746.
48. Gómez-Aparicio L, Zamora R, Castro J, Hódar JA (2008) Facilitation of tree saplings by nurse plants: microhabitat amelioration or protection against herbivores? J Veg Sci 19: 161–172.
49. Acácio V, Holmgren M, Rego F, Moreira F, Mohren GMJ (2009) Are drought and wildfires turning Mediterranean cork oak forests into persistent shrublands? Agrofor Syst 76: 389–400.
50. Chase JM (2007) Drought mediates the importance of stochastic community assembly. Proc Nat Ac Sci 104: 17430–17434.
51. Gómez-Aparicio L, Zamora R, Gómez JM, Hódar JA, Castro J, et al. (2004) Applying plant facilitation to forest restoration in Mediterranean ecosystems: a meta-analysis of the use of shrubs as nurse plants. Ecol. Appl. 14: 1128–1138.
52. Matías L, Zamora R, Castro J (2011) Repercussions of simulated climate change on the diversity of woody-recruit bank in a Mediterranean ecosystem. Ecosystems 14: 672–682.

14

Plant Diversity Impacts Decomposition and Herbivory via Changes in Aboveground Arthropods

Anne Ebeling[1]*, Sebastian T. Meyer[2], Maike Abbas[3], Nico Eisenhauer[1], Helmut Hillebrand[3], Markus Lange[4], Christoph Scherber[5], Anja Vogel[1], Alexandra Weigelt[6], Wolfgang W. Weisser[1,2]

1 Institute of Ecology, University of Jena, Jena, Germany, **2** Department of Ecology and Ecosystem Management, Center for Food and Life Sciences Weihenstephan, Technische Universität München, Freising, Germany, **3** Institute for Chemistry and Biology of the Marine Environment, Carl-von-Ossietzky-University Oldenburg, Wilhelmshaven, Germany, **4** Max Planck Institute for Biogeochemistry, Jena, Germany, **5** DNPW, Agroecology, Georg-August University Göttingen, Göttingen, Germany, **6** Department for Systematic Botany and Functional Biodiversity, University of Leipzig, Leipzig, Germany

Abstract

Loss of plant diversity influences essential ecosystem processes as aboveground productivity, and can have cascading effects on the arthropod communities in adjacent trophic levels. However, few studies have examined how those changes in arthropod communities can have additional impacts on ecosystem processes caused by them (e.g. pollination, bioturbation, predation, decomposition, herbivory). Therefore, including arthropod effects in predictions of the impact of plant diversity loss on such ecosystem processes is an important but little studied piece of information. In a grassland biodiversity experiment, we addressed this gap by assessing aboveground decomposer and herbivore communities and linking their abundance and diversity to rates of decomposition and herbivory. Path analyses showed that increasing plant diversity led to higher abundance and diversity of decomposing arthropods through higher plant biomass. Higher species richness of decomposers, in turn, enhanced decomposition. Similarly, species-rich plant communities hosted a higher abundance and diversity of herbivores through elevated plant biomass and C:N ratio, leading to higher herbivory rates. Integrating trophic interactions into the study of biodiversity effects is required to understand the multiple pathways by which biodiversity affects ecosystem functioning.

Editor: Mari Moora, University of Tartu, Estonia

Funding: The study was funded by the Deutsche Forschungsgemeinschaft (FOR 1451). The funders had no role in study design, data collection and analysis, decision to publish, or preparation of the manuscript.

Competing Interests: The authors have declared that no competing interests exist.

* Email: anne.ebeling@uni-jena.de

Introduction

Extensive studies have identified biodiversity loss as one of the main drivers of ecosystem change [1], showing that the relation between biodiversity and ecosystem processes is mainly positive [2,3]. For a long time biodiversity research has focused mainly on processes directly relating to the trophic level manipulated (e.g. aboveground productivity in plant diversity experiments [4,5]; but see [6]). More recent studies have shown that plant diversity and associated changes in plant species composition also affect ecosystem processes governed by higher trophic levels, such as decomposition and herbivory [7–10].

Decomposition of organic matter is a key process in ecosystems, replenishing the pool of plant available soil nutrients and releasing photosynthetically fixed carbon back to the atmosphere by the activity of several groups of soil organisms. Invertebrate macrofauna, such as isopods, diplopods and earthworms, fragment dead plant material and thereby pave the way for further microbial decay and mineralization. Consequently, changes in their density and diversity should lead to altered decomposition rates [11,12]. They may also affect nutrient cycling and plant productivity of grassland ecosystems. Herbivores can decrease plant productivity by decreasing plant performance, but contrary to that, they could also increase plant productivity by recycling nutrients and triggering compensatory growth [13,14]. Arthropod density and diversity were discussed as drivers of plant diversity effects on decomposition and herbivory, but empirical evidence is scarce [8,15]. Studies in experimental grasslands have generally reported higher diversity and abundances of herbivores, detritivores, pollinators, carnivores and parasitoid arthropods at higher plant diversity [16–23], although the strength of the effect differed between the groups.

Plant-animal, plant-ecosystem process, and animal-ecosystem process relationships have often been studied separately, but have rarely been linked. To understand how plants, arthropods, and arthropod-related processes such as decomposition and herbivory are linked, there is a strong need to conduct studies considering multiple trophic levels and ecosystem processes simultaneously. There is a large number of competing hypotheses linking consumer-resource dynamics in diverse communities (summarized in Fig. 1), applicable to the relationship between arthropod and plant communities. Hypotheses emphasizing effects of resources on consumers have focused on biomass (More Individuals Hypothesis- MIH, Productivity Hypothesis- PH), and stoichiometry (C:N- Mechanism 3 in Fig. 1), both variables being strongly affected by changing plant species richness [24,25] but also on

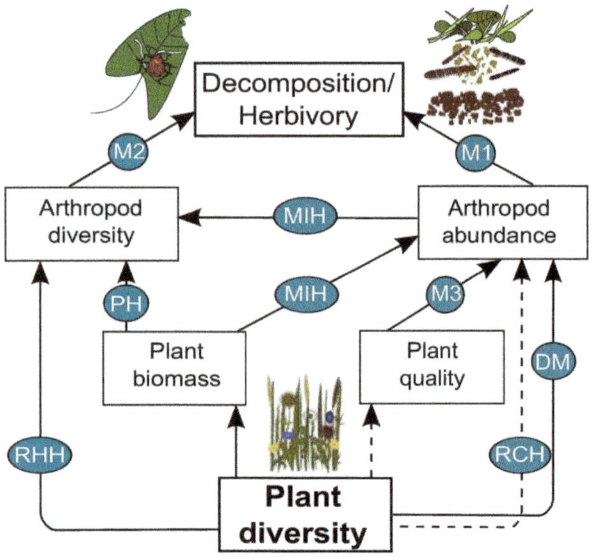

Figure 1. Possible mechanisms driving a positive plant diversity-ecosystem process relationship. The consumer-related ecosystem processes studied here are herbivory and decomposition. Based on existing literature we hypothesize different mechanisms driving the positive relationship between plant diversity and herbivory/decomposition, represented here by the different arrows. For details on the hypotheses see Introduction. Solid lines indicate expected positive correlations, dotted lines expected negative correlations. M1-3 = Mechanisms 1-3, DM = Dietary Mixing, MIH = More Individuals Hypothesis, PH = Productivity Hypothesis, RCH = Resource Concentration Hypothesis, RHH = Resource Heterogeneity Hypothesis. M1-3 are "unnamed" mechanisms, which are reported in the literature.

species richness per se (Dietary Mixing- DM, Resource Concentration Hypothesis- RCH, Resource Heterogeneity Hypothesis- RHH).

MIH and PH hypothesize a positive relationship between plant biomass and consumer diversity, but they differ in the underlying mechanisms. The MIH [26] applies to communities, which are limited in productivity and predicts higher consumer abundances in more productive sites as well as consumer diversity as an increasing function of total abundance. Following the PH, a higher overall resource level in diverse plant communities directly attracts more generalist consumer species [27]. The stoichiometry of resources is also reported to be a strong predictor of consumer abundances as higher food quality (e.g. higher N content of plants) increases arthropod fitness (e.g. fecundity) and hence their abundance [28]. Following the literature, the way and direction how plant species richness affects consumer abundance strongly depends on their food specialization. According to the RCH [29] we would expect lower abundances of specialized herbivorous pests in more diverse habitats due to lower densities of the respective host plant. For generalist herbivores the opposite pattern may be true (DM): the ability of some generalist herbivores (e.g., grasshoppers) to mix their diet to an optimal combination of nutrients or to dilute toxins could increase their fitness and hence their abundance, indicated by a positive relation between plant diversity and arthropod abundance [30]. For consumer species richness the RHH [31] predicts that due to a higher number of different resources with increasing plant diversity, the number of specialist consumer species increases.

While the theoretical framework of how plant diversity affects consumer (arthropod) abundance and diversity is quite elaborated, the effects of arthropod communities on herbivory and decomposition are less clear. Consumer effects on related processes could be driven by arthropod abundance (Mechanism 1 in Fig. 1) or species richness (Mechanism 2 in Fig. 1). If specialist herbivores dominate, then increasing herbivore diversity is likely to increase community-wide herbivory rates, while a higher number of herbivores will only increase herbivory on those plants that are attacked. By contrast, if herbivory is mainly due to generalists, then increasing herbivore abundance rather than diversity will increase herbivory. For decomposition, functional complementarity among decomposers has been shown in a number of studies [11], yet decomposition may well be limited by decomposer abundance. Thus, various chains of effects are possible emphasizing the need for empirical approaches. All the hypotheses and mechanisms (Fig. 1) are not mutually exclusive, however, and many of the hypothesized pathways can have interactive and direct or indirect effects. The relative importance of these pathways may change when considering herbivores as opposed to decomposers because some mechanisms like compensatory plant growth would only result from herbivory, but not from decomposition.

Here, we integrate consumers into our models of plant diversity effects on ecosystem processes by focusing on three components of plant communities (food quantity: aboveground biomass; food quality: C:N ratio of plants, and species richness- hereafter: plant diversity), two components of consumer communities (arthropod abundance, arthropod species richness), and two processes (decomposition and herbivory). We test (1) if plant diversity effects on arthropods and decomposition/herbivory found in former studies can be confirmed, (2) if effects of plant diversity on decomposition and herbivory are mediated by altered arthropod community composition in terms of overall abundance, and (3) which mechanisms link plants to arthropods (plant diversity, quantity or quality), and arthropods to decomposition and herbivory (abundance or species richness).

Material and Methods

Ethic statement

Arthropod sampling was conducted with the permission of the city council of Jena, Germany.

Experimental field site

The study was conducted in the framework of The Jena-Experiment (Thuringia, Germany, 50°55′ N, 11°35′ E; 130 m above sea level). This grassland biodiversity experiment uses sixty species native to regional floodplains. Plant communities were sown in 82 plots of 20×20 m with a gradient of species richness (1, 2, 4, 8, 16 and 60) and plant functional group richness (1–4 functional groups: grasses, small herbs, tall herbs, legumes) [32]. In 2009, the plot size was reduced to 100 m^2, and two of the monocultures were abandoned due to poor cover of the target species. The particular species mixtures were randomly selected with some constraints [32]. A gradient in abiotic soil properties was taken into account by arranging plots in four blocks perpendicular to the river Saale [32]. All plots are mown twice a year, a management regime common for such meadows. Plots are weeded twice a year to maintain the experimental diversity gradient. As sown and realized number of target species are highly correlated ($R^2 = 0.97$), we used the number of sown plant species for our analyses. The study site was established in May 2002 allowing for plenty of time for colonization by arthropods and making the established plant diversity gradient particularly well suited for examining consumer effects on ecosystem functioning [20].

Plant community analyses

In our path analyses we used aboveground plant biomass as measure of resource availability for arthropods, because herbivores directly feed on plants and decomposers on plant litter material, which is reported to be highly correlated with plant biomass [19]. In 2010, aboveground plant community biomass was harvested during peak standing biomass in late May and August on all plots, in the frame of the long-term measurement of aboveground productivity. This was done by clipping the vegetation three cm above the soil surface in two rectangles of 0.2×0.5 m. The harvested biomass was separated into sown species, cleaned from weeds, dried at 70°C for 72 h and weighed. As a general measure of plant quality we used plant molar C:N ratio, because it is the most practicable measure of plant quality across 60 different plant species, applicable to many different herbivore and decomposer species. Community carbon (C) and nitrogen (N) concentrations of the pooled, dried and milled plant material were measured separately for May and August 2010 with an elemental analyzer (Flash EA 112 Thermo).

Arthropod sampling

Arthropods were sampled in 2010 according to methods used in previous years with the permission of the city council of Jena. We sampled decomposers from May until September 2010, using two pitfall traps of 4.5 cm diameter per plot. The use of pit fall traps is quite common to sample litter- and surface-dwelling macro-arthropods like isopods and diplopods [33]. Traps were emptied every two weeks, but closed for two weeks during the mowing campaigns in June and September. In total, this resulted in 125 sampling days over the whole vegetation period. During the sampling periods, the field traps were filled with 3% formalin, and after emptying the traps, animals were stored in 70% ethanol. The resulting activity abundance data will be termed 'abundance' hereafter. Our data on decomposers comprise surface active decomposer macrofauna only. Less mobile decomposers, such as earthworms, may be important agents in surface litter decomposition [34], but are underrepresented in these samples.

Herbivorous arthropods were collected by suction sampling in June and July using a modified commercial vacuum cleaner (Kärcher A2500, Kärcher GmbH, Winnenden, Germany). Three subplots of 0.75 m×0.75 m were randomly chosen within each plot, covered quickly by a cage with gauze coat of the same size, and sampled until no arthropods were spotted anymore. The sampling was carried out between 9 a.m. and 4 p.m. within two 4-day sampling periods. Pit fall samples and suction samples were sorted to order level and most orders containing decomposers and herbivores were further identified to species level. For herbivores only Auchenorrhyncha, Coleoptera, and Heteroptera were identified to species level, and these orders covered 55% of the potential herbivores we sampled. Aphids and Diptera, which made together 30.2% of the sampled potential herbivores were not identified, because of their patchy distribution within a single plot (Aphids) and due to identification difficulties (Diptera).

Data of the different sampling campaigns were pooled.

Litter decomposition

We measured litter decomposition using the standard litterbag technique [11]. We placed one litterbag in each of the 80 plots of The Jena-Experiment between 17th June and 24th August 2009, as successfully done in previous studies [9,10,35]. The fact that we have a very high number of replicates within each plant diversity level (across plots) compensates for the low replication within plots (one litterbag each) since we were more interested in the effects of plant diversity than of those of specific plant mixtures. To allow access for decomposer macrofauna (e.g. earthworms, isopods, diplopods) we constructed litterbags with 4 mm mesh and anchored them to the soil surface. We filled each litterbag with ~3 g of wheat straw (N = 0.4%, C = 45.2%, C:N = 111.5) of ~3 cm length. At the end of the experiment, the remaining litter was cleaned, dried (70°C, 48 h) and weighed to determine mass loss. The use of wheat straw acts as a standard method to measure potential decomposition rates caused by decomposers and affected by microclimatic conditions, without including effects of litter quality [9,10,36]. This method focusses on microenvironmental effects on decomposition caused by changes in plant communities instead of combined effects of microenvironment and litter quality, composition or structure.

Herbivory

Arthropod herbivory was measured in the plant communities during peak biomass in August 2010. A maximum of 30 fully developed leaves were sampled randomly for each species from each plot where it occurred in the plant biomass samples. The minimum number of leaves for a species in a plot was one, and the average number was 21 per species and plot. On average 65% (SE±2.1%) of the sown species per plot occurred in the biomass sampled, therefore the measure can be regarded representative of the sown plant communities. The damaged surface area caused by herbivores of each leaf (in mm^2) was estimated visually by comparing the damaged leaf area to circular and square templates ranging in size from 1 mm^2 to 500 mm^2. The leaf area was measured with a leaf-area meter (LI-3000C Area Meter, LI-COR Biosciences, Lincoln, USA), and the proportional herbivory damage (herbivory rates) was calculated as the damaged area divided by measured leaf area. The community herbivory rates for each plot were calculated by summing the species-specific herbivory rates weighted by species-specific leaf biomass in the plot. Mammals were excluded from the experimental field site by a fence, thus only invertebrates were responsible for the observed herbivore damage. Slugs contributed to the measured leaf damage, but were not included in herbivore analyses. For details in the methods see [8].

Statistical analysis

We analyzed our data using (i) linear models and (ii) path analyses. Linear models served as affirmation of the results arising from the path analyses, as our sample size (N = 80) was somewhat lower than usually recommended for path analyses [37,38].

Linear models were used to test for effects of plant community properties on arthropods and ecosystem processes. These models were fitted using the statistical software package R (R development Core Team, http://www.R-project.org. version 2.13.1). Arthropod abundance, arthropod species richness (decomposer and herbivores), decomposition and herbivory were used as response variables. For each of the response variables, we fitted a model including block (factor), plant diversity (numeric), and functional group richness (numeric; number of plant functional groups occurring in the respective plot- see experimental design section) as explanatory variables. We omitted plant functional group richness from our further models, as it had no consistent effect on any of the response variables (see Table S1).

Path analyses were calculated in Amos v.20.0 (IBM SPSS Amos, http://www.spss.com/amos/) and were used to infer the hypothesized causal links between plant diversity and ecosystem processes, as ecosystem process responses are likely indirectly mediated by physical, structural or chemical attributes of plants and/or consumer groups studied. We constructed two models (one for decomposition and one for herbivory), based on our hypotheses extracted from the published literature (see Introduction, and Fig. 1), and did not perform any model simplification. Models included arthropod abundance and species richness, aboveground plant biomass as a measure of food quantity, and plant community C:N ratio as measure of food quality as endogenous variables. Because of our low overall sample size (N = 80), we did not construct latent variables but assumed direct paths among each pair of endogenous variables. As herbivores were sampled after the first biomass harvest we used herbivory, biomass and C:N data of the data collection in August 2010. For decomposition we used mean values of the first and second data collection for aboveground biomass and C:N ratio of plants.

For all analyses (linear models and path analyses) measures of plant diversity, decomposer abundance, herbivore abundance and decomposer species richness were log transformed, aboveground plant biomass was square-root transformed, and herbivory was logit transformed [39] to account for non-normality of errors and heteroscedasticity, and to linearize the relationships among variables. All data used in this publication are deposited in the project database (www.the-jena-experiment.de).

Results

Effects of plant diversity on the decomposer community and decomposition

Overall, the final dataset contained 1,316 macrofauna decomposer individuals belonging to 12 species (four Isopoda and eight Diplopoda species). Isopoda dominated the macrofauna decomposer community, representing on average 92.8% of the total community.

Decomposer abundance and species richness increased significantly with increasing plant diversity, but remained unaffected by the number of plant functional groups (Table S1, Fig. S1). The number of individuals increased from 78.6 per plot (SE±19.2) in monocultures to 244 per plot (SE±30.5) in 60-plant-species-mixtures. Mean decomposer species richness increased from 3.6 (SE±0.5) per plot in plant monocultures to 6.3 (SE±0.4) in 60-plant-species-mixtures. Decomposition rate increased with increasing plant species richness from 3 mg mg^{-1} d^{-1} (SE±0.0002) in monocultures up to 5 mg mg^{-1} d^{-1} (SE±0.0004) in 60-plant-species-mixtures. The number of plant functional groups, available in the respective plant communities (Table S1), did not significantly affect decomposition.

Mechanisms driving altered decomposition rates

A chi-squared test indicated that our path analysis model cannot be rejected as a potential explanation of the observed covariance matrix ($\chi^2_4 = 4.49$, $P = 0.344$). The model explained 27% of the variance in decomposition. The positive effect of plant diversity on decomposition (Table S2 and S3, Fig. 2a) was mediated by aboveground plant biomass, which increased with increasing plant diversity and which positively affected decomposer abundance (Fig. 3a) and species richness (marginally). Although decomposer abundances and species richness were strongly correlated (Fig. 3b; Table S1), only species richness increased litter decomposition (Fig. 3c; Table S1). While increasing plant diversity increased plant C:N ratio, there was no significant effect of plant C:N ratio on decomposers or decomposition. A chi-squared test for the same path analysis excluding species richness and abundance of decomposers indicated that it need to be rejected as a potential explanation of the observed covariance matrix ($\chi^2_6 = 15.60$, $P = 0.016$).

Effects of plant diversity on the herbivore community and herbivory rate

Overall, the final dataset contained 12,829 herbivores, belonging to 127 species (38 Auchenorrhyncha, 62 Coleoptera, 27 Heteroptera). On average the herbivore community was dominated by Coleoptera (48.6%), followed by Auchenorrhyncha (43.6%), and Heteroptera (7.8%). Plant diversity, but not plant functional group richness, strongly affected the abundance and species richness of herbivorous insects and herbivory rate (Table S1, Fig. S2). Herbivore abundance increased from 65.2 (SE±13.7) individuals per plot in monocultures to 251 (SE±36.7) individuals per plot in 60-plant-species-mixtures. Herbivore communities in monocultures contained on average 15.6 (SE±1.0) species per plot, whereas in 60-plant-species-mixtures 23.5 (SE±0.8) species per plot occurred. Plant diversity also led to a higher herbivory rate, nearly doubling from 2% (SE±0.4%) in monocultures to 3.5% (SE±0.9%) in 60-plant-species-mixtures.

Mechanisms driving altered herbivory

A chi-squared test indicated that the path analysis explained 22% of the variance in herbivory, and cannot be rejected as a potential explanation of the observed covariance matrix ($\chi^2_4 = 7.50$, $P = 0.112$). The positive relationship between plant diversity and herbivory rate (Table S4 and S5, Fig. 2b), was not driven by aboveground plant biomass, as there was no significant correlation between aboveground biomass and herbivore abundance and species richness. Herbivore abundance and species richness increased significantly with increasing plant diversity (Fig. S2). Increasing plant diversity increased plant C:N ratio, which, in turn, was positively correlated with herbivore abundance (Fig. 4a). Herbivore abundances increased herbivore species richness (Fig. 4b), but only higher abundances of herbivorous arthropods increased herbivory rate (Fig. 4c). A chi-squared test for the same path analysis excluding species richness and abundance of herbivores indicated that it need to be rejected as a potential explanation of the observed covariance matrix ($\chi^2_6 = 24.70$, $P = 0.001$).

Discussion

Plant diversity influenced the abundance and diversity of decomposers and herbivores, and in turn, these effects propagated to decomposition and herbivory rates. Effects of plant diversity on process rates could be better explained when including information on the consumer community into the explanatory model for both, decomposition and herbivory. Abundance of decomposers was positively associated with increased resource (plant) biomass, whereas herbivore abundance increased with higher C:N ratio of the plant material and plant species richness. However, herbivore richness increased directly with higher plant diversity, whereas decomposer richness increased indirectly (via plant biomass and decomposer abundance). Accordingly, we provide support for the More Individuals Hypothesis (MIH) to drive altered decomposer communities, whereas we explain changes in herbivore communities by Dietary Mixing (DM) and the Resource Heterogeneity Hypothesis (RHH). Decomposer species richness enhanced decomposition of litter material (Mechanism 2 in Fig. 1), whereas higher herbivory was mainly driven by the higher abundance of herbivores (Mechanism 1 in Fig. 1).

Decomposers and decomposition

Notably, the discussion of plant community effects (biomass and C:N ratio) on the decomposer community is based on to the use of standard litter material in our study instead of plot-specific litter

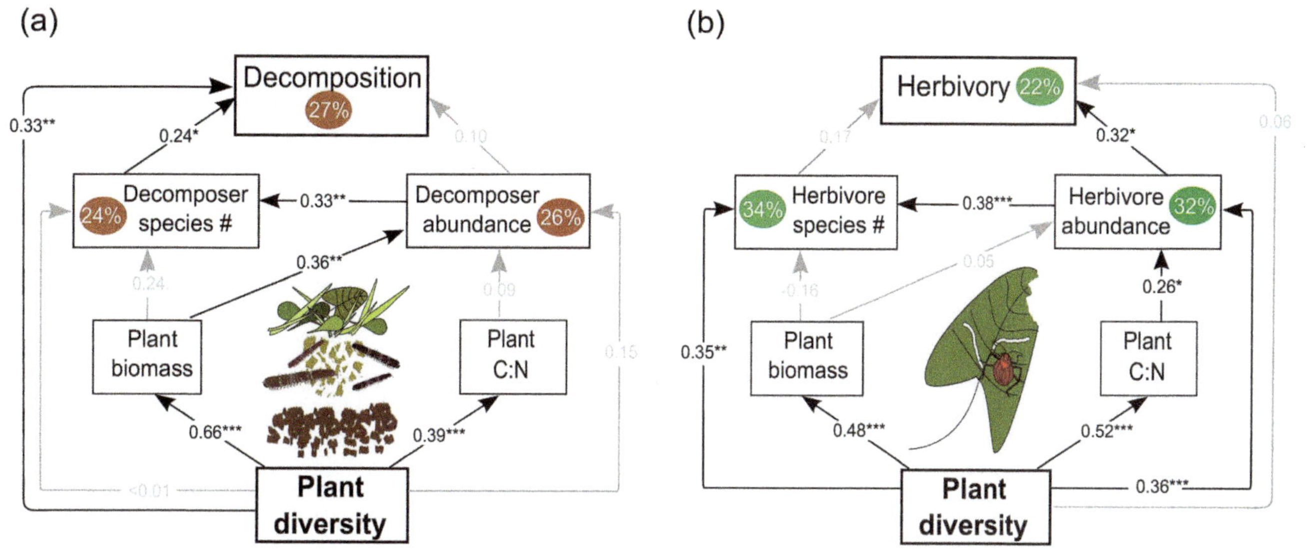

Figure 2. Path diagram explaining plant community effects on decomposition and herbivory. Models relate plant community variables (diversity, quantity and quality), species richness and abundance of (a) decomposer arthropods to decomposition, and (b) herbivorous insects to herbivory. Standardised path coefficients are given on top of the path arrows with significances indicated by *, $P<0.05$; **, $P<0.01$; ***, $P<0.001$. Non-significant paths are given in grey.

material. Consequently, we here discuss indirect effects between plant variables and decomposer communities (plants affect the environment for decomposers, which in turn leads to changes in the decomposer community) [9,10].

Decomposition was positively correlated with plant diversity as has been shown in some studies using standard litter [9,40–42] but not all [10,34]. Reich and colleagues [3] recently showed that plant diversity effects on plant biomass production increased over time, and it was suggested that this increase in plant diversity effects may be linked to delayed soil feedback effects of slowly assembling decomposer communities [43]. Indeed, the density and diversity of macro-decomposers and the density of meso-decomposers increased significantly with increasing plant diversity only four (aboveground) and six years (belowground) after establishment of The Jena-Experiment [20,23]. Consequently, the lack of plant diversity effects in other studies may be due to the short-term character of most previous experiments. Positive effects of plant diversity on decomposition processes can possibly be attributed to higher productivity of more diverse plant communities and thus elevated food availability for decomposers (see MIH in Fig. 1), as the amount of available litter material is positively correlated to aboveground biomass in grasslands [44].

Our path analysis identified aboveground biomass as an important driver of changes in the decomposer abundance, which supports the MIH. Plant biomass production increased with increasing plant diversity and this, in turn, increased decomposer species richness via abundance. Resulting increases in decomposition rates were mediated by increased decomposer species richness but not abundance, indicating a more diverse decomposer community to be functionally complementary in fragmenting litter material (according to M2 in Fig. 1).

Our results on decomposition are in accordance with a study of Heemsbergen et al. [36], who found functional dissimilarity among decomposer species being the best predictor for changes in leaf litter mass loss and soil respiration. Similar results are reported for benthic detritivores in streams, where decomposition rates also increase with detritivore richness [45]. In these systems, potential mechanisms involve increasing faciliation of resource use among species and reduction of inter-specific competiton [46]. Such facilitation can arise if the processing of dead organic matter by one species enhances the suitability of this resource for another species. The positive correlation between decomposer species richness and decomposition could also occur due to a sampling effect (higher chance of having a functionally dominant decomposer species in a species-rich community than in a species-poor one); however, we are not able to disentangle both possible effects here.

Even if we used wheat straw as litter material in our study we expected a negative effect of plant C:N ratio on the environment for decomposers (lower nutrient availability in litter and soil), thereby indirectly affecting their fitness (in our case: abundance). In contrast to our hypothesis we detected no significant effect of aboveground plant quality (C:N ratio) on decomposer abundance and species richness, maybe indicating that the C:N ratio of the fresh plant material is a poor predictor of the C:N ratio of the litter material.

In our path analysis a significant direct link between plant diversity and decomposition remained, indicating that part of the plant diversity effect on decomposition was independent of consumer-mediated effects. Potential environmental variables that could contribute to this link might include soil moisture of the top soil layer, vegetation density, and shade which are important predictors of soil dwelling decomposer community composition. All three variables are documented to increase with increasing plant diversity [10,47]. Further, our analysis was restricted to aboveground living decomposers, but we know that belowground decomposer groups are strongly affected by plant diversity [19,48]. Therefore, we suggest that also belowground decomposition might show an imprint of plant species richness via decomposer abundance or diversity. As data on soil dwelling decomposer species were not available from the the study year and could therefore not be included in the analyses this is a natural extension for future studies.

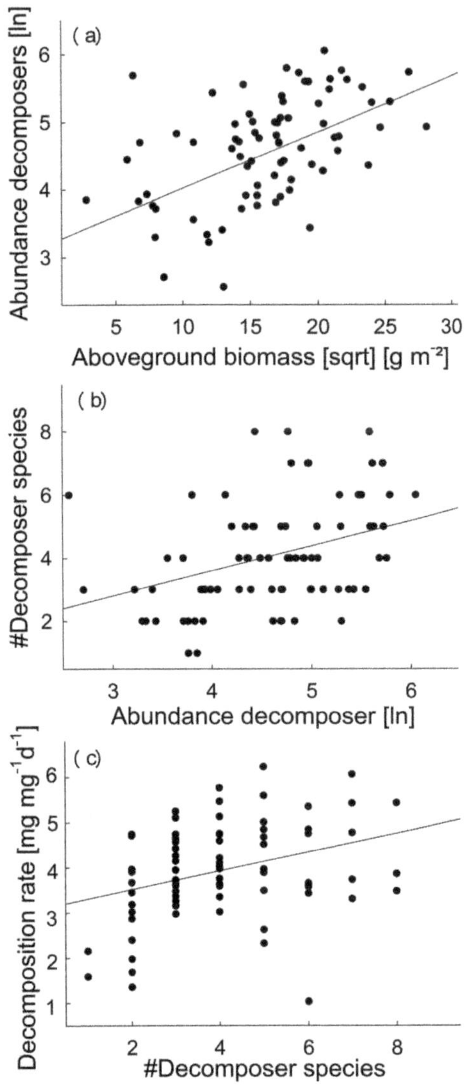

Figure 3. Pairwise correlations visualizing the significant links detected in the path analysis relating plants, decomposers and decomposition. We show the relationships between aboveground plant biomass and decomposer abundances (a), between decomposer abundances and their species richness (b) and decomposer species richness and decomposition (c). For statistics, see Table S2.

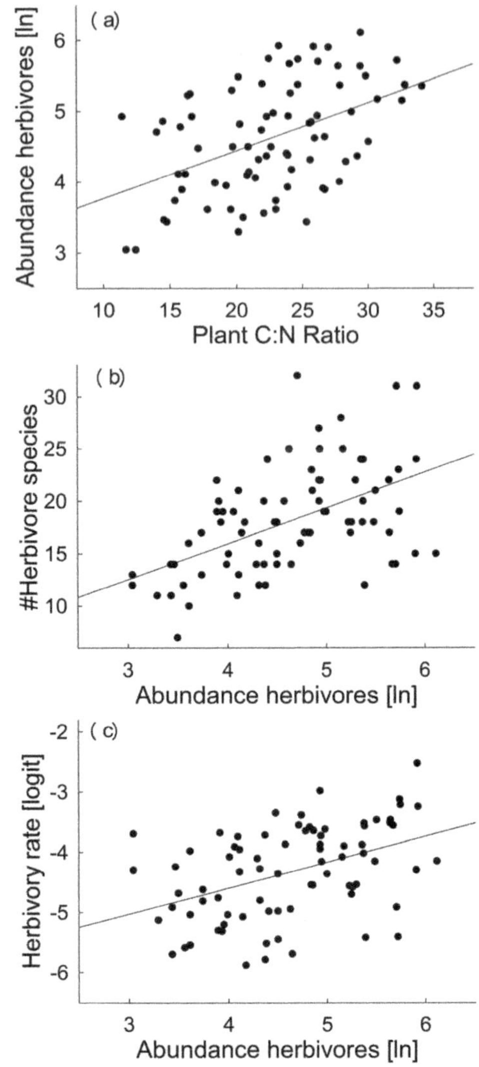

Figure 4. Pairwise correlations visualizing the significant links detected in the path analysis relating plants, herbivores and herbivory. We show the relationships between plant C:N ratio and herbivore abundance (a), herbivore species richness and their abundance (b), and between herbivore abundance and herbivory rate (c). For statistics, see Table S4.

Herbivores and herbivory

Plant diversity increased both herbivore abundance and richness. Plant diversity effects on herbivore richness could not be explained by variations in plant biomass or plant C:N, which indicates that it may be a direct link between resource diversity and consumer diversity. Such relationships are expressly predicted for specialist consumers, as more diverse resource assemblages enable the presence of a higher number of specialist consumer species (e.g. [16,17,21,31]; see RHH in Fig. 1). However, generalist consumers can also benefit from more diverse food plants, if these provide a nutritionally more balanced or temporally less variable food [17,30]. In contrast to findings of McNaughton et al. [49] and Borer et al. [22], who found higher aboveground plant biomass at higher plant species richness providing the food for a larger abundance of herbivores (see MIH in Fig. 1), increases in herbivore abundance in this study were not mediated by plant biomass.

Food plant quality (here C:N ratio) affected the herbivore abundance and thereby indirectly herbivory. As previously shown [25,50], plant C:N ratios increased with plant species richness. A higher food C:N ratio should lead to lower fitness and abundance of herbivores [28,51,52] if the plant C:N is above the consumers' threshold elemental ratios [53]. Consequently, community-wide herbivory was shown to decrease with poorer autotroph quality although individuals compensated for poor food quality by ingesting more [54]. If compensatory feeding in a low nutrition environment is effective, a neutral relationship between food quality and community herbivory might occur [51,55]. Yet, it is hard to imagine how compensatory feeding could lead to increased herbivore abundance at higher plant C:N as observed in our analysis, unless C:N ratio is correlated with other non-nutritional aspects of these trophic interactions. Increased investment in vertical growth and thus C-rich stem-material in the more species-rich plots is a promising candidate for such a relationship, which would increase the habitable volume for arthropods and thereby increase abundance. This interpretation is supported by the fact that the average height in summer of all plants occuring in a communitiy was a significant predictor of herbivory levels when modelling community herbivory based on plant functional traits [56].

Overall, plant diversity increased both herbivore abundance and richness, but only herbivore abundance then increased herbivory rates (confirming M1 in Fig. 1). The obvious interpretation of this relationship is that at higher plant diversity a larger abundance of herbivores consumes more plant biomass. Consequently, an increase of herbivory at higher plant diversity [8,57], can potentially be explained by an increase in herbivore abundance with increasing plant diversity, as documented in this study. Integrating data on feeding type or food specialization could help for a detailed understanding of the mechanisms responsible for the observed positive relation between herbivore abundance and herbivory rate. Herbivore species richness was indirectly linked to herbivory rates via its positive correlation to herbivore abundance.

Conclusion

Our results demonstrate the importance of interactions across trophic levels for ecosystem process rates of green and brown food webs. Decomposition was driven by arthropod species richness via arthropod abundance, and herbivory depended on the number of herbivores. Excluding data on consumer levels from our analysis (thus measuring only the directly plant related links from plant diversity and composition to rates of decomposition or herbivory) resulted in no explained variance (model rejection, see result section) for decomposition and herbivory. In particular, our results strongly support the conclusion that integrating trophic interactions into the study of biodiversity effects is required to understand the multiple pathways by which biodiversity affects ecosystem functioning [58]. In addition, we could show that the functional role of an important part of overall biodiversity – in our case the arthropod community – cannot always be deduced *a priori*, but must be unraveled using a mechanistic approach. While functional biodiversity research has shown the general importance of biodiversity for many ecosystem processes, understanding the functional role of biodiversity requires analyzing the mechanisms underlying BEF relationships. Given that decomposers and herbivores interact with communities of natural enemies exerting a top-down control in addition to the bottom-up control by the plant community [59] a consequent step forward would be to include higher trophic levels (predators, parasites, parasitoids) in the analysis of plant diversity effects on arthropod-mediated ecosystem processes.

Supporting Information

Figure S1 Effects of plant diversity on the abundance and species richness of decomposer arthropods, and decomposition.

Figure S2 Effects of plant diversity on the abundance and species richness of herbivorous arthropods, and herbivory rate.

Table S1 Results of linear models testing plant diversity effects on decomposing and herbivorous arthropods, and decomposition (mg mg^{-1} d^{-1}) and herbivory rate (%, logit transformed).

Table S2 Results of the path analysis linking plants, decomposers and decomposition.

Table S3 Standardized total effects of the structural equation model analysing plant diversity effects on decomposition.

Table S4 Results of the path analysis linking plants, herbivores and herbivory rate.

Table S5 Standardized total effects of the structural equation model analysing plant diversity effects on herbivory rate.

Acknowledgments

The study was funded by the Deutsche Forschungsgemeinschaft (FOR 1451). We thank the technical staff for their work in maintaining the field site and also many student helpers for weeding of the experimental plots and support during measurements. Hannah Loranger and Enrica De Luca are greatly acknowledged for providing data on herbivory rates and plant biomass respectively. We thank Jes Hines for helpful comments on this manuscript. Additional financial support was received by the AquaDiva @Jena project financed by the state of Thuringia, Germany.

Author Contributions

Conceived and designed the experiments: AE STM NE HH AV AW WWW. Performed the experiments: STM NE AV AE ML MA HH AW. Analyzed the data: AE STM CS. Contributed reagents/materials/analysis tools: STM CS NE AE HH AW WWW. Wrote the paper: AE STM MA NE HH ML AV AW WWW.

References

1. Hooper DU, Adair EC, Cardinale BJ, Byrnes JEK, Hungate BA, et al. (2012) A global synthesis reveals biodiversity loss as a major driver of ecosystem change. Nature 486: 105–108.
2. Cardinale BJ, Duffy JE, Gonzalez A, Hooper DU, Perrings C, et al. (2012) Biodiversity loss and its impact on humanity. Nature 486: 59–67.
3. Reich PB, Tilman D, Isbell F, Mueller K, Hobbie SE, et al. (2012) Impacts of biodiversity loss escalate through time as redundancy fades. Science (80-) 336: 589–592.
4. Cardinale BJ, Srivastava DS, Duffy JE, Wright JP, Downing AL, et al. (2006) Effects of biodiversity on the functioning of trophic groups and ecosystems. Nature 443: 989–992.
5. Marquard E, Weigelt A, Temperton VM, Roscher C, Schumacher J, et al. (2009) Plant species richness and functional composition drive overyielding in a six-year grassland experiment. Ecology 90: 3290–3302.
6. Naeem S, Thompson LJ, Lawler SP, Lawton JH, Woodfin RM (1994) Declining biodiversity can alter the performance of ecosystems. Nature 368: 734–737.
7. Scherber C, Milcu A, Partsch S, Scheu S, Weisser WW (2006) The effects of plant diversity and insect herbivory on performance of individual plant species in experimental grassland. J Ecol 94: 922–931.
8. Loranger H, Weisser WW, Ebeling A, Eggers T, Luca E De, et al. (2013) Invertebrate herbivory increases along an experimental gradient of grassland plant diversity. Oecologia:
9. Hector A, Beale AJ, Minns A, Otway SJ, Lawton JH (2000) Consequences of the reduction of plant diversity for litter decomposition: effects through litter quality and microenvironment. Oikos 90: 357–371.
10. Scherer-Lorenzen M (2008) Functional diversity affects decomposition processes in experimental grasslands. Funct Ecol 22: 547–555.
11. Hattenschwiler S, Gasser P (2005) Soil animals alter plant litter diversity effects on decomposition. Proc Natl Acad Sci U S A 102: 1519–1524.
12. Eisenhauer N (2012) Aboveground-belowground interactions as a source of complementarity effects in biodiversity experiments. Plant Soil 351: 1–22.

13. Gruner DS, Smith JE, Seabloom EW, Sandin SA, Ngai JT, et al. (2008) A cross-system synthesis of consumer and nutrient resource control on producer biomass. Ecol Lett 11: 740–755.
14. Allan E, Crawley MJ (2011) Contrasting effects of insect and molluscan herbivores on plant diversity in a long-term field experiment. Ecol Lett 14: 1246–1253.
15. Mulder CPH, Koricheva J, Huss-Danell K, Hogberg P, Joshi J (1999) Insects affect relationships between plant species richness and ecosystem processes. Ecol Lett 2: 237–246.
16. Koricheva J, Mulder CPH, Schmid B, Joshi J, Huss-Danell K (2000) Numerical responses of different trophic groups of invertebrates to manipulations of plant diversity in grasslands. Oecologia 125: 271–282.
17. Ebeling A, Klein AM, Schumacher J, Weisser WW, Tscharntke T (2008) How does plant richness affect pollinator richness and temporal stability of flower visits? Oikos 117: 1808–1815.
18. Haddad NM, Crutsinger GM, Gross K, Haarstad J, Knops JMH, et al. (2009) Plant species loss decreases arthropod diversity and shifts trophic structure. Ecol Lett 12: 1029–1039.
19. Scherber C, Eisenhauer N, Weisser WW, Schmid B, Voigt W, et al. (2010) Bottom-up effects of plant diversity on multitrophic interactions in a biodiversity experiment. Nature 468: 553–556.
20. Eisenhauer N, Milcu A, Sabais A, Bessler H, Brenner J, et al. (2011) Plant diversity surpasses plant functional groups and plant productivity as driver of soil biota in the long term. PLoS One 6: e16055.
21. Sabais ACW, Scheu S, Eisenhauer N (2011) Plant species richness drives the density and diversity of Collembola in temperate grassland. Acta Oecologica-International J Ecol 37: 195–202.
22. Borer ET, Seabloom EW, Tilman D, Novotny V (2012) Plant diversity controls arthropod biomass and temporal stability. Ecol Lett 15: 1457–1464.
23. Rzanny M, Voigt W (2012) Complexity of multitrophic interactions in a grassland ecosystem depends on plant species diversity. J Anim Ecol 81: 614–627.
24. Cardinale BJ, Wright JP, Cadotte MW, Carroll IT, Hector A, et al. (2007) Impacts of plant diversity on biomass production increase through time because of species complementarity. Proc Natl Acad Sci U S A 104: 18123–18128.
25. Abbas M, Ebeling A, Oelmann Y, Ptacnik R, Roscher C, et al. (2013) Biodiversity effects on plant stoichiometry. PLoS One 8: e58179.
26. Srivastava DS, Lawton JH (1998) Why more productive sites have more species: An experimental test of theory using tree-hole communities. Am Nat 152: 510–529.
27. Abrams PA (1995) Monotonic or unimodal diversity productivity gradients - what does competition theory predict. Ecology 76: 2019–2027.
28. Awmack CS, Leather SR (2002) Host plant quality and fecundity in herbivorous insects. Annu Rev Entomol 47: 817–844.
29. Root RB (1973) Organization of a plant-arthropod association in simple and diverse habitats - fauna of collards (Brassica-Oleracea). Ecol Monogr 43: 95–120.
30. Bernays EA, Bright KL (1993) Mechanisms of dietary mixing in grasshoppers - a review. Comp Biochem Physiol a-Physiology 104: 125–131.
31. Hutchinson GE (1959) Homage to Santa-Rosalia or why are there so many kinds of animals. Am Nat 93: 145–159.
32. Roscher C, Schumacher J, Baade J, Wilcke W, Gleixner G, et al. (2004) The role of biodiversity for element cycling and trophic interactions: an experimental approach in a grassland community. Basic Appl Ecol 5: 107–121.
33. Coleman DC, Crossley DA, Hendrix PF (2004) Fundamentals in soil ecology. Elsevier Academic Press.
34. Milcu A, Partsch S, Scherber C, Weisser WW, Scheu S (2008) Earthworms and legumes control litter decomposition in a plant diversity gradient. Ecology 89: 1872–1882.
35. Vogel A, Eisenhauer N, Weigelt A, Scherer-Lorenzen M (2013) Plant diversity does not buffer drought effects on early-stage litter mass loss rates and microbial properties. Glob Chang Biol 19: 2795–2803.
36. Heemsbergen DA, Berg MP, Loreau M, van Haj JR, Faber JH, et al. (2004) Biodiversity effects on soil processes explained by interspecific functional dissimilarity. Science (80-) 306: 1019–1020.
37. Arbuckle JL, Wothke W (1995) Amos 4.0 User's guide. Chicago: Small waters.
38. Grace J (2006) Structural equation modeling and natural systems. Cambridge: Cambridge University Press.
39. Warton DI, Hui FKC (2011) The arcsine is asinine: the analysis of proportions in ecology. Ecology 92: 3–10.
40. Bardgett RD, Shine A (1999) Linkages between plant litter diversity, soil microbial biomass and ecosystem function in emperate grasslands. Soil Biol Biochem 31: 317–321.
41. Spehn EM, Hector A, Joshi J, Scherer-Lorenzen M, Schmid B, et al. (2005) Ecosystem effects of biodiversity manipulations in european grasslands. Ecol Monogr 75: 37–63.
42. Cardinale BJ, Matulich KL, Hooper DU, Byrnes JE, Duffy E, et al. (2011) The functional role of producer diversity in ecosystems. Am J Bot 98: 572–592.
43. Eisenhauer N, Reich PB, Scheu S (2012) Increasing plant diversity effects on productivity with time due to delayed soil biota effects on plants. Basic Appl Ecol 13: 571–578.
44. Long ZT, Mohler CL, Carson WP (2003) Extending the resource concentration hypothesis to plant communities: Effects of litter and herbivores. Ecology 84: 652–665.
45. Jonsson M, Malmqvist B (2000) Ecosystem process rate increases with animal species richness: evidence from leaf-eating, aquatic insects. Oikos 89: 519–523.
46. Jonsson M, Malmqvist B (2003) Mechanisms behind positive diversity effects on ecosystem functioning: testing the facilitation and interference hypotheses. Oecologia 134: 554–559.
47. Rosenkranz S, Wilcke W, Eisenhauer N, Oelmann Y (2012) Net ammonification as influenced by plant diversity in experimental grasslands. Soil Biol Biochem 48: 78–87.
48. Hooper DU, Bignell DE, Brown VK, Brussaard L, Dangerfield JM, et al. (2000) Interactions between aboveground and belowground biodiversity in terrestrial ecosystems: Patterns, mechanisms, and feedbacks. Bioscience 50: 1049–1061.
49. McNaughton SJ, Oesterheld M, Frank DA, Williams KJ (1989) Ecosystem-level patterns of primary productivity and herbivory in terrestrial habitats. Nature 341: 142–144.
50. Van Ruijven J, Berendse F (2005) Diversity-productivity relationships: initial effects, long-term patterns, and underlying mechanisms. Proc Natl Acad Sci U S A 102: 695–700.
51. Berner D, Blanckenhorn WU, Korner C (2005) Grasshoppers cope with low host plant quality by compensatory feeding and food selection: N limitation challenged. Oikos 111: 525–533.
52. Ebeling A, Allan E, Heimann J, Köhler G, Scherer-Lorenzen M, et al. (2013) The impact of plant diversity and fertilization on fitness of a generalist. Basic Appl Ecol 14: 246–254.
53. Frost PC, Benstead JP, Cross WF, Hillebrand H, Larson JH, et al. (2006) Threshold elemental ratios of carbon and phosphorus in aquatic consumers. Ecol Lett 9: 774–779.
54. Hillebrand H, Borer ET, Bracken MES, Cardinale BJ, Cebrian J, et al. (2009) Herbivore metabolism and stoichiometry each constrain herbivory at different organizational scales across ecosystems. Ecol Lett 12: 516–527.
55. Schuldt A, Bruelheide H, Durka W, Eichenberg D, Fischer M, et al. (2012) Plant traits affecting herbivory on tree recruits in highly diverse subtropical forests. Ecol Lett 15: 732–739.
56. Loranger J, Meyer ST, Shipley B, Kattge J, Loranger H, et al. (2013) Predicting invertebrate herbivory from plant traits: polycultures show strong nonadditive effects. Ecology 94: 1499–1509.
57. Schuldt A, Baruffol M, Böhnke M, Bruelheide H, Härdtle W, et al. (2010) Tree diversity promotes insect herbivory in subtropical forests of south-east China. J Ecol 98: 917–926.
58. Duffy JE, Cardinale BJ, France KE, McIntyre PB, Thebault E, et al. (2007) The functional role of biodiversity in ecosystems: incorporating trophic complexity. Ecol Lett 10: 522–538.
59. Rzanny M, Kuu A, Voigt W (2013) Bottom-up and top-down forces structuring consumer communities in an experimental grassland. Oikos 122: 967–976.

15

Effects of Grazing Regimes on Plant Traits and Soil Nutrients in an Alpine Steppe, Northern Tibetan Plateau

Jian Sun[1], Xiaodan Wang[2]*, Genwei Cheng[2], Jianbo Wu[2], Jiangtao Hong[2], Shuli Niu[1]

1 Synthesis Research Centre of Chinese Ecosystem Research Network, Key Laboratory of Ecosystem Network Observation and Modelling, Institute of Geographic Sciences and Natural Resources Research, Chinese Academy of Sciences, Beijing, China, **2** The key laboratory of mountain surface processes and eco-regulation, Institute of Mountain Hazard and Environment, Chinese Academy of Sciences, Chengdu, China

Abstract

Understanding the impact of grazing intensity on grassland production and soil fertility is of fundamental importance for grassland conservation and management. We thus compared three types of alpine steppe management by studying vegetation traits and soil properties in response to three levels of grazing pressure: permanent grazing (M1), seasonal grazing (M2), and grazing exclusion (M3) in the alpine steppe in Xainza County, Tibetan Plateau. The results showed that community biomass allocation did not support the isometric hypothesis under different grassland management types. Plants in M1 had less aboveground biomass but more belowground biomass in the top soil layer than those in M2 and M3, which was largely due to that root/shoot ratios of dominant plants in M1 were far greater than those in M2 and M3. The interramet distance and the tiller size of the dominant clonal plants were greater in M3 than in M1 and M2, while the resprouting from rhizome buds did not differ significantly among the three greezing regimes. Both soil bulk density and soil available nitrogen in M3 were greater than in M1 at the 15–30 cm soil depth ($P=0.05$). Soil organic carbon and soil total nitrogen were greater in M3 than in M1 and M2 ($P=0.05$). We conclude that the isometric hypothesis is not supported in this study and fencing is a helpful grassland management in terms of plant growth and soil nutrient retention in alpine steppe. The extreme cold, scarce precipitation and short growing period may be the causation of the unique plant and soil responses to different management regimes.

Editor: Cheng-Sen Li, Institute of Botany, China

Funding: This research was jointly funded by the Knowledge Innovation Program of the Chinese Academy of Sciences (KZCX2-XB3-08), CAS Strategic Priority Research Program (XDA05050602), and the Open Fund of Key Laboratory of Ecosystem Network Observation and Modelling (110301A1PA). The funders had no role in study design, data collection and analysis, decision to publish, or preparation of the manuscript.

Competing Interests: The authors have declared that no competing interests exist.

* Email: wxd@imde.ac.cn

Introduction

Alpine grasslands make up the dominant ecosystem occupying approximately 94% of Northern Tibet [1]. The natural environment of the region is extremely harsh, and the alpine steppe, a fragile ecosystem, is extremely susceptible to the impacts of human activities [2]. It suffers from overgrazing, deforestation, and the harvesting of numerous herbs commonly used in traditional medicines [3–5]. Studies examining the response of above- and belowground biomass, the root/shoot ratio, the morphological characteristics of dominant plants, and the soil properties to human disturbance offer important insights that can contribute to adopting the most effective approach to grassland management in an alpine steppe, in which it is particularly difficult to recover from ecosystem degradation due to the region's long period of frost and relatively short growing season [3,6].

Although grazing and fencing, both of which have a substantial affect on vegetation traits and soil properties [7–9], are the most prevalent management regimes for grasslands worldwide, and although the effects of herbivores on soil properties in (sub)alpine ecosystems have recently been reported [10,11], knowledge about plants and soils in response to grassland management regimes (i.e., fencing and grazing) in the Tibetan Plateau remains limited due to an extremely difficult geographic situation [9,12]. With regard to soil properties, it has been documented that grazing depresses soil carbon storage by changing the plant biomass and composition of a Tibetan alpine meadow [13]. In contrast, Shi et al. [14] found grazing exclusion to decrease soil organic carbon storage in an alpine grassland of the Tibetan Plateau, while another report suggested that seasonal grazing might enrich soil nutrients [15]. Such conflicting results indicate that different grazing intensities may have varying impacts on soil properties.

Understanding the influence of different management types on grassland production is essential for improving grassland conservation and management [16]. Previous studies have yielded varying results for aboveground biomass changes. Grazing thus increased [17] or decreased [18] aboveground production in different cases. The fencing optimization hypothesis posits that fencing significantly enhances above- and belowground biomass by means of carbon reallocation to plant growth and promotes increases in soil nutrient concentrations [19,20]. Belowground biomass has also been shown to be affected by grazing, with some studies indicating that plants reduce the proportion of aboveground parts and allocate more biomass to belowground parts, so as to geminate and resist environmental stress (i.e., grazing pressure) [21]. Conversely, more belowground biomass was

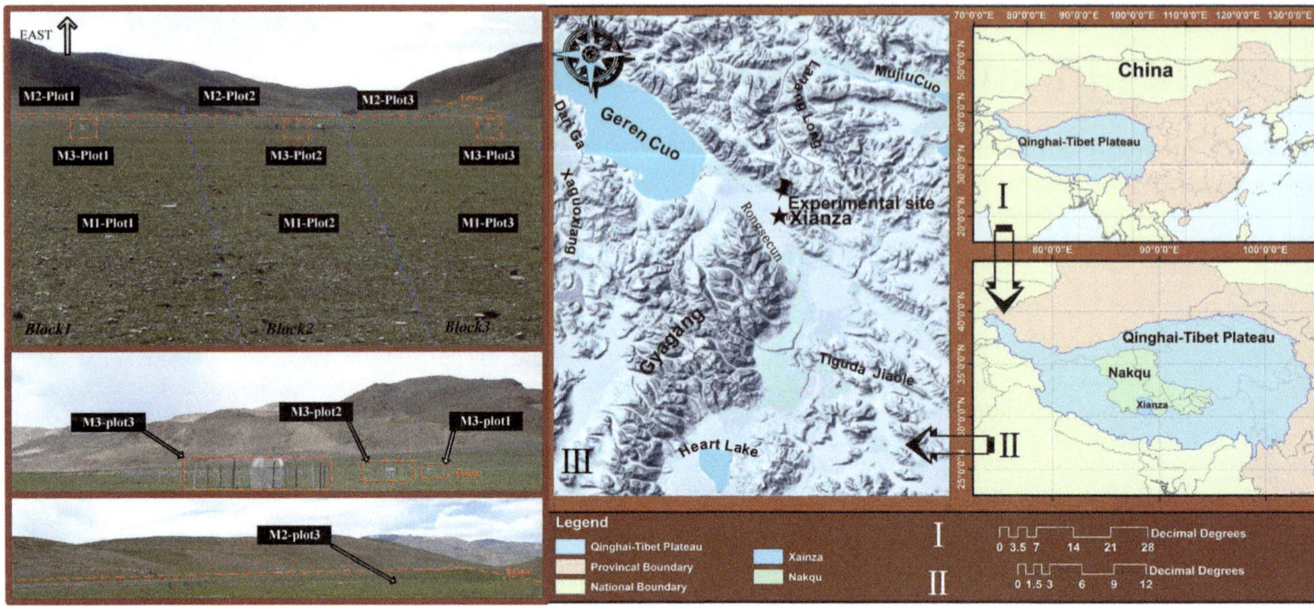

Figure 1. The experiment sites in alpine steppe, northern Tibetan Plateau.

allocated to the surface layer than the subsurface of the soil profile with increasing grazing pressure [22]. For instance, the root to shoot ratio in the grazing pattern was significantly higher than for the mowing and fencing patterns in temperate grassland [23].

The root:shoot ratio provides the basis for understanding the response or adaptive strategies of plants to environmental stress [24,25], and reflects how plants respond to different selection pressures [26,27]. Theory predicts the ratio of roots to shoots to respond isometrically across different individuals and community types under varying environmental conditions [28–30]. While biomass allocation has been widely examined from individual to community to ecosystem levels [31–35], few studies compare the biomass partitioning of dominant alpine species under different grassland management types.

Aside from biomass, the morphological characteristics of clonal plants are also regarded as a sensitive index for assessing the possible effect of grazing pressure [36,37]. The traditional theory is that a clone with a guerrilla-like foraging strategy should be able to locate and exploit favorable patches in which competitors are absent. In a habitat in which potential competitors are more homogeneously distributed, a phalanx clone exhibits a "consolidation" strategy and may be able to persist and monopolize locally available resources [38]. However, we know little about the responses of the morphological characteristics of clonal plants to grazing pressure in alpine steppes.

In this study, we examined plant traits and soil properties in response to different grassland management strategies in the Northern Tibetan Plateau. The specific aims are: (1) to reveal the impact of different grassland managements on plant traits and soil properties and to test the grazing optimization hypothesis; (2) to test the isometric partitioning theory on different alpine species under varying grassland management patterns; and (3) to explore the effects of grazing stress on clonal growth.

Materials and Methods

Study area

Our study was conducted in the Alpine Steppe Experiment Station (N 30°57′, E 88°42′, 4675 m a.s.l.) in Xainza County, Northern Tibet (Fig. 1). The annual mean air temperature of this region is 0°C, while the mean air temperature ranges from −10.1°C in January and 9.6°C in July. The average annual precipitation is 300 mm. There is no absolute frost-free season, and the frosty period can last up to 279.1 days [39]. The region's soil is classified as steppe soil with sandy loam [40]. Vegetation coverage of the alpine steppe is approximately 20% and is dominated by *Stipa purpurea* (Bunch grass) and *Carex moorcrofii* (sedge family, creeping rhizomes) (Fig. 2A). Other common species include *Stellera chamaejasme Linn.* (Bunch weeds), *Oxytropis glacialis* (Bunch weeds), and *Leontopodium alpinum* (Fig. 2B). All of these plants are perennial herbs.

Grazing treatments and experimental plots

There were three grazing treatments in this study. i.e, M1 (grazing): The pasture is always used for grazing. M2 (seasonal grazing): The pasture has been used for grazing in the non-growing seasons but not in the growing season (from May to September) since 2007. M3 (grazing exclusion): Herbivores have been excluded from the pasture since 2010.

The study area was divided into three blocks. Each block contained three plots, the eastern plot having a grazing intensity of M2, the middle plot a grazing intensity of M3, and the western plot a grazing intensity of M1 (Fig. 1). The dimensions of each M3 fenced plot were 6×6 m; the dimensions for each M2 plot were 200×500 m. Three M1 plots were located outside of the fenced area in each block, with each plot having dimensions of 200×500 m. In each plot, a quadrat (30 cm×30 cm) was used to measure biomass and take samples. The estimated average livestock density in both M1 and M2 was approximately 120 sheep units per km^2. (The "sheep unit" is the most frequently used unit of measurement for evaluating the carrying capacity of pasture areas in China and includes figures for cattle, which are converted

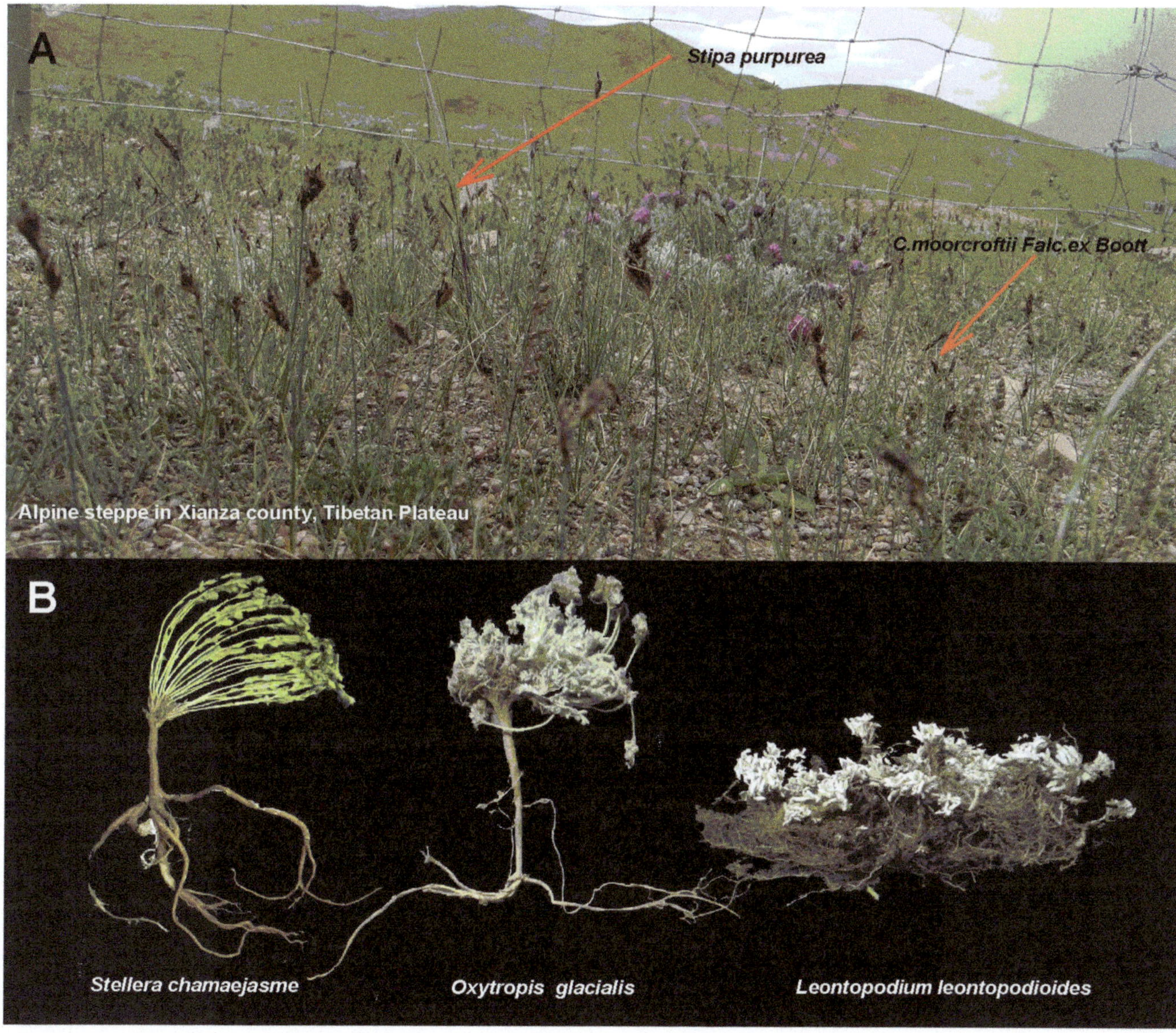

Figure 2. The vegetation composition diagram is in the study area. The Vegetation coverage of the alpine steppe was about 20%, which are dominated by *Stipa purpurea* and *Carex moorcrofii* (panel A) and the accompanying species are *Stellera chamaejasme Linn.*, *Leontopodium alpinum*, *Oxytropis microphylla* (panel B) and so on.

to sheep; thus, e.g., one cow is equivalent to five sheep based on the average weight of a cow) [41].

Ethics statement

No specific permits were required for the samples collected from any of the sites, and the field studies did not involve endangered or protected species.

Biomass measure and sample taking

We established a 30 cm×30 cm quadrat in each of the sampling plots to harvest aboveground biomass (AGB) in late August of 2012. After harvesting the aboveground biomass, we took soil samples by auger with a 5-cm core diameter at two depths (0–15 cm and 15–30 cm) to estimate belowground biomass (BGB). According to previous research, most belowground biomass is located in the top 30 cm layer [42]. Soil samples were also taken at depths of 0–15 cm and 15–30 cm in order to measure soil elements. After being air-dried and sieved (2 mm mesh), the soil samples were carefully handpicked to extract the surface organic materials and fine roots for an analysis of soil chemical properties. Soil nutrients (including soil organic carbon, soil total nitrogen content, available nitrogen, total phosphorus and available phosphorus) were determined using standard protocol [43].

In order to estimate the root:shoot ratio of the dominant species (*S. purpurea*, *C. moorcrofii*, *O. glacialis*, and *L. alpinum*), we took three soil blocks in each sampling plot of different management regimes. The patch was excavated with a spade with a diameter of 20 cm to a depth of 30 cm, which was selected based on the root morphological traits of the species [42]. The root samples obtained from the sites were immediately placed in a cloth bag and then soaked in water to remove the residual soil by means of a 0.5 mm sieve. The roots of each sample were carefully separated from soil and other belowground materials first. Then individuals of the all species were carefully separated from other plant roots. Finally, we

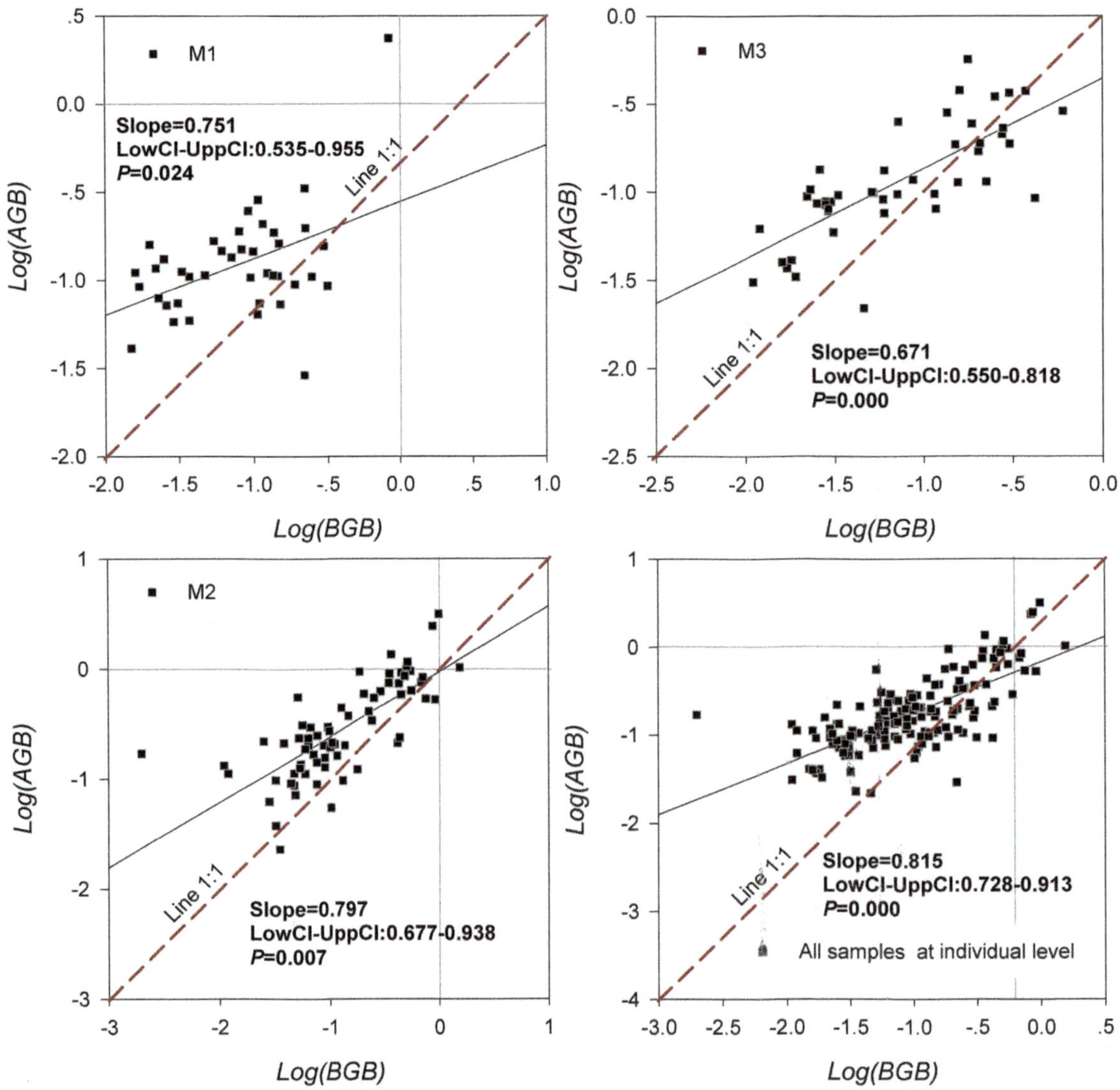

Figure 3. Relationships between AGB and BGB under grazing (M1), seasonal grazing (M2), and grazing exclusion (M3) regimes.

separated the roots and shoots of each individual. Individual biomass (shoots and roots) and community biomass (AGB and BGB) was oven-dried at 65°C until a constant weight was achieved. Individual biomass data were used to analyze allometric functions under different management regimes (with a total of 40, 45, and 68 individuals in M1, M2, and M3, respectively).

The plant morphology of *C. moorcrofii* is regarded as a fairly good indicator that reflects the different grassland management regimes [44]. We thus sampled *C. moorcrofii.* from other 30 cm×30 cm quadrats at each of the 9 plots. The samples were then taken back to the laboratory, where the morphological characteristics (plant height, interramet distance, number of rhizome buds, and tillering number) were immediately measured. Special attention was paid to avoid destroying the structural integrity of the plant throughout this process.

Statistical analysis

We performed a standardized major axis (SRMA) regression to examine whether AGB and BGB scale isometrically at an individual level across all samples, and then explored their relationships under different management types. The regression relationship of the form $LogAGB = \alpha + \beta LogBGB$ was used to describe the allometric relationship between AGB and BGB, where x is BGB, y is AGB, α is the intercept, and β is the scaling slope [25,28,45,46]. The scaling slope and y-intercept of the allometric function were determined using the software package titled "Standardized Major Axis Tests & Routines Version 2.0" [47]. If a 95% confidence interval of the scaling slope covered 1.0, the relationship between AGB and BGB was considered to be isometric.

One-way ANOVA was used to test differences in soil properties and biomes (AGB, BGB, and R/S) among the different treatments, and the Tukey test was used to distinguish differences at a $P = 0.05$ level.

Results

Biomass partitioning

The relationship between BGB and AGB under different grazing pressures was characterized by the linear function $LogAGB = \alpha + \beta LogBGB$ (Fig. 3). The slopes of the allometric

Table 1. Aboveground biomass (AGB), belowground biomass (BGB), and their ratios (R/S) under different grassland managements.

Management Types	Category	M1		M2		M3	
		Mean	STDEV	Mean	STDEV	Mean	STDEV
AGB (g·cm^{-2})		48.09	12.39	51.09	35.48	62.25	3.21
	0–15 cm	312.92	65.6	238.07	53.95	247.16	38.96
BGB (g·cm^{-2})	15–30 cm	11.95	7.64	6.77	5.64	14.72	10.27
	Ratio(0–15 cm/15–30 cm)	26.19	8.59	35.17	9.57	16.79	3.79
	0–30 cm	324.87	58.18	244.84	59.59	261.88	48.67
R/S		7.37	3.58	5.79	2.23	4.2	0.67

Note: The differences of biomes under the different grassland management type were insignificant at 0.05 level.

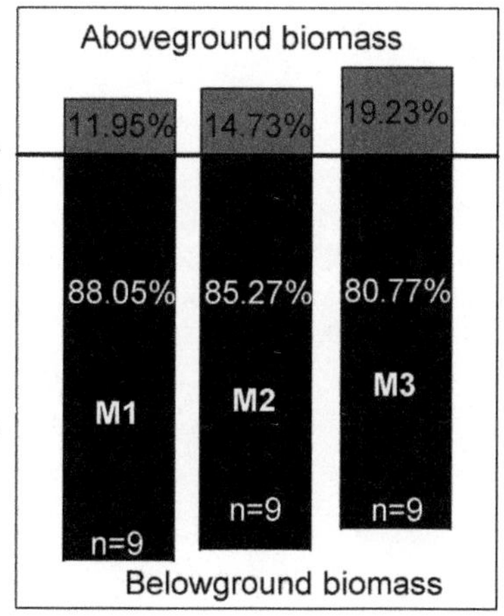

Figure 4. Biomass fractions (% of total biomass) in herbaceous plants in M1, M2 and M3 of Tibetan Plateau.

relationship were 0.751 (for M1), 0.797 (for M2), 0.671 (for M3), and 0.815 (for all samples at an individual level), respectively. All of the slopes differed significantly from 1 at $P = 0.05$.

Responses of AGB, BGB, and R/S to different management types

AGB in M1 was slightly but not significantly lower than in M2 and M3 (Table 1). In contrast, BGB of M1, M2, and M3 amounted to 324.87, 244.84, and 261.88 g·m^{-2}, respectively. The ratio of root to shoot in M1 was also higher than in M2 and M3. The root biomass was mainly distributed in a soil depth of 0–15 cm, while the root biomass in a soil depth of 15–30 cm only accounted for 2.77–5.62% of total BGB. The ratios of BGB in the 0–15 cm-layer to BGB in the 15–30 cm-layer of M1, M2, and M3 were 26.19, 35.17, and 16.79, respectively.

To investigate the characteristics of biomass allocation further, we analyzed the dry matter fraction of M1, M2, and M3 in the Tibetan Plateau (Fig. 4). The results indicated that the roots of total plants in the Tibetan Plateau amounted to 88.05% (in M1), 85.27% (in M2), and 80.77% (in M3).

Traits of clonal plants under different management types

C. moocroftii was tallest in M3 and shortest in M1 ($P<0.05$, Fig. 5A). Both interramet distance (Fig. 5C) and the tiller number (Fig. 5D) in M3 were significantly larger than in M1 and M2. No significant differences were found in the number of rhizome buds among M1, M2, and M3 treatments ($P<0.05$, Fig. 5B).

Soil properties in response to management types

At a soil depth of 0–15 cm, soil bulk density, soil organic carbon, total nitrogen, and available nitrogen were non-significant among M1, M2, and M3 (Table 2). At a soil depth of 15–30 cm, both soil bulk densiy and soil available nitrogen in M3 were greater than in M1 at a $P = 0.05$ level. Soil organic carbon and soil total nitrogen in M3 were greater than in M1, and M2 ($P = 0.05$).

Discussion

Biomass partitioning

Based on the results of our SMA analysis, biomass allocation at the community level did not agree with the isometric hypothesis under different grassland management types. Plants consistently sense changes in their environment and often allocate a greater

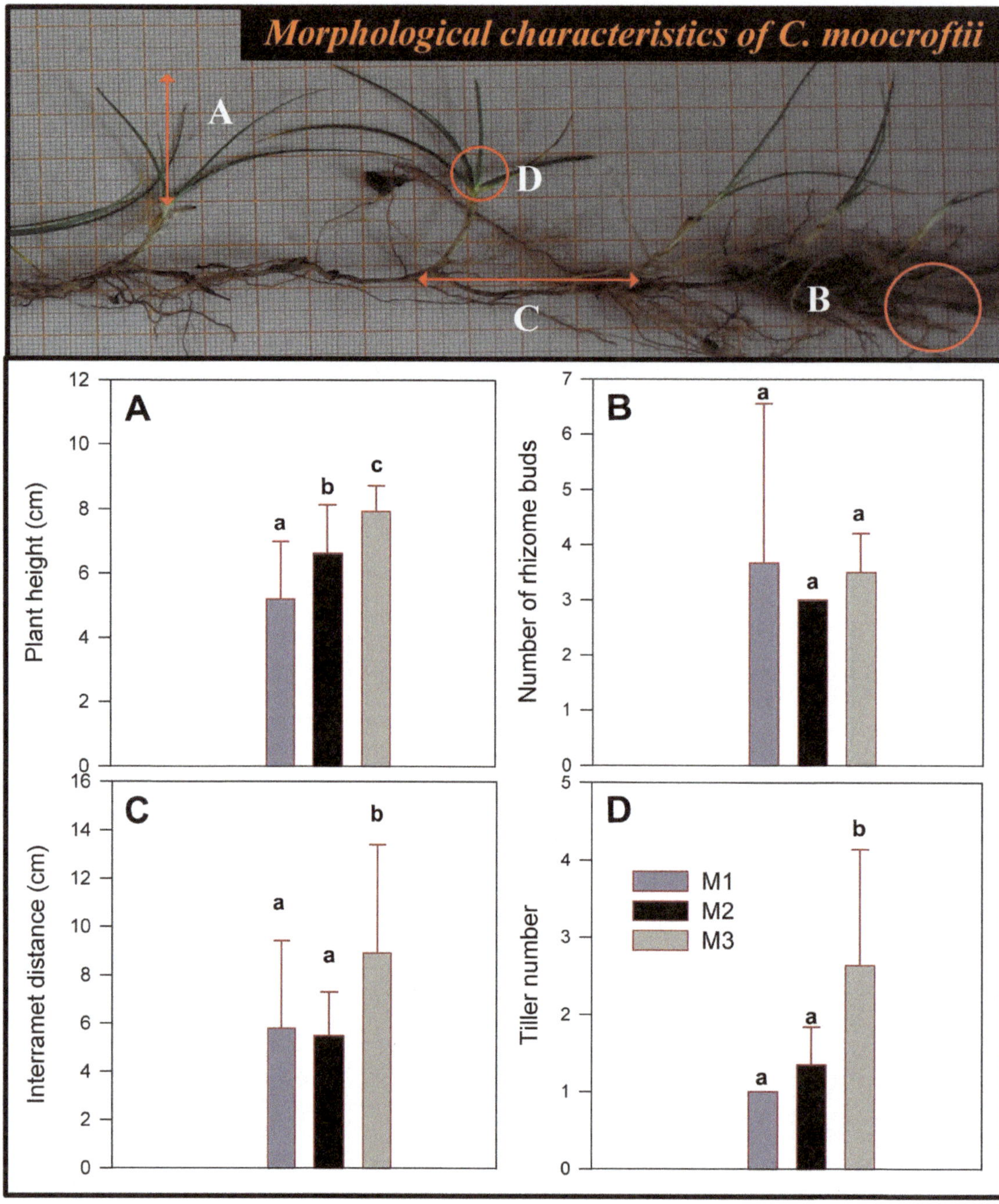

Figure 5. The morphological characteristics of *C. moocroftii* in the different grassland management types. A, represent plant height, B, number of rhizome buds, C, interramet distance, and D tiller number.

proportion of their biomass to the root system when water or mineral elements are scarce [48]. Moreover, roots have also been found to store carbohydrates in alpine grasslands [49]. In the alpine steppe, greater plant biomass allocation to belowground reflects the plants' response to the harsh alpine environment (limited precipitation and low temperatures) for survival. Another reason may be the relatively narrow variation in plant size in that region. Moreover, many species in the Tibetan alpine grasslands do not have typical hierarchical branching structures [42], which did not meet the assumption that stem length scales isometrically with respect to root length. This characteristic is clearly reflected in the root morphology structure shown in Fig. 2B and Fig. 5. Biomass allocation in these areas may thus support the allometric biomass partitioning hypothesis, rather than the isometric allocation hypothesis.

Responses of AGB, BGB, and R/S to different management types

AGB in the grazing plots was found to be lower than in the non-grazing plots, which is in agreement with the findings of Medina-Roldan et al. [50]. We suggested that alpine grassland production might be decreased to some degree by grazing as a result of the

Table 2. Soil properties under the different grassland managements.

Management Types	Depth	M1		M2		M3	
		Mean	STDEV	Mean	STDEV	Mean	STDEV
Soil bulk density (g·cm^{-3})	**0–15 cm**	1.71	0.09	1.48	0.08	1.60	0.11
Soil organic carbon (g·kg^{-1})		10.09	3.38	12.72	2.45	11.02	0.85
Soil total nitrogen(g·kg^{-1})		1.12	0.26	1.54	0.40	1.19	0.11
Soil available Nitrogen(mg·kg^{-1})		62.96	24.14	84.65	17.24	81.61	7.14
Soil total phosphorous(g·kg^{-1})		0.34	0.03	0.33	0.04	0.28	0.02
Soil available phosphorous(mg·kg^{-1})		3.91	0.72	3.83	0.16	3.39	0.29
Soil bulk density(g·cm^{-3})	**15–30 cm**	1.76 **a**	0.04	1.58 **ab**	0.09	1.48 **b**	0.14
Soil organic carbon (g·kg^{-1})		5.21 **a**	0.87	5.81 **a**	0.25	10.16 **b**	2.58
Soil total nitrogen(g·kg^{-1})		0.64 **a**	0.13	0.57 **a**	0.04	1.07 **b**	0.27
Soil available Nitrogen(mg·kg^{-1})		32.36 **a**	4.87	40.57 **ab**	8.51	68.01 **b**	17.37
Soil total phosphorous(g·kg^{-1})		0.27	0.01	0.24	0.04	0.37	0.09
Soil available phosphorous(mg·kg^{-1})		2.95	0.54	3.25	0.61	3.33	0.11

Note: The bold lower case represents the result of variance analysis at 0.05 level.

consumption effect of the livestock. The higher BGB in M1 than M2 and M3 in our alpine steppe conflict with the results of Cheng et al. [51], who found grazing to significantly decrease belowground biomass in the Loess Plateau. Gao et al. [52] also reported significantly less belowground biomass under conditions of heavy grazing in comparison with a non-grazing site in Inner Mongolia. The conflicting results between our study and previous studies in other areas probably stem from differences in environmental conditions. In the extreme cold region of alpine grassland, plants reduce the proportion of aboveground parts and allocate larger amounts of biomass to belowground parts in order to germinate and resist grazing pressures [21]. More BGB in the top soil layer of M1 than M2 and M3 suggests that grazing results in more belowground biomass being allocated to the topsoil [53], possibly due to the alpine environment inhibiting root growth in ungrazed plots. We did not find an ameliorating effect of reduced grazing (M2) or grazer exclusion (M3), suggesting that the effects of grazing cessation may only be detectable in the long term, as proposed by Steffens et al. [54].

The root:shoot ratio (R/S) in M1 was much larger than in M2 and M3, suggesting that plants allocate more to BGB than AGB in order to maximize resources for optimal growth in the grazed environment [21]. We compared the dry matter fraction of M1, M2, and M3 in the Tibetan Plateau (Fig. 4) with dry matter fractions in Argentina, Bolivia, Ecuador, the Arctic, and the Alps [49], and we found that the proportion of BGB was much higher in our study sites than for plants in a semi-arid grassland ecosystem (Bolivia), mountain grassland ecosystem (Argentina), and a humid mountain grassland ecosystem (Ecuadorian Andes). The main reason may be associated with the comparatively slow depletion of carbohydrates in roots, resulting from low respiration rates in the extremely cold winter, and the slower root turnover in colder environments [55]. Meanwhile, long-term grazing results in a higher BGB fraction in the Tibetan Plateau than for plants in other cold alpine environments and the Arctic.

The clonal plant in response to different management types

Clonal plants defend against large herbivores by regulating their morphological characteristics [56]. Clonal plant *C. moocroftii* in M3 was notably taller than in M1 and M2 because it was not bitten by livestock, and they can plastically respond to environmental heterogeneity by placing ramets and changing tiller size in favorable sites [57,58]. We found the interramet distance and tiller number in M3 to be greater than for plants in M1 and M2. Tillers may produce more daughter tillers as a result of increased nutrient availability in grazing exclusion. *C. moocroftii* thus responds to grazing by decreasing the tiller number [37]. The results also indicate that *C. moocroftii* may have both phalanx and guerrilla strategies. The phalanx strategy involves the production of a compact structure of closely spaced ramets (i.e., M1), and the guerrilla strategy involves the production of loosely arranged and more widely spaced ramets in order to seek out more soil nutrients in a relatively suitable habitat (i.e., M3).

Clonal growth by means of lateral roots is regarded as a typical trait of opportunistic species growing in disturbed habitats, as such growth potentially produces a large number of buds on lateral roots [59]. However, our results implied that different grassland management systems did not affect resprouting from rhizome buds, which suggests that reprouting from rhizome buds is not a major adjustment stragegy in response to grazing intensity. This finding is probably due to the short time period since the implementation of fencing and seasonal grazing.

Soil properties in response to different management types

Grazing intensity is one of the most important factors influencing soil properties [50,60]. It has been documented that the trampling action of grazing animals impacts the soil by increasing bulk density [61] and mechanical resistance, and reducing porosity, water infiltration, and aggregate stability [62], and native perennial cover and litter cover, with the consequence of changing the soil nutrient concentrations [63]. In our study, bulk density was found to increase with increasing grazing intensity, and all soil organic carbon, total nitrogen, and available nitrogen contents in M3 were slightly higher than in M1 and M2. We thus report that grazing exclusion may improve soil N availability. Arevalo et al. [64] found an increase in soil phosphorus content in response to increased goat grazing pressure in pastures. In contrast, our results indicated that soil total phosphorous and available phosphorous contents in M1 did not differ significantly from those of M2 and M3. The extreme cold, limited precipitation and short growing period may lead to no obvious change in phosphorus content.

Conclusions

By studying plant traits and soil nutrients in response to grazing intensity in an alpine steppe, we found that the isometric hypothesis is not supported in this particular region. The extreme cold, limited amount of precipitation and short growing period may lead to the plants' and soils' unique to different management types. Overall, the study suggests that the implementation of grazing exclusion played a positive role in the sustainable development of the alpine steppe region.

Acknowledgments

The authors are grateful to two anonymous reviewers all gave very helpful editorial comments.

Author Contributions

Conceived and designed the experiments: XDW GWC. Performed the experiments: JS JBW JTH. Analyzed the data: JS. Contributed reagents/materials/analysis tools: JS GWC JTH. Wrote the paper: JS SLN.

References

1. Lu XY, Fan JH, Yan Y, Wang XD (2013) Responses of Soil CO_2 Fluxes to Short-Term Experimental Warming in Alpine Steppe Ecosystem, Northern Tibet. Plos One 8.
2. Gao QZ, Li Y, Wan YF, Qin XB, Jiangcun WZ, et al. (2009) Dynamics of alpine grassland NPP and its response to climate change in Northern Tibet. Climatic Change 97: 515–528.
3. Sun J, Cheng GW, Li WP, Sha YK, Yang YC (2013) On the Variation of NDVI with the Principal Climatic Elements in the Tibetan Plateau. Remote Sensing 5: 1894–1911.
4. Wang GX, Cheng GD, Shen YP, Qian J (2003) Influence of land cover changes on the physical and chemical properties of alpine meadow soil. Chinese Science Bulletin 48: 118–124.
5. Sun J, Cheng GW, Li WP (2013) Meta-analysis of relationships between environmental factors and aboveground biomass in the alpine grassland on the Tibetan Plateau. Biogeosciences 10: 1707–1715.
6. Yu HY, Luedeling E, Xu JC (2010) Winter and spring warming result in delayed spring phenology on the Tibetan Plateau. Proceedings of the National Academy of Sciences of the United States of America 107: 22151–22156.
7. Zheng SX, Ren HY, Li WH, Lan ZC (2012) Scale-Dependent Effects of Grazing on Plant C: N: P Stoichiometry and Linkages to Ecosystem Functioning in the Inner Mongolia Grassland. Plos One 7.
8. Wu GL, Liu ZH, Zhang L, Chen JM, Hu TM (2010) Long-term fencing improved soil properties and soil organic carbon storage in an alpine swamp meadow of western China. Plant and Soil 332: 331–337.
9. Wang XD, Yan Y, Cao YZ (2012) Impact of historic grazing on steppe soils on the northern Tibetan Plateau. Plant and Soil 354: 173–183.
10. Haynes AG, Schutz M, Buchmann N, Page-Dumroese DS, Busse MD, et al. (2014) Linkages between grazing history and herbivore exclusion on decomposition rates in mineral soils of subalpine grasslands. Plant and Soil 374: 579–591.
11. Risch AC, Haynes AG, Busse MD, Filli F, Schutz M (2013) The Response of Soil CO2 Fluxes to Progressively Excluding Vertebrate and Invertebrate Herbivores Depends on Ecosystem Type. Ecosystems 16: 1192–1202.
12. Fan YJ, Hou XY, Shi HX, Shi SL (2013) Effects of grazing and fencing on carbon and nitrogen reserves in plants and soils of alpine meadow in the three headwater resource regions. Russian Journal of Ecology 44: 80–88.
13. Sun DS, Wesche K, Chen DD, Zhang SH, Wu GL, et al. (2011) Grazing depresses soil carbon storage through changing plant biomass and composition in a Tibetan alpine meadow. Plant Soil and Environment 57: 271–278.
14. Shi XM, Li XG, Li CT, Zhao Y, Shang ZH, et al. (2013) Grazing exclusion decreases soil organic C storage at an alpine grassland of the Qinghai-Tibetan Plateau. Ecological Engineering 57: 183–187.
15. Whitford WG, Steinberger Y (2012) Effects of seasonal grazing, drought, fire, and carbon enrichment on soil microarthropods in a desert grassland. Journal of Arid Environments 83: 10–14.
16. Yang XH, Guo XL, Fitzsimmons M (2012) Assessing light to moderate grazing effects on grassland production using satellite imagery. International Journal of Remote Sensing 33: 5087–5104.
17. Hilbert DW, Swift DM, Detling JK, Dyer MI (1981) Relative Growth-Rates and the Grazing Optimization Hypothesis. Oecologia 51: 14–18.
18. Lkhagva A, Boldgiv B, Goulden CE, Yadamsuren O, Lauenroth WK (2013) Effects of grazing on plant community structure and aboveground net primary production of semiarid boreal steppe of northern Mongolia. Grassland Science 59: 135–145.
19. Wu GL, Du GZ, Liu ZH, Thirgood S (2009) Effect of fencing and grazing on a Kobresia-dominated meadow in the Qinghai-Tibetan Plateau. Plant and Soil 319: 115–126.
20. Pandey CB, Singh JS (1992) Influence of Rainfall and Grazing on Belowground Biomass Dynamics in a Dry Tropical Savanna. Canadian Journal of Botany-Revue Canadienne De Botanique 70: 1885–1890.
21. Ma WL, Shi PL, Li WH, He YT, Zhang XZ, et al. (2010) Changes in individual plant traits and biomass allocation in alpine meadow with elevation variation on the Qinghai-Tibetan Plateau. Science China-Life Sciences 53: 1142–1151.
22. Burke IC, Lauenroth WK, Riggle R, Brannen P, Madigan B, et al. (1999) Spatial variability of soil properties in the shortgrass steppe: The relative importance of topography, grazing, microsite, and plant species in controlling spatial patterns. Ecosystems 2: 422–438.
23. Liu M, Liu GH, Wu X, Wang H, Chen L (2014). Vegetation traits and soil properties in respones to utilization patterns of grassland in Hulun Buir City, Inner Mongolia, China. Chinese Geographical Science, 24(4):471–478.
24. Xie JB, Tang LS, Wang ZY, Xu GQ, Li Y (2012) Distinguishing the Biomass Allocation Variance Resulting from Ontogenetic Drift or Acclimation to Soil Texture. Plos One 7.
25. Yang YH, Luo YQ (2011) Isometric biomass partitioning pattern in forest ecosystems: evidence from temporal observations during stand development. Journal of Ecology 99: 431–437.
26. Baruch Z (1994) Responses to Drought and Flooding in Tropical Forage Grasses.1. Biomass Allocation, Leaf Growth and Mineral Nutrients. Plant and Soil 164: 87–96.
27. Sanchez-Vilas J, Bermudez R, Retuerto R (2012) Soil water content and patterns of allocation to below- and above-ground biomass in the sexes of the subdioecious plant Honckenya peploides. Annals of Botany 110: 839–848.
28. Enquist BJ, Niklas KJ (2002) Global allocation rules for patterns of biomass partitioning in seed plants. Science 295: 1517–1520.
29. Niklas KJ (2005) Morphogenesis and biomechanics: The roles of mechanical perturbation and other environmental stresses in plant development and evolution. Evolving Form and Function: Fossils and Development: 25–41.
30. Cheng DL, Niklas KJ (2007) Above- and below-ground biomass relationships across 1534 forested communities. Annals of Botany 99: 95–102.
31. McConnaughay KDM, Coleman JS (1999) Biomass allocation in plants: Ontogeny or optimality? A test along three resource gradients. Ecology 80: 2581–2593.
32. Shipley B, Meziane D (2002) The balanced-growth hypothesis and the allometry of leaf and root biomass allocation. Functional Ecology 16: 326–331.
33. Reich PB, Oleksyn J, Wright IJ, Niklas KJ, Hedin L, et al. (2010) Evidence of a general 2/3-power law of scaling leaf nitrogen to phosphorus among major plant groups and biomes. Proceedings of the Royal Society B-Biological Sciences 277: 877–883.
34. McCarthy MC, Enquist BJ (2007) Consistency between an allometric approach and optimal partitioning theory in global patterns of plant biomass allocation. Functional Ecology 21: 713–720.
35. Enquist BJ (2002) Universal scaling in tree and vascular plant allometry: Toward a general quantitative theory linking plant form and function from cells to ecosystems. Tree Physiology 22: 1045–1064.
36. Tolvanen A, Schroderus J, Henry GHR (2001) Age- and stage-based bud demography of Salix arctica under contrasting muskox grazing pressure in the High Arctic. Evolutionary Ecology 15: 443–462.

37. Jonsdottir IS (1991) Effects of Grazing on Tiller Size and Population-Dynamics in a Clonal Sedge (Carex-Bigelowii). Oikos 62: 177–188.
38. Cheplick GP (1997) Responses to severe competitive stress in a clonal plant: Differences between genotypes. Oikos 79: 581–591.
39. Lu XY, Fan JH, Yan Y, Wang XD (2011) Soil water soluble organic carbon under three alpine grassland types in Northern Tibet, China. African Journal of Agricultural Research 6: 2066–2071.
40. Wang XD, Li MH, Liu SZ, Liu GC (2006) Fractal characteristics of soils under different land-use patterns in the and and semiarid regions of the Tibetan Plateau, China. Geoderma 134: 56–61.
41. Xie YZ, Wittig R (2004) The impact of grazing intensity on soil characteristics of Stipa grandis and Stipa bungeana steppe in northern China (autonomous region of Ningxia). Acta Oecologica-International Journal of Ecology 25: 197–204.
42. Wang LA, Niu KC, Yang YH, Zhou P (2010) Patterns of above- and belowground biomass allocation in China's grasslands: Evidence from individual-level observations. Science China-Life Sciences 53: 851–857.
43. Bao SD (2000) Soil chemical anlysis of agriculture. Beijing: Chinese Agriculture Press.
44. Hu JY (2007) Morphological and structural traits and their ecological adaptation of two dominant species in alpine grassland in Tibetan Plateau. Master's degree thesis. Institute of botany, Chinese academy of sciences.
45. Yang YH, Fang JY, Ji CJ, Han WX (2009) Above- and belowground biomass allocation in Tibetan grasslands. Journal of Vegetation Science 20: 177–184.
46. Li HT, Han XG, Wu JG (2006) Variant scaling relationship for mass-density across tree-dominated communities. Journal of Integrative Plant Biology 48: 268–277.
47. Warton DI, Wright IJ, Falster DS, Westoby M (2006) Bivariate line-fitting methods for allometry. Biological Reviews 81: 259–291.
48. Hermans C, Hammond JP, White PJ, Verbruggen N (2006) How do plants respond to nutrient shortage by biomass allocation? Trends in Plant Science 11: 610–617.
49. Patty L, Halloy SRP, Hiltbrunner E, Korner C (2010) Biomass allocation in herbaceous plants under grazing impact in the high semi-arid Andes. Flora 205: 695–703.
50. Medina-Roldan E, Paz-Ferreiro J, Bardgett RD (2012) Grazing exclusion affects soil and plant communities, but has no impact on soil carbon storage in an upland grassland. Agriculture Ecosystems & Environment 149: 118–123.
51. Cheng J, Wu GL, Zhao LP, Li Y, Li W, et al. (2011) Cumulative effects of 20-year exclusion of livestock grazing on above- and belowground biomass of typical steppe communities in arid areas of the Loess Plateau, China. Plant Soil and Environment 57: 40–44.
52. Gao YZ, Giese M, Lin S, Sattelmacher B, Zhao Y, et al. (2008) Belowground net primary productivity and biomass allocation of a grassland in Inner Mongolia is affected by grazing intensity. Plant and Soil 307: 41–50.
53. Rodriguez MA, Brown VK, Gomezsal A (1995) The Vertical-Distribution of Belowground Biomass in Grassland Communities in Relation to Grazing Regime and Habitat Characteristics. Journal of Vegetation Science 6: 63–72.
54. Steffens M, Kolbl A, Totsche KU, Kogel-Knabner I (2008) Grazing effects on soil chemical and physical properties in a semiarid steppe of Inner Mongolia (PR China). Geoderma 143: 63–72.
55. Gill RA, Jackson RB (2000) Global patterns of root turnover for terrestrial ecosystems. New Phytologist 147: 13–31.
56. Smit C, Bakker ES, Apol MEF, Olff H (2010) Effects of cattle and rabbit grazing on clonal expansion of spiny shrubs in wood-pastures. Basic and Applied Ecology 11: 685–692.
57. Yu FH, Dong M, Krusi B (2004) Clonal integration helps Psammochloa villosa survive sand burial in an inland dune. New Phytologist 162: 697–704.
58. Gao Y, Xing F, Jin YJ, Nie DD, Wang Y (2012) Foraging responses of clonal plants to multi-patch environmental heterogeneity: spatial preference and temporal reversibility. Plant and Soil 359: 137–147.
59. Sosnova M, van Diggelen R, Klimesova J (2010) Distribution of clonal growth forms in wetlands. Aquatic Botany 92: 33–39.
60. Gao YH, Schumann M, Chen H, Wu N, Luo P (2009) Impacts of grazing intensity on soil carbon and nitrogen in an alpine meadow on the eastern Tibetan Plateau. Journal of Food Agriculture & Environment 7: 749–754.
61. Evans CRW, Krzic M, Broersma K, Thompson DJ (2012) Long-term grazing effects on grassland soil properties in southern British Columbia. Canadian Journal of Soil Science 92: 685–693.
62. Zhou ZC, Gan ZT, Shangguan ZP, Dong ZB (2010) Effects of grazing on soil physical properties and soil erodibility in semiarid grassland of the Northern Loess Plateau (China). Catena 82: 87–91.
63. Yates CJ, Norton DA, Hobbs RJ (2000) Grazing effects on plant cover, soil and microclimate in fragmented woodlands in south-western Australia: implications for restoration. Austral Ecology 25: 36–47.
64. Arevalo JR, Chinea E, Barquin E (2007) Pasture management under goat grazing on Canary Islands. Agriculture Ecosystems & Environment 118: 291–296.

16

A Comparative Metagenome Survey of the Fecal Microbiota of a Breast- and a Plant-Fed Asian Elephant Reveals an Unexpectedly High Diversity of Glycoside Hydrolase Family Enzymes

Nele Ilmberger[1], Simon Güllert[1], Joana Dannenberg[1], Ulrich Rabausch[1], Jeremy Torres[1], Bernd Wemheuer[2], Malik Alawi[3], Anja Poehlein[2], Jennifer Chow[1], Dimitrij Turaev[4], Thomas Rattei[4], Christel Schmeisser[1], Jesper Salomon[5], Peter B. Olsen[5], Rolf Daniel[2], Adam Grundhoff[6], Martin S. Borchert[5], Wolfgang R. Streit[1]*

1 Universität Hamburg, Biozentrum Klein Flottbek, Abteilung für Mikrobiologie & Biotechnologie, Hamburg, Germany, **2** Georg-August-University Göttingen, Institute of Microbiology and Genetics, Göttingen, Germany, **3** University Medical Center Hamburg-Eppendorf, Bioinformatics Service Facility, Hamburg, Germany, **4** University of Vienna, CUBE - Division for Computational Systems Biology, Department of Microbiology and Ecosystem Science, Vienna, Austria, **5** Novozymes A/S, Microbial Discovery, Bagsværd, Denmark, **6** Heinrich Pette Institute, Leibniz Institute for Experimental Virology, Hamburg, Germany

Abstract

A phylogenetic and metagenomic study of elephant feces samples (derived from a three-weeks-old and a six-years-old Asian elephant) was conducted in order to describe the microbiota inhabiting this large land-living animal. The microbial diversity was examined via 16S rRNA gene analysis. We generated more than 44,000 GS-FLX+454 reads for each animal. For the baby elephant, 380 operational taxonomic units (OTUs) were identified at 97% sequence identity level; in the six-years-old animal, close to 3,000 OTUs were identified, suggesting high microbial diversity in the older animal. In both animals most OTUs belonged to Bacteroidetes and Firmicutes. Additionally, for the baby elephant a high number of Proteobacteria was detected. A metagenomic sequencing approach using Illumina technology resulted in the generation of 1.1 Gbp assembled DNA in contigs with a maximum size of 0.6 Mbp. A KEGG pathway analysis suggested high metabolic diversity regarding the use of polymers and aromatic and non-aromatic compounds. In line with the high phylogenetic diversity, a surprising and not previously described biodiversity of glycoside hydrolase (GH) genes was found. Enzymes of 84 GH families were detected. Polysaccharide utilization loci (PULs), which are found in Bacteroidetes, were highly abundant in the dataset; some of these comprised cellulase genes. Furthermore the highest coverage for GH5 and GH9 family enzymes was detected for Bacteroidetes, suggesting that bacteria of this phylum are mainly responsible for the degradation of cellulose in the Asian elephant. Altogether, this study delivers insight into the biomass conversion by one of the largest plant-fed and land-living animals.

Editor: Stefan Bereswill, Charité-University Medicine Berlin, Germany

Funding: This work was in part funded by the ministry of education and research "Bundesministerium für Bildung und Forschung" (BMBF; http://www.bmbf.de/) within the network program ExpresSys and the grant FKZ 0315586F to WRS. JT was funded by the German Academic Exchange program (DAAD; https://www.daad.de/de/index.html). The funders had no role in study design, data collection and analysis, decision to publish, or preparation of the manuscript.

Competing Interests: Some of the authors are affiliated with Novozymes A/S.

* Email: wolfgang.streit@uni-hamburg.de

Introduction

The microbiota of mammalian intestines is the main driver of plant cell wall degradation as mammalian genomes do not encode for a significant number of genes linked to structural polysaccharide, i.e. cellulose, degradation [1]. Thereby herbivores can gain 70% of their energy from microbial polysaccharide breakdown [1–3]. Within the herbivores two groups are distinguished corresponding to the location of the main fermentation: the foregut fermenters (mainly ruminants) and the hindgut fermenters. In foregut fermenters the fermentation takes place in pregastric chambers, the rumen. In hindgut fermenters the main fermentation chamber is the colon or the caecum (e.g. elephants, their caecum is up to 1.5 m long) and therein the dry matter content is significantly higher than in the rumen [4]. In addition, ruminants grind the lignocellulosic material mechanically by chewing regularly during the fermentation process. Ruminants therefore have a very effective way of digesting lignocellulose. Hindgut fermenters in contrast are able to digest faster; this is an advantage at high body size as more food can be ingested [5]. Recently, the intestinal or fecal microbiota of different animals has been investigated using next generation sequencing (NGS) technology. Among them were cow, reindeer, wallaby, yak, giant panda, buffalo, swine, Iberian lynx and termite [6–16]. These studies delivered a significant amount of novel sequence data giving insight into the phylogeny, the metabolism and the genetic potential of the intestinal microbiota. Furthermore, these studies have led to a better understanding of fecal and gut microbial communities and they have underlined the importance of these communities for host survival, fitness, physiology and nutrient utilization [17–19]. In addition these studies have suggested a core and a variable gene set, influenced by host traits, environment, type of diet and other not yet identified factors [20–22]. In general and as expected, the herbivorous microbiomes encoded for high numbers of carbohydrate active enzymes (CAZymes). For many of the studied systems Clostridia were most abundant and were identified as main cellulose degraders [11,23]. Interestingly, in termites Spirochaetes and Fibrobacteres are supposed to be the main contributors to cellulose digestion [16]. For these microbes the absence of cellulosomes and the release of at least some of the cellulases in the ruminal fluid were proposed [16]. Furthermore, some Bacteroidetes have been described as being highly abundant in different fecal/intestinal samples [6,8–10]. Recent metagenomic and genomic studies have highlighted the presence of polysaccharide utilization loci (PULs) from Bacteroidetes that include cellulase genes [8,10]. PULs were primarily described as starch degradation operon composed of genes encoding proteins designated SusA to SusG. These proteins coordinately cleave carbohydrates and transport sugar into the cell [1]. PULs have been identified in different gut samples and genomes of *Bacteroides* sp. [3,8,24–26].

Adult Asian elephants can reach a weight of 5,000 kg and consume about 150 kg plant materials per day. The nourishment is composed of highly fibrous plant material, mainly grass, fruits, leaves, twigs, roots and bark. As our understanding of the microbiota inhabiting elephants is poor, we were interested in characterizing the microbial population in the feces of this large herbivorous animal. Within the current paper we deliver evidence that the microbiome of Asian elephants is highly diverse and that microbes of the Bacteroidetes phylum are presumably the main cellulose degraders. Thereby we show that elephants are generalists rather than specialists with respect to the degradation of plant biomass.

Materials and Methods

DNA isolation

No specific permissions were required for these activities. The study did not involve endangered or protected species. Coordinates are: Longitude/Latitude 9.941572/9.941572. Fresh feces samples derived from a six-years-old female Asian elephant ("Kandy", *2003 in Hamburg) and from a three-weeks-old male elephant ("Assam", *2012 in Hamburg), both living in the zoo 'Hagenbecks Tierpark' in Hamburg (Germany), were collected from the zoo staff directly after defecation. The samples were transported, directly and on ice, to the laboratory in Hamburg Klein Flottbek for further analysis. The elephants were not treated with antibiotics. The older Asian elephant was mainly fed with grass, hay, leaves and twigs, with additional fruits and vegetables. In contrast, the three-weeks-old male elephant was breast-fed. DNA isolation was performed with the QIAamp DNA Stool kit from Qiagen (Hilden, Germany) as described previously [27].

Metagenome sequencing

Libraries were prepared with the NEB DNA Ultra Kit following the manufacturer's protocol. Illumina sequencing for the six-years-old elephant was performed using a HiSeq 2000 instrument (1.5 lanes, paired-end run (2×100 bases)). For the three-weeks-old animal a HiSeq 2500 instrument (one lane, paired-end run (2×100 bases)) was used for sequencing at the HPI in Hamburg. *De novo* assembly was performed with the Velvet assembly program version 1.2.08 [28]. For the investigation of the sequences the IMG server (https://img.jgi.doe.gov/cgi-bin/mer/main.cgi) was used. To further analyze the possible biological processes linked to the individual genes and ORFs mainly the KEGG [29], the COG [30] and the Pfam [31] databases were employed using a cut off of 10^{-5}.

Amplification and sequencing of 16S rRNA genes

To assess the microbial diversity, variable regions of the 16S rRNA genes were amplified as previously published [32] but with minor modifications.

The V3–V5 region was amplified using the following primer set: V3for 5′-TCTC ATCCCTGCGTGTCTCCGACTCAGACGCTCGACACCTACGGGNGGCWGCAG-3′ and V5rev 5′-CCTATCCCCTGTGTGCCTTGGCAGTCTCAGCCGTCAATTCMTTTRAGTTT-3′. The primers contained Roche 454 pyrosequencing adaptors, keys and one unique MID per sample (underlined). To assess the archaeal diversity, the V4–V6 region was amplified using the primer set: A519F 5′-CCATCTCATCCCTGCGTGTCTCCGACT CAGATATCGCGAGCAGCMGCCGCGGAA- 3′ and A1041R 5′-CCTATCCCCTGTGT GCCTTGGCAGTCTCAGGGCCATGCACCWCCTCTC-3′. The PCR reaction (50 µl) contained 0.5 U of Phusion High-Fidelity DNA Polymerase (Thermo Scientific, Germany), 10 µl 5x Phusion GC Buffer, 200 µM of each *d*NTP, 2.5% DMSO, 1.5 mM $MgCl_2$, 4 µM of each primer, and 20 ng isolated DNA. PCR cycling conditions were: initial denaturation at 98°C for 3 min, followed by 28 cycles of denaturation at 98°C for 30 s, annealing at 61°C for 30 s (archaeal primer set: 66°C), and extension at 72°C for 25 s. The final extension was conducted at 72°C for 5 min. Negative controls were performed with H_2O instead of template DNA. The obtained PCR products were purified via Gel/PCR DNA Fragments Extraction Kit (Geneaid Biotech, Taiwan) as recommended by the manufacturer. Three separate PCR reactions were conducted for each sample. After gel extraction, the reaction products were pooled in equal amounts. The 16S rRNA gene sequencing was performed at the Göttingen Genomics Laboratory using a Roche GS-FLX+454 pyrosequencer and titanium chemistry (Roche, Branford, USA).

Processing and analysis of 16S rRNA gene data sets

Via pyrosequencing generated raw sequences were processed according to [33], with the following modifications: After raw data extraction, reads shorter than 300 bp and those possessing long homopolymer stretches (>8 bp) or primer mismatches (>5 bp) were removed. The sequences were denoised employing Acacia [34]. Chimeric sequences were removed using UCHIME in reference mode with the most recent SILVA SSU database as reference dataset (SSURef 115 NR) [35]. The processed 16S rRNA gene sequences were uploaded to the SILVA NGS (SILVA next-generation sequencing) server for taxonomic classification [36]. Microbial taxonomy was determined using default settings with two adjustments: The cluster sequence identity threshold was decreased to 0.97 and the maximal taxonomic depth was increased to 30. Rarefaction curves were calculated employing the QIIME 1.8 software package [37].

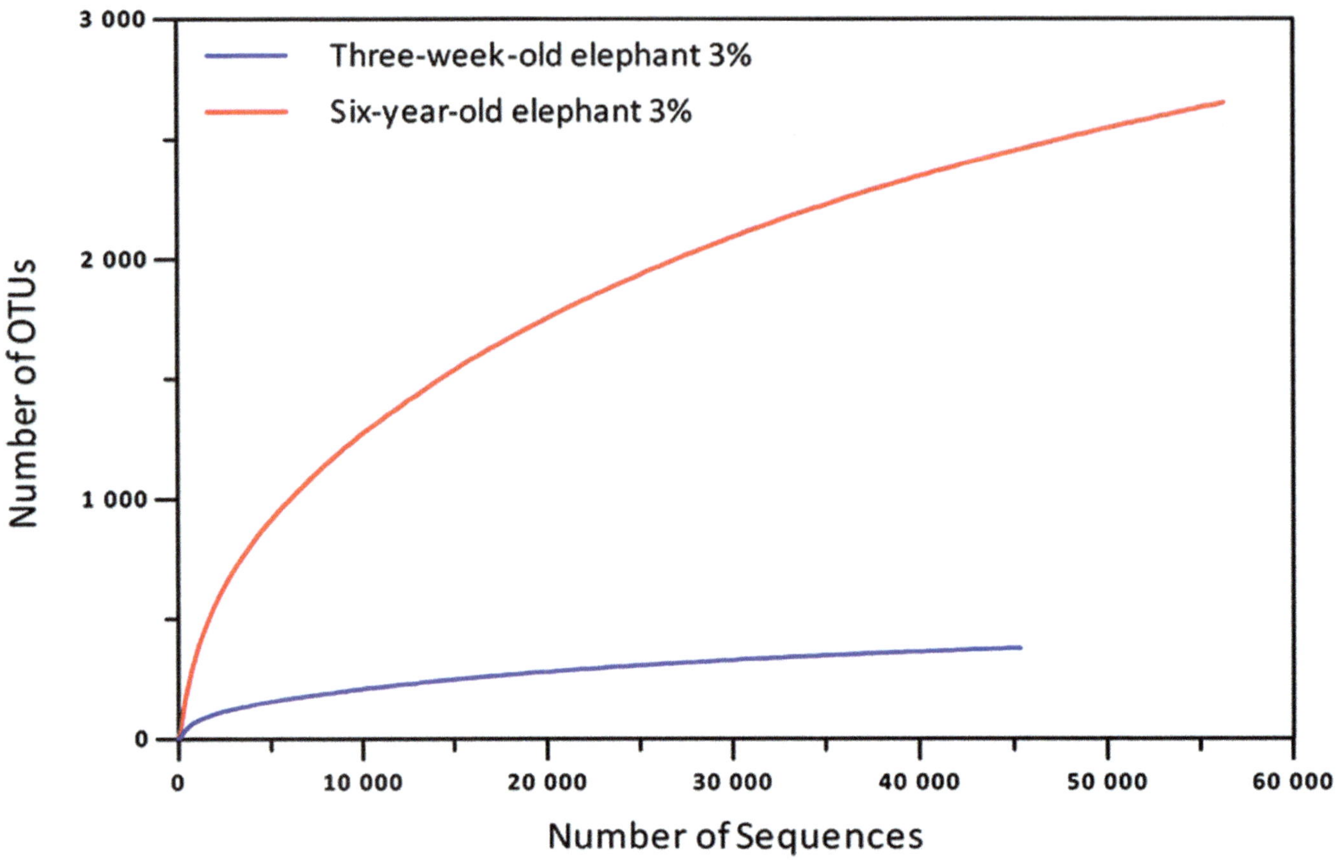

Figure 1. Rarefaction curves calculated for the feces sample of the three-weeks-old and the six-years-old Asian elephant at 3% genetic distance of 16S rRNA genes. The curve for the six-years-old elephant comprises 8,014 archaeal sequences which were clustered to 54 OTUs. The sequences were denoised employing Acacia. Chimeric sequences were removed using UCHIME in reference mode with the most recent SILVA SSU database as reference dataset (SSURef 115 NR).

Table 1. Overall numbers of sequences and contigs generated for the two elephant feces samples.

Parameter	Six-years-old elephant	Three-weeks-old elephant
16S rRNA gene sequences		
No. of GS-FLX+454 sequences	56,124	44,508
Feces Microbiome DNA		
No. of Illumina reads	415,784,264	200,946,703
No. of bases assembled	929,519,943	138,494,152
No. of assembled sequences	260,535	50,765
Coverage	42	73
Mean contig size (bp)	3,566	5,801
N50 (bp)	5,751	22,052
Largest contig (bp)	344,979	597,113
Gene count	1,068,385	173,134
GC mean %	43,96	47,71
Protein coding genes	1,052,245	171,134
with COG	629,485	111,920
with Pfam	818,373	136,417
with KEGG	210,308	42,640

A)

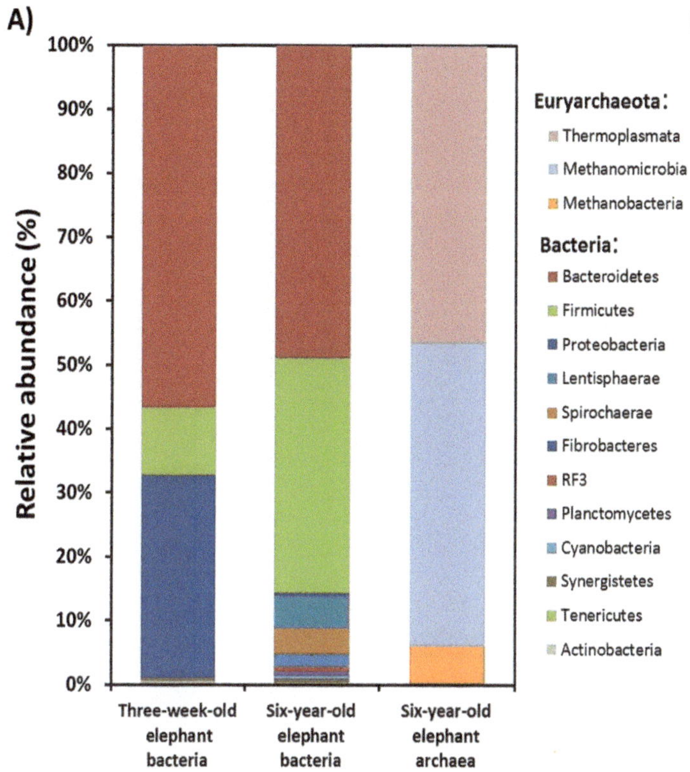

B)

Archaeal / Bacterial Class	Relative abundance (%) Three-week-old elephant	Six-year-old elephant
Methanobacteria	0,0	6,1
Methanomicrobia	0,0	47,4
Thermoplasmata	0,0	46,5
Actinobacteria	0,1	0,0
Coriobacteriia	0,6	0,1
Bacteroidia	34,4	48,8
Flavobacteria	21,9	0,0
Sphingobacteriia	0,3	0,0
Fibrobacteria	0,0	2,2
Bacilli	0,1	0,1
Clostridia	9,2	31,6
Erysipelotrichia	0,1	2,6
Negativicutes	1,4	2,1
Lentisphaeria	0,0	4,9
Planctomycetacia	0,0	0,7
Alphaproteobacteria	0,0	0,1
Betaproteobacteria	0,1	0,1
Epsilonproteobacteria	0,0	0,0
Gammaproteobacteria	31,6	0,3
Spirochaetes	0,0	4,4
Synergistia	0,4	0,5
Mollicutes	0,0	0,3
RF3	0,0	0,7
Other	0,0	0,5
No relative	0,0	0,2

Figure 2. Relative abundances of different phyla and classes in the two elephant feces samples. A: Relative abundance of phyla in the feces of the three-weeks-old and the six-years-old Asian elephant based on 16S rRNA gene sequences. For the three-weeks-old elephant no Archaea were observed. **B:** Phylogenetic comparison on class level between both elephants. Heat map colors indicate the abundances of the respective 16S rRNA genes.

Comparison with other fecal metagenomes and statistical analysis

An additional set of metagenomes of animals' microbiotas was used for comparative analysis by the IMG/M ER webpage of the US Department of Energy Joint Genome Institute (https://img.jgi.doe.gov/cgi-bin/mer/main.cgi) [38]. The phylogenetic distribution of the metagenomes was analyzed. A phylogenetic tree was automatically created with the public Metagenomics RAST server [39].

Binning

The assembly of both metagenomic samples was performed with the Ray Meta assembler [40], version 2.3.1, using a k-mer length of 31 and default parameters. Scaffolds of $\geq$2 kb length and a mode k-mer coverage depth of $>$5 were binned based on unsupervised tetranucleotide frequencies using MetaWatt2.0 [41]. To assess bin taxonomy, homogeneity and completeness we identified bacterial and archaeal phylogenetic marker genes using AMPHORA2 [42]. Using project-specific scripts we estimated the most recent taxonomic level supported by most high-confidence AMPHORA2 markers (confidence score $\geq$0.9) for each bin. At these levels the phylogenetic markers determined the taxonomic affiliation and completeness of each bin. The tRNA genes, rRNA genes and protein-coding genes were predicted in each bin using RNAmmer [43], tRNAscan-SE [44] and GeneMarkS [45], respectively.

Transmission electron microscopy (TEM)

Slices were prepared with the microtome Reichert-Jung Ultracut E. Fixation was performed in 2% glutaraldehyde in 75 mM cacodylate buffer (pH 7.0). After washing it was supplied with 2% agar in 75 mM cacodylate buffer (pH 7.0) and further fixed with 1% OsO_4 in 50 mM cacodylate buffer (pH 7.0). After washing with 75 mM cacodylate buffer (pH 7.0) water was removed with acetone and the sample was infiltrated with Spurr resin (Polysciences, Warrington, PA, USA). TEM pictures were observed with the LEO 906 E, the camera Gatan 794 and the software Digital micrograph.

DNA sequences obtained and GenBank submissions

This project has been deposited at GenBank using the BioProject number PRJNA240141. The sequences derived from Illumina and 454 sequencing were deposited in the NCBI Short Read Archive, the study accession number is SRP040073. Assembled sequence data with predicted gene models and annotation is available from www.jgi.doe.gov. (DOE Joint Genome Institute), the IMG Project Id is 50566.

Results and Discussion

Population structure of the elephant feces samples

Recent metagenomic research has demonstrated a close correlation between host diet and intestinal microbiome and it has highlighted the metabolic diversity within intestinal and fecal samples of mammals and insects [6,8,9,11,15,16,46]. These intriguing findings inspired us to analyze the fecal microbiota of

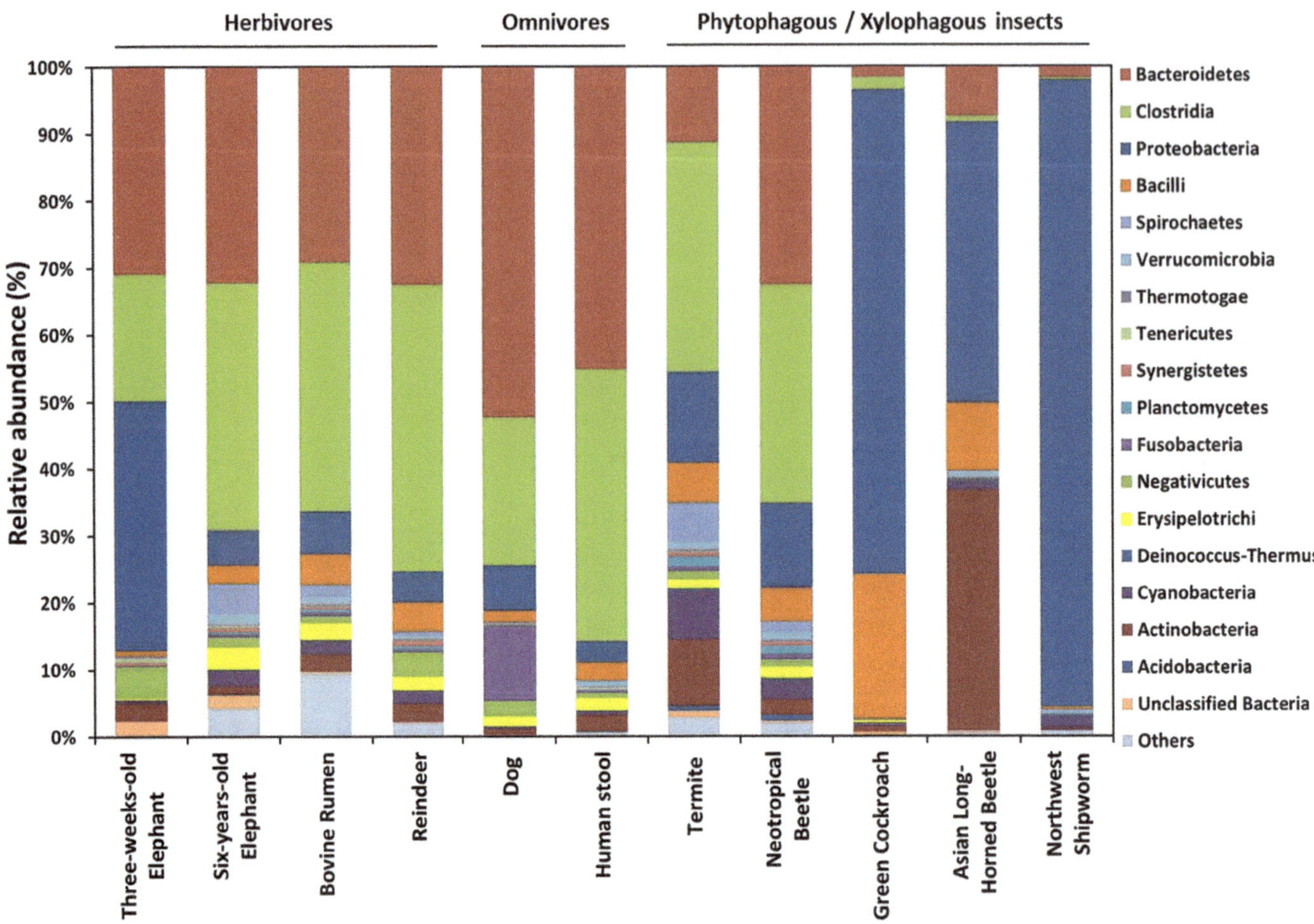

Figure 3. Phylogenetic analysis of the elephant feces in comparison with other fecal and intestinal metagenome data sets. Data indicate the phylogenetic relation based on gene similarities in the metagenome sequences. The percent of sequences assigned to each phylum according to IMG/M ER is shown based on the total number of obtained sequences of each data set. Sequence data for the metagenomes were extracted from the IMG/M ER web page of the US Department of Energy Joint Genome Institute and the respective bioprojects (IMG Genome IDs: Six-years-old Elephant (this study): 3300001598; three-weeks-old Elephant (this study): 3300001919; Green Cockroach: 2228664000; Termite: 3300001544; Dog: 2019105001; Reindeer: 2088090000; Neotropical Beetle: 3300000114; Asian Long-Horned Beetle: 2084038013; Bovine Rumen: 2061766007; Northwest Shipworm: 2189573029; Human stool: 7000000038).

one of the largest land-living herbivores (only the African elephant is larger than the Asian elephant), which is known to be an exception in the world of large foregut fermenters as digesting faster than the high body size would allow [5]. To estimate the diversity of the microbes and their individual tasks that meet the energy need of elephants, a detailed phylogenetic analysis using NGS technologies was performed. Thereby we analyzed the community of a plant-fed female specimen (six-years-old) and of a three-weeks-old breast-fed male animal. Both animals were living in a local zoo and they were not treated with antibiotics. For the baby elephant 44,508 partial 16S rRNA sequences with an average length of 536 bp were generated and 56,124 sequences with an average length of 523 bp for the six-years-old elephant (TABLE 1). Rarefaction curves suggested that both datasets had the required amount of sequence data to assess operational taxonomic unit (OTU) richness and that both datasets had a similar coverage of biodiversity (FIGURE 1).

The feces sample of the six-years-old elephant revealed an almost 10-fold higher diversity than that of the baby animal (FIGURE 1, 2). For the three-weeks-old animal the phylogenetic analysis suggested the presence of approximately 380 OTUs based on a 97% sequence identity cut off for bacterial 16S rRNA genes. However, in the feces of the older animal 2,656 OTUs (calculated max. 3,487 OTUs) were identified (FIGURE 1, TABLE S1). The Shannon index for the baby elephant was 4.3 and for the six-years-old elephant it was 8.4. Furthermore the Chao1 index for species richness was 449 for the baby and 3,281 for the six-years-old elephant (TABLE S1), leading to the overall conclusion that the microbiome of the older elephant was much more diverse than that of the baby.

The baby elephant's feces sample was dominated by bacteria belonging to the Bacteroidetes (FIGURE 2). More than 50% of the microbes were Flavobacteriales (22%) or Bacteroidales (34%). The sample of the baby elephant furthermore comprised a high number of Proteobacteria (32%) and Firmicutes (11%) (FIGURE 2). On genus level most abundant bacteria were *Myroides* (22%), the S24-7 group of Bacteroidetes (14%) and *Pseudomomas* and *Psychrobacter* species (12% each) (supporting information S1). With the archaeal primer set no PCR product was observed.

The bacterial population of the six-years-old animal was dominated by Firmicutes (36%) and Bacteroidetes (47%). Furthermore Spirochaetes (4%), Fibrobacteres (2%) and Lentisphaeria (5%) were detected. Interestingly no bacteria belonging to these phyla were detected in the three-weeks-old elephant's

Table 2. Selected and recent metagenome studies published from insect or mammalian fecal and gut samples.

Microbiome	OTUs detected	Assemble DNA (Mbp)	Sequencing strategy	No. of GH families identified	Total no. of CAZymes	No. of GH per Mbp assembled DNA	f)Total no. of cellulolytic enzymes	Reference
Iberian lynx	n.d.	18.3	454 GS FLX	42	372	20.3	39	[15]
Tammar wallaby	236	85.2	454 GS FLX	53	557	6.5	175	[9]
Reindeer[a]	1,182	26	454 GS FLX	30	d)5,000	e)9.9	d)400	[8]
Giant panda	85	37	Illumina	44	440	11.9	124	[11]
Bovine	1,000	b)1,930	Illumina	60	27,755	14.4	5,670	[36]
Bovine	161–259	c)104	454 GS20	35	3,828	c)36.8	1,017	[7]
Termite	216	71	454 GS20+ Sanger	45	703	9.9	259	[16]
Elephant gut (six-years-old)	2,656	929.5	Illumina	82	11,038	11.9	2,074	This study
Elephant gut (three-weeks-old)	380	138.5	Illumina	70	1,873	13.5	242	This study

n.d., not determined; a), in 2 samples; b), 286 Gb resulted in approx. 2 Gb of assembled DNAs >1 kb scaffolds; c),unassembled; d), in unassembled reads; e), per 503 Mbp unassembled DNA; f), total gene count with respect to the GH1, GH3, GH 5, GH6, GH8, GH9, GH44, GH45, GH48, GH51, GH74, GH94 family enzymes. Total gene counts, size of assembled DNAs, OTUs and numbers of carbohydrate active enzymes were extracted from the indicated references.

sample. On genus level uncultured *Lachnospiraceae* (10%), *Ruminococcaceae* (10%), *Prevotellaceae* (11%) and the RC9 gut group of *Rickenellaceae* (13%) dominated (supporting information S2). Archaea were found in this sample. About 50% of these belonged to the Methanomicrobiales and the remainder belonged mostly to the Thermoplasmatales, i.e. Candidatus *Methanomethylophilus* (FIGURE 2). The respective organisms most likely contribute to elephant's methane production during feed fermentation [47].

To further verify these data, we analyzed the sequences obtained by Illumina sequencing for the presence of rRNA gene fragments or complete genes. 102 16S rRNA gene fragments were found for the baby elephant and 338 16S rRNA gene fragments for the six-years-old elephant. The results from this analysis largely confirmed the results of the 16S rRNA amplicon sequencing. Additionally, binning was performed with the metagenomic sequence data. For the baby elephant 234 bins were observed, for the six-years-old elephant 1,401. Thereof 131, or 935, respectively, were unassigned (TABLE S2, S3). The results largely confirm the 16S rRNA analysis, the higher microbial richness of the six-years-old elephant and the phylogenetic affiliations. For the older elephant Firmicutes and Bacteroidetes were dominant with 39 and 41% of the bins, respectively. Spirochaetes and Fibrobacteres were present at lower levels and 6% of the bins were assigned to Archaea (TABLE S3). For the baby elephant Firmicutes (35%), Bacteroidetes (32%) and Proteobacteria (28%) were dominant (TABLE S2). This is, in comparison with the 16S rRNA analysis, an overrepresentation of Firmicutes and an underrepresentation of Bacteroidetes. This difference nevertheless can easily be explained by the different methods.

Overall and as expected the microbiota of the six-years-old elephant resembled those of most herbivorous mammals like reindeer, wallaby and cow [6,8,9] with Firmicutes, Bacteroidetes and Proteobacteria being dominant (FIGURE 3). Surprisingly, the number of OTUs was at least two-fold higher than those of other herbivores (FIGURE 3; TABLE 2 and references given herein). Since Asian elephants have a rather diverse diet in the zoo compared to wild living animals this relatively high bacterial diversity may reflect an adaptation to this feeding manner. Furthermore the nourishment is composed of different plants and the microbiome is limited to the small zoo population. Thus, the results altogether do not represent the microbiome of wild living animals. It has been shown e.g. for pandas that the communities of wild and captive animals differ greatly [11].

As the baby elephant was breast-fed, the microbes inhabiting the intestine of this animal were in part determined by the mother's milk and of those bacteria attached to the mother transmitted by direct contact [48]. We speculate that especially the relatively high level of Proteobacteria was a result of the breast-feeding (FIGURE 3). This observation fitted well with the fact that in human breast milk more than 60% of the bacteria were Proteobacteria [49]. While the observed differences in OTU richness of the microbial communities of both elephants can most likely be explained by the differing nourishment, it should be noted that baby elephants also eat feces from older animals. Because of this it is likely that a significant fraction of the microbes in its feces originated from other animals from the herd.

Nevertheless, only 13 OTUs were present in both elephants' microbiomes. These were seven Firmicutes, three Bacteroidetes and three Proteobacteria (TABLE S4). Interestingly, most of these OTUs were considerably higher represented in one of the samples (TABLE S4). Altogether the two communities are distinct with only few overlaps. Though of course analyzing only two

Table 3. Gene count, relative gene count and relative coverage of genes for families of carbohydrate-active enzymes discovered in the sequences of the elephant feces samples according to CAZy.

	Baby elephant			Six-years-old elephant		
[1]GH family	Gene count	Relative count	Relative coverage	Gene count	Relative count	Relative coverage
1	33	0.01761	0.00979	103	0.00933	0.00667
2	220	0.11745	0.09455	917	0.08311	0.082006
3	148	0.07901	0.08355	804	0.07287	0.07784
4	16	0.00854	0.00430	57	0.00517	0.00405
5	19	0.0101	0.01064	517	0.04686	0.04808
8	4	0.00213	0.00134	85	0.00770	0.01198
9	9	0.00481	0.00406	119	0.01078	0.01199
10	6	0.00320	0.00836	258	0.02338	0.02915
11	–	–	–	20	0.00181	0.00142
13	124	0.06620	0.07574	845	0.07658	0.07423
15	1	0.00053	0.00019	1	0.00009	0.00008
16	22	0.01174	0.01399	198	0.01794	0.01552
17	–	–	–	5	0.00045	0.00022
18	51	0.02723	0.03378	86	0.00779	0.00824
19	1	0.00053	0.00150	3	0.00027	0.00011
20	108	0.05766	0.05088	266	0.02411	0.01881
23	69	0.03684	0.06012	294	0.02664	0.03324
24	12	0.00641	0.00685	26	0.00236	0.00191
25	30	0.01602	0.03030	229	0.02075	0.02255
26	8	0.00427	0.00341	103	0.00933	0.00931
27	12	0.00641	0.00591	158	0.01432	0.01366
28	26	0.01388	0.01502	242	0.02193	0.02112
29	79	0.04218	0.04032	376	0.03408	0.02950
30	17	0.00908	0.00727	95	0.00861	0.00975
31	35	0.01869	0.01000	318	0.02882	0.02962
32	21	0.01121	0.00879	119	0.01078	0.01409
33	31	0.01655	0.01639	149	0.01350	0.01003
35	36	0.01922	0.01745	123	0.01115	0.01155
36	32	0.01708	0.01637	310	0.02809	0.02396
37	3	0.00160	0.00112	17	0.00154	0.00144
38	12	0.00641	0.00653	81	0.00734	0.00645
39	3	0.00160	0.00374	89	0.00807	0.00554
42	9	0.00481	0.00262	37	0.00335	0.00247
43	98	0.05232	0.03978	894	0.08102	0.09836
44	–	–	–	7	0.00063	0.00116
45	–	–	–	7	0.00063	0.00084
46	–	–	–	1	0.00009	0.00004
50	2	0.00107	0.00343	26	0.00236	0.00184
51	24	0.01281	0.00653	239	0.02166	0.02139
53	6	0.00320	0.00632	88	0.00798	0.00976
54	–	–	–	13	0.00118	0.00136
55	2	0.00107	0.00030	8	0.00073	0.00052
57	11	0.00587	0.00893	115	0.01042	0.00941
63	7	0.00374	0.00434	57	0.00517	0.00318
64	–	–	–	2	0.00018	0.00008
65	7	0.00374	0.00574	39	0.00353	0.00285
66	3	0.00160	0.00068	10	0.00091	0.00069
70	–	–	–	1	0.00009	0.00002

Table 3. Cont.

	Baby elephant			Six-years-old elephant		
[1)]GH family	Gene count	Relative count	Relative coverage	Gene count	Relative count	Relative coverage
73	36	0.01922	0.01851	202	0.01831	0.01721
74	–	–	–	11	0.00100	0.00084
76	12	0.00641	0.00510	9	0.00082	0.00081
77	28	0.01495	0.01611	193	0.01749	0.01614
78	41	0.02189	0.03159	413	0.03743	0.02897
79	3	0.00160	0.00264	–	–	–
81	–	–	–	3	0.00027	0.00009
84	9	0.00481	0.00250	19	0.00172	0.00165
88	33	0.01762	0.01190	54	0.00489	0.00342
92	79	0.042178	0.04649	213	0.01930	0.01900
93	–	–	–	2	0.00018	0.00009
94	5	0.00267	0.00140	89	0.00807	0.00906
95	47	0.02509	0.01727	164	0.01486	0.01587
97	44	0.02349	0.02573	197	0.01785	0.02091
98	1	0.00053	0.00021	23	0.00208	0.00330
99	–	–	–	2	0.00018	0.00012
102	3	0.00160	0.00084	4	0.00036	0.00010
103	3	0.00160	0.01026	2	0.00018	0.00005
104	4	0.00214	0.00193	–	–	–
105	25	0.01335	0.00548	126	0.01142	0.01231
106	13	0.00694	0.01614	97	0.00879	0.00832
108	6	0.00320	0.00165	14	0.00127	0.00081
109	22	0.01175	0.01299	125	0.01133	0.00876
110	15	0.00801	0.00812	30	0.00272	0.00237
113	–	–	–	11	0.00100	0.00036
115	7	0.00374	0.00223	74	0.00671	0.00891
116	8	0.00427	0.00147	26	0.00236	0.00139
117	4	0.00214	0.00079	9	0.00082	0.00055
120	6	0.00320	0.00197	45	0.00408	0.00468
123	10	0.00534	0.00288	46	0.00417	0.00276
125	17	0.00908	0.01145	35	0.00317	0.00313
126	–	–	–	2	0.00018	0.00004
127	11	0.00587	0.00802	88	0.00798	0.00639
128	1	0.00053	0.00298	15	0.00136	0.00121
129	1	0.00053	0.00024	7	0.00063	0.00034
130	22	0.01175	0.01020	127	0.01151	0.01197

[1)]GH families according to the CAZy database http://www.cazy.org; Searches for glycoside hydrolases were performed with pfam HMMs, named in accordance with the CAZy nomenclature scheme. - no count observed, GH families not listed were not detected in any of the samples; the relative counts indicate the relative number in comparison to all GHs in the respective sample, and the relative coverage indicates the respective coverage in relation to the coverage of all GHs in the respective sample; GH families with cellulases are in bold.

individuals, these data give hint to the great changes the intestinal microbiota is subjected to during animal's development.

Metagenome survey reveals remarkable metabolic richness

In this study metagenomic sequence data was generated for the intestinal microbiota of a six-years-old Asian elephant and of a three-weeks-old baby elephant. The reads observed from Illumina sequencing were assembled to 929,519,943 bp in 260,535 contigs for the six-years-old animal. The largest contig was 344,979 bp. The average contig length was 3,566 bp (TABLE 1). For the baby elephant 138,494,152 bp were assembled. The average contig length was 2,728 bp. The largest contig was 597,113 bp (TABLE 1).

To broadly characterize the sequence data, general traits were analyzed. Approximately 1,052,245 protein-coding genes could be identified for the-six-years-old elephant and 171,134 for the baby elephant. Of these, a total of 210,308 for the former and 42,640

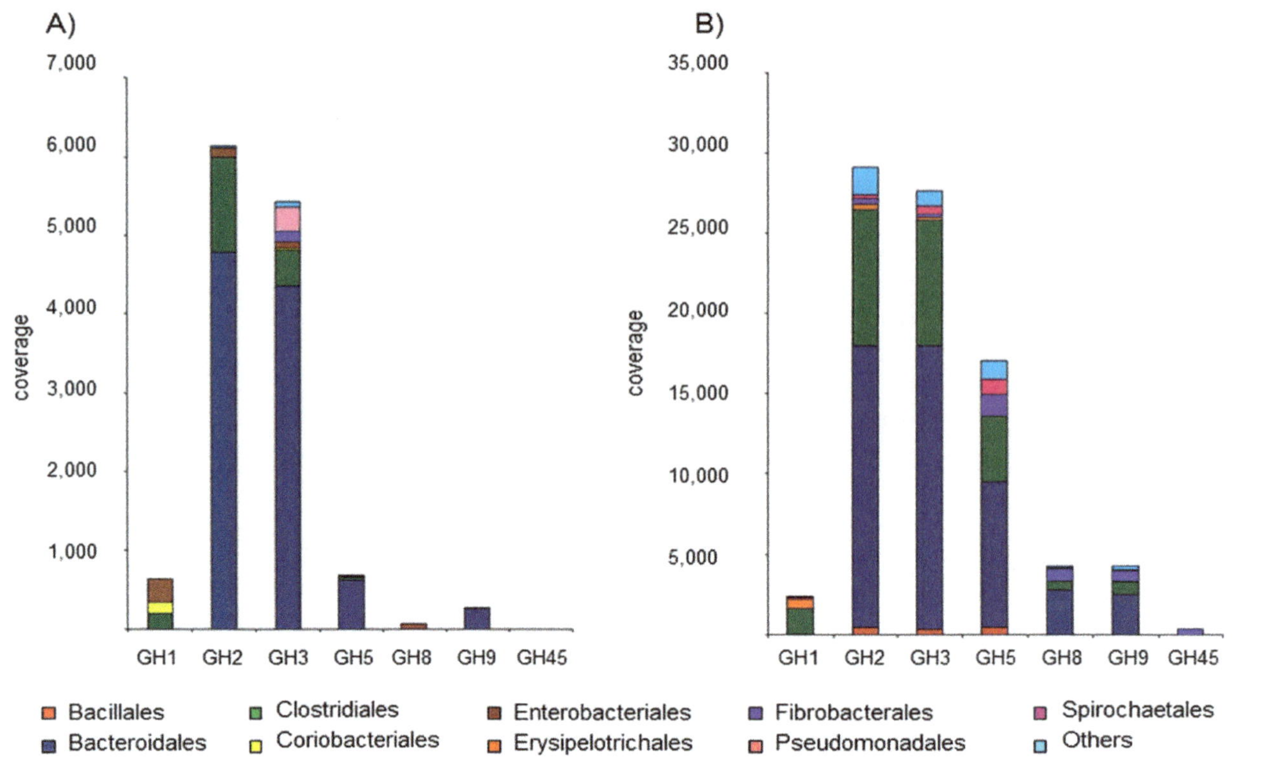

Figure 4. Overall coverage of selected cellulolytic GH family genes in the feces samples in relation to their phylogenetic affiliation. **A**) three-weeks-old elephant, **B**) six-years-old elephant.

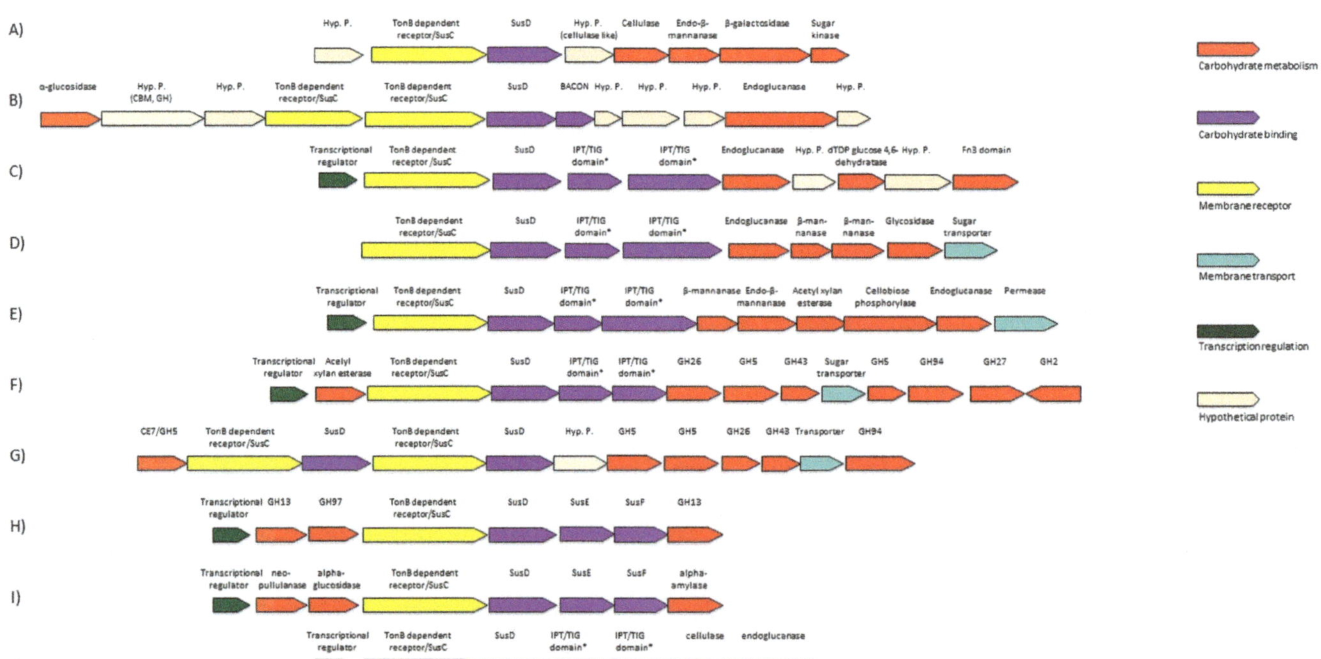

Figure 5. Physical map of selected putative polysaccharide utilization loci (PULs). A)–E) show regions from the sequence data derived from the microbiome of the six-years-old elephant. The different contigs are available from www.jgi.doe.gov (DOE Joint Genome Institute) under the IMG Project Id: 50566 with the scaffold id numbers: EMG_10007792, EMG_10000304, EMG_10002947, EMG_10003848 and EMG_10000174. **F–J)** indicate PULs from the literature or databases. **F)** from Tammar wallaby foregut [9], **G)** from Svalbard reindeer rumen [8,21], **H)** *Bacteroides thetaiotaomicron* [3], **I)** *B. faecis* CAG:32 (GenBank: FR891562.1), **J)** *Prevotella* sp. Sc00026 (GenBank JX424618.1). *indicates IPT/TIG domain containing hypothetical proteins which have also been annotated as SusE and SusF proteins.

for the latter were similar to putative proteins in the KEGG database (TABLE 1). A small fraction (1.0/0.6% respectively) of the putative proteins was derived from eukaryotes. Although the available sequences do not allow a complete analysis, when taking the coverage of the obtained sequences into account, the 1.1 Gbp assembled DNA represented a significant fraction of the metagenome and the sequences therefore gave a good first estimation of the communities' metabolic potential. The amount of assembled sequence data was in general significantly higher than in similar studies (TABLE 2). Only for the cow rumen metagenome a two-fold higher amount of assembled sequence data was reported [6]. Furthermore the N50 of both datasets (5,751 and 22,052 bp, for the baby and the six-years-old elephant, respectively) were average or high in comparison with similar studies. E.g. in a study with Iberian Lynx the N50 was 4,370 bp [15] and for a study with bovine rumen it was 24 kb [6].

The KEGG analysis suggested that the metabolic potential of the microbiomes of the fecal samples is highly diverse and versatile. Genes of many of the classical catabolic pathways linked to the degradation of diverse polysaccharides and proteins, but also for the degradation of aromatic compounds, were identified. In both samples about 25% of the putative proteins were involved in carbohydrate metabolism and glycan biosynthesis (data not shown). A high number of genes coding for proteases (9,998 and 1,396 for the six-years-old and the three-weeks-old animal, respectively) were observed. Further 6,657/843 genes encoding putative esterolytic and/or lipolytic enzymes were detected, suggesting high hydrolytic activity for esters of fatty acids within both microbial communities.

Unique GH diversity in the fecal samples

Since the older elephant was fed on a large variety of different plant-based polysaccharides we focused on the analysis of genes and enzymes linked to carbohydrate catabolism. For a more detailed analysis we compared our set of predicted CAZymes (carbohydrate active enzymes) to entries in the CAZy database. The CAZy database contains a large set of validated CAZymes and it provides a sequence-based family classification of enzymes that are involved in the modification and breakdown of polysaccharides [50]. One of the most striking findings was the observation that the metagenomes of both elephants exhibited extraordinary high diversity of GHs when compared to other fecal samples of herbivores (TABLE 2). Our analyses identified a total of 11,038 putative genes for CAZymes from 82 different GH families in the six-years-old elephant's microbiome (TABLE 2). In the three-weeks-old elephant we detected 1,873 GH genes from 70 GH families. The majority of the identified GH genes were predicted to represent full-length genes. Furthermore, more than 50% of the enzymes showed less than 50% identity to the nearest neighbor (data not shown).

Interestingly, the diversity of GH family enzymes observed in both samples in this study was in general 2-fold higher than reported for any of the previously studied samples (TABLE 2) and with respect to the study on the cow rumen the GH diversity still was 20% higher [6]. This observation is in line with the observed higher phylogenetic diversity in the feces sample of the older elephant. When comparing GH diversity and OTU richness for the baby elephant, a very high GH diversity was observed (TABLE 2). Within the older elephant's fecal metagenome enzymes belonging to GH2, GH3, GH5, GH43 and GH78 families were predominant. Altogether 4,039 genes were linked to these five GH families, being equivalent to one third of all observed CAZymes in this study (TABLE 3). With respect to the overall occurrence of GH2, GH3, GH5 and GH43 family enzymes this observation fitted well with reports on other herbivores such as cow rumen, reindeer and others summarized in TABLE 2. GH2 and GH3 family enzymes encompass ß-galactosidases, ß-glucosidases, exoglucanases but also xylosidases. These enzymes are involved in the breakdown of a large variety of oligosaccharides. GH5 enzymes are endoglucanases involved in cellulose breakdown. Similar to microbiota of other herbivores the six-years-old elephant's data set contained a large number of putative genes matching enzymes of GH families specific for the metabolism of xylo-oligosaccharides (TABLE 3). The most abundant were GH10 mainly acting as ß-1,4-xylanases and the GH43 family enzymes acting mainly as ß-xylosidases (TABLE 3).

Further, we observed a high number of GH78 and GH13 family enzymes. GH78 family enzymes are mainly α-L-rhamnosidases and involved in rhamnose removal from polysaccharides or other molecules including polyphenols. GH78 has only been reported to be predominant in the cow rumen metagenome [6]. Since GH13 family enzymes are mainly involved in starch breakdown the relatively high number of GH13 enzymes suggests that starch depolymerization plays a major role for nutrient uptake in the elephant's intestinal tract. This is likely since the elephants are also fed with fruits and vegetables containing high amounts of starch.

With respect to those enzymes involved in cellulose breakdown our data analysis suggested that within the feces microbiome of the six-years-old animal at least 10 GH families can be identified which are partially involved in cellulose hydrolysis (e.g. GH1, GH3, GH5, GH8, GH9, GH44, GH45, GH51, GH74 and GH94). The predominant ones were GH5 (517) and GH9 (119) cellulases and endoglucanases. Further, 239 GH51 endoglucanases/arabinofuranosidases could be identified (TABLE 3). A total number of 84 GH94 enzymes were found. GH94 enzymes are acting as cellobiose or cellodextrin phosphorylases. Thus altogether 2,074 candidate enzymes were identified to be involved in cellulose degradation in the feces sample of the six-years-old elephant and 242 in the feces of the three-weeks-old elephant, respectively.

The overall occurrence of cellulolytic GH genes per assembled Mbp of metagenome DNA in the elephant feces sample was comparable to those numbers reported for the cow rumen and the reindeer [6,8] (TABLE 2).

The observation that the baby elephant microbiome already encoded a large GH diversity was intriguing. It however may suggest that the mother transfers already very early a part of its microbiome to the baby. This is in line with earlier reports [48]. Baby elephants occasionally eat feces from older elephants what might explain the relatively high content of GH family enzymes in the three-weeks-old animal. Nevertheless, the majority of GH family enzymes in the three-weeks-old elephant were ß-galactosidases (GH2). The rather frequent occurrence of many GH2 family enzymes is most likely linked to the breast-feeding and the high contents of lactose in the mothers' milk.

Bacteroidales constitute the main cellulolytic organisms within the elephant feces microbiome

Interestingly, when considering protein sequence identities, the GH5, GH9 and other GH families associated with cellulolytic enzyme activities belonged in majority to Bacteroidales (FIGURE 4). In both elephants most of these sequences belonged to *Bacteroidaceae*, followed by *Lachnospiraceae* for the older and *Porphyromonadaceae* for the baby elephant (data not shown). This partially differs from other herbivores such as cow rumen in which cellulose degrading enzymes mainly derived from *Clostridium* or *Ruminococcus* species [1]. Nevertheless, a significant role of

Bacteroidetes in carbohydrate degradation was already suggested for the intestines of other animals like wallaby and reindeer [8,9]. Furthermore electron microscopic examinations confirmed the absence of significant numbers of cellulosome-producing bacteria in the elephants' fecal samples (data not shown). Instead small coccoid cells were identified in close proximity to plant cell walls (FIGURE S1). Furthermore searches within the metagenome data set failed to identify high numbers of cellulosome associated proteins. For the older elephant 4 genes encoding a protein with a dockerin and a GH domain were discovered, for the baby elephant none. Because of these observations and the high coverage of GH5 and GH9 genes for Bacteroidales (FIGURE 4) we speculate that these are the main cellulose degraders in the elephant gut. Members of this phylum are known to degrade a variety of carbohydrates but only rarely cellulose [3]. They are well known to degrade starch and other carbohydrates with proteins encoded in an operon designated as PUL (polysaccharide utilization locus) typically containing the genes *susA* to *susG*. We identified 1,383 putative *susD* genes for the six-years-old elephant and 733 for the breast-fed animal. Interestingly, about 25 putative *sus* operons were identified which included a putative cellulase gene suggesting a possible role of PULs during cellulose breakdown. These clusters had a size from 14 kb to 25 kb. Next to cellulases these operons included a variety of different glycoside hydrolases like mannanases, galactosidases and arabinosidases indicating the flexibility of Bacteroidetes. PULs have been identified in a growing number of herbivore microbiomes and in aquatic environments [8,9,51]. The observation here, however, that 1,383 *susD* homologues (733 for the baby elephant) were identified suggests a high importance of these for polysaccharide degradation. Some of the clusters comprising a *susD* gene are exemplarily shown in FIGURE 5. When comparing these to clusters from other feces samples or those deriving from single strains it becomes obvious that most of the operons show high synteny (FIGURE 5). Interestingly, the operon organization in the elephant derived clusters resembles *Prevotella* species rather than *Bacteroides faecis* or the model organism *B. thetaiotaomicron* (FIGURE 5).

Besides the Bacteroidetes, the Fibrobacteres show high abundance of GH5 genes with more than 8% for the six-years-old elephant (FIGURE 4), compared to rather low abundance on phylogenetic level (2%). In contrast to the six-years-old elephant the baby elephant did not show any residing Spirochaetes or Fibrobacteres. Microorganisms of both phyla are known for their ability to digest lignocellulose and to contribute to this task in termites [16]. *Fibrobacter succinogenes* has been described as cellulolytic bacterium able to grow on crystalline cellulose. Nevertheless its mode of cellulose degradation is not elucidated. It does not produce multiprotein complexes like some *Ruminococcus* or *Clostridium* species or secrete cellulases in the surrounding medium [52]. *F. succinogenes* instead produces cellulose-binding proteins that putatively mediate close proximity of the bacterium to the substrate [46]. Therefore and as a relatively high numbers of GH5 genes have high sequence similarity to Fibrobacteres it is likely that they also play a significant role in cellulose degradation in the elephant gut.

Similar to other samples like the termite gut microbiome [16] and the cow rumen [6], exoglucanase genes were absent, suggesting that these plant-cell-wall (GH6) and *Clostridium*-derived (GH48) cellulolytic enzymes play virtually no role during cellulose degradation in the elephant feces microbiome.

Concluding remarks

Elephants are the largest land-living animals, but our knowledge on their microbiome is limited. The data presented here give a first insight into the phylogeny of the elephant gut microbiome (i.e. fecal samples). The phylogenetic analysis of the feces of the six-years-old Asian elephant indicated that these animals host a very diverse community dominated by Bacteroidetes and Firmicutes (approx. 3,000 OTUs at 97% identity level). Furthermore, metagenome sequencing revealed a very high GH diversity of altogether 84 families. Bacteroidetes thereby seem to have a predominant role in biomass degradation. Altogether these findings distinguish the elephant feces microbiota from that of other animals like cow (Clostridiales and *Prevotellaceae* predominant) or termite (Spirochetes and Fibrobacteres predominant). Thus elephants do have a unique and flexible microbiome that meets their high energy need and allows them to digest a wide range of plant-based biopolymers. In this context it is possible to speculate that the elephant is rather a generalist and not a specialist regarding the breakdown of plant biomass. This speculation is in line with the nutrition of the six-years-old animal, which is based on leaves, twigs, hay, grass, vegetables and fruits. While both studied animals were zoo animals, in nature elephants also nurture from a wide variety of plant-derived biomass.

Supporting Information

Figure S1 TEM picture of a typical cell found in the feces of the six-years-old Asian elephant.

Table S1 Description of diversity and richness of the fecal samples of the three-weeks-old and the six-years-old Asian elephant based on 16S rRNA gene analysis. The data for the six-years-old includes Bacteria and Archaea, the data for the three-weeks-old elephant only Bacteria.

Table S2 Bins observed for the baby elephant.

Table S3 Bins observed for the six-years-old elephant.

Table S4 OTUs observed in the 16S rRNA gene datasets of the six-years-old and the three-weeks-old elephant and their frequencies.

Supporting Information S1 16S rRNA gene analysis of the baby elephant.

Supporting Information S2 16S rRNA gene analysis of the six-years-old elephant.

Author Contributions

Conceived and designed the experiments: NI MSB WRS. Performed the experiments: NI SG JD BW. Analyzed the data: MA AP CS JS PBO TR DT. Contributed reagents/materials/analysis tools: RD AG MSB. Contributed to the writing of the manuscript: NI SG UR JT JC RD AG WRS.

References

1. Björnhag G (1994) The Digestive System in Mammals: Food, Form and Function: Cambridge University Press: 287–309.
2. Flint HJ, Bayer EA, Rincon MT, Lamed R, White BA (2008) Polysaccharide utilization by gut bacteria: potential for new insights from genomic analysis. Nature Reviews Microbiology 6: 121–131.
3. Flint HJ, Scott KP, Duncan SH, Louis P, Forano E (2012) Microbial degradation of complex carbohydrates in the gut. Gut Microbes 3: 289–306.
4. Bayané A, Guiot S (2011) Animal digestive strategies versus anaerobic digestion bioprocesses for biogas production from lignocellulosic biomass. Reviews in Environmental Science and Bio/Technology 10: 43–62.
5. Clauss M, Frey R, Kiefer B, Lechner-Doll M, Loehlein W, et al. (2003) The maximum attainable body size of herbivorous mammals: morphophysiological constraints on foregut, and adaptations of hindgut fermenters. Oecologia 136: 14–27.
6. Hess M, Sczyrba A, Egan R, Kim TW, Chokhawala H, et al. (2011) Metagenomic discovery of biomass-degrading genes and genomes from cow rumen. Science 331: 463–467.
7. Brulc JM, Antonopoulos DA, Miller ME, Wilson MK, Yannarell AC, et al. (2009) Gene-centric metagenomics of the fiber-adherent bovine rumen microbiome reveals forage specific glycoside hydrolases. Proceedings of the National Academy of Sciences 106: 1948–1953.
8. Pope PB, Mackenzie AK, Gregor I, Smith W, Sundset MA, et al. (2012) Metagenomics of the Svalbard reindeer rumen microbiome reveals abundance of polysaccharide utilization loci. PLoS One 7 (6): e38571.
9. Pope PB, Denman SE, Jones M, Tringe SG, Barry K, et al. (2010) Adaptation to herbivory by the Tammar wallaby includes bacterial and glycoside hydrolase profiles different from other herbivores. Proceedings of the National Academy of Sciences 107: 14793–14798.
10. Dai X, Zhu Y, Luo Y, Song L, Liu D, et al. (2012) Metagenomic insights into the fibrolytic microbiome in yak rumen. Public Library of Science One 7 (7): e40430.
11. Zhu L, Wu Q, Dai J, Zhang S, Wei F (2011) Evidence of cellulose metabolism by the giant panda gut microbiome. Proceedings of the National Academy of Sciences 108: 17714–17719.
12. Yang S, Ma S, Chen J, Mao H, He Y, et al. (2010) Bacterial diversity in the rumen of Gayals (*Bos frontalis*), Swamp buffaloes (*Bubalus bubalis*) and Holstein cow as revealed by cloned 16S rRNA gene sequences. Molecular Biology Reports 37: 2063–2073.
13. Singh KM, Ahir VB, Tripathi AK, Ramani UV, Sajnani M, et al. (2012) Metagenomic analysis of Surti buffalo (*Bubalus bubalis*) rumen: a preliminary study. Molecular Biology Reports 39: 4841–4848.
14. Lamendella R, Santo Domingo J, Ghosh S, Martinson J, Oerther D (2011) Comparative fecal metagenomics unveils unique functional capacity of the swine gut. BioMed Central Microbiology 11: 103–119.
15. Alcaide M, Messina E, Richter M, Bargiela R, Peplies J, et al. (2012) Gene sets for utilization of primary and secondary nutrition supplies in the distal gut of endangered Iberian lynx. Public Library of Science One 7: e51521.
16. Warnecke F, Luginbuhl P, Ivanova N, Ghassemian M, Richardson TH, et al. (2007) Metagenomic and functional analysis of hindgut microbiota of a wood-feeding higher termite. Nature 450: 560–565.
17. Jacobs JP, Braun J (2014) Immune and genetic gardening of the intestinal microbiome. FEBS Letters doi: 10.1016/j.febslet.2014.02.052.
18. Buffie CG, Pamer EG (2013) Microbiota-mediated colonization resistance against intestinal pathogens. Nature Reviews Immunology 13: 790–801.
19. Varel VH (1987) Activity of Fiber-Degrading Microorganisms in the Pig Large Intestine. Journal of Animal Science 65: 488–496.
20. Ley RE, Hamady M, Lozupone C, Turnbaugh PJ, Ramey RR, et al. (2008) Evolution of Mammals and Their Gut Microbes. Science 320: 1647–1651.
21. Schmitt S, Tsai P, Bell J, Fromont J, Ilan M, et al. (2012) Assessing the complex sponge microbiota: core, variable and species-specific bacterial communities in marine sponges. The ISME Journal 6: 564–576.
22. Chassard C, Lacroix C (2013) Carbohydrates and the human gut microbiota. Current Opinion in Clinical Nutrition & Metabolic Care 16: 453–460.
23. Leser TD, Amenuvor JZ, Jensen TK, Lindecrona RH, Boye M, et al. (2002) Culture-Independent Analysis of Gut Bacteria: the Pig Gastrointestinal Tract Microbiota Revisited. Applied and Environmental Microbiology 68: 673–690.
24. Xu J, Bjursell MK, Himrod J, Deng S, Carmichael LK, et al. (2003) A genomic view of the human-*Bacteroides thetaiotaomicron* symbiosis. Science 299: 2074–2076.
25. Sonnenburg ED, Zheng H, Joglekar P, Higginbottom SK, Firbank SJ, et al. (2010) Specificity of polysaccharide use in intestinal bacteroides species determines diet-induced microbiota alterations. Cell 141: 1241–1252.
26. Mackenzie AK, Pope PB, Pedersen HL, Gupta R, Morrison M, et al. (2012) Two SusD-Like proteins encoded within a polysaccharide utilization locus of an uncultured ruminant *Bacteroidetes* phylotype bind strongly to cellulose. Applied and Environmental Microbiology 78: 5935–5937.
27. Ilmberger N, Meske D, Juergensen J, Schulte M, Barthen P, et al. (2012) Metagenomic cellulases highly tolerant towards the presence of ionic liquids-linking thermostability and halotolerance. Appl Microbiol Biotechnol 95: 135–146.
28. Zerbino DR, Birney E (2008) Velvet: Algorithms for *de novo* short read assembly using de Bruijn graphs. Genome Res 18: 821–829.
29. Nakaya A, Katayama T, Itoh M, Hiranuka K, Kawashima S, et al. (2013) KEGG OC: a large-scale automatic construction of taxonomy-based ortholog clusters. Nucleic Acids Research 41: D353–D357.
30. Tatusov RL, Natale DA, Garkavtsev IV, Tatusova TA, Shankavaram UT, et al. (2001) The COG database: new developments in phylogenetic classification of proteins from complete genomes. Nucleic Acids Research 29: 22–28.
31. Finn RD, Tate J, Mistry J, Coggill PC, Sammut SJ, et al. (2008) The Pfam protein families database. Nucleic Acids Research 36: D281–D288.
32. Wemheuer B, Guellert S, Billerbeck S, Giebel HA, Voget S, et al. (2013) Impact of a phytoplankton bloom on the diversity of the active bacterial community in the southern North Sea as revealed by metatranscriptomic approaches. Federation of European Microbiological Societies Microbiology Ecology 87: 378–389.
33. Wemheuer B, Taube R, Akyol P, Wemheuer F, Daniel R (2013) Microbial diversity and biochemical potential encoded by thermal spring metagenomes derived from the Kamchatka Peninsula. Archaea 2013: 136714–136725.
34. Bragg L, Stone G, Imelfort M, Hugenholtz P, Tyson GW (2012) Fast, accurate error-correction of amplicon pyrosequences using Acacia. Nat Methods 9: 425–426.
35. Pruesse E, Peplies J, Glockner FO (2012) SINA: accurate high-throughput multiple sequence alignment of ribosomal RNA genes. Bioinformatics 28: 1823–1829.
36. Quast C, Pruesse E, Yilmaz P, Gerken J, Schweer T, et al. (2013) The SILVA ribosomal RNA gene database project: improved data processing and web-based tools. Nucleic Acids Res 41: D590–596.
37. Caporaso JG, Kuczynski J, Stombaugh J, Bittinger K, Bushman FD, et al. (2010) QIIME allows analysis of high-throughput community sequencing data. Nat Methods 7: 335–336.
38. Markowitz VM, Chen IM, Chu K, Szeto E, Palaniappan K, et al. (2012) IMG/M: the integrated metagenome data management and comparative analysis system. Nucleic Acids Research 40: D123–129.
39. Meyer F, Paarmann D, D'Souza M, Olson R, Glass EM, et al. (2008) The metagenomics RAST server - a public resource for the automatic phylogenetic and functional analysis of metagenomes. BioMed Central Bioinformatics 9: 386–399.
40. Boisvert S, Raymond F, Godzaridis É, Laviolette F, Corbeil J (2012) Ray Meta: scalable de novo metagenome assembly and profiling. Genome Biology 13: R122.
41. Strous M, Kraft B, Bisdorf R, Tegetmeyer HE (2012) The Binning of Metagenomic Contigs for Microbial Physiology of Mixed Cultures. Frontiers in Microbiology 3: 410.
42. Wu M, Scott AJ (2012) Phylogenomic Analysis of Bacterial and Archaeal Sequences with AMPHORA2. Bioinformatics 28: 1033–1034.
43. Lagesen K, Hallin P, Rødland EA, Stærfeldt HH, Rognes T, et al. (2007) "RNAmmer: consistent and rapid annotation of ribosomal RNA genes. Nucleic Acids Research 35: 3100–3108.
44. Lowe TM, Eddy SR (1997) tRNAscan-SE: a program for improved detection of transfer RNA genes in genomic sequence. Nucleic Acids Research 25: 955–964.
45. Besemer J, Lomsadze A, Borodovsky M (2001) GeneMarkS: a self-training method for prediction of gene starts in microbial genomes. Implications for finding sequence motifs in regulatory regions. Nucleic Acids Research 29: 2607–2618.
46. Engel P, Martinson VG, Moran NA (2012) Functional diversity within the simple gut microbiota of the honey bee. Proceedings of the National Academy of Sciences 109: 11002–11007.
47. Crutzen PJ, Aselmann I, Seiler W (1986) Methane production by domestic animals, wild ruminants, other herbivorous fauna, and humans. Tellus 388: 271–284.
48. Collado MC, Cernada M, Baüerl C, Vento M, Pérez-Martínez G (2012) Microbial ecology and host-microbiota interactions during early life stages. Gut Microbes 3: 352–365.
49. Ward TL, Hosid S, Ioshikhes I, Altosaar I (2013) Human milk metagenome: a functional capacity analysis. BioMed Central Microbiology 13: 116–127.
50. Lombard V, Golaconda Ramulu H, Drula E, Coutinho PM, Henrissat B (2014) The carbohydrate-active enzymes database (CAZy) in 2013. Nucleic Acids Research 42: D490–D495.
51. Kabisch A, Otto A, Konig S, Becher D, Albrecht D, et al. (2014) Functional characterization of polysaccharide utilization loci in the marine Bacteroidetes *Gramella forsetii* KT0803. ISME J. doi: 10.1038/ismej.2014.4.
52. Suen G, Weimer PJ, Stevenson DM, Aylward FO, Boyum J, et al. (2011) The Complete Genome Sequence of *Fibrobacter succinogenes* S85 reveals a cellulolytic and metabolic specialist. Public Library of Science ONE 6: e18814.

17

Comparing the Relative Importance of Water-Borne Cues and Direct Grazing for the Induction of Defenses in the Brown Seaweed *Fucus vesiculosus*

Carla R. Flöthe[1]*, Uwe John[1], Markus Molis[2]

1 Section Ecological Chemistry, Alfred-Wegener-Institut, Helmholtz-Zentrum für Polar- und Meeresforschung, Bremerhaven, Germany, 2 Section Functional Ecology, Alfred-Wegener-Institut, Helmholtz-Zentrum für Polar- und Meeresforschung, Bremerhaven, Germany

Abstract

Some seaweed species have been shown to release water-borne cues after herbivore attack, for example, to attract natural enemies of the herbivore. These cues may also be sensed by neighboring seaweeds and used to adjust their defenses in anticipation of a possible herbivore attack. Several studies indicated information transfer between seaweed individuals in the past, including the brown seaweed *Fucus vesiculosus*. Previous work showed induction of defenses in *F. vesiculosus* in response to water-borne cues released by isopod-grazed conspecifics. In contrast, another study on induced responses after exposure to cues from isopod-grazed neighbors using the same seaweed species yielded contradictory results. This study reassessed the ability of *F. vesiculosus* individuals to sense water-borne cues released by isopod-grazed neighbors in a series of experiments that monitored *F. vesiculosus* palatability in response to direct grazing by *Idotea baltica* and water-borne cues from isopod-grazed neighbors relative to unmanipulated seaweed pieces. Two-choice feeding assays were conducted with both fresh and reconstituted seaweed pieces. Direct grazing by *I. baltica* induced a chemical defense in *F. vesiculosus*, confirming results of previous studies. In contrast, evidence for increased herbivore resistance in seaweed pieces that were located downstream of isopod-grazed *F. vesiculosus* could not be provided. The lack of defense induction in response to grazing of conspecific neighbors may be explained by the environmental conditions and the scattered distribution of *F. vesiculosus* individuals in the intertidal zone of Helgoland, which may render resource investment in the emission and/or response to water-borne cues at this site unprofitable.

Editor: Tilmann Harder, University of New South Wales, Australia

Funding: Financial support was provided by the PACES research program of the Alfred-Wegener-Institut Helmholtz-Zentrum für Polar- und Meeresforschung. CRF was financially supported by the Rosa Luxemburg Foundation. The funders had no role in study design, data collection and analysis, decision to publish, or preparation of the manuscript.

Competing Interests: The authors have declared that no competing interests exist.

* Email: carla.floethe@awi.de

Introduction

The ability of seaweeds to induce defenses in response to attack by herbivores is well known and a meta-analysis revealed that brown and green seaweeds frequently induce chemical anti-herbivory defenses [1]. The induction of defenses is suggested to be associated with the release of info-chemicals in some seaweed species, which may act as an indirect defense mechanism by attracting and directly influencing the foraging behavior of natural enemies of the attacking herbivore [2]. As the induced anti-herbivory defense does not necessarily spread to adjacent undamaged thallus parts [3] and almost all seaweeds have a limited internal transport system, these info-chemicals may also optimize within-plant signaling to ensure a systemic defense without receiving damage at all thallus parts [4].

At the same time, emitted info-chemicals become 'public' information and may also be sensed by other seaweeds. Three decades ago, researchers discovered that vascular plants apparently 'listen' to their neighbors, since herbivore attack resulted in increased resistance to herbivory not only in the attacked plant but also in plants growing nearby [5]. Thus, 'communication' between herbivore-damaged and undamaged neighbors may allow induction or preparation of defenses prior to attack [6]. Toth and Pavia [4] initially demonstrated that the brown seaweed *Ascophyllum nodosum* sensed water-borne cues released by periwinkle-grazed conspecific neighbors and increased secondary metabolite production without being subjected to direct grazing. Further studies reported that water-borne cues induced defenses in several green, red and various brown seaweed species [7–13]. Other studies failed to detect induction of anti-herbivory traits after exposure to info-chemicals (e.g. [14]) and 'communication' between adjacent seaweeds was suggested to be dependent on both the attacked seaweed species [8] and the attacking herbivore species [10]. However, the detection of defenses induced by water-borne cues may be complicated by highly variable temporal defense patterns. *Ascophyllum nodosum*, for example, was shown to use an oscillating temporal defense pattern in response to water-borne cues from grazed neighbors [15]. Although info-chemical triggered anti-herbivory traits may have been overlooked in some macroalgae due to temporally variable defenses, 'communication' via water-borne cues may indeed not occur in all seaweed species for several reasons. First, emitters of info-chemicals may be faced with metabolic costs in terms of production, transport, storage and release (e.g. [16]) and maintenance costs may arise for the synthesis of enzymes involved in these processes (e.g. [17]). Thereby, water-borne cues may incur particularly high costs because they are released into the environment and may have to be renewed constantly. Second, once water-borne cues are released, they may not only attract the herbivore's predator, but also other members of the food web, such as other herbivore species or ineffective, but competitively superior natural enemies (e.g. [17,18]). Third, neighboring seaweeds are likely to compete for resources [19]. Helping adjacent seaweeds may shift the competitive balance between emitter and receiver towards a disadvantage for the emitter, which already received damage but possibly helps its competitor to successfully escape herbivore attack.

In this study, induced anti-herbivory traits were investigated in *Fucus vesiculosus*, a common brown seaweed in the intertidal zone of North Sea shores [20]. *Fucus vesiculosus* was shown to sense cues from conspecific neighbors grazed by the isopod *Idotea baltica* in some studies [10,13], while Yun and colleagues [8] reported contradictory results about the ability of *F. vesiculosus* individuals to 'communicate' with each other. They showed that the exposure to water-borne cues from *I. baltica*-grazed neighbors reduced the palatability of reconstituted *F. vesiculosus* pieces, but not that of live seaweed pieces. As *F. vesiculosus* was shown to use an oscillating temporal defense pattern in response to direct grazing by isopods and periwinkles [21,22] and seaweed defenses triggered by water-borne cues may also show temporal variability [15], this inconsistency may have been caused by a temporal mismatch between feeding assays and single defense 'pulses' of water-borne cue-exposed *F. vesiculosus*.

A series of laboratory experiments was used in this study to (1) reassess whether *F. vesiculosus* responds to water-borne cues from conspecific neighbors grazed by the isopod *I. baltica* and (2) initially test for temporal variation in the efficacy of water-borne cue-exposed *F. vesiculosus* to deter grazers.

Materials and Methods

Organisms used in this study were collected during low tide in the mid rocky intertidal at Kringel, Helgoland, NE Atlantic (54°10′60″N, 7°53′15″E) in July 2012. At this site, the intertidal is dominated by perennial canopy-forming brown seaweeds, such as toothed wrack (*Fucus serratus*) and bladder wrack (*F. vesiculosus*), as well as irish moss *Mastocarpus stellatus*. The isopod *I. baltica* is a littoral mesograzer species [23], but can also be found among drift algae of the genera *Fucus* and *Ascophyllum* [24]. All isopods used for this study were taken from an *I. baltica* culture that was fed with customary fish food and *Ascophyllum nodosum*, and maintained in an aerated 200 L flow-through tank with a 12/12 h light/dark cycle within a constant temperature room at 15°C. Every year, new individuals from drift algae collected in the Helgoland Bight were introduced into the culture.

This study was conducted in compliance with the legal requirements of the state order no. 791-4-37 issued by the government of the Land Schleswig-Holstein on 24 April 1981 that declared the rocky shores below the high tide limit of Helgoland Island a nature reserve and allows ecologists to access sites to accomplish field research.

Experimental set-up and design

An induction experiment was run in a laboratory of the Biologische Anstalt Helgoland using a flow-through aquaria system in which *F. vesiculosus* was exposed to seawater that previously flowed over grazed conspecifics. Twenty aquaria (25 L volume; 480×230×260 mm) were divided with a polyethylene mesh (mesh size 2 mm) into equally sized upstream and downstream compartments (Fig. 1). Aquaria were irradiated by two fluorescent tubes (Osram Lumilux Daylight L 36W/865) in a 12:12 h light-dark cycle at a mean (± SD) photon flow rate of 125 (±22) μmol m^{-2} s^{-1} (PAR). Filtered North Sea water was pumped over a cotton filter into two 200 L tanks from where aquaria were individually supplied at a mean (± SD) flow rate of 104 (±19) ml/min. Filtered seawater with a mean (± SD) temperature of 19.6±0.7°C flowed through upstream and downstream compartments. To avoid animal escapes from the set-up, each aquarium was covered with a 3 mm thick transparent acrylic plate.

On 27 July 2012, 8 apical pieces with a mean (± SD) wet mass of 1.77 (±0.44) g lacking visual feeding scars were cut from each of 140 *F. vesiculosus* individuals (=1,120 pieces in total). An algal individual was defined as the tissue stemming from a single holdfast. Within 30 minutes all algal pieces were transported to the laboratory, where macroscopic epibionts were gently removed with a soft sponge. To identify genetically identical *F. vesiculosus* pieces in the set-up, seaweed pieces stemming from the same individual were marked with colored threads. Four pieces of each of 7 seaweed individuals were allocated to each compartment of each of 10 treatment and control aquaria (=56 pieces per aquarium; 112 per replicate) (Fig. 1). Treatment and corresponding control pieces

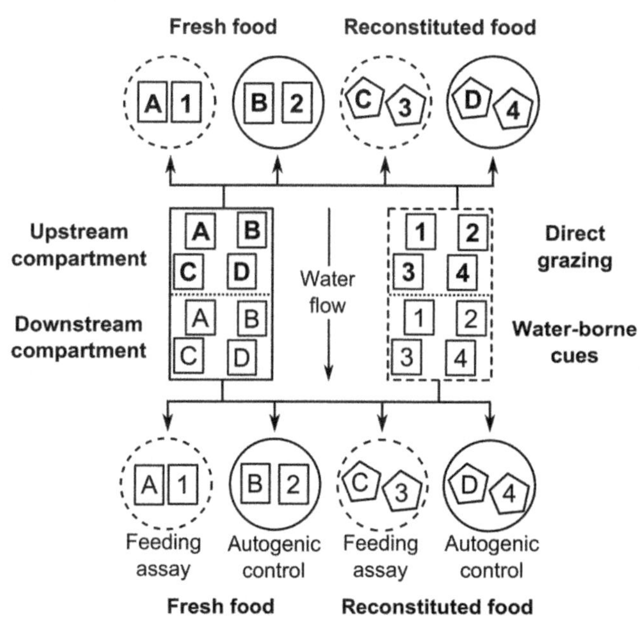

Figure 1. Schematic illustration showing the allocation of *F. vesiculosus* pieces (small rectangles) for a single replicate at one of several sampling time points. Upstream and downstream compartments were separated by a polyethylene mesh and contained 28 pieces each (only 4 shown). From these, 4 pieces were allocated to feeding arenas (circles) at each of the sampling time points. Arrow indicates direction of water flow. Dashed lines indicate containers with herbivores. Solid lines indicate containers without grazers. Letters and numbers indicate control and treated (direct grazing/water-borne cues) pieces of *F. vesiculosus*, respectively. Bold letters and numbers indicate *F. vesiculosus* pieces located in upstream compartments, while all other characters indicate seaweed pieces located in downstream compartments. Pentagons indicate reconstituted food items used for feeding assays with naïve herbivores.

Table 1. Results of two-tailed paired *t*-tests comparing isopod consumption between different 3 day intervals of the induction phase (n = 10).

Time interval since start of induction [d]	1–3	4–6	7–9	10–12	13–15	16–18	19–21	22–24	25–27
1–3	-	0.851	0.056	**0.012**	0.559	0.093	0.077	**0.047**	0.505
4–6		-	**<0.001**	**<0.001**	0.536	0.063	0.065	**0.027**	0.501
7–9			-	**0.004**	0.794	0.842	0.903	0.274	0.183
10–12				-	0.467	0.285	0.314	0.763	**0.031**
13–15					-	0.794	0.861	0.429	0.765
16–18						-	0.826	0.404	0.181
19–21							-	0.253	0.476
22–24								-	0.202
25–27									-

Significant p-values, i.e. $\alpha \leq 0.05$, in bold.

originated from the same *F. vesiculosus* individual in the field to make sure that potential differences in palatability between both pieces detected in a feeding assay at a specific point in time during the induction phase were not confounded by inter-individual variation in, for example, morphology [25] or chemical composition [26]. Afterwards, seaweed pieces were separately anchored with cable ties (width 1.8 mm) to a polyethylene mesh (mesh size 2 mm) which was placed on the bottom of the aquarium to prevent floating of *F. vesiculosus* pieces.

Two sequential experimental phases were applied: acclimation and induction. During acclimation, algal pieces remained in the set-up for 4 days without grazers to reduce potential effects by cutting the algae as well as putative induced defensive traits, which may have been attained by unknown grazing histories in the field. According to prior studies, 3 days are sufficient to reduce anti-herbivory defenses in *F. vesiculosus* (e.g. [27]). At the end of the acclimation phase, the wet mass of *F. vesiculosus* pieces was determined by blotting them dry with paper towels for 20 seconds and weighing them to the nearest 0.001 g (Sartorius CPA323S, Sartorius, Göttingen, Germany). This was the standard procedure to measure the wet mass of all food items in this study. Afterwards, a feeding assay was started to confirm equal palatability of designated treated and control pieces after acclimation (see details below).

The same day, the induction phase was started by adding 6 *I. baltica* to upstream compartments of all treatment aquaria. Upstream compartments of control aquaria as well as all downstream compartments remained without herbivores. Treatment and control aquaria of each replicate were paired and pairs randomly arranged in the set-up. At the end of the acclimation phase and from day 12 on, 4 genetically identical *F. vesiculosus* pieces of each upstream (direct grazing) and downstream (water-borne cues) compartment were allocated to feeding assays every three days during the 27 day induction phase. The first of these 4 pieces from each compartment was transferred to a feeding arena with naïve grazers. The second piece was allocated to a feeding arena without consumers to determine autogenic wet mass changes during the feeding assay. The third and the fourth piece

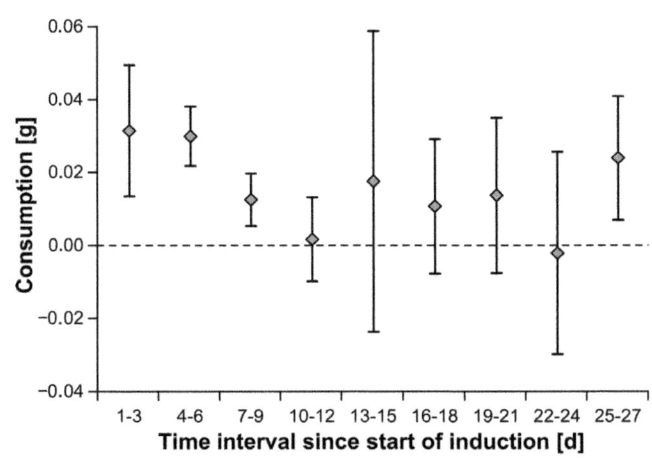

Figure 2. Mean and 95% confidence intervals of *Fucus vesiculosus* consumption by *Idotea baltica* during 3 day intervals in the induction phase (n = 10). Stippled line marks level of no consumption. Confidence intervals overlapping with stippled line indicate intervals when consumption was statistically not significantly different from zero. Different letters indicate significant difference in consumption.

Table 2. Results of the 2-factorial RM-ANOVA for the induction phase, comparing feeding rates affected by treatment (control vs. water-borne cues/direct grazing) and time.

	Water-borne cues				Direct grazing			
	df	MS	F	p	df	MS	F	p
Fresh food								
Time	5	0.005	0.72	0.612	5	9.36×10^{-4}	0.42	0.830
Error	54	0.007			54	2.21×10^{-3}		
Treatment	1	0.003	0.33	0.569	1	0.04	15.85	0.000
Treatment×Time	5	0.004	0.46	0.804	5	0.01	2.53	0.039
Error	54	0.008			54	2.67×10^{-3}		
Reconstituted food								
Time	5	0.02	2.16	0.072	5	0.03	3.97	0.004
Error	54	0.01			54	0.01		
Treatment	1	0.01	0.47	0.494	1	0.10	12.17	0.001
Treatment×Time	5	0.01	0.32	0.899	5	0.02	2.49	0.042
Error	54	0.02			54	0.01		

Consumption was assessed in two-choice feeding assays using either fresh or reconstituted *Fucus vesiculosus* (n = 10).

of each compartment were stored at −80°C and used in feeding assays with artificial food pellets within 4 weeks (Fig. 1).

At the same time, one isopod was removed from the upstream compartment of each treatment aquarium to apply a comparable grazing pressure to remaining seaweed pieces. Furthermore, the wet mass of three *F. vesiculosus* pieces from each upstream compartment was measured at the start of the induction phase and in 3 day intervals thereafter to monitor herbivore consumption during the induction phase. The pieces were chosen randomly for each interval and means were calculated from these pieces as a replicate measure of consumption for statistical analysis (see formula in subchapter 'feeding assays').

Feeding assays

Fresh algae. Herbivore preferences for control or treated *F. vesiculosus* were determined in 72 h two-choice feeding assays at several sampling time points during the induction phase. Transparent 8 L plastic aquaria (325×175×185 mm) were used as feeding arenas. Each feeding arena contained naïve herbivores, which were allowed to choose between a *F. vesiculosus* pieces that was either previously grazed or exposed to water-borne cues before and a control piece of genetically identical *F. vesiculosus*. To avoid grazer adaptations, naïve consumers, which were not in contact with *F. vesiculosus* before, were used in feeding assays. One male *I. baltica* was used as consumer in each feeding assay. At the beginning and the end of the feeding assays, the wet mass of the *F. vesiculosus* pieces in the feeding arenas (= assayed alga) was measured. A second seaweed piece was removed from the same compartment, from where the assayed algae originated and allocated to a feeding arena without grazers to correct herbivore consumption rates for non-feeding related (autogenic) changes in seaweed wet mass. The risk of underestimating error variance and, thus, of committing a type I error was reduced by using the same number of autogenic controls and assayed algae [28]. The following formula was used to calculate the consumption of each assayed *F. vesiculosus* piece (adopted from [29]):

$$\text{consumption} = T_{start} \times \left(C_{end}/C_{start}\right) - T_{end}$$

where T_{start} and T_{end} represent the wet mass of an assayed algal piece before and after the feeding assay, respectively, and C_{start} and C_{end} represent the wet mass of the corresponding autogenic control alga and the beginning and the end of the feeding assay, respectively.

Reconstituted food. Additional feeding assays with reconstituted food were run to determine whether induced changes in *F. vesiculosus* palatability were due to chemical traits. As described above, the third and the fourth *Fucus vesiculosus* piece, which were removed from each compartment, were stored at −80°C for a period of maximum four weeks prior to the preparation of reconstituted food. Frozen seaweed pieces were lyophilized at −30°C and 0.37 mbar for 24 hours (Christ Beta 1-8 LD plus, Martin Christ Gefriertrocknungsanlagen, Osterode am Harz, Germany). Afterwards, all pieces were ground for 10 seconds at a frequency of 25 Hz with a mixer mill (Schwingmühle MM 400, Retsch Laborgeräte, Haan, Germany). Subsequently, 0.4 g of this powder was mixed with 3.6 ml of molten agar (a blend of 0.02 g agar per one ml of boiling distilled water) after agar had cooled to 45°C to minimize putative thermal destruction of chemical seaweed compounds. This mixture was applied to a polystyrene mosquito mesh (mesh size 1.5 mm) and flattened between two PVC plates coated with wax paper (method adopted from [30]). Uniform thickness of food pellets was guaranteed by placing a 1 mm thick plastic template between both PVC plates. A food

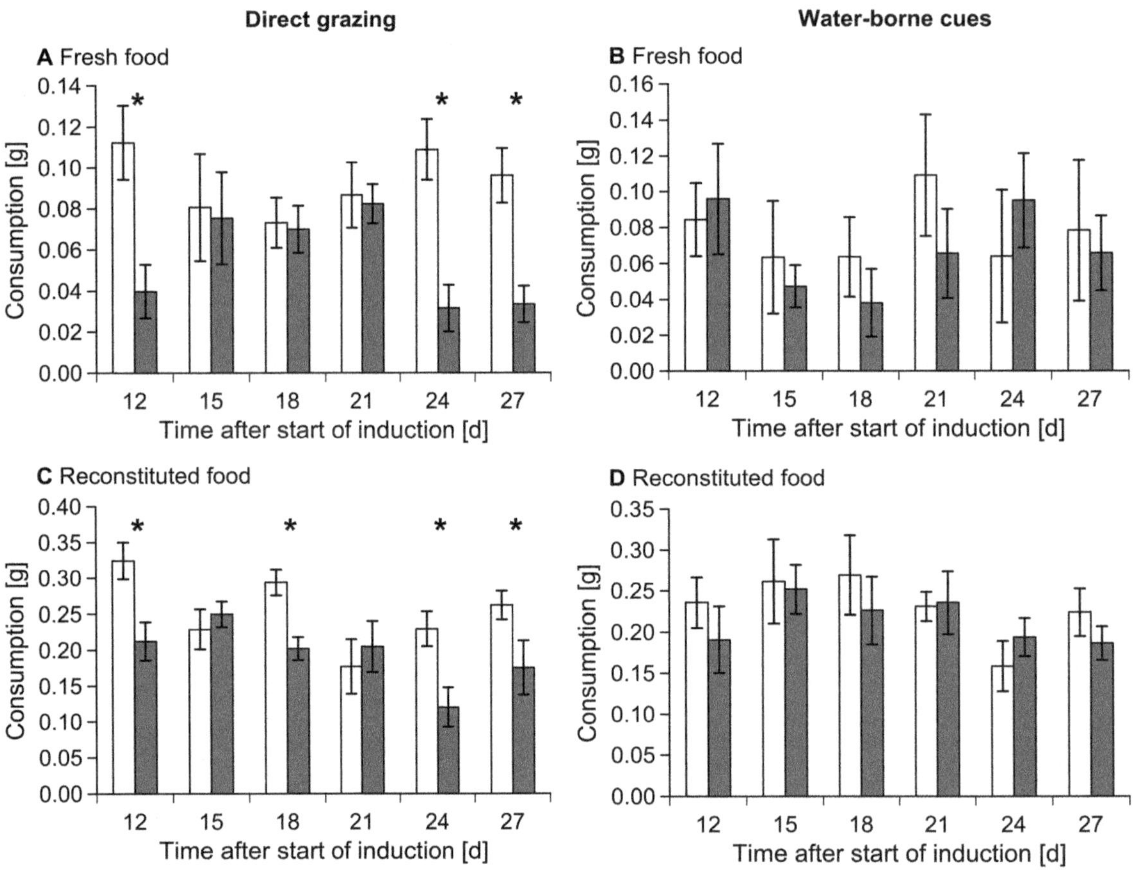

Figure 3. Mean ± SE consumption of *Fucus vesiculosus* by *Idotea baltica* in two-choice feeding assays during the induction phase (n = 10). Isopod consumption of fresh and reconstituted *Fucus vesiculosus* pieces that were previously grazed by *Idotea baltica* (A and C) or located downstream of isopod-grazed conspecifics (B and D) (grey bars) and seaweed pieces which were not exposed to grazing or located downstream of grazed *F. vesiculosus* before (controls; open bars). Asterisks indicate significant results of one-tailed paired *t*-tests comparing distribution of differences between control and grazed pieces against the null-hypothesis of no difference.

item of 2×2 cm area was cut from each pellet after solidification. Each food item was placed in a glass Petri dish (Ø 10 cm, 2 cm height) and transferred to a separate feeding arena. Each feeding arena contained one Petri dish with a food item made from a control *F. vesiculosus* piece and a second Petri dish with a food item made from a *F. vesiculosus* pieces that was either previously grazed or exposed to water-borne cues before. Food items were placed in Petri dishes within feeding arenas to permit correct allocation of fragments that occasionally broke off from food items. Both food items in each feeding arena were weighed and naïve grazers (one male *I. baltica*) were introduced to feeding arenas containing food items made from the third *F. vesiculosus* piece which was removed from each compartment. Autogenic wet mass changes during feeding assays were determined by using reconstituted food pellets made from the fourth *F. vesiculosus* piece which was removed from each compartment (Fig. 1). No grazers were added to feeding arenas that assessed autogenic wet mass changes. Feeding assays were terminated and food items re-weighed after 3 days or when ≥50% of one food item was consumed, whichever came first.

Statistical analysis

Consumption of inducers during induction phase. Consumption of *F. vesiculosus* by *I. baltica* during the induction phases was analyzed by two-tailed paired *t*-tests, comparing consumption between different 3 day intervals of the induction phase.

Feeding assays. Herbivore consumption rates from feeding assays performed directly after the acclimation phase were analyzed by one-tailed paired *t*-tests.

Repeated-measures analyses of variances (RM-ANOVAs) were applied to test for the effect of direct grazing by *I. baltica* and isopod grazing on conspecific neighbors on the palatability of *F. vesiculosus* (within-subjects measure: 2 levels, fixed) at different times during the induction phase (between-subjects measures: 6 levels, fixed). RM-ANOVA was used because treatments were not independent and standard ANOVA cannot be properly applied when two food types are simultaneously offered to the same individual consumer [31]. As the within-subject factor had only two levels (control vs. treatment), testing for sphericity is not applicable [32].

Due to ambiguous selection of an appropriate error term for post-hoc tests involving within-subject by between-subject interactions, no post-hoc tests were computed for time×grazing interactions [33]. Instead, one-tailed paired *t*-tests (due to experimental evidence that direct grazing and water-borne cues induce anti-herbivory defenses in *F. vesiculosus*, e.g. [10]) were performed for each time separately. Normal distribution of differences in the consumption of treated and control seaweed pieces was confirmed using the Kolmogorov-Smirnov test.

Table 3. Results of one-tailed paired *t*-tests comparing controls and seaweed pieces that were previously exposed to direct grazing by *I. baltica*.

Time [d]	Fresh algae		Reconstituted food	
	t	p	*t*	p
0	0.18	0.429	0.30	0.385
12	3.64	**0.003**	3.44	**0.004**
15	0.14	0.447	−0.55	0.495
18	0.18	0.430	3.28	**0.005**
21	0.25	0.403	−0.46	0.456
24	3.74	**0.002**	3.41	**0.004**
27	3.89	**0.002**	1.85	**0.049**

Consumption by conspecific consumers was assessed in feeding assays using either fresh or reconstituted food pieces of *F. vesiculosus* (n = 10). Time = days after start the of the induction phase. Significant p-values, i.e. $\alpha \leq 0.05$, in bold.

Results

Consumption of inducers during induction phases

During the induction phase isopods consumed on average 7.7% of initial *F. vesiculosus* wet mass. Inducer consumption varied significantly during the induction phase (Table 1, Fig. 2). With 36 *t*-tests performed in total, the probability of finding 7 tests below $p<0.05$ by chance is 0.001 [34].

Feeding assays

At the end of the acclimation phase, the palatability of *F. vesiculosus* pieces in designated control and treatment aquaria was neither significantly different when tested in assays using fresh algae (one-tailed paired *t*-test: direct grazing $t_9 = 0.18$, $p = 0.429$ and water-borne cues $t_9 = 0.29$, $p = 0.388$) nor reconstituted food (one-tailed paired *t*-test: direct grazing $t_9 = 0.30$, $p = 0.385$ and water-borne cues $t_9 = 0.16$, $p = 0.438$).

RM-ANOVA results did not indicate a significant preference of isopods for fresh and reconstituted *F. vesiculosus* pieces that were located downstream of ungrazed control *F. vesiculosus* pieces compared to seaweed pieces that were located downstream of isopod-grazed conspecifics during the induction phase. There was no significant interaction between treatment and time (Table 2, Fig. 3).

In contrast, isopods significantly preferred ungrazed *F. vesiculosus* to previously grazed seaweed pieces in both fresh and reconstituted food feeding assays. In addition, a significant interaction between isopod grazing and time was found in assays using fresh and reconstituted *F. vesiculosus* pieces (Table 2).

One-tailed paired *t*-tests, that analyzed feeding preferences at each time, revealed that *I. baltica* significantly preferred fresh control *F. vesiculosus* to previously grazed algae pieces 12, 24, and 27 days after the onset of grazing (Table 3, Fig. 3A). Reconstituted food made from previously ungrazed seaweed pieces was preferred to reconstituted food made from previously grazed *F. vesiculosus* 12, 18, 24, and 27 days after the start of the induction phase (Table 3, Fig. 3C).

Discussion

The induction experiment did not suggest induction of defenses in response to neighbor grazing in *F. vesiculosus*. However, the same experiment clearly showed a significant reduction in palatability of directly grazed upstream seaweed pieces and provided strong evidence for induced defenses in these *F. vesiculosus* pieces. The finding that bioassays conducted with reconstituted food and fresh algae showed very similar results confirms the induction of chemical anti-herbivory traits. Furthermore, *F. vesiculosus* showed a pulsating temporal defense pattern. Because different genetic individuals were used for feeding assays at each sampling time point, the observed temporal pattern may have been caused by genetic differences between *F. vesiculosus* individuals. However, pulsating defenses have been repeatedly shown for *F. vesiculosus* in the recent past [21,22] and, in addition, were also found in the related seaweed species *Ascophyllum nodosum* [15]. Therefore, it is most likely that short-term variations in seaweed palatability are an intrinsic feature of the *F. vesiculosus* anti-herbivory defense that may entail economic, ecologic and/or evolutionary benefits [21].

The observed lack of a response to water-borne cues released by isopod-grazed neighbors is therefore surprising because fast-swimming isopods were shown to frequently switch between *F. vesiculosus* individuals [35,36] and may be suggested to imply a relatively high risk of attack for nearby seaweed individuals. Although this result partially coincides with findings reported by Yun and colleagues [8], it deviates from several studies reporting on induction of anti-herbivory defenses in Baltic *F. vesiculosus* located downstream of isopod-grazed conspecifics [10,13]. Therefore, this study may provide evidence for geographic variation in the ability to emit and/or receive water-borne cues in response to herbivory between genetically distinct *F. vesiculosus* populations originating from Helgoland (North Sea) and the Baltic Sea [37,38]. Inter-population differences in the ability of seaweeds to induce anti-herbivory defenses were, for instance, reported in European and North American populations of a closely related seaweed (*Ascophyllum nodosum*) [39]. Long and colleagues [40] suggested that this variation may be related to a difference in consumer pressure between European and North American sites. Similarly, *F. vesiculosus* individuals from North Sea and Baltic Sea populations may have evolved different abilities to 'communicate' with each other due to either different grazing pressures of *I. baltica* [41] and/or different abiotic conditions between the North and Baltic Seas.

The Baltic Sea is a semi-enclosed virtually tideless habitat [42] where *F. vesiculosus* forms dense and almost mono-specific stands [10]. In such an environment, where water-flow is low and adjacent seaweeds are in very close proximity, the benefits of seaweed-seaweed communication (e.g. increased efficacy of the anti-herbivory defense at the population level [43] or increased inclusive fitness by warning close relatives growing nearby [44]) may outweigh costs affiliated with the production of water-borne cues.

In contrast, the study site on Helgoland represents a relatively exposed habitat with a mean tidal range of about 2.3 m [45] where *F. vesiculosus* forms only moderately dense stands (C. Flöthe, personal observation). In such a high-flow environment where *F. vesiculosus* individuals are as well not in direct contact to each other, water movement may dilute or wash away chemical cues to such an extent that the transmission of these cues between single *F. vesiculosus* individuals is strongly impeded or even becomes impossible. In addition, the release of water-borne cues is suggested to incur a metabolic cost [17] and may cause even higher costs when info-chemical are used by 'unauthorized' receivers, such as additional herbivores or unwanted predators [46]. Therefore, it may be assumed that the benefits of seaweed-seaweed 'communication' do not outweigh costs affiliated with the production of water-borne cues at the study site. Thus, investing resources in growth and reproduction rather than the synthesis and emission of water-borne cues may offer greater benefits to Helgoland *F. vesiculosus*.

However, straightforward generalizations based on the results obtained in this study should be made carefully. A high degree of variation was shown to exist in the anti-herbivory response of *F. vesiculosus* (e.g. [10]) and 'communication' between *F. vesiculosus* individuals in response to another herbivore species at the same site cannot be excluded. Furthermore, even if defense induction in response to water-borne cues released by isopod-grazed conspecifics may not occur in wave-exposed Helgoland *F. vesiculosus*, it may occur at the same place in systems with calm water conditions, such as tide pools [47].

Acknowledgments

We thank Monique Bartels, Viktor Koch, and Willy Petzold for assistance in the laboratory.

Author Contributions

Conceived and designed the experiments: CRF MM UJ. Performed the experiments: CRF. Analyzed the data: CRF. Wrote the paper: CRF MM UJ.

References

1. Toth GB, Pavia H (2007) Induced herbivore resistance in seaweeds: a meta-analysis. J Ecol 95: 425–434.
2. Coleman R, Ramchunder S, Davies K, Moody AJ, Foggo A (2007) Herbivore-induced infochemicals influence foraging behaviour in two intertidal predators. Oecologia 151: 454–463.
3. Hemmi A, Honkanen T, Jormalainen V (2004) Inducible resistance to herbivory in *Fucus vesiculosus*: duration, spreading and variation with nutrient availability. Mar Ecol Prog Ser 273: 109–120.
4. Toth GB, Pavia H (2000) Water-borne cues induce chemical defense in a marine alga (Ascophyllum nodosum). PNAS 97: 14418–14420.
5. Baldwin IT, Schultz JC (1983) Rapid changes in tree leaf chemistry induced by damage: evidence for communication between plants. Science 221: 277–279.
6. Baldwin IT, Halitschke R, Paschold A, von Dahl CC, Preston CA (2006) Volatile signaling in plant-plant interactions: "talking trees" in the genomics era. Science 311: 812–815.
7. Yun HY, Cruz J, Treitschke M, Wahl M, Molis M (2007) Testing for the induction of anti-herbivory defences in four Portuguese macroalgae by direct and water-borne cues of grazing amphipods. Helgol Mar Res 61: 203–209.
8. Yun HY, Engelen AH, Santos RO, Molis M (2012) Water-borne cues of a non-indigenous seaweed mediate grazer-deterrent responses in native seaweeds, but not vice versa. PLoS One 7: e38804.
9. Toth GB (2007) Screening for induced herbivore resistance in Swedish intertidal seaweeds. Mar Biol 151: 1597–1604.
10. Rohde S, Molis M, Wahl M (2004) Regulation of anti-herbivore defence by *Fucus vesiculosus* in response to various cues. J Ecol 92: 1011–1018.
11. Macaya EC, Rothäusler E, Thiel M, Molis M, Wahl M (2005) Induction of defenses and within-alga variation of palatability in two brown algae from the northern-central coast of Chile: Effects of mesograzers and UV radiation. J Exp Mar Biol Ecol 325: 214–227.
12. Díaz E, Güldenzoph C, Molis M, McQuaid C, Wahl M (2006) Variability in grazer-mediated defensive responses of green and red macroalgae on the south coast of South Africa. Mar Biol 149: 1301–1311.
13. Haavisto F, Välikangas T, Jormalainen V (2010) Induced resistance in a brown alga: phlorotannins, genotypic variation and fitness costs for the crustacean herbivore. Oecologia 162: 685–695.
14. Sotka EE, Taylor RB, Hay ME (2002) Tissue-specific induction of resistance to herbivores in a brown seaweed: the importance of direct grazing versus waterborne signals from grazed neighbors. J Exp Mar Biol Ecol 277: 1–12.
15. Flöthe CR, Molis M (2013) Temporal dynamics of inducible anti-herbivory defenses in the brown seaweed *Ascophyllum nodosum* (Phaeophyceae). J Phycol 49: 468–474.
16. Dicke M, Sabelis MW (1992) Costs and benefits of chemical information conveyance: proximate and ultimate factors. In: Roitberg BD, Isman MB, editors. Insect chemical ecology: an evolutionary approach. New York: Chapman & Hall. pp. 122–155.
17. Dicke M, Sabelis MW (1989) Does it pay plants to advertise for bodyguards? Towards a cost-benefit analysis of induced synomone production. In: Lambers H, Konings H, Cambridge ML, editors. Causes and consequences of variation in growth rate and productivity of higher plants. The Hague: SPB Academic Publishing. pp. 341–358.
18. Kalberer NM, Turlings TCJ, Rahier M (2001) Attraction of a leaf beetle (*Oreina cacaliae*) to damaged host plants. J Chem Ecol 27: 647–661.
19. Edwards MS, Connell SD (2012) Competition, a major factor structuring seaweed communities. In: Wiencke C, Bischof K, editors. Seaweed Biology. Heidelberg: Springer-Verlag Berlin Heidelberg. pp. 135–156.
20. Lüning K (1985) Meeresbotanik - Verbreitung, Ökophysiologie und Nutzung der marinen Makroalgen. Stuttgart, Germany: Thieme Verlag. 383 p.
21. Flöthe CR, Molis M, Kruse I, Weinberger F, John U (2014) Herbivore-induced defence response in the brown seaweed *Fucus vesiculosus* (Phaeophyceae): temporal pattern and gene expression. Eur J Phycol. DOI: 10.1080/09670262.2014.946452.
22. Flöthe CR, Molis M, John U (2014) Induced resistance to periwinkle grazing in the brown seaweed *Fucus vesiculosus* (Phaeophyceae): molecular insights and seaweed-mediated effects on herbivore interactions. J Phycol 50: 564–576.
23. Orav-Kotta H, Kotta J (2004) Food and habitat choice of the isopod *Idotea baltica* in the northeastern Baltic Sea. Hydrobiologia 514: 79–85.
24. Franke HD, Gutow L, Janke M (2007) Flexible habitat selection and interactive habitat segregation in the marine congeners *Idotea baltica* and *Idotea emarginata* (Crustacea, Isopoda). Mar Biol 150: 929–939.
25. Hoffmann A (2011) Untersuchungen zur phänotypischen Plastizität in Braunalgen [Diploma thesis]. Karlsruhe, Germany: Karlsruher Institut für Technologie. 64 p.
26. Tuomi J, Ilvessalo H, Niemelä P, Sirén S, Jormalainen V (1989) Within-plant variation in phenolic content and toughness of the brown alga *Fucus vesiculosus* L. Bot Mar 32: 505–510.
27. Rohde S, Wahl M (2008) Temporal dynamics of induced resistance in a marine macroalga: time lag of induction and reduction in *Fucus vesiculosus*. J Exp Mar Biol Ecol 367: 227–229.
28. Roa R (1992) Design and analysis of multiple-choice feeding-preference experiments. Oecologia 89: 509–515.
29. Cronin G, Hay M (1996) Within-plant variation in seaweed palatability and chemical defenses: optimal defense theory versus the growth-differentiation balance hypothesis. Oecologia 105: 361–368.
30. Hay ME, Kappel QE, Fenical W (1994) Synergisms in plant defenses against herbivores: interactions of chemistry, calcification, and plant quality. Ecology 75: 1714–1726.
31. Peterson C, Renaud P (1989) Analysis of feeding preference experiments. Oecologia 80: 82–86.
32. Quinn GGP, Keough MJ (2002) Experimental design and data analysis for biologists. Cambridge: Cambridge University Press.
33. Winer BJ, Brown DR, Michels KM (1991) Statistical principles in experimental design. New York: McGraw-Hill.
34. Moran MD (2003) Arguments for rejecting the sequential Bonferroni in ecological studies. Oikos 100: 403–05.
35. Jormalainen V, Tuomi J (1989) Reproductive ecology of the isopod *Idotea baltica* (Pallas) in the northern Baltic. Ophelia 30: 213–223.
36. Jormalainen V, Honkanen T, Heikkilä N (2001) Feeding preferences and performance of a marine isopod on seaweed hosts: cost of habitat specialization. Mar Ecol Prog Ser 220: 219–230.
37. Johannesson K, André C (2006) Life on the margin: genetic isolation and diversity loss in peripheral marine ecosystem, the Baltic Sea. Mol Ecol 15: 2013–2029.
38. Tatarenkov A, Jönsson RB, Kautsky L, Johannesson K (2007) Genetic structure in populations of *Fucus vesiculosus* (Phaeophyceae) over spatial scales from 10 m to 800 km. J Phycol 43: 675–685.

39. Long JD, Hamilton RS, Mitchell JL (2007) Asymmetric competition via induced resistance: specialist herbivores indirectly suppress generalist preference and populations. Ecology 88: 1232–40.
40. Long JD, Toth G, Pavia H (2013) Proximate and ultimate causes for transatlantic variation in seaweed defenses. Mar Ecol Prog Ser 493: 83–89.
41. Kersen P, Kotta J, Bučas M, Kolesova N, Deķere Z (2011) Epiphytes and associated fauna on the brown alga Fucus vesiculosus in the Baltic and the North Seas in relation to different abiotic and biotic variables. Mar Ecol 32: 87–95.
42. Yun HY, Rohde S, Linnane K, Wahl M, Molis M (2010) Seaweed-mediated indirect interaction between two species of meso-herbivores. Mar Ecol Prog Ser 408: 47–53.
43. Bruin J, Sabelis MW, Dicke M (1995) Do plants tap SOS signals from their infested neighbours? TREE 10: 167–170.
44. Heil M, Karban R (2010) Explaining evolution of plant communication by airborne signals. Trends Ecol Evol 25: 137–144.
45. Eschweiler N, Molis M, Buschbaum C (2009) Habitat-specific size structure variations in periwinkle populations (*Littorina littorea*) caused by biotic factors. Helgol Mar Res 63: 119–127.
46. Bruin J, Dicke M (2001) Chemical information transfer between wounded and unwounded plants: backing up the future. Biochem Syst Ecol 29: 1103–1113.
47. Macaya EC, Thiel M (2008) In situ tests on inducible defenses in *Dictyota kunthii* and *Macrocystis integrifolia* (Phaeophyceae) from the Chilean coast. J Exp Mar Biol Ecol 354: 28–38.

18

Demographic Outcomes and Ecosystem Implications of Giant Tortoise Reintroduction to Española Island, Galapagos

James P. Gibbs[1]*, Elizabeth A. Hunter[2], Kevin T. Shoemaker[3], Washington H. Tapia[4], Linda J. Cayot[5]

1 Department of Environmental and Forest Biology, State University of New York College of Environmental Science and Forestry, Syracuse, New York, United States of America, **2** Warnell School of Forestry and Natural Resources, University of Georgia, Athens, Georgia, United States of America, **3** Department of Ecology and Evolution, Stony Brook University, Stony Brook, New York, United States of America, **4** Department of Applied Research, Galapagos National Park Service, Puerto Ayora, Galapagos, Ecuador, **5** Galapagos Conservancy, Fairfax, Virginia, United States of America

Abstract

Restoration of extirpated species via captive breeding has typically relied on population viability as the primary criterion for evaluating success. This criterion is inadequate when species reintroduction is undertaken to restore ecological functions and interactions. Herein we report on the demographic and ecological outcomes of a five-decade-long population restoration program for a critically endangered species of "ecosystem engineer": the endemic Española giant Galapagos tortoise (*Chelonoidis hoodensis*). Our analysis of complementary datasets on tortoise demography and movement, tortoise-plant interactions and Española Island's vegetation history indicated that the repatriated tortoise population is secure from a strictly demographic perspective: about half of tortoises released on the island since 1975 were still alive in 2007, *in situ* reproduction is now significant, and future extinction risk is low with or without continued repatriation. Declining survival rates, somatic growth rates, and body condition of repatriates suggests, however, that resources for continued population growth are increasingly limited. Soil stable carbon isotope analyses indicated a pronounced shift toward woody plants in the recent history of the island's plant community, likely a legacy of changes in competitive relations between woody and herbaceous plants induced by now-eradicated feral goats and prolonged absence of tortoises. Woody plants are of concern because they block tortoise movement and hinder recruitment of cactus–a critical resource for tortoises. Tortoises restrict themselves to remnant cactus patches and areas of low woody plant density in the center of the island despite an apparent capacity to colonize a far greater range, likely because of a lack of cactus elsewhere on the island. We conclude that ecosystem-level criteria for success of species reintroduction efforts take much longer to achieve than population-level criteria; moreover, reinstatement of endangered species as fully functioning ecosystem engineers may often require large-scale habitat restoration efforts in concert with population restoration.

Editor: Cheryl A. Nickerson, The Biodesign Institute, Arizona State University, United States of America

Funding: The 2010 field surveys of tortoise-plant interactions as well as this synthesis of the accumulated information was funded by the Galapagos Conservancy in collaboration with the Galapagos National Park Service, the Prometeo Program of Ecuador's National Secretariat for Higher Education, Science, Technology and Innovation, and U.S. National Science Foundation grant DEB 1258062. The funders had no role in study design, data collection and analysis, decision to publish, or preparation of the manuscript.

Competing Interests: The authors have declared that no competing interests exist.

* Email: jpgibbs@esf.edu

Introduction

Restoration of extirpated species and populations via captive breeding and reintroduction has become a widespread practice in biodiversity conservation [1–2]. As such projects proliferate there is growing need to define and document reintroduction success [2–3]. Documentation of a viable, self-sustaining population is essential for declaring success of any reintroduction project [4]. Yet reintroduction efforts are often motivated not just by population viability considerations but also by broader concerns about restoring ecological functions and interactions, thereby highlighting the need to document the role of reintroduced species as agents of ecosystem change [4].

Species reintroduction can generate profound and complex ecosystem changes, especially if the focal species can be characterized as an "ecosystem engineer" [5–7]. For example, carnivore reintroduction to Yellowstone National Park in the western United States likely triggered a trophic cascade whereby predation pressure on large herbivores has led to the restoration of plant communities [6],[8]. Yet ecosystem-level effects of species reintroduction may manifest only after the focal species crosses a threshold of effective density far higher than needed to ensure population persistence alone [9–10] and thereby result in separate time frames and planning horizons for achieving population-level versus ecosystem-level objectives of species reintroduction programs.

Despite the promise of species reintroduction as a tool for restoring populations and in some cases ecosystems, prolonged absence of certain species can result in habitat degradation and thereby lead to failed reintroduction efforts despite repeated attempts [1],[4],[11–12]. In some cases, management interventions may be required to render release sites hospitable to a focal species before the species can assume its former ecological role [13]. For example, invasive species removal may facilitate successful repatriation in cases where a species' former ecological role has been supplanted by an invasive competitor [14]. For these reasons assessments of the success and feasibility of reintroduction projects must carefully consider ecological changes wrought in the absence of the focal species.

Reintroduction of tortoise populations is increasingly proposed not only to conserve biodiversity [15] but also as tools for ecosystem restoration [16–18]. Giant tortoises, once widespread on all continents except Antarctica, are ecosystem engineers that manipulate the distribution and abundance of other organisms through direct effects of herbivory, disturbance and seed dispersal on plant communities and subsequent indirect impacts on animal communities [14],[19–20]. The effects of giant tortoises on terrestrial ecosystems of oceanic islands (to which giant tortoises are currently restricted) are potentially on par with those of megaherbivores as drivers of savanna structure and function, a system for which the role of ecosystem engineers as agents of ecosystem change is far better understood [21–22]. Consequently, giant tortoises have been proposed as generic tools for island ecosystem restoration [17].

Herein we examine the demographic and ecological outcomes of the repatriation of Galapagos giant tortoises (*Chelonoidis hoodensis*) to Española Island (Fig. 1). *C. hoodensis* is a CITES Appendix I and IUCN Red List "critically endangered" species [23] that once numbered in the thousands of individuals but dropped to just 15 individuals by 1960 due to historical exploitation of tortoises throughout the archipelago [24]. Between 1963 and 1974, all survivors (12 females and 3 males) were brought into captivity and to date >1500 of their captive-raised offspring have since been released onto the island [25]. Successful reproduction of repatriated *C. hoodensis* was first documented in 1990 [24] suggesting the repatriated population might become self-sustaining.

Concerns remain, however, about current and future habitat quality. Specifically, the island's dense and extensive woody plant cover is hypothesized to inhibit tortoise movement (thereby excluding tortoises from colonizing new areas) and to outcompete grasses and juvenile cactus that are critical to tortoises as forage [16]. The present-day dominance of woody plants on Española Island may be a legacy of both the absence of giant tortoises and the presence of large numbers of feral goats on the island for ¾ of a century (1905 to 1978), which might have shifted competitive relationships between woody and herbaceous plants by altering soil water availability in ways that favor woody plants [26–28]. Furthermore, sudden eradication of introduced livestock (in this case, goats) in the absence of native herbivores (in this case, giant tortoises) may have exacerbated woody plant incursion [28]. A particular concern relates to tortoise interactions with a large-seeded, arboreal prickly pear cactus (*Opuntia megasperma*). The cactus is restricted to just three islands and is listed by the IUCN as a "vulnerable" species due to extensive population reduction caused by feral goats. The cactus serves as a critical source of food, water, and shade for tortoises [29]. Although tortoises have been observed to injure and crush young *O. megasperma* individuals, in aggregate tortoise impacts on cactus populations, including seed dispersal [20], appear to be marginally positive on Española Island [16].

We integrated complementary datasets on tortoise demography, tortoise-plant interactions and vegetation history on Española Island to comprehensively assess the reintroduced tortoise population's trajectory in the context of past, current and future ecological conditions. We first examined a 32-year-long tortoise mark-recapture dataset for evidence that the population is self-sustaining. We then assessed the relationship between tortoise density and habitat features (including cactus and woody plant density) in the context of historical changes in woody plant cover on the island. Last, we simulated how different management scenarios, including cessation of the repatriation program and propagation of cactus and control of woody plants, might affect the long-term stability of the tortoise population. We demonstrate that the first step of restoring a viable population of giant tortoises has been achieved; however, significant habitat restoration challenges remain before the role of tortoises as self-sustaining, island-wide ecosystem engineers is fully realized.

Materials and Methods

Study area and the tortoise monitoring program

Española Island ($-1°$ $38'$ N; $-89°$ $67'$ W) occupies the southeastern sector of the Galapagos Archipelago and is small (60.48 km^2, <1% of the archipelago), low elevation (max. 206 m), flat, and arid (annual precipitation varies between 10–600 mm with average annual temperatures of 23.8°C). Dominant herbaceous plants include *Galactia striata* and *G. tenuiflora, Phaseolus mollis, Rhynchosia minima* and *Paspalum* spp. Woody plants, primarily *Cordia lutea* and *Prosopis juliflora*, occur at high densities throughout the island (Fig. 2). The arboreal *Opuntia* cactus is almost entirely restricted to the island's central region [30], a 1250-hectare area that encompasses the three sites used for releasing repatriates from the *ex situ* breeding program: El Caco, Las Tunas, and Gardner Bay (Fig. 2a). Releases began with 17 captive-bred tortoises in 1975 and continued with 2- to 5-year-old tortoises introduced every year from 1975–1994 and sporadically thereafter from 1997–2007 for a total of 1482 tortoises repatriated [29].

Since the reintroduction program was initiated all repatriated tortoises as well as individuals encountered in the field during surveys have been given unique identification markings. Up to 1991, a unique combination of notches (made with machete or metal file) was carved onto the tortoises' marginal scutes. Thereafter a duplicate marking system was introduced consisting of numbers branded on the rear costal scutes combined with insertion of passive integrated transponder (PIT) tags (AVID Identification Systems Inc., Norco, California, USA) into the left rear leg of each tortoise. Locations of individual tortoises have been registered since 2002 with hand-held Global Positioning System (GPS) devices, thereby enabling tracking movements of marked individuals between survey bouts during later phases of the tortoise monitoring program. Most (96%) tortoises captured were weighed with a calibrated spring scale. Curved carapace length, measured as the distance from the notch at the anterior of the carapace over the tops of the vertebral scutes to the tip of the supracaudal scute, was used as the measure of tortoise size. Sex and age were assessed on the basis of external morphology. Native-hatched tortoises (offspring of repatriates), first encountered in significant numbers after 2000, were identified on the basis of lack of notches or PIT tags and the distinct appearance of their scutes: more compressed growth segments, lighter coloration, and more scute abrasion given size.

Figure 1. Adult male Galapagos giant tortoise (*Chelonoidis hoodensis*) resting beneath an adult arboreal prickly pear cactus (*Opuntia megasperma*) surrounding by woody plants (mostly muyuyo or *Cordia lutea*), Española Island, May 2010 (image: J. P. Gibbs).

Fieldwork was carried out on an endangered species and on public land in strict accordance with permit conditions of the Galapagos National Park Directorate and following the protocol approved by State University of New York, College of Environmental Science and Forestry's Institutional Animal Care & Use Committee (Protocol Number: 121201).

Tortoise-plant interactions

Because of the hypothesized importance of woody plants as a potential factor structuring tortoise-plant interactions and thereby affecting tortoise population recovery, we quantified the extent of woody vegetation on Española Island over both space and time. Spatial extent was determined by delineating woody vegetation from cloud-free Quickbird satellite imagery (December 29, 2006, 0.6 meters resolution) using a supervised classification calibrated with woody plant density estimates from field plots (see below, IDRISI Andes 5.0). We aggregated these data to obtain a map of percent woody cover by summing the number of 0.6×0.6 m cells classified as woody in each 24 m cell. Temporal trend in woody plant cover, emphasizing possible changes associated with goat impacts and tortoise near-absence over the last 100 years, was inferred from soil carbon isotope analysis [31]. To do so, we first determined diagnostic $\delta^{13}C$ signatures of the island's dominant woody plants (*Bursera*, *Cordia*, *Croton*, *Parkinsonia*, *Prosopis*) and herbaceous plants (*Cyperus* and *Paspalum*) as well as cactus (*Opuntia*) ($n = 10$ of each species). We then sampled soil at 10 cm increments through the soil horizon from soil pits dug at Gardner Bay and in the Caco area. All analyses were made at the Environmental Stable Isotope Laboratory at SUNY-ESF in Syracuse, New York. Accuracy and precision of stable isotope measurements were verified using National Institute of Standards and Technology (NIST) reference materials (M. A. Teece, pers. comm.). Radiocarbon dating of two soil samples at two points in the soil profile of the Caco pit was performed by Beta Analytic Inc., Miami, Florida, to enable calibration of soil depth and soil sample age. We tested trends in the dominance of woody plants in the soil record using linear models of $\delta^{13}C$ by depth for two time periods (recent history versus earlier, see Results) using a general linear mixed model with soil pit treated as a random effect.

To assess specific hypotheses about tortoise-plant interactions we performed two studies. The first was re-censusing cactus in June 2014 in a 60 ha macroplot in the Tunas tortoise release area first censused in May 2004 [16] to examine cactus population trends in an area where tortoises have long been established at high density. The second involved measuring tortoise and cactus abundance as well as woody vegetation extent on 96 field plots within the central "tortoise zone" in May-June 2010. The sampling frame was established by buffering by 250 m all tortoise occurrences recorded during previous surveys. We then randomly placed one plot within every 250×250 m grid cell. Each plot comprised five 20-m-radius circular sub-plots with a central sub-plot and four sub-plots directly adjacent at the cardinal directions. Within each sub-plot we recorded total numbers of tortoises, tortoise droppings, juvenile cactus (no bark on trunk), subadult cactus (some bark on trunk), and adult cactus (trunk entirely bark-covered). Density of woody plants was estimated using the point-quarter method [32] for two basal diameter classes: ≥ 5 cm (mature plants) and ≤ 0.5 cm (recruits). All measurements were subsequently averaged at the plot-level for analysis of interactions among tortoises, cactus, and woody vegetation via binary classification trees [33]. We also modeled tortoise density as a function of cactus and woody plant density, as well as distance from release site, using a linear negative binomial regression model. The model was run in WinBUGS, using 30,000 MCMC iterations (treating the first 10% as a burn-in and capturing every 10th observation to reduce serial autocorrelation).

After 30 years of repatriation, tortoises remain restricted to the central part of the island (Fig. 2a); therefore, we evaluated whether tortoises have the intrinsic capacity to move further afield if suitable habitat were available. To do so, observed annual displacement rates (km per year) of individuals from 2000 to

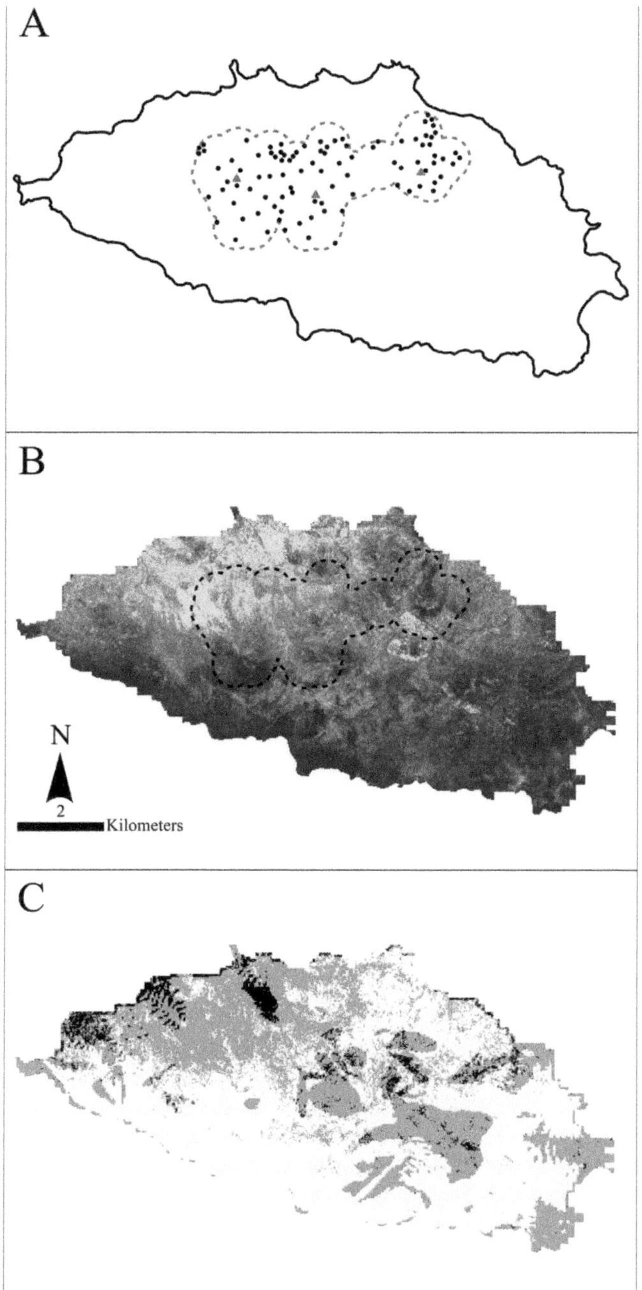

Figure 2. Study areas on Española Island. (A): gray dashed line is "tortoise zone" or maximum extent of tortoises observed on the island since the repatriation program was initiated; black circles are 2010 survey plot locations for tortoises, cactus, and woody plants; gray triangles are tortoise introduction sites (West: Tunas, central: Cacos, East: Gardner). Woody plant cover on Española Island (B; darker shades indicate higher percent woody plant cover), classified using cloud-free Quickbird satellite imagery (December 29, 2006, 0.6 meters resolution) into areas with and without woody vegetation using a supervised classification calibrated with plot woody plant density estimates (IDRISI Andes 5.0) (percent cover is the percent vegetated cells per 24 m×24 m area). Potentially highly suitable cactus restoration areas (C) with low slopes (<2.06 degrees, bottom quartile of available slopes) and low woody plant cover (<52% cover woody plants, bottom quartile of available woody plant cover); black areas have both low slopes and low woody cover, gray areas have either low slope or low woody cover.

2003 were used as a proxy for intrinsic capacity for range expansion. We tested for differences among age/sex classes (juvenile and adult, female and male) in mean displacement rates with a linear mixed effects model and then examined whether the observed expansion in range area (computed as a minimum convex polygon, or MCP, over all known occurrence locations) differed from a null expectation associated with random movement. We generated the null distribution for range expansion by simulating 500 replicate movement scenarios in which observed starting locations were randomly permuted (thereby de-linking starting locations from observed movement distances and directions). Finally, we tested the hypothesis that tortoises near the edge of the tortoise zone would be more likely to move away from the current range than tortoises situated far from the edge by regressing the direction of tortoise movement relative to the nearest edge with the distance of the starting location from the nearest edge. All analyses were run in the R statistical computing language (raster package and adehabitatHR package) [35–36].

Last, to estimate carrying capacity (K) for tortoises of the central zone they now occupy (Fig. 2a), we used the regression model of tortoise density (described previously) to compute the expected tortoise density at release sites, extrapolating the resulting estimate to the rest of the tortoise zone (assuming tortoises had reached carrying capacity at distance = 0 from release sites and that mean cactus and woody plant densities computed from plot-level data pertained to the entire tortoise zone). In addition, we used the same model to evaluate island-wide carrying capacity under three potential restoration scenarios: (1) a cactus regeneration scenario in which the cactus density of the tortoise zone is restored to the entire island, (2) a woody plant removal scenario in which woody plant density is reduced by half across the entire island, and (3) a combined cactus regeneration/woody plant removal scenario. The potential effect of these restoration scenarios was expressed as the percent increase in the tortoise carrying capacity over the current estimated carrying capacity.

Tortoise demography

We assessed the demographic outcome of the tortoise repatriation program in two ways. The first was direct estimation of tortoise population size in 2010 (after 34 years of repatriations) from mean tortoise densities measured on the 96 field plots in the tortoise zone (see above). The second involved estimation of abundance, vital rates, growth rates and body condition based on a capture history database assembled from surveys between 1975 and 2007 (recapture data were available every year except 1995, 1996 and 1998), during which an average of 120 tortoises was recaptured per year (range: 16 to 377). Trends in somatic growth (cm/year) and body condition (deviations from expected mass-length relationships) were assessed using generalized additive mixed models (GAMMs, mgcv [34]) that included the categorical covariates sex/age (juvenile, adult, male, female or unknown) and individual (as a random effect in order to account for repeated measures [37]) as well as year (1975–2007) and body size (curved carapace length). Growth was determined from change in curved carapace length between subsequent measurements of tortoises (2.5% percent of observations were removed from both extremes of the growth rate estimates distribution as well as recaptures >3 years apart to guard against anomalies caused by errors in field recording or data entry).

We also extracted annual abundance and survival rates from the capture history database using a Bayesian open-population model (Cormack-Jolly-Seber analogue) in conjunction with the known numbers of tortoises released each year. We estimated survival (φ) separately for juveniles (<8 y.o.) and subadults/adults (≥8 y.o.).

To accommodate temporal dynamics in survival rate (e.g., environmental stochasticity, trends in survival, density dependence), survival rates were estimated separately for each 4-year time period (subadults/adults), 2-year time periods (juveniles), and 1-year time period (year-of-release). Mean capture probability was estimated separately for juveniles/subadults (<15 y.o.) and reproductive adults (≥15 y.o.) with inter-annual variance in capture probability modeled as a logit-normal random effect. Because our modeling framework explicitly tracked the status of each released tortoise (i.e., alive or dead) each year, we were able to estimate annual abundance of released tortoises (excluding native-born individuals) at two distinct release areas (Tunas and Caco versus Gardner Bay) as the sum of living individuals occupying each of these release areas (Fig. 2a) (due to proximity and lack of barriers to movement of tortoises between the release sites of El Caco and Las Tunas these were treated as a single release site for analysis).

We estimated population parameters with Markov-Chain Monte Carlo (MCMC) in WinBUGS (version 1.4) via the R2WinBUGS package in R [34]. Uninformative uniform prior probability distributions were assigned to all parameters and the initial 5,000 MCMC samples were discarded as a burn-in. We performed a further 10,000 MCMC iterations and saved every 10th iteration to reduce serial autocorrelation among samples [38]. Convergence of the Markov chains to the stationary posterior distribution was tested with the Gelman–Rubin diagnostic [38]. We summarized posterior distributions for all parameters with the mean of all MCMC samples as a point estimate and the 2.5 and 97.5 percentiles of the MCMC samples as a 95% credible interval [38]. A detailed description of our demographic and population projection model is provided in Appendix S1 with accompanying R and WinBUGS code in Appendix S2.

We projected over the next 100 years the abundance of giant tortoises on Española Island on the basis of estimates of survival and temporal variation in survival derived from the capture-recapture analysis, carrying capacity estimates from tortoise-habitat relationships (see above), and fecundity estimates extrapolated from observations of native-born tortoises (see Appendix S3). We used a female-only model, because females are generally presumed to be the limiting sex for most turtles and tortoises [39]. Survival rates of tortoises aged 1–2 were not estimable (all tortoises were released at age 3 and above) and were assigned a range of values typical of long-lived turtle species [40]. Initial female abundance was approximated using the estimated total number of released individuals that were alive in 2007 (derived from capture-recapture analyses, and ignoring native-born individuals). We used a "ceiling" type density dependence model in which the maximum population growth rate (R_{max}) was defined by the vital rates estimated for this study (a reasonable assumption given the population was likely below carrying capacity for much of the study period). All parameter values used in our population models, and the information sources for these parameter values, are presented in Table 1.

Total fertility rate (total number of yearlings produced per reproductive adult) was computed as the product of four parameters: mean proportion of females in the population, mean number of eggs produced per female, mean hatching success rate, and mean neonate (hatchling) survival. We used outside sources of information to estimate the first two fertility components (Table 1) and an approximate-likelihood technique to estimate the remaining fertility parameters (see Appendix S1 for details): on the basis of the increase in the proportion of native-born individuals observed over time we concluded that the product of egg viability and hatchling survival on Española Island was between 0.023 (equivalent to both parameters set at 0.15) and 0.066 (equivalent to both parameters set at 0.26). Recruitment was modeled as a Poisson process [38]. Lacking information on the covariance structure of the four components of fertility, we conservatively assumed that all four parameters were perfectly correlated, which may underestimate population stability. All population projections consisted of 100 separate population simulations for 150 years each starting in the year 2011. For all simulations we recorded total abundance of tortoises and total abundance of adults (the ecological effects of tortoises are largely driven by adults; see below). Last, we estimated extirpation probability as the proportion of simulation runs reaching the quasi-extinction threshold (75 adults or fewer at any point during the simulation) after 150 years. All population projection models were coded in the R statistical computing language, and the code is provided in Appendix S3.

Results

Spatial and temporal pattern in woody plant extent on Española Island

Extent of woody plants on Española Island showed dramatic variation across the island in 2006 (Fig. 2b). Regression tree analysis indicated that the fraction of a given 24 m×24 m area on the island dominated by woody plants was highest (0.96) within 1.1 km of the coast; farther from the coast, woody cover was more extensive on lower slopes (<3.2 degrees, 0.86) than steeper ones and lower (0.58) in the tortoise zone than outside (0.68). In terms of temporal variation in woody cover, extent of woody plants apparently increased strongly and recently in terms of the history of the island's plant community (Fig. 3). Radiocarbon dating from the Caco soil pit samples at 40 and 60 cm yielded dates of 510+/−40 BP and 1100+/−40 BP, respectively; thus our 0.8–1.1 m soil pits captured >1,000 years of vegetation history on the island. Stable carbon isotope signatures ($\delta^{13}C$ values) were highly distinct between herbaceous and woody plants sampled: all woody plants were consistently more negative (−26 to −30‰) and all herbaceous plants consistently less negative (−13 to −16‰; Fig. 3) with cactus (−15‰) grouping among the complement of herbaceous plants sampled. In recent history (soil depths of ≦40 cm or within 510 YBP), $\delta^{13}C$ values have become more negative with time ($t = 5.49$, $p < 0.001$) indicating a shift towards greater dominance by woody plants (Fig. 3); prior to 510 YBP, $\delta^{13}C$ values indicated no consistent trend ($t = 0.38$, $p = 0.71$).

Tortoise-habitat relationships

In 2010 tortoise density was highest (1.8 tortoises/ha) in locations where adult cactus density was highest (>4.0 cactus/ha) (Fig. 4a). Where adult cactus density was lower (<4.0 cactus/ha), tortoises occurred at much lower densities (0.66/ha) and only if large diameter woody plant density was also low (<10.7 stems/100 m^2). No tortoises occurred on plots with low adult cactus densities when combined with high densities of large diameter woody plants. Tortoise dropping density (reflecting tortoises in transit) was greatest (15.2/ha) where woody plant density was lowest (Fig. 4b) yet was still high (13.0/ha) in areas of high woody plant extent if adult cactus were present; otherwise there were dramatically fewer droppings (4.1/ha) suggesting that tortoises avoid woody plant areas particularly if no cactus is present.

Tortoise density in the tortoise zone declined with distance from reintroduction release points (Tunas, Caco and Gardner), increased with adult cactus density, and decreased with woody plant density. The credible interval of the effect of distance from introduction points did not overlap with zero (95% C.I.: −5.31 to −2.42), but it did for the cactus (95% C.I.: −0.14 to 2.33) and

Table 1. Summary of input parameters used to simulate the population dynamics of Española giant tortoises during and after repatriation to Española Island, Galapagos.

Parameter	Lower-Upper Bound	Source
Age at first reproduction	17–17	Age of oldest repatriated tortoises at year of first appearance of native-born individuals on the island [24]
Proportion of population female	0.3–0.7	Not known; wide interval centered on 0.5
Carrying capacity	424–3921	Estimated from tortoise density relationships to distance from introduction sites, and cactus and woody plant density (see Methods)
Fecundity	4–10	Mean annual eggs produced per female [68]
Annual survival, ages 1–3	0.6–0.9	Not known; estimated using rates typical for turtles [40] with wide interval to accommodate uncertainty
Hatchling survival (ages 0–1 years)	0.15–0.26	Estimated via simulation to obtain values of a fraction of native-born animals similar to that observed in the field
Mean hatching success rate	0.15–0.26	Estimated via simulation to obtain values of a fraction of native-born animals similar to observed in the field

Parameters estimated from the 32-year capture-recapture dataset include juvenile and adult survival, temporal variation in survival, and survival at year-of-release (see text).

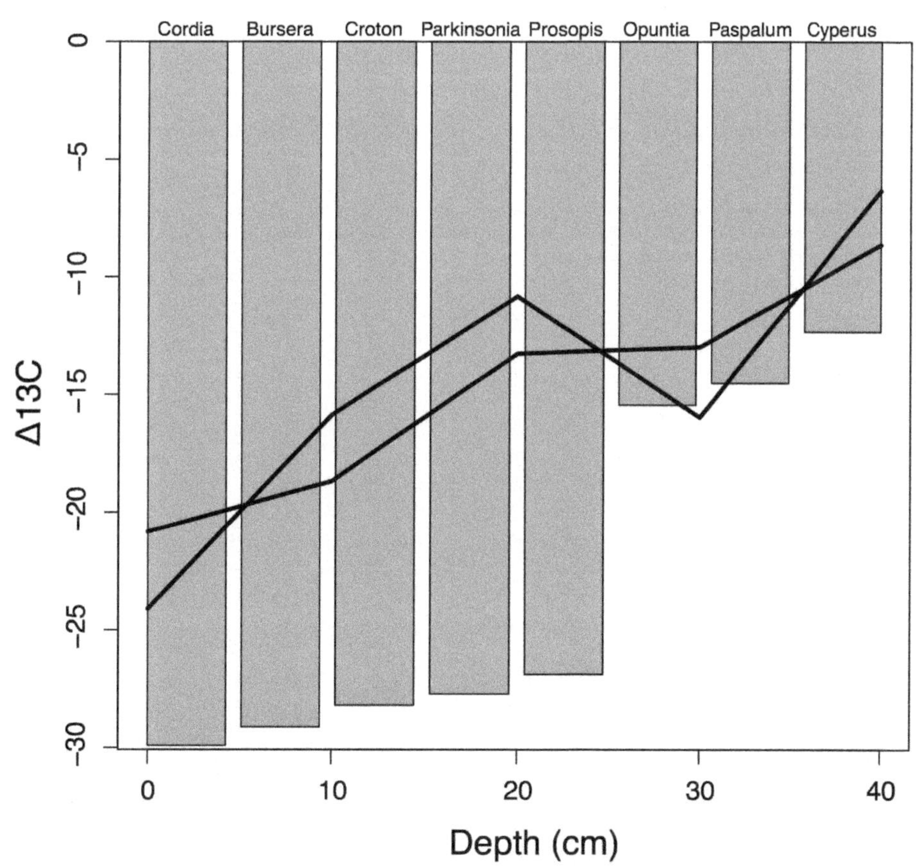

Figure 3. Carbon isotope signatures ($\delta^{13}C$ values) of soils at increasingly greater depths in two soil pits (Gardner and El Caco) on Española Island, Galapagos. Radiocarbon dating indicated that depths of 40 cm are approximately 500 years BP; therefore, samples below the 40 cm depth are grouped to show pre-disturbance averages. Points are $\delta^{13}C$ values of the island's dominant woody plants (*Bursera, Cordia, Croton, Parkinsonia, Prosopis*) and herbaceous plants (*Cyperus* and *Paspalum*) as well as cactus (*Opuntia*) (averaged across 10 samples of each, with standard errors for all means too miniscule to be evident in this graphic).

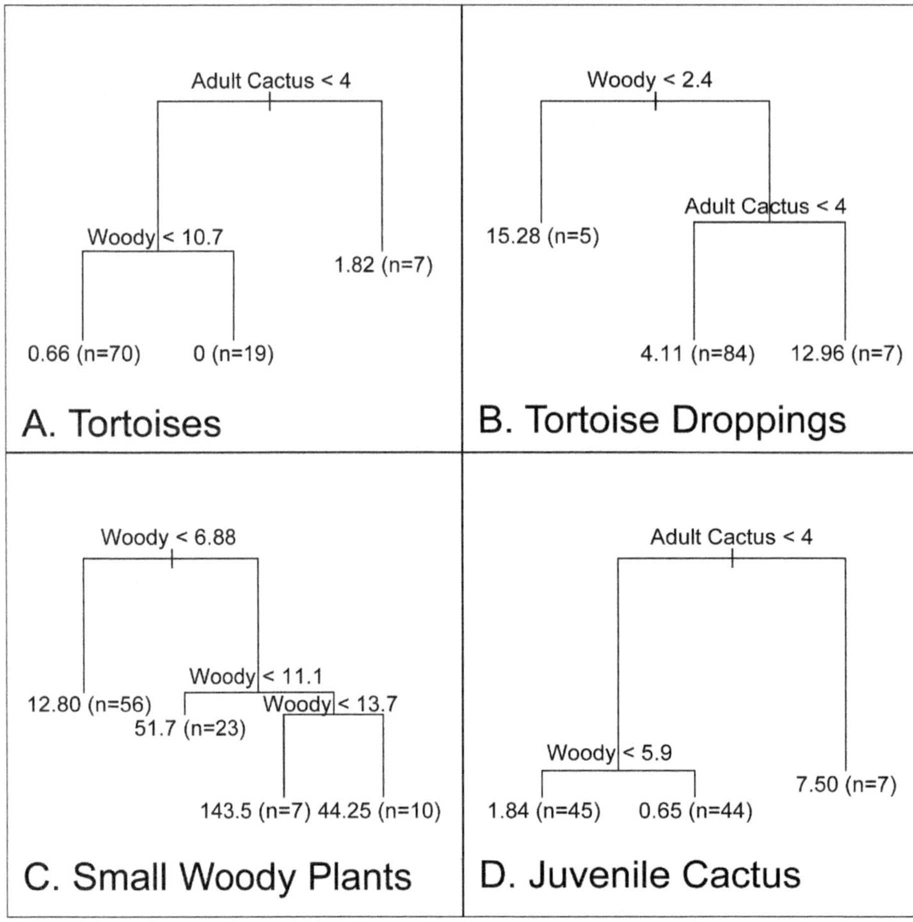

Figure 4. Relationships derived from binary classification trees between *density of tortoises* (A) and *tortoise droppings* (B) with thresholds of adult cactus density and large diameter woody plant density; between *small diameter woody plants* (C) with thresholds of large diameter woody plant density; and between *density of juvenile cactus* (D) with thresholds of large diameter woody plant density and adult cactus density. Density of tortoises, tortoise droppings, and cactus are expressed as number/hectare; woody plants as stems/100 m^2.

woody plant (95% C.I.: −2.45 to 2.04) effects. Using these relationships, we estimated a carrying capacity of 2081 tortoises (95% C.I.: 424 to 3921) within the tortoise zone. Under the three potential restoration scenarios, the change in the tortoise carrying capacity over the current carrying capacity was estimated to be a 7% increase under the cactus regeneration scenario, a 52% increase under the woody plant removal scenario, and a 62% increase under the cactus regeneration and woody plant removal scenario.

The area observed to be used by tortoises (MCP) expanded little during the period for which GPS locations were available (2000 and after; Fig. 5a). During this same period random movements by tortoises would have generated a substantial expansion of the currently occupied core area (Fig. 5b); in fact, the observed final MCP was entirely outside the statistical distribution of plausible final MCPs produced by random movement (Fig. 5c). This disparity suggests that tortoise movements were not random with respect to the edge of the observed core area, which was also confirmed by our final movement analysis: tortoises near to the edges of the core area were much more likely to move away from the current range edge than tortoises situated more centrally within the core area (Fig. 5d).

For cactus populations in the tortoise zone, we estimated densities of 1.7 juvenile cactus/ha, 0.4 subadult cactus/ha, and 1.0 adult cactus/ha for a total of 2142 juvenile cacti (95% C.I.: 1303.8–2980.3), 457.5 subadult cacti (201.7–713.2) and 1310.2 adult cacti (627.4–1992.9). Density of juvenile cactus was highest (7.5/ha) where adult cactus densities were >4 individuals/ha, next highest yet dramatically lower (1.8/ha) where adult cactus densities were <4 individuals/ha and density of large diameter woody plants was <5.9 stems/100 m^2, and virtually absent (0.65/ha) where adult cactus reached densities of <4 individuals/ha and density of large diameter woody plants was >5.9 stems/stems/100 m^2 (Fig. 4d). Density of small diameter woody plants (woody plant regeneration) was extremely high (143 stems/100 m^2) at intermediate densities of large diameter woody plants (11–13 stems/100 m^2) and not associated with tortoise density (Fig. 4c).

Over a 10-year period in the Tunas tortoise release zone macroplot [16], total individual cacti censused increased from 315 to 396 individuals (+26%): stage structure (adults versus juveniles and subadults) did not vary between census periods ($G_{adj} = 2.38$, df = 1, P = 0.12) despite juveniles and subadults increasing in abundance (194 to 266 individual juveniles and subadults, +37%) more rapidly than adult cacti (121 to 130 adults, or +7%).

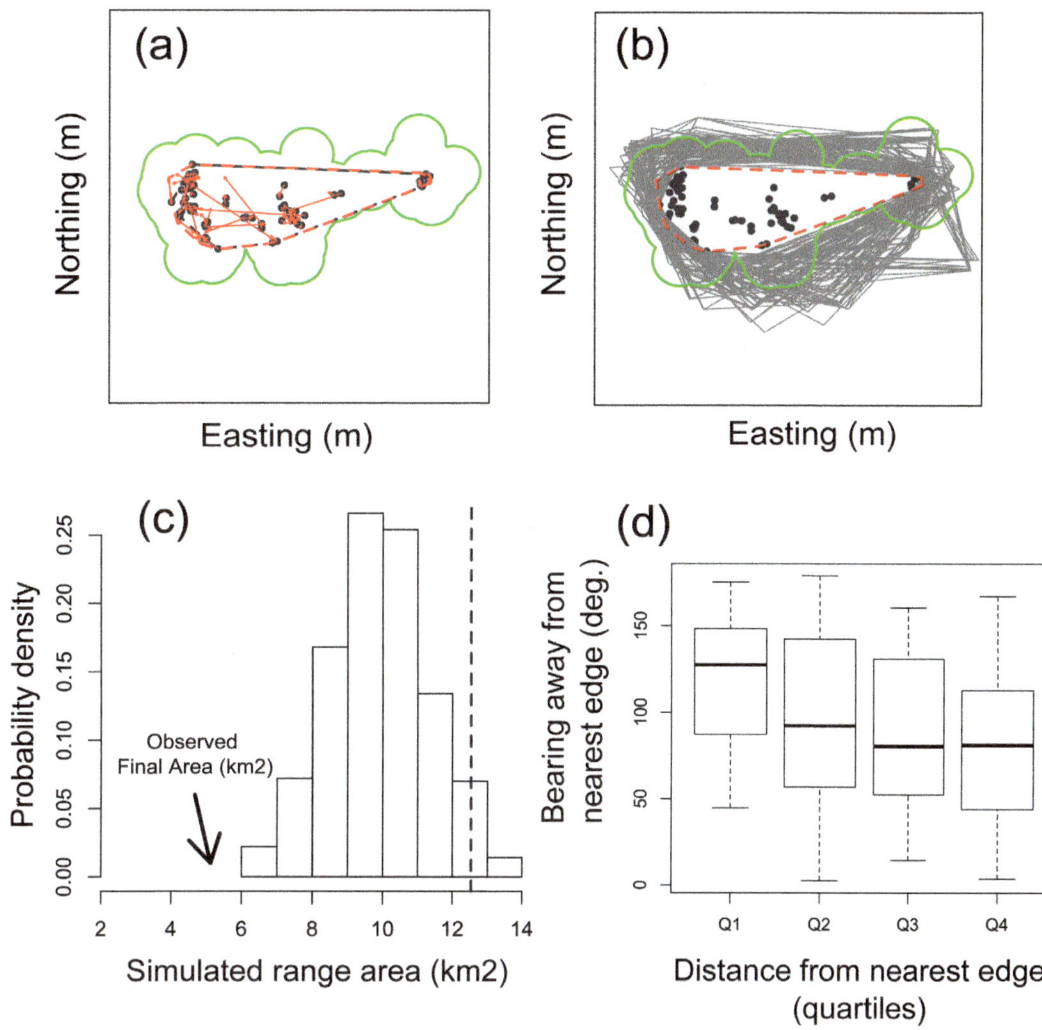

Figure 5. Movement patterns of giant tortoises reintroduced to Española Island, Galapagos. Closed circles in (A) and (B) represent the initial locations of all tortoises captured from 2000 to 2003 that were later recaptured. Arrows in (A) represent observed movement distances and directions. Minimum convex polygons (MCP) enclose the initial locations (grey) and initial plus final locations (red; differences represent the observed range expansion resulting from the observed movements). The grey polygons in (B) depict hypothetical MCPs (n = 500) for which initial locations have been decoupled from observed movement distances and directions. The observed final range area was smaller than all resulting hypothetical MCPs (C). In fact, those tortoises nearest to the range margins exhibited a strong tendency to move away from the nearest edge (D).

Demographic reconstruction of the reintroduction program and future projections

Our recapture history comprised 1459 individuals from among cohorts initially released in El Caco (*n* = 780), Las Tunas (*n* = 370), and Gardner Bay (*n* = 309). Individuals were captured on average 2.2 times each (range: 1 to 21). Annual somatic growth rates (cm per year in curved carapace length) averaged 2.99 cm±2.19 cm (SD), with 35.6% of deviance explained by sex and age, carapace length, and year (Table 2): juveniles grew an average of 4.46 cm/year (±1.18 SD, n = 175), unknowns (intermediate between juveniles and adults) 3.23 cm/year (±2.19 SD, n = 361), males 2.25 cm/year (±2.27, n = 278) and females 2.31 cm/year (±2.01, n = 212). Our suite of independent variables accounted for virtually all variation (98%) in tortoise body mass: after controlling for the inherent size-mass relationship, males and unknowns were relatively lighter than females, and tortoise condition declined with time (Table 2).

Annual probability of capture for juvenile giant tortoises averaged 0.38 (CI 0.17–0.58), and was slightly lower for adult giant tortoises (Fig. S1–1 in Appendix S1). Mean annual survival was high for both juvenile (mean = 0.958, CI 0.86–0.99) and adult giant tortoises (mean = 0.977, CI 0.93–0.99), with mean juvenile survival approximately 1% lower than adult survival (Fig. S1-1 in Appendix S1). Adult survival rate varied temporally from 0.94 to 0.99, juvenile survival rate varied temporally from 0.88 to 0.99, and survival at year-of-release varied from 0.29 to 0.98, with generally lower survival rates after 1994 (Fig. 6a). Maximum population growth rate (R_{max}, or the discrete population growth rate excluding density dependent effects) was estimated at 1.01 on the basis of our vital rate (survival and fecundity) estimates (popbio package in R [34]). With only 6 years of data for which to estimate survival and abundance at the "Gardner" release area, credible intervals for all survival terms were much more diffuse (Fig. 6b) and most survival information was derived from the much larger and longer-term records from the Tunas and Caco release site. We estimate that more than half of tortoises released to the island since 1975 were still alive in 2007 (Fig. 6a). Notably, the independent estimate of tortoise population size in 2010 derived from plot-

Table 2. Summary of generalized additive mixed models fitted to giant tortoises (n = 909) on Española Island, Galapagos between 1975 and 2007 to explore relationships between (A) somatic growth rates (cm/year) and age/sex while controlling for time (year) and individual length (carapace length) (35.6% of deviance explained) and (B) body mass and age/sex while controlling for time (year) and individual length (carapace length) (97.8% of deviance explained).

(A) Somatic growth							
Parametric coefficients					**Nonlinear effects (nonparametric)**		
Parameter	**Estimate**	**Standard Error**	***t***	***P***	**df**	***F***	***P***
Constant	27.665	18.323	1.510	0.131			
Sex/Age (Female vs. Juvenile)	0.649	0.177	3.662	0.0003			
Sex/Age (Female vs. Male)	0.527	0.145	3.626	0.0003			
Sex/Age (Female vs. Unknown)	−0.209	0.172	−1.211	0.226			
Year	−0.0122	0.00918	−1.330	0.183			
Carapace length					5.466	49.550	0.001
(B) Mass-length							
Parameter	**Estimate**	**Standard Error**	***t***	***P***	**df**	***F***	***P***
Constant	173144.0	16114.663	10.745	0.001			
Sex/Age (Female vs. Juvenile)	29.967	162.393	0.184	0.853			
Sex/Age (Female vs. Male)	−447.683	144.858	−3.0912	0.00202			
Sex/Age (Female vs. Unknown)	−1190.300	175.237	−6.793	0.001			
Month							
Year	−78.221	8.073	−9.689	0.001			
Carapace length					8.992	8413	0.001

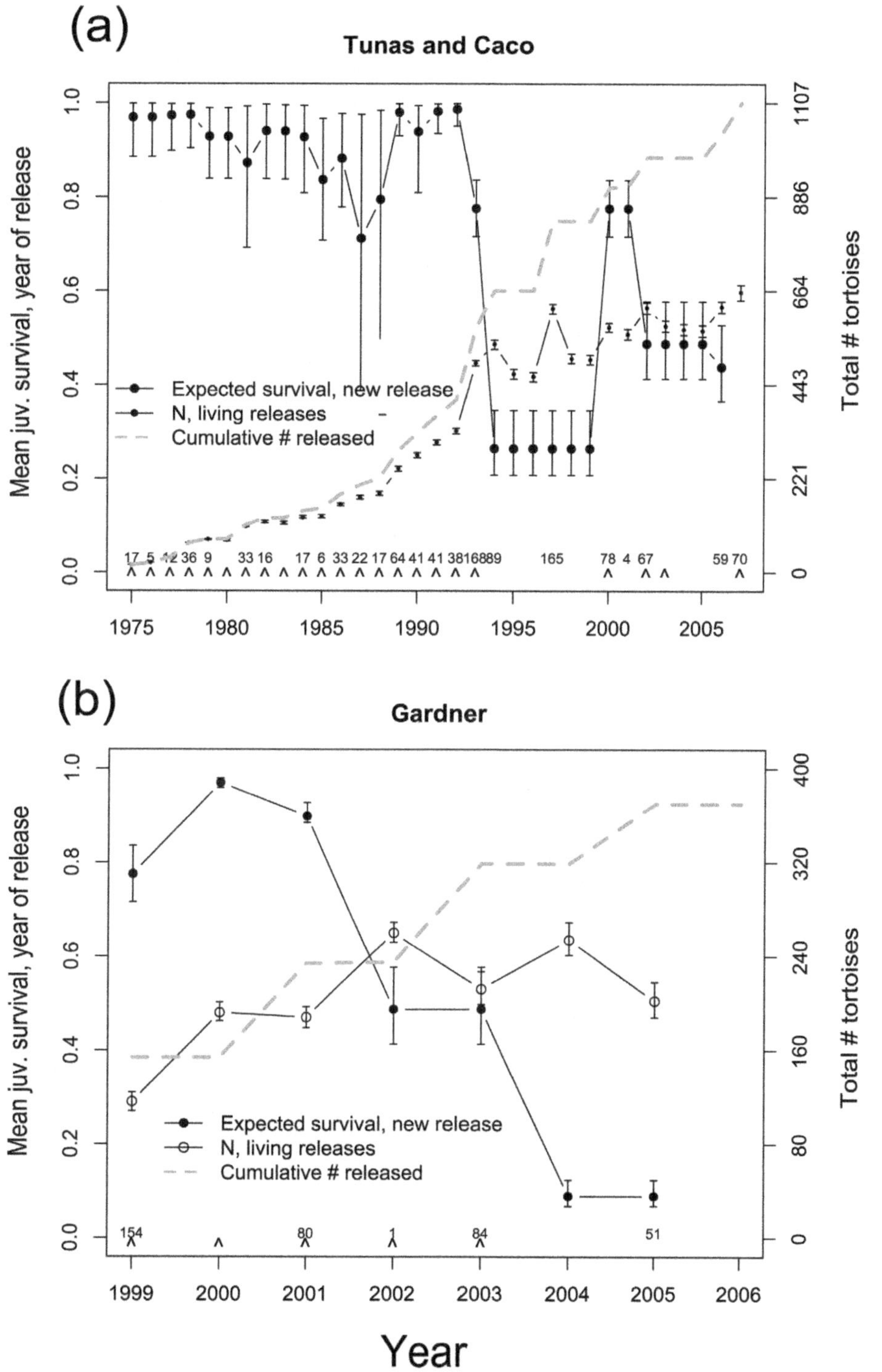

Figure 6. Estimates derived from a 32-year capture-recapture data set of abundance and vital rates of giant tortoises reintroduced to Española Island, Galapagos depicted by release sites (see Fig. 1): Tunas and Caco combined (A) and Gardner (B). Closed circles depict expected survival rates for newly released juveniles, the dashed line cumulative number of tortoises released, and open circles the estimated number of released individuals remaining. Release years are denoted with carets located above the x axis, and the numbers above the carets indicate the total number of tortoises released during each repatriation event.

based counts of tortoises for the Tunas and Caco area of 0.613 tortoises/ha (or total population of 769.5 tortoises, 439.7–1099.2, 95% confidence limits) corresponded well to the population size estimate derived from the mark-recapture analysis: 864 tortoises surviving in 2007 (Fig. 6a & b).

Population projections built from vital rates estimated from the mark-recapture data suggest that the benefits of continuing tortoise repatriation efforts to Española in terms of population viability would be very low. Continued repatriation for 25 more years, termination of repatriation, and termination of repatriation coupled with one-time removal of 50 adults (for possible translocation to other islands) yielded nearly equal and negligible extinction risk estimates over the 100-year time frame (Figs. 7a–c). In terms of restoring the role of giant tortoises as ecosystem engineers, however, the scenarios' outcomes contrasted in a substantive manner: carrying capacity was projected to be reached under the continued repatriation scenario by 2075 versus by 2100 for repatriation termination and by 2115 for removal of 50 adults in 2012 with termination of repatriation. This pattern was even more pronounced when adult (female) abundance was examined in isolation: adult abundance plateaus at ca. 1000 adult female tortoises (ca. 2000 total adults) by the year 2045 under the continued repatriation scenario (Fig. 7d), whereas adult abundance only reached half this level by the end of the simulation period under scenarios in which repatriation is terminated (Fig. 7e, f).

Discussion

The science of species reintroduction has typically focused on the initial steps of a reintroduction project: how to physically release the animals [41], how to improve genetic mixing of the small initial population [42], how suitable receiving habitat is for a release [43], and how to prevent excessive dispersal away from the intended release site [44–45]. Although these considerations are important for initial re-establishment of any population, how to generate wide-scale re-delivery of ecosystem-level processes via the introduced species is more complicated and rarely examined. In the case of the giant tortoise repatriation on Española Island in Galapagos, we determined that the re-introduced tortoise population was likely now self-sustaining over the long-term, but realization of ecological goals for the tortoise re-introduction program (notably, full occupation of the island by tortoises and the restoration of the island's once-dominant herbaceous and graminoid vegetation) will require considerably more time and perhaps further intervention in the form of habitat management.

More specifically, our analysis indicated that the repatriated Española tortoise population is secure from a strictly demographic perspective: extinction risk was very low (<1%) with or without continued repatriation, and population growth rates were weakly positive. We are confident in our population reconstruction in part because the estimate of tortoise population size derived from plot-based counts of tortoises yielded an estimate of population size that corresponded to the final abundance estimate derived in a

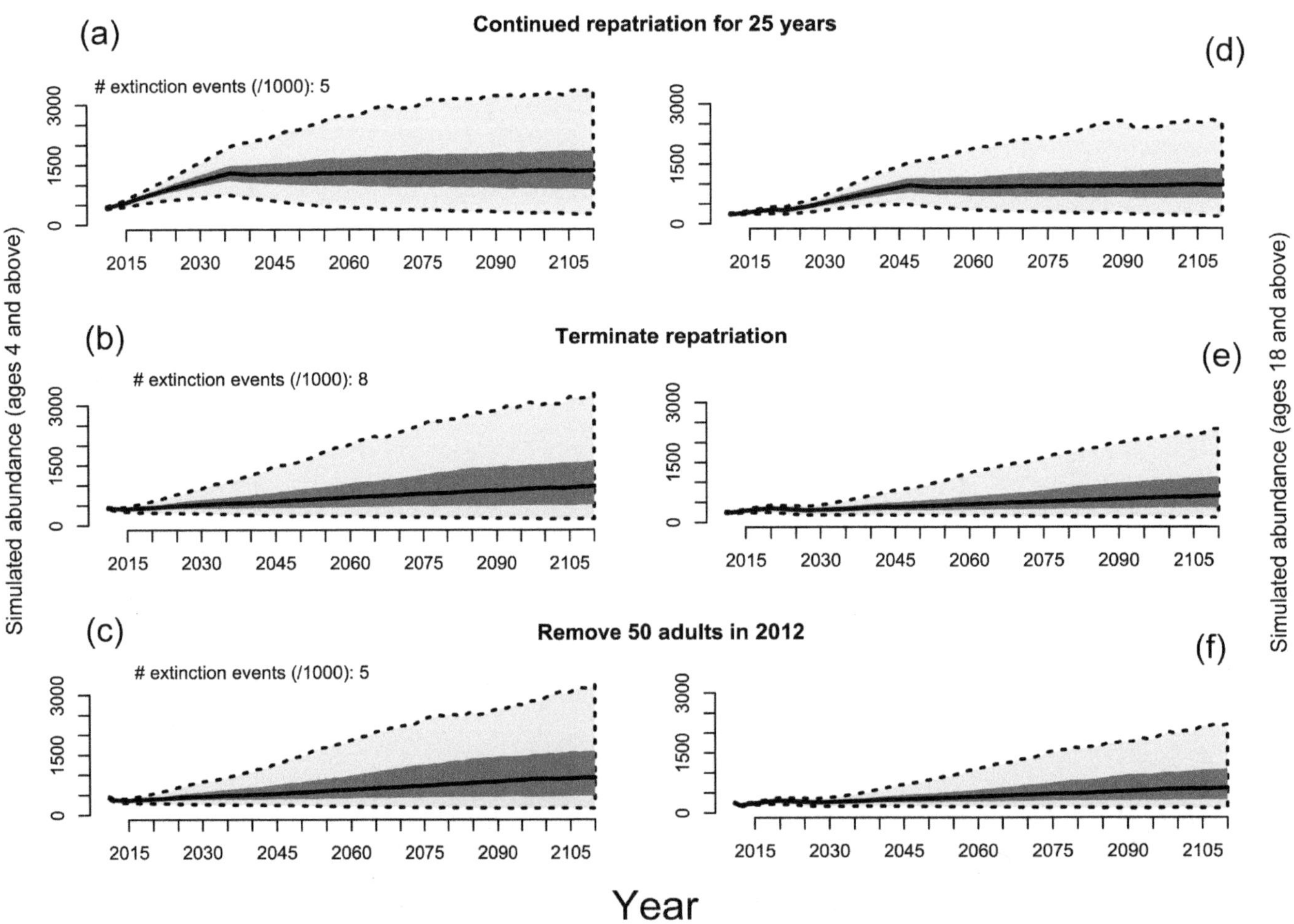

Figure 7. Summary of projected 100-year population dynamics for the repatriated giant tortoise population on Española Island, Galapagos, for (A, B, C) all female tortoises ≥4 years old, and (D, E, F) only adult female tortoises ≥18 years. Vital rates for the projection models were developed largely from a 32-year-capture-recapture database and carrying capacity (K) was determined using tortoise densities measured at field plots and modeled as a function of vegetation characteristics and distance from release sites. We modeled three management scenarios: (A, D) continued release of captive-raised tortoises for 25 years, (C, E) termination of tortoise repatriation, and (C, F) removal of 50 reproductive adult females at the beginning of the simulation (e.g., to accelerate repatriation efforts on other islands).

completely independent manner from analysis of capture-recapture data. This analysis was made possible by a rigorous monitoring program conducted in parallel with and since the inception of the reintroduction effort 32 years ago, and because ancillary sources of information were available (e.g., plot-based analyses of vegetation characteristics and tortoise densities) from which we could estimate carrying capacity.

Our demographic analyses for the Española population imply that approximately half of tortoises released on the island since 1975 were still alive in 2007, and that estimated fecundity rates were generally sufficient to offset mortality, resulting in modestly positive population growth even if the repatriation program were to be terminated (Fig. 6). Similar, high intrinsic population growth rates were reported for Aldabran giant tortoises, whose population rebounded from very low numbers in the early 20th century (when they were in the hundreds or low thousands [46]) to over 100,000 tortoises by the 1970s after cessation of severe harvesting pressures, which implies a discrete annual population growth of at least 1.07 [46–47]. The Aldabra example and our case study of Española tortoises together illustrate that giant tortoise conservation efforts starting from very few individuals have the potential for relatively rapid success if the deterministic factors that caused population decline originally are ameliorated.

The apparent decline in the survival rate of release cohorts we observed since ca. 1994 (Fig. 6a), along with decreasing growth rates and body condition, may be evidence, however, that resources on Española island are becoming limiting to continued population growth. In other words, the population may now be at or near carrying capacity of the area occupied. Two caveats may apply to this conclusion. First, declines in apparent survival could be due to emigration of young tortoises from the core area that is the focus on tortoise monitoring, perhaps motivated by density-dependent food competition among tortoises in a growing population. This said, recent surveys outside the tortoise zone [48] have revealed very little sign (droppings, bedding sites, trails) of tortoises elsewhere on the island. Second, the reduction we observed in growth rates and condition over time could be simple artifacts of ontogenetic changes in carapace length in an aging population given that this is a prominently saddle-backed species whose distinct morphology is manifested after sexual maturity [49]. However, the magnitude of relative change of carapace length (the measure of size used in this study) versus other measures of body size is indistinguishable in male and very modest in female saddle-backed tortoises between younger and older individuals [50]).

Population growth is generally not limited by environmental constraints during early phases of reintroduction efforts because densities are below carrying capacity [51–52], with population limiting factors not manifesting their effects until population densities increase. With high estimated adult, subadult and juvenile annual survival rates of ca. 98% throughout the reintroduction effort on Española Island, and in the absence of opportunity for migration (as seems to be the case for this island population where suitable habitat is limited to a small sector of the island), low survival of hatchlings and very young individuals is likely to limit population growth as the population approaches carrying capacity. This was evidenced by the generally low estimated survival ($\leq$50%) of tortoises released since 1994, after which population growth rates tended to be reduced (Fig. 6a). While juvenile survival rates estimated for giant tortoises in this study are not atypical for turtles and tortoise populations experiencing stable growth [40], higher juvenile survival rates are likely required to boost population growth rates [53].

We estimated a low probability of extinction for the Española tortoise population under all of the future management scenarios we assessed. We have used an admittedly short time-frame (100 years) for considering extinction risk for such a long-lived animal (generation time is estimated at ca. 50 to 60 years on the basis of our vital rate estimates) yet modeling time frames beyond 100 years are of little relevance to managers and conservation planners [54]. Clearly, projecting the success of reintroduction efforts can be difficult for long-lived species such as tortoises, whose reproduction can be delayed for a decade or more and for which high adult annual survival often obscures critical population dynamics [53]. For example, few reintroduction projects for long-lived species like tortoises and turtles have monitored populations long enough to estimate adult survival (with some exceptions [41]), let alone to estimate carrying capacity and to project density-dependent declines in population growth rates, yet such estimates are critical for confirming reintroduction success and estimating the long-term ecological effects of reintroduction efforts.

Despite evident achievement of population viability through this repatriation program, significant constraints evidently remain on the tortoise population's ability to achieve high densities and extensive spatial distribution. More specifically, the tortoises' preferred habitat, areas with high adult cactus density and low woody plant density, is currently limited across the island, and consequently the tortoises' movements are restricted to a small portion of the island despite an intrinsic movement capacity to expand to a far larger area than the species currently occupies. Our plot-based analysis of tortoise-cactus-woody-plant interactions supports the general hypothesis that woody plants have negative impacts on tortoises. This likely occurs because woody plants block tortoise movement thereby increasing travel costs and restricting access to habitat. Woody plants also reduce cactus recruitment, likely through light and water competition and perhaps by mediating tortoise movement thereby limiting dispersal and germination of cactus seed [16].

The current predominance of woody plants appears to be anomalous in the context of the history of vegetation on the island given that temporal trends in stable carbon isotopes of soils indicated a pronounced shift toward woody plants in the recent history of Española Island's plant community. This inference relied on the clear separation we observed in $\delta^{13}C$ values for woody and herbaceous plants but also relies on a reasonable but impossible-to-validate assumption that the plant species sampled in the present also were those that dominated in the past. The cause of the recent shift towards woody plants cannot be determined from data available to us because in addition to the herbivory by goats and centuries-long effective absence of tortoises, climate change may have altered plant communities and the $\delta^{13}C$ values expressed in the soil profile. This said, other studies in the Galapagos have indicated a change in the historical plant community composition with the reduction in giant tortoise populations and invasion of exotic herbivores [55]. Moreover in other semi-arid ecosystems, including ones that share some of the woody plant genera that dominate Española Island, the loss of a native herbivore and the introduction of an invasive herbivore has resulted in woody plant encroachment to an extent that the ecosystem cannot evidently recover without intervention [28]. Given the current widespread dominance of woody plants on Española Island, it is possible that the system has already reached a new, stable and woody-plant-dominated state that may require extensive woody plant removal or other habitat management to transition to a more mixed woody plant-cactus-grassland state [56].

The status of cactus on the island also remains of particular concern, not just because the species itself is of conservation concern due to its highly restricted distribution and reduced abundance due to historical goat impacts, but also because tortoises are heavily dependent on it. There are currently only an estimated 1310 adult cacti in the zone occupied by tortoises on Española Island, or just 1–2 cactus per tortoise. Moreover, our density estimates for cactus on Española Island of 1.7 juvenile cactus/ha, 0.4 subadult cactus/ha, and 1.0 adult cactus/ha (3.1 total stems/ha) are anomalously low for Galapagos in comparison to, for example, those on other comparable arid islands (Pinzón and Santa Fe Islands), which have stem densities of about 300–440/ha, that is, two orders of magnitude higher [57]. Our surveys also revealed a stage structure among cacti that is dramatically skewed to adults, thereby indicating a constraint on recruitment of young cacti into the adult population. Cactus populations elsewhere in Galapagos show a typical juvenile-biased size distribution [58], with a clear dominance of younger age classes typical of non-declining plant populations.

The tortoise re-introduction does appear to have a net positive impact on the cactus population, itself of conservation concern due to destruction by goats now removed. In this study we observed that the cactus population is increasing in one of the primary tortoise release sites, Las Tunas, confirming the suggestion [16] that tortoises have a modest yet positive impact on cactus populations. However, the population growth rate of adult cacti observed (about 7% over the last decade) is considerably less than the per decade rate at which the tortoise population has grown over the last four decades (about 200 tortoises or 33% population growth per decade). Notably tortoise population growth is comparable to the growth rate of subadult and juvenile cacti observed herein, or 37%. Only adult cactus substantially produce resources needed by tortoises (fruits, pads, shade) and the disparity of population growth rates of tortoises versus adult cacti indicate that ratios of tortoises to adult cacti will increase with time; however, as long as adult cactus survival remains high we predict a slowly improving status for the cactus population resulting from interactions with reintroduced tortoises, which foster recruitment of young cacti that will eventually transition to adult status producing resources (pads, fruits and shade) of use to tortoises.

Given that tortoises are now restricted to a small portion of Española Island and that tortoise habitat has likely degraded, it comes as little surprise that carrying capacity estimated for tortoises on Española appears to be quite low: approximately 2100 tortoises for the 1250 ha "tortoise zone" or 1.7 tortoises/hectare. This is striking given that tortoise densities on other islands in the archipelago have been reported as high as 8 tortoises/hectare [59]. Therefore, continued reintroductions of juvenile tortoises to the island would seem ill-advised. Currently juveniles have low survival rates and are therefore recruited infrequently into the population and likely play only a small role in sustaining it.

Instead of further repatriations, a more useful approach to improve the long-term outlook for this species to expand its ecological role on the island would be to improve habitat in such a way that the tortoises, now with significant numbers of older repatriates evidently reproducing, can colonize more of the island. More specifically, woody plant density is evidently a limiting factor to population growth given that a 50% reduction in woody plant density was projected to correspond to a 52% increase in the tortoise carrying capacity. Achieving this scale of restoration via manual removal of woody plants (the only currently available option) on half the island would, however, be extraordinarily expensive and logistically difficult given the remoteness of the island. Prescribed burns are not currently used as a habitat management tool in Galapagos given the unclear role of natural fires as an historical agent of natural disturbance in the archipelago [60], thus mechanical or chemical means may be required to reduce woody plant extent to levels more consistent with the recent history of the plant community on the island (Fig. 3).

Given the depleted and evidently declining status of tree cactus on the island combined with its importance to tortoises, a special focus on cactus management may be warranted given the near absence of cactus on much of the uncolonized parts of the island. Large-scale reseeding of cactus, now feasible given the availability of new technologies (e.g., [61]), could increase cactus recruitment, growth and survival dramatically. Given light competition between cactus and woody plants, cactus out-planting would best coincide with woody plant control (see Fig. 2c for potential cactus restoration areas with low slopes and low woody plant cover). Because tortoises are suspected to not only disperse cactus seeds and thereby increase recruitment but also mechanically flatten small cactus and thereby reduce cactus recruitment [19], a phased approach involving clearing of woody plants in new zones of the island unoccupied by tortoises and distant from occupied areas coupled with cactus out-plantings would be most effective. Timing of tortoise colonization could be controlled by establishing corridors to cactus groves in areas cleared of woody plants once sufficient number of out-planted cactus had grown to "tortoise-proof" size (subadult and older stages), that is, a decade or more after planting [62].

The reintroduction of an ecosystem engineer can have indirect benefits for other species of conservation concern [6],[8]. We have demonstrated that in the case of the Española tortoise in addition to recruitment benefits conferred to the endangered Opuntia cactus that is a primary tortoise food source, woody plant reduction could also aid the Waved Albatross (*Phoebastria irrorata*), a critically endangered seabird that effectively only nests on Española Island and that is likely limited by access to open-ground take-off and landing sites [48],[63]. Tortoises may be able to reduce or at least control woody plant density near albatross sites, but the current tortoise zone only marginally overlaps with that of the Waved Albatross, which are restricted to the southern part of the island [64]. Española tortoises and Waved Albatross may have a long co-evolutionary relationship: the absence of actual nest-building by albatross and their habit of rolling their eggs may be a response to the threat of trampling by tortoises [65]. The potential cross-species benefits of this tortoise reintroduction supports the more general claim that re-establishment of ecosystem engineers should be prioritized in conservation and ecological restoration [66–67]. For these benefits to be realized in the case of Española Island, however, we conclude that reinstatement of Española giant tortoises as fully functioning ecosystem engineers and not simply a viable population will likely require large-scale habitat restoration efforts in concert with further population restoration.

Acknowledgments

Mark Teece kindly oversaw the stable carbon isotope analyses as part of a larger archipelago-wide analysis and he and Steven Blake provided many useful comments on the manuscript. The mark-recapture data set upon which many components of this study were based was generated over 32 years during many arduous site visits by many park guards of the Galapagos National Park Service, biologists from the Charles Darwin Research Station (now Charles Darwin Foundation) and cooperating scientists. The 2010 field surveys of tortoise-plant interactions as well as this synthesis of the accumulated information was made possible by support from the Galapagos Conservancy in collaboration with the Galapagos National Park Service, the Prometeo Program of Ecuador's National Secretariat for Higher Education, Science, Technology and Innovation, and U.S. National Science Foundation grant DEB 1258062.

Author Contributions

Conceived and designed the experiments: JG WT LC. Performed the experiments: JG WT LC. Analyzed the data: JG EH KS. Contributed reagents/materials/analysis tools: EH KS JG. Contributed to the writing of the manuscript: JG EH KS WT LC.

References

1. Griffith B, Scott JM, Carpenter JW, Reed C (1989) Translocation as a species conservation tool: status and strategy. Science 245: 477–480.
2. Seddon PJ, Armstrong DP, Maloney RF (2007) Developing the science of reintroduction biology. Conservation Biology 21: 303–312.
3. Dodd CK, Seigel RA (1991) Relocation, repatriation, and translocation of amphibians and reptiles: are they conservation strategies that work? Herpetologica 47: 336–350.
4. Armstrong DP, Seddon PJ (2008) Directions in reintroduction biology. Trends in Ecology and Evolution 23: 20–25.
5. Wilmers CC, Crabtree RL, Smith DW, Murphy KM, Getz WM (2003) Trophic facilitation by introduced top predators: grey wolf subsidies to scavengers in Yellowstone National Park. Journal of Animal Ecology 72: 909–916.
6. Beyer HL, Merrill EH, Varley N, Boyce MS (2007) Willow on Yellowstone's northern range: evidence for a trophic cascade? Ecological Applications 17: 1563–1571.
7. Johnson BE, Cushman JH (2007) Influence of a large herbivore reintroduction on plant invasions and community composition in a California grassland. Conservation Biology 21: 515–526.
8. Ripple WJ, Beschta RL (2012) Trophic cascades in Yellowstone: the first 15 years after wolf reintroduction. Biological Conservation 145: 205–213.
9. Hunter EA, Gibbs JP (2013) Densities of ecological replacement herbivores required to restore plant communities: a case study of giant tortoises on Pinta Island, Galapagos. Restoration Ecology: 10.1111/rec.12055.
10. Ripple WJ, Beschta RL (2007) Restoring Yellowstone's aspen with wolves. Biological Conservation 138: 514–519.
11. Davidson AD, Friggens MT, Shoemaker KT, Hayes CL, Erz J, et al. (2014) Population dynamics of reintroduced Gunnison's prairie dogs in the southern portion of their range. In press. Journal of Wildlife Management.
12. Byers JE, Cuddington K, Jones CG, Talley TS, Hastings A, et al. (2006). Using ecosystem engineers to restore ecological systems. Trends in Ecology and Evolution 21: 493–500.
13. Field KJ, Tracy CR, Medica PA, Marlow RW, Corn PS (2007) Return to the wild: translocation as a tool in conservation of the Desert Tortoise (*Gopherus agassizii*). Biological Conservation 136: 232–245.
14. Hamann O (1993) On vegetation recovery, goats and giant tortoises on Pinta Island, Galapagos, Ecuador. Biodiversity and Conservation 2: 138–151.
15. Truett J, Phillips M (2009) Beyond historic baselines: restoring Bolson tortoises to Pleistocene range. Ecological Restoration 27: 144–151.
16. Gibbs JP, Márquez C, Sterling EJ (2008) The role of endangered species reintroduction in ecosystem restoration: tortoise-cactus interactions on Española Island, Galapagos. Restoration Ecology 16: 88–93.
17. Hansen DM, Donlan CJ, Griffiths CJ, Campbell KJ (2010) Ecological history and latent conservation potential: large and giant tortoises as a model for taxon substitutions. Ecography 33: 272–284.
18. Griffiths CJ, Hansen DM, Jones CG, Zuël N, Harris S (2011) Resurrecting extinct interactions with extant substitutes. Current Biology 21: 762–765.
19. Gibbs JP, Sterling EJ, Zabala FJ (2010) Giant tortoises as ecological engineers: a long-term quasi-experiment in the Galapagos Islands. Biotropica 42: 208–214.
20. Blake S, Wikelski M, Cabrera F, Guezou A, Silva M, et al. (2012) Seed dispersal by Galapagos tortoises. Journal of Biogeography 82: 310–321.
21. Asner GP, Levick SR (2012) Landscape-scale effects of herbivores on treefall in African savannas. Ecology Letters 15: 1211–1217.
22. Cromsigt JPGM, Te Beest M (2014) Restoration of a megaherbivore–landscape-level impacts of white rhinoceros in Kruger National Park, South Africa. Journal of Ecology 102: 1365–2745.
23. Tortoise & Freshwater Turtle Specialist Group (1996) *Chelonoidis nigra* ssp. *hoodensis*. In: IUCN 2013. IUCN Red List of Threatened Species. Version 2013.2. <www.iucnredlist.org>. Downloaded on 28 January 2014.
24. Márquez C, Morillo G, Cayot LJ (1991) A 25-year management program pays off: repatriated tortoises on Española reproduce. Noticias de Galapagos 50: 17–18.
25. Cayot LJ (2008) The restoration of giant tortoise and land iguana populations in Galapagos. Galapagos Research 65: 39–43.
26. Walker BH, Ludwig D, Holling CS, Peterman RM (1981) Stability of semi-arid savanna grazing systems. Journal of Ecology 69: 473–498.
27. Archer S (1995) Tree-grass dynamics in a *Prosopis*-thornscrub savanna parkland: reconstructing the past and predicting the future. Ecoscience 2: 83–99.
28. Browning DM, Archer SR (2011) Protection from livestock fails to deter shrub proliferation in a desert landscape with a history of heavy grazing. Ecological Applications 21: 1629–1642.
29. Cayot LJ (1987) Ecology of giant tortoises (Geochelone elephantopus) in the Galapagos Islands. Syracuse University, Syracuse, New York.
30. Márquez C, Vargas H, Snell H, Mauchamp A, Gibbs J, et al. (2003) ¿Por qué tan pocas Opuntia en la Isla Española-Galapagos? Ecología Aplicada 2: 21–29.
31. Bai E, Boutton TW, Liu F, Wu XB, Archer SR (2012) Spatial patterns of soil $\delta^{13}C$ reveal grassland-to-woodland successional processes. Organic Geochemistry 42: 1512–1518.
32. Mueller-Dombois D, Ellenberg H (1974) Aims and methods of vegetation ecology. New York: Wiley.
33. De'ath G, Fabricius KE (2000) Classification and regression trees: a powerful yet simple technique for ecological data analysis. Ecology 81: 3178–3192.
34. R Development Core Team (2012) R: A language and environment for statistical computing. R Foundation for Statistical Computing, Vienna, Austria. ISBN 3-900051-07-0. Available: http://www.R-project.org/.
35. Hijmans RJ (2013) raster: Geographic data analysis and modeling. R package version 2.1–49. http://CRAN.R-project.org/package=raster.
36. Calenge C (2006) The package adehabitat for the R software: a tool for the analysis of space and habitat use by animals. Ecological Modelling 197: 516–519.
37. Crawley MJ (2007) The R Book. John Wiley & Sons, West Sussex, England.
38. Bolker BM (2008) Ecological models and data in R. Princeton, NJ: Princeton University Press.
39. Pearse D, Avise JC (2001) Turtle mating systems: behavior, sperm storage, and genetic paternity. Journal of Heredity 92: 206–211.
40. Pike DA, Pizzatto L, Pike BA, Shine R (2008) Estimating survival rates of uncatchable animals: the myth of high juvenile mortality in reptiles. Ecology 89: 607–611.
41. Tuberville TD, Norton TM, Todd BD, Spratt JS (2008) Long-term apparent survival of translocated gopher tortoises: A comparison of newly released and previously established animals. Biological Conservation 141: 2690–2697.
42. Milinkovitch MC, Monteyne D, Gibbs JP, Fritts TH, Tapia W, et al. (2004) Genetic analysis of a successful repatriation programme: giant Galapagos tortoises. Proceedings of the Royal Society of London 271: 341–345.
43. Schadt S, Revilla E, Wiegand T, Knauer F, Kaczensky P, et al. (2002) Assessing the suitability of central European landscapes for the reintroduction of Eurasian lynx. Journal of Applied Ecology 39: 189–203.
44. Reinert HK, Rupert R (1999) Impacts of translocation on behavior and survival of timber rattlesnakes, *Crotalus horridus*. Journal of Herpetology 33: 45–61.
45. Yott A, Rosatte R, Schaefer JA, Hamr J, Fryxell J (2010) Movement and spread of a founding population of reintroduced elk (*Cervus elaphus*) in Ontario, Canada. Restoration Ecology 19: 70–77.
46. Bourn D, Gibson C, Augeri D, Wilson CJ, Church J, et al. (1999) The rise and fall of the Aldabran giant tortoise population. Proceedings of the Royal Society B: Biological Sciences 266: 1091–1100.
47. Swingland I, Coe M (1979) The natural regulation of giant tortoise populations on Aldabra Atoll: Recruitment. Philosophical Transactions of the Royal Society of London Series B, Biological Sciences 286: 177–188.
48. Gibbs JP, Woltz HW (2010) A pilot survey of the central colony of the waved albatross *Phoebastria irrorata* on Española Island. Galapagos Research 67: 18–21.
49. Chiari Y, Claude J (2011) Study of the carapace shape and growth in two Galapagos tortoise lineages. Journal of Morphology 272: 379–386.
50. Chiari Y, Hyseni C, Fritts TH, Glaberman S, Marquez C, et al. (2009) Morphometrics parallel genetics in a newly discovered and endangered taxon of Galapagos tortoise. PLoS One 4: e6272.
51. Nicoll MAC, Jones CG, Norris K (2003) Declining survival rates in a reintroduced population of the Mauritius kestrel: evidence for non-linear density

dependence and environmental stochasticity. Journal of Animal Ecology 72: 917–926.

52. Armstrong DP, Davidson RS, Perrott JK, Roygard J, Buchanan L (2004) Density-dependent population growth in a reintroduced population of North Island saddlebacks. Journal of Animal Ecology 74: 160–170.
53. Congdon JD, Dunham AE, van Loben Sels RC (1993) Delayed sexual maturity and demographics of Blanding's turtles (*Emydoidea blandingii*): implications for conservation and management of long-lived organisms. Conservation Biology 7: 826–833.
54. Shoemaker KT, Breisch AR, Jaycox JW, Gibbs JP (2014) Disambiguating the minimum viable population concept: Response to Reed and McCoy. In Press. Conservation Biology.
55. Froyd CA, Coffey EED, van der Knaap WO, van Leeuwen JFN, Tye A, et al. (2014) The ecological consequences of megafaunal loss: giant tortoises and wetland biodiversity. Ecology Letters 17: 144–154.
56. Scheffer M, Carpenter S, Foley JA, Folke C, Walker B (2001) Catastrophic shifts in ecosystems. Nature 413: 591–596.
57. Racine CH, Downhower JF (1974) Vegetative and reproductive strategies of *Opuntia* (Cactaceae) in the Galapagos Islands. Biotropica 6: 175–186.
58. Hicks DJ, Mauchamp A (2000) Population structure and growth patterns of *Opuntia echios* var. *gigantea* along an elevational gradient in the Galapagos Islands. Biotropica 32: 235–243.
59. Márquez C, Wiedenfeld D, Snell H, Fritts T, MacFarland C, et al. (2004) Estado actual de las poblaciones de tortugas terrestres gigantes (*Geochelone* spp., Chelonia: Testudinae) en las Islas Galapagos. Ecologia Aplicada 3: 98–111.
60. Trueman M, Hobbs RJ, Van Niel K (2013). Interdisciplinary historical vegetation mapping for ecological restoration in Galapagos. Landscape Ecology 28: 519–532.
61. Liu M, Zhiyang L, Ren W (2014) Research on the effect of waterboxx technology on *Haloxylon ammodendron* afforestation in arid and semiarid areas. Advanced Science, Engineering and Medicine 6: 236–239.
62. Hicks DJ, Mauchamp A (1996) Evolution and conservation biology of the Galapagos opuntias (Cactaceae). Haseltonia 4: 89–102.
63. Douglas H (1998) Changes in the distribution and abundance of waved albatrosses at Isla Española, Galapagos Islands, Ecuador. Condor 100: 737–740.
64. Anderson DJ, Huyvaert KP, Apanius V, Townsend H, Gillikin CL, et al. (2002) Population size and trends of the waved albatross *Phoebastria irrorata*. Marine Ornithology 30: 63–69.
65. Anderson DJ (2008) Waved Albatross. In T. De Roy, M. Jones, and J. Fitter, editors. Albatross: their world, their ways. Richmond Hill, Ontario, Canada: Firefly Books Ltd.
66. Soulé ME, Estes JA, Berger J, Del Rio CM (2003) Ecological effectiveness: conservation goals for interactive species. Conservation Biology 17: 1238–1250.
67. Hobbs RJ, Cramer VA (2008) Restoration ecology: interventionist approaches for restoring and maintaining ecosystem function in the face of rapid environmental change. Annual Review of Environment and Resources 33: 39–61.
68. Márquez C, Cayot LJ, Rea S (1999) La Crianza de tortugas gigantes en cautiverio: Un manual operativo. Quito, Ecuador: A & B Editores.

19

Intraspecific Trait Variation Driven by Plasticity and Ontogeny in *Hypochaeris radicata*

Rachel M. Mitchell[1]*, Jonathan D. Bakker[1,2]

1 School of Environmental and Forest Sciences, University of Washington, Seattle, Washington, United States of America, 2 Smithsonian Environmental Research Center, Edgewater, Maryland, United States of America

Abstract

The importance of intraspecific variation in plant functional traits for structuring communities and driving ecosystem processes is increasingly recognized, but mechanisms governing this variation are less studied. Variation could be due to adaptation to local conditions, plasticity in observed traits, or ontogeny. We investigated 1) whether abiotic stress caused individuals, maternal lines, and populations to exhibit trait convergence, 2) whether trait variation was primarily due to ecotypic differences or trait plasticity, and 3) whether traits varied with ontogeny. We sampled three populations of *Hypochaeris radicata* that differed significantly in rosette diameter and specific leaf area (SLA). We grew nine maternal lines from each population (27 lines total) under three greenhouse conditions: ambient conditions (control), 50% drought, or 80% shade. Plant diameter and relative chlorophyll content were measured throughout the experiment, and leaf shape, root:shoot ratio, and SLA were measured after five weeks. We used hierarchical mixed-models and variance component analysis to quantify differences in treatment effects and the contributions of population of origin and maternal line to observed variation. Observed variation in plant traits was driven primarily by plasticity. Shade significantly influenced all measured traits. Plant diameter was the only trait that had a sizable proportion of trait variation (30%) explained by population of origin. There were significant ontogenetic differences for both plant diameter and relative chlorophyll content. When subjected to abiotic stress in the form of light or water limitation, *Hypochaeris radicata* exhibited significant trait variability. This variation was due primarily to trait plasticity, rather than to adaptation to local conditions, and also differed with ontogeny.

Editor: Eric Gordon Lamb, University of Saskatchewan, Canada

Funding: The authors have no support or funding to report.

Competing Interests: The authors have declared that no competing interests exist.

* Email: rmm57@duke.edu

Introduction

Recent and growing interest in plant functional traits has deepened our understanding of critical ecological mechanisms. For example, interspecific trait-based approaches have identified differing processes of community assembly across spatial scales [1] and in temperate and tropical ecosystems [2], shaped understanding of abiotic drivers such as altitude [3,4], climate [5,6], temperature [7,8], and resource gradients [9] on community assembly and diversity, and identified global patterns of plant-trait expression [10]. In addition, trait-based approaches have provided a mechanistic understanding of correlations between biodiversity and ecosystem function [11–13].

Although trait-based approaches have improved our understanding of ecological mechanisms, the role of intraspecific trait variation - variation from individual to individual within a species - is less well understood. Intraspecific variation in plant functional traits can be large, though it has often been ignored or its contribution underestimated [14–16]. Intraspecific variation can structure communities [17] and influence ecosystem function [18]. For example, intraspecific variation in plant functional traits has been shown to structure communities along environmental gradients [19], to drive diversity in forest ecosystems [20], and to enhance community resistance to plant invasion [21].

Changes in intraspecific variation in plant functional traits have been observed in response to a wide range of environmental conditions (e.g., flooding [19], fire [22], drought [23] and temperature [24]), but few studies have examined the mechanisms controlling these changes (but see [25–27]). Two primary processes are hypothesized to drive variation in trait expression: adaptation to local conditions, creating distinct genetic ecotypes [28], and phenotypic plasticity in response to prevailing abiotic conditions [29]. Those studies that have investigated sources of intraspecific variation have, at times, found contrasting results, depending on the species and functional trait being investigated [30–32]. For example, Firn et al. found significant plasticity in specific leaf area (SLA; leaf area per unit mass) for an exotic, invasive grass but not for co-occurring native grasses [33]. In *Quercus suber*, genetically distinct populations displayed significantly different mean values for SLA, suggesting local adaptation to climate regimes, but populations also demonstrated high phenotypic plasticity [34]. In contrast, Robson et al. found that differences in SLA in *Fagus sylvatica* were primarily driven by genetic provenance [35].

If observed intraspecific trait variation is largely driven by the genetics of a given population, abiotic or biotic filters can selectively eliminate individuals with traits that exceed the bounds of the filter. For example, species or populations with low variability in traits related to drought stress (e.g., leaf size) may be unable to cope with reduced water availability under shifting climate. However, if variation in critical plant traits is largely a plastic response, independent of genetic diversity, even rare species may respond to filtering by altering their trait expression and thus persist in the community. However, plasticity is not without cost, as very plastic species may express phenotypes unsuited to abiotic filters, or experience costs associated with the trait of plasticity [36,37].

In addition to being influenced by adaptation and plasticity, trait expression also changes due to ontogeny. Determining how traits vary ontogenetically is important because traits influence ecosystem function [11], and temporal differences in plant trait expression may facilitate species coexistence and biodiversity maintenance [38]. Smith *et al.* found that SLA decreased with age in *Populus tremuloides* [39], while Cornelissen et al. found poor correlation between greenhouse seedlings and field-measured adults for 90 woody plant species, indicating strong ontogenetic differences [40]. The effects of ontogeny can differ depending on the trait being measured. For example, wood density, but not leaf mass per unit area, was strongly affected by ontogeny in tropical *Nothofagus pumilio* [24]. Despite evidence of correlation between trait variability and ontogenetic state, this source of variability is rarely accounted for, and often explicitly avoided by use of standard collection protocols [41].

Examination of the mechanisms controlling intraspecific variation is confounded by the fact that variation is both adaptive and with cost [37,42,43]. One species, population within a species, or genetic lineage may be inherently more plastic than others for some traits. In a widely distributed species, some populations may exhibit greater plasticity in response to abiotic stress due to high environmental heterogeneity selecting for phenotypically plastic genotypes, or through genetic correlation to other traits that are under selection [44]. For example, Balaguer *et al.* [45] found that populations of *Quercus coccifera* growing in homogenous conditions displayed significantly less phenotypic plasticity in light-responsive traits than those growing in more heterogeneous conditions. Pratt & Mooney found a similar pattern for growth and flower production in *Artemisia californica* in response to precipitation variability [46].

We used *Hypochaeris radicata*, a globally widespread weed, as a model species for examining trait variation in terms of both trait means (average trait values) and trait dispersion (variability of trait values). *H. radicata* occurs in a number of ecosystems, from undisturbed grasslands to highly disturbed urban environments. Because it persists across such a wide variety of conditions, it likely exhibits considerable trait plasticity [47]. However, persistence under differing abiotic conditions may lead to distinct population-level trait differences through ecotypic adaptation or transgenerational plastic responses [14,28,38,48].

We collected and germinated seeds from three populations, exposed the seedlings to abiotic stress (shade, drought) in the greenhouse, and measured a range of traits on each individual. We used these data to experimentally test three hypotheses. First, we hypothesized that abiotic stress would significantly impact trait expression for all populations, leading to a convergence in trait values for plants experiencing abiotic stress. Second, we hypothesized that population of origin and maternal line of origin, and not random variation between individuals, would explain the majority of observed trait variation that was not explained by abiotic conditions. Third, we hypothesized that over the lifetime of a plant, individuals, maternal lines, and populations within a treatment would become more similar in terms of both trait means and variances in order to accommodate stressful abiotic conditions.

Materials and Methods

Ethics Statement

No special permits were required for this research. Permission to conduct research at Glacial Heritage Preserve was granted by Thurston County. Permission to conduct research at Smith Prairie was granted by the non-profit owners, Pacific Rim Institute for Environmental Stewardship. Permission to collect samples at Union Bay Natural Area was granted by the University of Washington. This research did not include any populations of endangered or threatened species. Data used in this manuscript are available through the TRY database and can be accessed by submitting a data request at http://www.try-db.org.

Study Species

Hypochaeris radicata is a short-lived perennial herb species common in intact and degraded grassland ecosystems around the globe, including Europe, North America, Australia, New Zealand and Japan. It forms basal rosettes with oblong-lanceolate leaves and produces several upright, branching flower stalks during a long flowering period – June through October in the Pacific Northwest of the USA. *H. radicata* is self-incompatible [49,50] and its seeds are wind dispersed up to a few hundred meters [51].

Experimental Design

Three populations (Glacial Heritage (GL), Smith Prairie (SP), and Union Bay Natural Area (UB)) spanning a 200 km range in Washington state, USA, were sampled for this experiment. Populations occurred in exotic-dominated grasslands, but experienced different abiotic conditions in terms of annual precipitation, soil type, and land-use history (Table S1). Specific leaf area and rosette diameter were measured on 20–25 plants (maternal lines) per population, and one mature seed-head was collected from each plant. Measured plants were at least 3 m apart to prevent sampling from closely related individuals. Previous analyses detected significant differences in trait variability for SLA and rosette diameter [16]; follow-up analyses using PermDISP and PERMANOVA (see "Statistical Analysis", below) indicated significant differences in trait dispersion and mean values for diameter (Fig. 1A, $P<0.0001$) and significantly different dispersion for SLA (Fig. 1B, $P=0.05$) but no significant differences in seed mass (Fig. 1C).

The single seed head collected per maternal line did not contain enough seeds to allow individual maternal lines to be replicated in each abiotic treatment. Instead, maternal lines were subjected to Principal Components Analysis (PCA), and the first principal component (PC1) was used to select maternal lines spanning the range of traits expressed in each population. Thus, for each population, the three highest, three most central, and three lowest maternal lines (hereafter referred to as "high", "central", and "low") along PC1, which explained 49% of the variation, were identified (Fig. 1D). One maternal line from each group was then randomly assigned to an abiotic treatment (Figure S1).

Seeds from the selected maternal lines (3 populations ×9 maternal lines/population) were cold-stratified for 10 days and germinated at 25°C. Twenty germinants per maternal line (540 total) were planted individually into 164 ml "conetainers" (Stuewe & Sons, Oregon, USA) filled with Sunshine Mix #3 potting soil (Sun Gro Horticulture, Massachusetts, USA). Germinants were allowed to acclimate for seven days before treatments were applied.

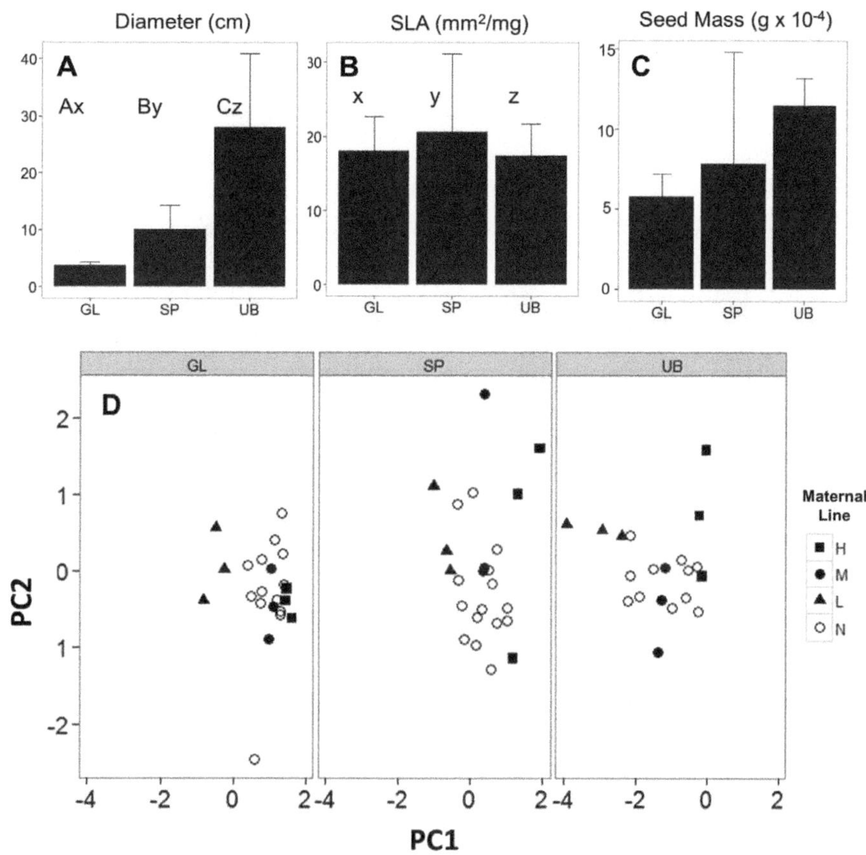

Figure 1. Mean (+ SD) of field-based measurements of three traits for the populations of origin (GL: Glacial Heritage; SP: Smith Prairie; UB: Union Bay). Significant PERMANOVA results ($\alpha = 0.05$) are indicated by different uppercase letters and significant PERMDISP results are indicated by different lowercase letters. D) Results of a Principal Components Analysis of the three traits measured on individuals from each population. Populations were analyzed together, but are shown in separate graphs for clarity. PC1 explained 49% of observed variation. Filled symbols indicate maternal "high" (H), "medium" (M) and "low" (L) maternal lines used in the greenhouse experiment, while open symbols (N) indicate maternal lines measured in the field but not included in the experiment.

Plants were grown in the greenhouse under one of three treatments: ambient conditions, 80% shade (80% of light is blocked), and drought conditions (50% of winter precipitation). The shade and drought treatment levels were selected to mimic marginal conditions where *H. radicata* is known to survive. Beginning on day seven, the high, low and central maternal lines were randomly assigned to the three treatments, and individuals were randomly arranged on the bench to account for edge effects. The ambient and shade treatments received weekly top-watering equivalent to average October-November rainfall across the three populations (53 ml per week per conetainer). The drought treatment received half as much (26 ml per week per conetainer) and visibly induced plant stress; individuals wilted between waterings. Each treatment received weekly 10-10-10 NPK fertilizer plus micronutrients per manufacturer specifications.

Mortality and rosette diameter (a measure of competitive ability similar to height [16]) were measured weekly, beginning on day seven. Relative chlorophyll content (correlated with photosynthesis and leaf life-span [52]) was measured weekly beginning on day 14 (plant leaves were too small on day seven for data collection) using a SPAD 502 plus chlorophyll meter (Spectrum Technologies, Illinois, USA). The length and width of the largest leaf on each individual were measured after five weeks to quantify whether leaves were more lanceolate or ovate, which can have consequences for a number of functions, including light interception [53]. Plants were destructively sampled after five weeks. One leaf per individual was harvested, scanned, dried at 60°C, and weighed using a Mettler AE 163 analytical scale (Mettler Toledo, Ohio, USA). One-sided leaf area was calculated from the scanned images using ImageJ version 1.45 (http://rsb.nih.gov/ij). SLA, which is correlated with resource acquisition and leaf life-span [54,55], was calculated as leaf area (mm^2) divided by leaf dry mass (mg). Above and belowground biomass were harvested separately, dried at 60°C, weighed, and used to calculate root:shoot ratio, a measure of resource allocation.

Statistical Analysis

Differences between trait means for the populations of origin measured in the field were quantified using PERMANOVA, a non-parametric permutational form of ANOVA [56]. Differences in trait dispersion were quantified using PermDISP, a permutational analog to the Levene's test [57]. These analyses were conducted using the PERMANOVA+ add-on to PRIMER-E [58], with Euclidean distances, type III sums of squares, and 9,999 permutations. Populations were considered to differ if $P<0.05$.

For plants grown under greenhouse conditions, measured traits were checked for normality, and corrective measures were taken when traits were non-normal. Correlation between dependent variables was checked and, when response variables were strongly correlated, only one representative variable was tested (the weakest

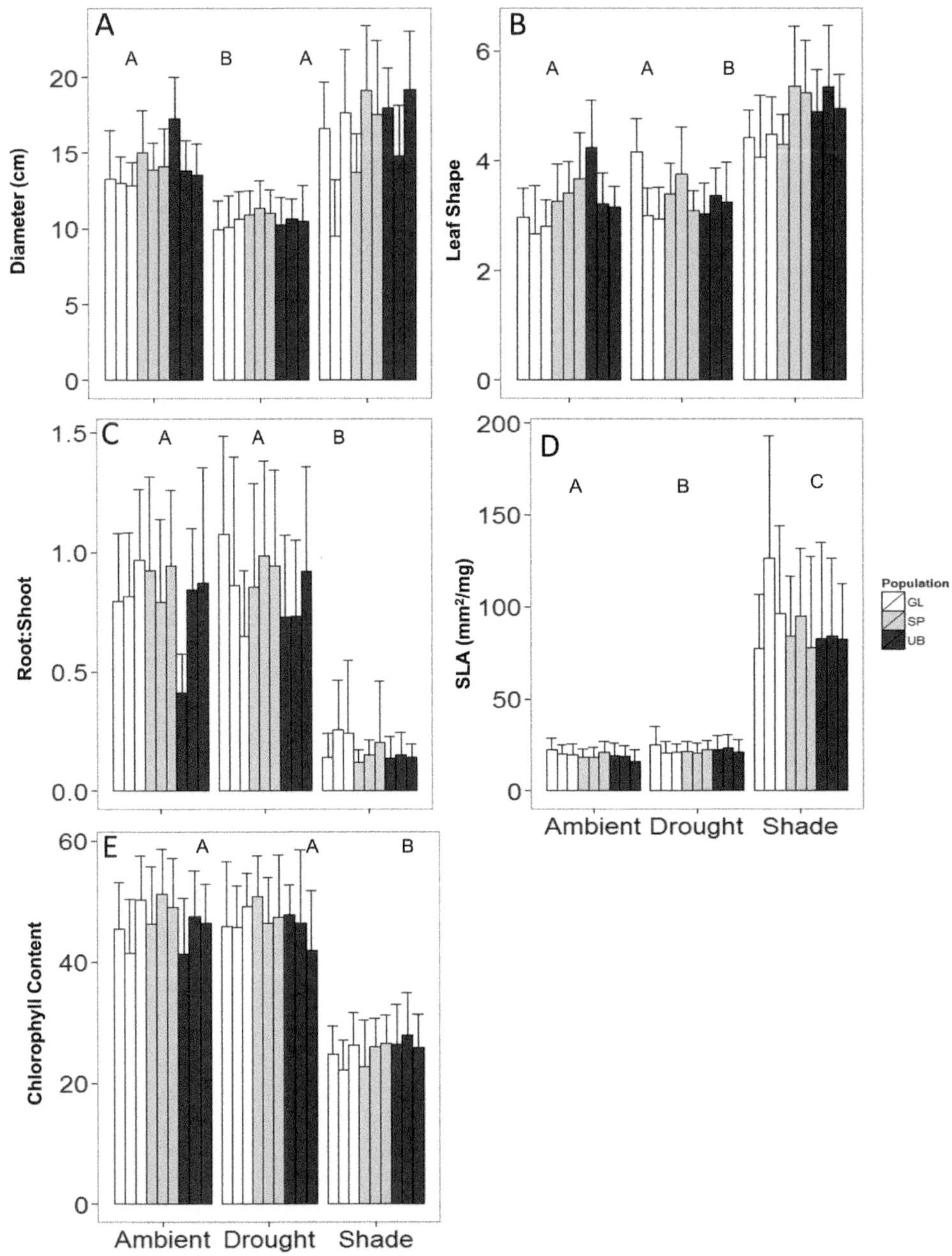

Figure 2. Mean (+ SD) values of A) final diameter, B) leaf shape (length:width ratio), C) root:shoot ratio, D) SLA, and E) final relative chlorophyll content for each maternal line (individual bar, arranged from low to high based on scores from PC1) in each population (unique color; defined in Table S1) for each treatment. Significant treatment differences ($\alpha = 0.05$) in trait means are indicated by different letters.

correlation among tested variables was $r^2 > 0.65$, see Fig S2). Five traits were selected for analysis: diameter, leaf shape (leaf length:width ratio; larger numbers indicate more lanceolate leaves and lower numbers indicate more ovate leaves), root:shoot ratio, SLA, and final relative chlorophyll content. SLA and diameter were log transformed to correct for non-normality and heteroscedacity.

We used hierarchical linear mixed effects models and variance component analysis to quantify the influence of abiotic treatment and ecotypic factors on observed traits. Population and maternal line were designated as random effects, with maternal line nested within populations. The variance attributable to population of origin, maternal line, and individual were expressed as proportions of the total random-effects variance. Abiotic treatment (ambient, shade, or drought) was treated as a fixed effect. Models were fit with the "lme4" package in R (version 2.14.2) and *P*-values were generated using the "languageR' package. Significant treatment effect was followed by pairwise comparisons using TukeyHSD through the "multcomp" package (version 1.2–19), with attention focused on the differences between the ambient treatment and each of the shade and drought treatments.

Ontogenetic changes were examined by analyzing changes over time in diameter and relative chlorophyll content. These traits were selected because they did not require repeated, destructive

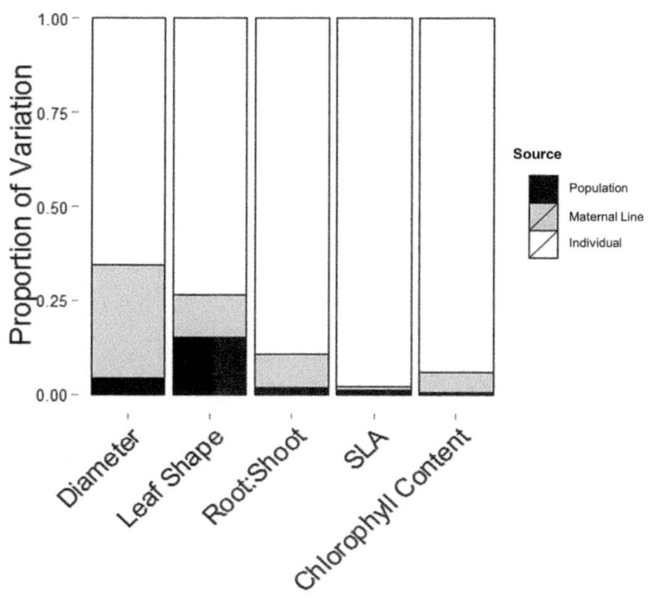

Figure 3. Proportion of total random effects variance explained by population of origin, maternal line, and individual plant (residual) for five measured traits.

measurements. Individual plants were nested within maternal lines within populations in these models, and time was included as a fixed effect. The fixed effects (abiotic treatment and time) and their interaction were iteratively added, beginning with the null model (including only random effects) and selecting the best model using Akaike's Information Criterion (AIC). Significant treatment by time effects were followed by separate analyses of each date to quantify the variance attributable to population of origin, maternal line, and individuals.

Results

Response to Abiotic Stress

Mortality was low, even for plants grown under abiotic stress: only 8 of 540 individuals died. However, the abiotic treatments affected plant trait expression (Fig. 2, See Table S2 for GLMM results). Mean final diameter was larger for plants in the shade treatment (16.4 cm) than the ambient treatment (14.1 cm), but smaller in the drought treatment (10.6 cm) (Fig. 2A). Plants in the shade treatment had significantly more lanceolate leaves (4.8) than those in the ambient (3.3) and drought treatments (3.3) (Fig. 2B), and also had significantly lower root:shoot ratios (0.17 vs 0.82 and 0.86, respectively) (Fig. 2C). Both the shade and drought treatments increased mean SLA (88.2 and 22.1 mm^2/mg, respectively) compared to those in the ambient treatment (19.4 mm^2/mg) (Fig. 3D). Relative chlorophyll content was significantly lower for plants in the shade treatment (25.6 SPAD units) compared with those in the drought (46.8 units) or ambient treatments (46.7 SPAD units) (Fig. 2E).

Contribution of Population and Maternal Line to Trait Variation

A surprisingly small amount of variation was explained by population of origin or maternal line for most measured traits (Fig. 3). Diameter was the only trait that showed an appreciable maternal contribution to observed variation; 30% of the total random effects variation was attributable to maternal lines, while 5% was attributable to populations (Fig. 3). Variation in leaf shape showed nearly equal contributions from population (15%) and maternal sources (12%). Visual inspection of the mean values for each maternal line showed no clear relationship with the maternal plant's score on PC1 (Fig. 1).

Ontogenetic Effects

There were significant interactions between treatment and time for both plant diameter and relative chlorophyll content (Fig. 4A and C). Treatment groups were indistinguishable on days 7 and 14. On day 21, plants in the shade treatment had significantly smaller diameter than those in the drought and ambient treatments ($P = 0.03$). On days 28 and 35, diameter in the drought treatment was significantly lower than the shade or ambient treatments ($P \leq 0.0001$; Fig 4A). Individuals accounted for an increasing amount of the variation in plant diameter, while variation between populations decreased over time (Fig. 4B). Variation between maternal lines remained fairly consistent.

Shading rapidly and strongly affected relative chlorophyll content, as indicated by significant treatment effects already on day 14, while drought did not ($P<0.0001$; Fig 4C). Variation between individuals increased between days 14 and 21, and remained high throughout the experiment (Fig. 4D). Variation between maternal lines and populations decreased over the course of the experiment.

Discussion

The roles of plasticity and adaptation in shaping trait expression

We set out to determine whether observed intraspecific variation was driven by plasticity or by adaptation to local abiotic conditions. Our results indicate that, for *H. radicata*, the observed trait variation is largely attributable to plasticity. Neither shading nor drought stress were strong enough to cause appreciable mortality, confirming that *H. radicata* can tolerate a very wide range of abiotic conditions. However, plants grown under these conditions exhibited markedly different functional traits. For example, SLA values previously reported for *H. radicata* individuals have ranged from 1.7–46.1 mm^2/mg [16,59,60]; in this study, SLA ranged from 6.6–239.1 mm^2/mg. Much of the increased range of SLA in this study was due to increases in response to shading (Fig. 2D), a response which has been demonstrated in the past [61]. The inclusion of shade-grown individuals, which have significantly higher SLA than sun-grown individuals, would dramatically shift the trait distribution for this species. Inclusion of such plants would also likely influence the overall trait distribution of the plant community, and may in turn impact estimations of critical ecosystem processes like litter production and nutrient cycling [13] through differing nutrient contributions and breakdown times in the decomposition pool [62]. Standard trait collection procedures [41,63] focus trait measurements exclusively on sun-exposed plants and therefore do not account for the full trait variation of species that can tolerate varying levels of shade. The omission of individuals in partial or full-shade from trait measurements may hinder our understanding of trait mediated ecosystem processes and responses. We suggest it would be more appropriate to sample the full range of conditions experienced by a species.

The variance explained by population of origin and maternal line was surprisingly small for every measured trait. Only leaf shape and diameter had significant variation attributable to differences across populations and maternal lines respectively, and even this amount of variation (30%) was dwarfed by treatment

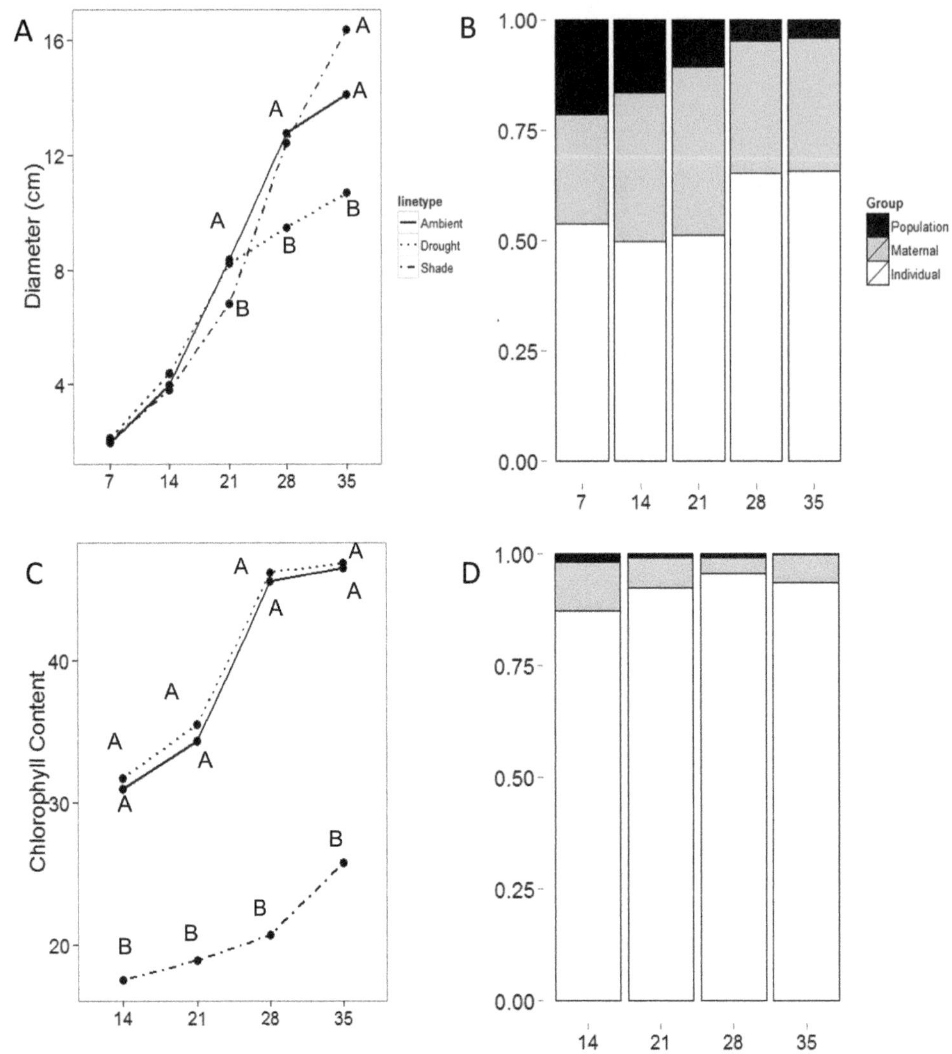

Figure 4. Change in average trait values in response to treatments (A and C; in each time period, letters indicate significantly different groups) and (B and D) the amount of variation attributed to population of origin, maternal line, and individual plant for diameter (top) and relative chlorophyll content (bottom).

effects. Virtually none of the observed variation could be attributed to population or maternal line for SLA (Fig. 3), confirming that this trait is highly plastic, a finding which supports previous analyses of this trait [61]. Furthermore, the position of each maternal line on PC1 had little correlation with the ranking of each maternal line in the greenhouse, suggesting that individuals within maternal lines exhibited significant variability and plasticity, and are not necessarily constrained by maternal genetics. The very small amount of variation attributable to population or maternal sources indicates that, for the traits measured in this experiment, plastic responses to prevailing conditions are the primary driver of observed variation.

The fact that all of the measured traits displayed significant plasticity has important implications for how traits are measured, and how they are incorporated into larger, trait-based models. Although some species are shade-intolerant, many species experience variable light conditions through interactions with neighboring plants in the community. Trait values based on fully sun-exposed leaves may inaccurately represent the trait contributions of such species, and thus may misrepresent community-level trait values, and the ecosystem functions dependent on those trait values. In addition, attempts to model species presence or response to abiotic change may be hindered by a reliance on trait means derived from sun-exposed individuals. If species, or populations within species, are variable and highly plastic for traits critical to survival, such taxa may be present in a wider range of habitat types, and be better able to cope with a wider range of conditions [64,65] than expected from trait values measured on individuals growing in a subset of the taxa's possible growing conditions.

The role of ontogeny in shaping trait expression

Variation in trait values over the lifespan of a plant is an additional, and often unaccounted for, source of intraspecific trait variability. This source of variation may be underappreciated as collection techniques typically focus on "mature" plants and tissues [41] yet maturity can be difficult to distinguish in the field. Because plants of differing ontogenies may appear "mature" in the field, intraspecific trait variation measured in field populations could be partially attributed to differing ontogeny between individuals, rather than to genetic or environmental conditions.

Furthermore, ontogenetic differences in trait variability may be important in the structure and function of communities. For example, ontogenetic trait differences may lead to temporal niche partitioning, facilitating biodiversity maintenance through limiting similarity [66]. Ontogenetic differences leading to higher palatability can also alter plant fitness by influencing interactions with herbivores and predators [67]. Furthermore, temporal variation in plant traits may exacerbate or mediate the impacts of abiotic filters. Young plants face similar pressures during establishment, and tend to exhibit similar trait values, in both this study, and others (e.g., [68]). This lack of trait variability in young plants may increase susceptibility to stress, while increased variability in older plants may buffer the population against abiotic change. Incorporating trait values of plants at different ontogenetic stages could provide a clearer understanding of community and population dynamics, and more accurate extrapolation to ecosystem processes and functions.

Conclusion

Taken together, these results indicate that plasticity in response to abiotic conditions and ontogeny, and not adaptation and genetically distinct ecotypes, drive trait expression in *H. radicata*. However, the generality of this conclusion needs to be explored. *Hypochaeris radicata* is globally widespread, and thus may be inherently more plastic than species with restricted distributions [69,70]. To determine whether plasticity is a characteristic trait of widespread or dominant species, this research should be replicated with other widespread species and with species from restricted ranges. In addition, further research should examine how sensitive these conclusions are to the identity of the trait being examined.

The treatments applied during this experiment represented strong abiotic filters for this species, but are within the range of conditions experienced by *H. radicata*. Studies involving light or moisture gradients could assess whether the importance of population of origin or plasticity changes depending on the strength of the abiotic filter, a question which was not addressed in this study. Studies could also examine other abiotic filters, such as soil fertility. Finally, the additive and interactive effects of multiple abiotic filters on trait expression could be tested in both field and greenhouse settings.

Despite growing interest in understanding and incorporating intraspecific variation in plant functional traits into larger community- and ecosystem-level studies, the factors driving observed variation remain poorly understood. Our results suggest that observed intraspecific variation is largely driven by trait plasticity rather than genetic factors. We also found strong evidence of variation in plant traits due to ontogeny, a source of trait variation that is often ignored.

Supporting Information

Figure S1 Diagram illustrating selection of populations and maternal lines.

Figure S2 Correlation Matrix of Traits.

Table S1 Table of Abiotic Conditions at Sample Locations.

Table S2 Results of General Linear Mixed Models.

Acknowledgments

We thank Ty Robinson for assistance in the field, and Loretta Fisher for assistance in the lab. We also thank Janneke Hille Ris Lambers, Soo-Hyung Kim, Sarah Reichard, Regina Rochefort, and Martha Groom for their advice and guidance.

Author Contributions

Conceived and designed the experiments: RMM JDB. Performed the experiments: RMM. Analyzed the data: RMM. Contributed reagents/materials/analysis tools: RMM JDB. Wrote the paper: RMM JDB.

References

1. Messier J, McGill BJ, Lechowicz MJ (2010) How do traits vary across ecological scales? A case for trait-based ecology. Ecol Lett 13: 838–848. doi:10.1111/j.1461-0248.2010.01476.x.
2. Myers JA, Chase JM, Jiménez I, Jørgensen PM, Araujo-Murakami A, et al. (2013) Beta-diversity in temperate and tropical forests reflects dissimilar mechanisms of community assembly. Ecol Lett 16: 151–157. doi:10.1111/ele.12021.
3. Long W, Zang R, Schamp BS, Ding Y (2011) Within- and among-species variation in specific leaf area drive community assembly in a tropical cloud forest. Oecologia 167: 1103–1113. doi:10.1007/s00442-011-2050-9.
4. Swenson NG, Anglada-Cordero P, Barone Ja (2011) Deterministic tropical tree community turnover: evidence from patterns of functional beta diversity along an elevational gradient. Proc R Soc B 278: 877–884. doi:10.1098/rspb.2010.1369.
5. Swenson NG, Weiser MD (2010) Plant geography upon the basis of functional traits: an example from eastern North American trees. Ecology 91: 2234–2241.
6. Thuiller W, Lavorel S, Midgley G, Lavergne S (2004) Relating plant traits and species distributions along bioclimatic gradients for 88 Leucadendron taxa. Ecology 85: 1688–1699.
7. Laughlin DC, Fulé PZ, Huffman DW, Crouse J, Laliberté E (2011) Climatic constraints on trait-based forest assembly. J Ecol 99: 1489–1499. doi:10.1111/j.1365-2745.2011.01885.x.
8. Hudson JMG, Henry GHR, Cornwell WK (2011) Taller and larger: shifts in Arctic tundra leaf traits after 16 years of experimental warming. Glob Chang Biol 17: 1013–1021. doi:10.1111/j.1365-2486.2010.02294.x.
9. Fonseca C, McC Overton J, Collins B, Westoby M (2000) Shifts in trait-combinations along rainfall and phosphorus gradients. J Ecol 88: 964–977.
10. Díaz S, Cabido M, Casanoves F (1998) Plant functional traits and environmental filters at a regional scale. J Veg Sci 9: 113–122.
11. Diaz S, Hodgson JG, Thompson K, Cabido M, Cornelissen JHC, et al. (2004) The plant traits that drive ecosystems: Evidence from three continents. J Veg Sci 15: 295–304. doi:10.1111/j.1654-1103.2004.tb02266.x.
12. Díaz S, Cabido M (2001) Vive la différence: plant functional diversity matters to ecosystem processes. 16: 646–655.
13. William KC, Johannes HCC, Kathryn A, Ellen D, Valerie TE, et al. (2008) Plant species traits are the predominant control on litter decomposition rates within biomes worldwide. Ecol Lett 11: 1065–1071.
14. Albert CH, Thuiller W, Yoccoz NG, Douzet R, Aubert S, et al. (2010) A multi-trait approach reveals the structure and the relative importance of intra- vs. interspecific variability in plant traits. Funct Ecol 24: 1192–1201. doi:10.1111/j.1365-2435.2010.01727.x.
15. Albert CH, Thuiller W, Yoccoz NG, Soudant A, Boucher F, et al. (2010) Intraspecific functional variability: extent, structure and sources of variation. J Ecol 98: 604–613. doi:10.1111/j.1365-2745.2010.01651.x.
16. Mitchell RM, Bakker JD (2014) Quantifying and comparing intraspecific functional trait variability: a case study with Hypochaeris radicata. Funct Ecol 28: 258–269. doi:10.1111/1365-2435.12167.
17. Siefert A (2012) Incorporating intraspecific variation in tests of trait-based community assembly. Oecologia 170: 767–775. doi:10.1007/s00442-012-2351-7.
18. Lecerf A, Chauvet E (2008) Intraspecific variability in leaf traits strongly affects alder leaf decomposition in a stream. Basic Appl Ecol 9: 598–605. doi:10.1016/j.baae.2007.11.003.
19. Jung V, Violle C, Mondy C, Hoffmann L, Muller S (2010) Intraspecific variability and trait-based community assembly. J Ecol 98: 1134–1140. doi:10.1111/j.1365-2745.2010.01687.x.
20. Clark JS (2010) Individuals and the variation needed for high species diversity in forest trees. Science (80-) 327: 1129–1132. doi:10.1126/science.1183506.

21. Crutsinger GM, Souza L, Sanders NJ (2008) Intraspecific diversity and dominant genotypes resist plant invasions. Ecol Lett 11: 16–23. doi:10.1111/j.1461-0248.2007.01118.x.
22. Moreira B, Tavsanoglu C, Pausas JG (2012) Local versus regional intraspecific variability in regeneration traits. Oecologia 168: 671–677. doi:10.1007/s00442-011-2127-5.
23. Jung V, Albert CH, Violle C, Kunstler G, Loucougaray G, et al. (2014) Intraspecific trait variability mediates the response of subalpine grassland communities to extreme drought events. J Ecol 102: 45–53. doi:10.1111/1365-2745.12177.
24. Fajardo A, Piper FI (2011) Intraspecific trait variation and covariation in a widespread tree species (Nothofagus pumilio) in southern Chile. New Phytol 189: 259–271. doi:10.1111/j.1469-8137.2010.03468.x.
25. Grassein F, Till-Bottraud I, Lavorel S (2010) Plant resource-use strategies: the importance of phenotypic plasticity in response to a productivity gradient for two subalpine species. Ann Bot 106: 637–645. doi:10.1093/aob/mcq154.
26. Wellstein C, Chelli S, Campetella G, Bartha S, Galiè M, et al. (2013) Intraspecific phenotypic variability of plant functional traits in contrasting mountain grasslands habitats. Biodivers Conserv 22: 2353–2374. doi:10.1007/s10531-013-0484-6.
27. Freschet GT, Bellingham PJ, Lyver PO, Bonner KI, Wardle Da. (2013) Plasticity in above- and belowground resource acquisition traits in response to single and multiple environmental factors in three tree species. Ecol Evol: n/a–n/a. doi:10.1002/ece3.520.
28. Joshi J, Schmid B, Caldeira MC, Dimitrakopoulos PG, Good J, et al. (2001) Local adaptation enhances performance of common plant species. Ecol Lett 4: 536–544. doi:10.1046/j.1461-0248.2001.00262.x.
29. Schlichting C (1986) The evolution of phenotypic plasticity in plants. Annu Rev Ecol Syst 17: 667–693.
30. Hansen C, Garcia MB, Ehlers BK (2013) Water availability and population origin affect the expression of the tradeoff between reproduction and growth in Plantago coronopus. J Evol Biol 26: 1–10. doi:10.1111/jeb.12114.
31. Grassein F, Till-Bottraud I, Lavorel S (2010) Plant resource-use strategies: the importance of phenotypic plasticity in response to a productivity gradient for two subalpine species. Ann Bot 106: 637–645. doi:10.1093/aob/mcq154.
32. Bonser SP, Ladd B, Monro K, Hall MD, Forster M (2010) The adaptive value of functional and life-history traits across fertility treatments in an annual plant. Ann Bot 106: 979–988. doi:10.1093/aob/mcq195.
33. Firn J, Prober SM, Buckley YM (2012) Plastic traits of an exotic grass contribute to its abundance but are not always favourable. PLoS One 7: e35870. doi:10.1371/journal.pone.0035870.
34. Ramírez-Valiente JA, Sánchez-Gómez D, Aranda I, Valladares F (2010) Phenotypic plasticity and local adaptation in leaf ecophysiological traits of 13 contrasting cork oak populations under different water availabilities. Tree Physiol 30: 618–627. doi:10.1093/treephys/tpq013.
35. Robson TM, Sánchez-Gómez D, Cano FJ, Aranda I (2012) Variation in functional leaf traits among beech provenances during a Spanish summer reflects the differences in their origin. Tree Genet Genomes 8: 1111–1121. doi:10.1007/s11295-012-0496-5.
36. Dewitt TJ, Sih A, Wilson DS (1998) Costs and limits of phenotypic plasticity. Trends Ecol Evol 13: 77–81.
37. Auld JR, Agrawal A, Relyea R (2010) Re-evaluating the costs and limits of adaptive phenotypic plasticity. Proc R Soc B 277: 503–511. doi:10.1098/rspb.2009.1355.
38. Violle C, Enquist BJ, McGill BJ, Jiang L, Albert CH, et al. (2012) The return of the variance: intraspecific variability in community ecology. Trends Ecol Evol 27: 244–252. doi:10.1016/j.tree.2011.11.014.
39. Smith EA, Collette SB, Boynton TA, Lillrose T, Stevens MR, et al. (2011) Developmental contributions to phenotypic variation in functional leaf traits within quaking aspen clones. Tree Physiol 31: 68–77. doi:10.1093/treephys/tpq100.
40. Cornelissen JHC, Cerabolini B, Puyravaud JP, Maestro M, Werger MJA, et al. (2003) Functional traits of woody plants: correspondence of species rankings between field adults and laboratory-grown seedlings? J Veg Sci 14: 311–322.
41. Cornelissen J, Lavorel S, Garnier E, Diaz S, Buchmann N, et al. (2003) A handbook of protocols for standardised and easy measurement of plant functional traits worldwide. Aust J Bot 51: 335–380.
42. Bradshaw AD (2006) Unravelling phenotypic plasticity-Why should we bother? New Phytol 170: 639–641. doi:10.1111/j.1469-8137.2006.01758.x.
43. Nicotra AB, Atkin OK, Bonser SP, Davidson AM, Finnegan EJ, et al. (2010) Plant phenotypic plasticity in a changing climate. Trends Plant Sci 15: 684–692. doi:10.1016/j.tplants.2010.09.008.
44. Valladares F, Gianoli E, Gómez JM (2007) Ecological limits to plant phenotypic plasticity. New Phytol 176: 749–763. doi:10.1111/j.1469-8137.2007.02275.x.
45. Balaguer L, Martinez-Ferri E, Valladares F, Perez-Corona ME, Baquedano FJ, et al. (2001) Population divergence in the plasticity of the response of Quercus coccifera to the light environment. Funct Ecol 15: 124–135. doi:10.1046/j.1365-2435.2001.00505.x.
46. Pratt JD, Mooney KA (2013) Clinal adaptation and adaptive plasticity in Artemisia californica: implications for the response of a foundation species to predicted climate change. Glob Chang Biol 19: 2454–2466. doi:10.1111/gcb.12199.
47. Davidson AM, Jennions M, Nicotra AB (2011) Do invasive species show higher phenotypic plasticity than native species and, if so, is it adaptive? A meta-analysis. Ecol Lett 14: 419–431. doi:10.1111/j.1461-0248.2011.01596.x.
48. Herman JJ, Sultan SE (2011) Adaptive transgenerational plasticity in plants: case studies, mechanisms, and implications for natural populations. Front Plant Sci 2: 102. doi:10.3389/fpls.2011.00102.
49. Pico F, Ouborg NJ, Van Groenendael JM (2004) Influence of selfing and maternal effects on life-cycle traits and dispersal ability in the herb *Hypochaeris radicata* (Asteraceae). Bot J Linn Soc 146: 163–170.
50. Ortiz MA, Talavera S, Garcia-Castano JL, Tremetsberger K, Stuessy T, et al. (2006) Self-Incompatibility and Floral Parameters in Hypochaers sec. Hypocaheris (Asteraceae). Am J Bot 93: 234–244.
51. Mix C, Arens PFP, Rengelink R, Smulders MJM, Van Groenendael JM, et al. (2006) Regional gene flow and population structure of the wind-dispersed plant species *Hypochaeris radicata* (Asteraceae) in an agricultural landscape. Mol Ecol 15: 1749–1758. doi:10.1111/j.1365-294X.2006.02887.x.
52. Poorter L, Bongers F (2006) Leaf traits are good predictors of plant performance across 53 rain forest species. Ecology 87: 1733–1743.
53. Nicotra A, Leigh A, Boyce C (2011) The evolution and functional significance of leaf shape in the angiosperms. Funct Plant … 38: 535–552.
54. Reich PB, Walters MB, Ellsworth D (1992) Leaf life-span in relation to leaf, plant and stand characteristics among diverse ecosystems. Ecol Monogr 62: 365–392.
55. Westoby M (1998) A leaf-height-seed (LHS) plant ecology strategy scheme: 213–227.
56. Anderson MJ (2001) A new method for non-parametric multivariate analysis of variance. Austral Ecol 26: 32–46. doi:10.1046/j.1442-9993.2001.01070.x.
57. Anderson MJ (2004) PERMDISP: a FORTRAN computer program for permutational analysis of multivariate dispersions (for any two-factor ANOVA design) using permutation. Department of Statistics, University of Auckland, Auckland, New Zealand.
58. Clarke K, Gorley R (2006) Primer v6: User Manual/Tutorial. Plymouth: PRIMER-E.
59. Mokany K, Ash J (2008) Are traits measured on pot grown plants representative of those in natural communities? J Veg Sci 19: 119–126. doi:10.3170/2007-8-18340.
60. Kleyer M, Bekker RM, Knevel IC, Bakker JP, Thompson K, et al. (2008) The LEDA Traitbase: a database of life-history traits of the Northwest European flora. J Ecol 96: 1266–1274. doi:10.1111/j.1365-2745.2008.01430.x.
61. Poorter H, Niinemets Ü, Poorter L, Wright IJ, Villar R (2009) Causes and consequences of variation in leaf mass per area (LMA): a meta-analysis. New Phytol 182: 565–588. doi:10.1111/j.1469-8137.2009.02830.x.
62. Cornwell WK, Cornelissen JHC, Amatangelo K, Dorrepaal E, Eviner VT, et al. (2008) Plant species traits are the predominant control on litter decomposition rates within biomes worldwide. Ecol Lett 11: 1065–1071.
63. Pérez-Harguindeguy N, Díaz S, Garnier E, Lavorel S, Poorter H, et al. (2013) New handbook for standardised measurement of plant functional traits worldwide. Aust J Bot: 167–234.
64. Sultan SE (2000) Phenotypic plasticity for plant development, function and life history. Trends Plant Sci 5: 537–542.
65. Richards CL, Bossdorf O, Muth NZ, Gurevitch J, Pigliucci M (2006) Jack of all trades, master of some? On the role of phenotypic plasticity in plant invasions. Ecol Lett 9: 981–993. doi:10.1111/j.1461-0248.2006.00950.x.
66. Diaz S, Cabido M (2001) Vive la difference: plant functional diversity matters to ecosystem processes. Trends Ecol Evol 16: 646–655.
67. Boege K (2010) Induced responses to competition and herbivory: natural selection on multi-trait phenotypic plasticity. Ecology 91: 2628–2637.
68. Mediavilla S, Escudero A (2004) Stomatal responses to drought of mature trees and seedlings of two co-occurring Mediterranean oaks. For Ecol Manage 187: 281–294. doi:10.1016/j.foreco.2003.07.006.
69. Sultan SE, Barton K, Wilczek AM (2009) Contrasting patterns of transgenerational plasticity in ecologically distinct congeners. Ecology 90: 1831–1839. doi:doi:10.1890/08-1064.1.
70. Gitzendanner M, Soltis P (2000) Patterns of genetic variation in rare and widespread plant congeners. Am J Bot 87: 783–792.

Herbivore Impacts on Marsh Production Depend upon a Compensatory Continuum Mediated by Salinity Stress

Jeremy D. Long*, Laura D. Porturas

Biology Department and Coastal & Marine Institute Laboratory, San Diego State University, San Diego, California, United States of America

Abstract

Plant communities are disturbed by several stressors and they are expected to be further impacted by increasing anthropogenic stress. The consequences of these stressors will depend, in part, upon the ability of plants to compensate for herbivory. Previous studies found that herbivore impacts on plants can vary from negative to positive because of environmental control of plant compensatory responses, a.k.a. the Compensatory Continuum Hypothesis. While these influential studies enhanced our appreciation of the dynamic nature of plant-herbivore interactions, they largely focused on the impact of resource limitation. This bias limits our ability to predict how other environmental factors will shape the impact of herbivory. We examined the role of salinity stress on herbivory of salt marsh cordgrass, *Spartina foliosa*, by an herbivore previously hypothesized to influence the success of restoration projects (the scale insect, *Haliaspis spartinae*). Using a combination of field and mesocosm manipulations of scales and salinity, we measured how these factors affected *Spartina* growth and timing of senescence. In mesocosm studies, *Spartina* overcompensated for herbivory by growing taller shoots at low salinities but the impact of scales on plants switched from positive to neutral with increasing salinity stress. In field studies of intermediate salinities, scales reduced *Spartina* growth and increased the rate of senescence. Experimental salinity additions at this field site returned the impact of scales to neutral. Because salinity decreased scale densities, the switch in impact of scales on *Spartina* with increasing salinity was not simply a linear function of scale abundance. Thus, the impact of scales on primary production depended strongly upon environmental context because intermediate salinity stress prevented plant compensatory responses to herbivory. Understanding this context-dependency will be required if we are going to successfully predict the success of restoration efforts and the ecological consequences of anthropogenic disturbances.

Editor: Cristina Armas, Institute of Ecology and Biodiversity, Chile

Funding: This research was supported by grants from the United States National Science Foundation Biological Oceanography program (OCE-0825846 awarded to JL and Geoff Trussell), California Sea Grant (NA10OAR4170060), and CSU COAST (awarded to LP). The funders had no role in study design, data collection and analysis, decision to publish, or preparation of the manuscript.

Competing Interests: The authors have declared that no competing interests exist.

* Email: jlong@mail.sdsu.edu

Introduction

Environmental stress can disturb communities directly by reducing survivorship or forcing migrations, or indirectly by altering the traits responsible for determining how species interact [1–5]. An important plant trait that can be affected by stress is the ability to compensate for herbivory [6]. In a foundational study, Maschinski and Whitham (1989) proposed the Compensatory Continuum Hypothesis whereby plant compensatory responses can vary in sign and strength, largely via environmental factors. The discovery that compensatory responses to herbivory can be shaped by the environment profoundly influenced our understanding of plant-herbivore interactions [7,8]. However, most studies focused on the effect of resource limitation on compensatory responses [8]. Indeed, a well-cited revision of the Compensatory Continuum Hypothesis, the Limiting Resource Model, focused on the primacy of the level and identity of the limiting resource [9]. The roles of other environmental stressors on herbivory remain unclear. If compensatory responses generally fall along a continuum determined by environmental stress, then understanding this complexity will become important to successfully predicting the consequences of anthropogenic stressors like pollution and climate change.

Although there is emerging evidence that compensatory responses vary with multiple environmental contexts, we have an incomplete appreciation for the role of environmental stress other than resource limitation. For example, overcompensation appears more commonly for monocot herbs when nutrients are not limiting but overcompensation is more common for dicots under nutrient limitation [8]. Water-stress can also determine compensatory responses and the consequences of herbivory ([10–13] but see [14–16] for exceptions). In most of these cases, water stress leads to reduced compensatory responses. Unfortunately, the consequences of other common stressors, like salinity, are poorly known. Although the immediate consequences of salinity stress may largely result from changes in water uptake, salinity stress can create unique problems for plants that influence their ability to compensate for herbivory [6,17]. For example, water and salinity stress may have very different impacts on natural communities because the former increases soluble nitrogen and nutritional quality of plants [14,15,18], whereas the latter can decrease nutritional quality [19,20]. Also, salinity stress may influence compensatory responses because of negative, salt-specific effects on hormonal regulation, photosynthesis, and cellular function [17]. Furthermore, salinity stress might directly suppress herbivore performance via increases in plant tissue salt concentrations [21]. Thus, previous studies may fail to predict the impact of salinity stress on plant compensatory responses and their consequences because of stress-specific effects on plant physiology.

Because of their distinct gradients in environmental stress and plant distributions [22,23] and their susceptibility to anthropogenic stressors like climate change, salt marshes represent important habitats where salinity stress may interact with herbivory to influence primary production. Three observations support this hypothesis. First, experimental salinity stress increased the abundance of marsh planthoppers ([24,25] but see [20] for an exception). Second, palatability of marsh plants depended upon whether plants were collected from high salinity versus low salinity sites [26]. Third, salt addition stress magnified the negative impact of snail herbivory on *Spartina* biomass [27], perhaps because salt impaired the ability of *Spartina* to tolerate herbivory or increase resistance after attack [28]. Taken together, these studies suggest that salinity stress may shape the impact of herbivores on marsh plants and that the synergy of these factors may contribute to marsh dieback [27]. Given that climate change is expected to alter soil salinities via changes in sea level and precipitation [29,30], there is a pressing need to understand the impact of salinity stress on marsh herbivory.

We examined the impact of salinity stress on top-down control of Pacific cordgrass, *Spartina foliosa*, by the specialist scale insect, *Haliaspis spartinae*. Scale insects are common pests that can reduce the growth and survival of plants [31–33]. Although no study has experimentally tested the impact of *H. spartinae* on cordgrass, it has been implicated in the failure of some restoration projects to achieve desired outcomes [34,35]. These authors noted that, "In 1992, where the highest densities of *Haliaspis* occurred, cordgrass was shorter than in previous years and senesced unusually early in the season" [34]. Because cordgrass height is one measure of restoration success (in part because endangered birds require tall plants to make nests [35]), it was hypothesized that the scales may prevent successful restoration of these sensitive habitats [34,35]. Thus, an improved understanding of scale impacts and how these change with environmental conditions could lead to improved restoration outcomes.

We tested for a causative relationship between scales and plant performance, and the role of salinity stress on scale herbivory, by manipulating scale presence and salinity in a series of mesocosm and field studies. We predicted that 1) scales would exert the strongest negative impacts on *Spartina* at intermediate salinities where compensatory responses would be impaired, and 2) further stressing *Spartina* with high salinities would weaken scale impacts.

Methods

We conducted three major experiments to examine the impact of scale insects (*Haliaspis spartinae*, hereafter scales) on *Spartina foliosa* stems (hereafter *Spartina*). First, we used a field experiment to demonstrate that scales can negatively influence *Spartina* performance. Second, we used a mesocosm experiment to examine the effect of reduced salinities on *Spartina*-scale interactions. Finally, we conducted a field experiment to determine the effect of elevated soil salinities on these interactions. The latter two experiments were conducted because 1) the heaviest infestations of scales on *Spartina* (# per stem) occur in the high marsh and 2) high soil salinities in this zone suppress *Spartina* growth [36]. Thus, we hypothesized that top-down control of *Spartina* by scales may be influenced by soil salinity.

Infestations of *Haliaspis*, a *Spartina* specialist, appear highest in Southern California marshes. We have observed heavy infestations of *Haliaspis* from South San Diego Bay to Upper Newport Bay, California. Throughout this range, there is variation within and between marshes in *Haliaspis* densities. Although *Haliaspis* is believed to be indigenous to these marshes, there is some evidence that the distribution and abundance of this scale is expanding. For example, in the late 1980s and early 1990s, *Haliaspis* was abundant in human-constructed marshes but rare or absent in more natural marshes [34,37–39]. Our recent surveys suggest that *Haliaspis* has now expanded into some of these natural marshes at extremely high densities.

The two field experiments were conducted in the high zone of Sweetwater Marsh in South San Diego Bay. Sweetwater is a tidal, hypersaline marsh with soil salinities that frequently exceed 50 ppt. This marsh also has the highest maximum densities of scales per stem that we have encountered during our surveys of southern California marshes (maximum number of scales per stem >3000). Within Sweetwater Marsh, the highest scale densities per stem are observed in the high marsh zone. Salt marsh perennials including *Sarcorcornia virginica*, *Frankenia grandifolia*, *Jaumea carnosa*, *Batis maritima*, and *Suaeda californica* dominate this habitat. This research was permitted by the U.S. Fish and Wildlife Service (SUP 81680-12002) – the organization that manages Sweetwater Marsh.

Effects of scale removal (2011 Field experiment)

To examine the impact of scales on cordgrass performance, we manipulated scale presence on *Spartina* at Sweetwater Marsh (32.6412°N, 117.1142°W). On 20 May 2011, we randomly selected 20 *Spartina* stems separated by at least 1 m and that contained high densities of mature scales relative to nearby stems (70±19 scales per stem, mean ± SE). Stems were randomly assigned to Scale or No Scale treatments, and scales were removed from half of the stems (No Scale stems, N = 10) by brushing the adaxial surface of leaves with a soft toothbrush. This procedure effectively removes scales because they cluster on the adaxial surface of leaves. As a procedural control, we brushed the abaxial surface of Scale stems. For 20 weeks, we maintained treatments by manually removing scales from No Scale stems with a toothbrush every 1–2 weeks. Because scales have a short larval dispersal period followed by a sessile juvenile and adult period, manual removals effectively create Scale/No Scale treatments [40].

Every 1–2 weeks, we recorded Scale density (# per stem), and *Spartina* senescence, stem height, and seed presence. *Spartina* senescence was determined when the stem's meristem was completely brown. As a conservative estimate, all plants that had not senesced by our final sampling date (7 October 2011) were assumed to have senesced one week later. Stem height was determined by recording the maximum vertical distance from shoot base to the tip of the longest leaf. Seed presence was recorded as a binomial variable (present/absent). After 20 weeks (7 October 2011), we collected shoots, recorded seed presence, dried the shoots, and recorded final dry shoot biomass. We observed early senescence of many plants in the Scale treatment. These were included in all performance measurements. On 19 August 2011, we observed leaves that had been grazed entirely across their width. Although we did not identify this grazer, this type of damage is consistent with large, vertebrate grazers (e.g. birds). One Scale stem and two No Scale stems contained such damage. To exclude further vertebrate grazing, we placed tomato cages lined with chicken wire (opening ~3 cm) around all stems.

We compared two measures of growth of Scale and No Scale plants using two-tailed, two-sample t-tests. First, we calculated the proportional change in shoot height during the 20-week experiment. Second, we compared final dry shoot biomass. We used a two-tailed Fisher's Exact Test to compare the proportion of starting plants that ever produced seeds because 9 of 10 Scale plants did not produce seeds. We compared days to senescence using a two-tailed, two-sample t-test.

Effects of scale removal and salinity reduction (2011 Mesocosm experiment)

We examined the impact of reduced salinities on *Spartina*-scale interactions using mesocosms because logistical hurdles prevented us from reducing salinities in the field. A limitation of this design is that our mesocosm experiment may have been confounded by other factors. We sought to minimize differences between the field and our mesocosm experiment by conducting mesocosm research outdoors at SDSU's Coastal & Marine Institute Laboratory that is located 14 km from Sweetwater Marsh. This lab is located adjacent to San Diego Bay (i.e. the same water body from which all organisms were collected). Thus, ambient environmental conditions during our mesocosm studies were likely similar to field conditions.

We collected mud cores (~15 cm diameter × 15 cm deep) from Sweetwater Marsh that each contained a single *Spartina* stem with scales. We transferred these to the marine lab, and planted each stem-containing mud core into a 2.6 L plastic pot (15 cm height) with holes in the bottom for drainage. Stems were randomly assigned to each of four different treatments created by manipulating Scales (Scale, No Scale) and Salinity (Freshwater, Saltwater). While we recognize that freshwater exposure of southern California *Spartina* is extremely rare, this treatment was included because previous studies showed that salt marsh plants grow optimally in reduced salinity [36]. Most studies of plant stress define stress as any condition that reduces the optimal performance of plants (e.g. [41]). Additionally, ambient soil salinities as low as 15 ppt have been reported for *Spartina foliosa*, especially during high rainfall years [42].

All treatments had 8 replicates, with the exception of the Scale-Saltwater treatment, which had 7. Scales were removed from No Scale stems with a toothbrush and scale treatments were maintained weekly. Stems from each treatment were randomly assigned to shallow pools of a specified salinity treatment ($N = 2$). Salinity was manipulated by filling the pools to the soil-air interface with either freshwater or saltwater. The saltwater pools were filled with water from the flow-through seawater system that takes in water from San Diego Bay (~32 ppt). Freshwater pools were filled with water from a tap. Twice per week, we measured the pool salinity with a refractometer (Model RHS-10ATC, Aquatic Eco-Systems Inc.), drained and refilled each pool, and then re-measured pool salinity. We chose to measure the salinity of the pool water rather than the porewater (soil salinity) so that we would minimize disturbance to the soil in the pots. Thus, the magnitude by which our freshwater mesocosm reduced soil salinity is currently unclear. During the water changes, each pot was allowed to drain outside of the pools for ~30 minutes. All pools were kept outside in full sunlight.

We maintained these treatments for six months (April–October 2011). Every week, we recorded Scale density (# per stem starting 3 May 2011), and *Spartina* senescence and stem height (see above for details). Because the replicates of each treatment were divided amongst two replicate pools, we calculated an average for all replicates within each pool at each time point. During measurements, each pot was allowed to drain outside of the pools for 1–4 hours. We used stem height to calculate the average proportional growth at each time point. These data were analyzed with a Repeated Measures ANOVA. We visually inspected our data and the resulting covariance matrix to confirm that they met the assumptions of this test. However, we were concerned that our data violated the sphericity assumption, perhaps because of the high number of repeated measurements, so we applied a Greenhouse-Geisser correction to this analysis ($\varepsilon = 0.061$). We compared the number of days to senescence using ANOVA. Both ANOVA analyses examined the effect of the fixed factors of Scales (Scale, No Scale) and Salinity (Freshwater, Saltwater) on the response variable.

Effects of scale removal and salinity addition (2012 Field experiment)

To examine the influence of elevated salinity on the top-down control of *Spartina* by scales, we conducted a manipulative experiment at Sweetwater Marsh during 2012. We selected two sites in March 2012 [hereafter North (32.6412°N, 117.1142°W) and South (32.6388°N, 117.1099°W)]. These sites were selected because of their different elevations (+1.6 m and +2.1 m above Mean Lower Low Water, respectively). Also, we observed negative effects of scales on *Spartina* at the North site during the 2011 experiment. At each site, we randomly selected 1×1 m plots that contained *Spartina* and were separated by at least 1 m. Plots were randomly assigned to a Salinity treatment (Ambient or Elevated) and a Scale treatment (Scale or No Scale). We replicated each of four treatments at both sites ($N = 7$ or 8; North or South, respectively).

For two months, we pre-treated Elevated Salinity plots with 750 g of solar evaporated sea salt (produced by South Bay Salt Works, Chula Vista, CA) every 2 weeks during low tide. On 26 May 2012, we increased salt additions in Elevated Salinity plots to 1500 g every 2 weeks to create salinities near the maximum levels we have recorded in the field. We collected a single soil core (~5 cm deep) from each plot for soil salinity analysis on seven dates (31 March, 14 April, 28 April, 12 May, 8 June, 22 June, 6 July, 9 July). Soil salinity was not sampled after 9 July so as to minimize further disturbance to plots. On each of the first seven dates, we sampled plots prior to that day's salt addition so they reflected salinity differences persisting two weeks after the previous salt addition. In contrast, soil samples from the last date that soil salinities were measured (9 July) were collected 3 d after salt addition. This measurement provided better insight into soil salinity dynamics because the greatest impact of salt additions on soil salinity was observed 3 d after salt additions. We measured soil salinity using a modification of an existing protocol [43]. Briefly, soil cores were weighed, dried for 2 days at 60°C, and then reweighed. Then, we rehydrated cores with the original water volume plus 15 ml deionized water and measured the salinity of the slurry supernatant with a refractometer (Model RHS-10ATC, Aquatic Eco-Systems Inc.). We used these measurements to determine the soil salinity of the original samples.

We marked a single target *Spartina* stem near the center of each Scale and No Scale plot with flagging tape. Scales were removed from No Scale stems using toothbrushes. All stems were initially measured on 26 May 2012, and stems were re-measured and treatments were maintained every 2 weeks thereafter. We measured scale density and recorded the date of senescence. Because previous experiments suggested that plant growth did not extend past September, we stopped this experiment on 14 September 2012.

Salinities were compared using Repeated Measures ANOVA. We excluded the soil salinity data from 6 July 2012 from this analysis because we lost 6 samples during processing and these were not distributed evenly across treatments. We excluded one Enhanced Salinity, Scale plot from the North site because a sample was lost on 6/8/12. We also conducted a three factor ANOVA to examine the effects of Site (North, South), Scales (Scale, No Scale), and Salt (Ambient, Elevated) on the number of days to senescence. We excluded one plant assigned to the North, Scale, Ambient plot because scales never colonized it.

Results

Effects of scale removal (2011 Field experiment)

We observed dispersal of juvenile scales beginning in late June and ending in late July 2011 (Fig. S1). During this time, scale densities on Scale plants increased from 154±62 to 771±184 scales per stem (mean ± SE). Although we occasionally observed recruitment onto No Scale plants, the maximum density of scales we observed on No Scale stems was only 124±15 scales per stem. In contrast, we observed an order of magnitude greater maximum scale densities on Scale stems 1531±358 scales per stem. Thus, we effectively maintained differences in scale densities between our treatments.

Scales suppressed *Spartina* shoot height growth by 69% (Fig. 1a; $t = 3.381$, $p = 0.003$) and final dry shoot biomass by 37% (Fig. 1b; $t = 2.531$, $p = 0.021$). By 7 October 2011, 5 No Scale stems, but only 1 Scale stem, produced seeds. However, the proportion of stems producing seeds was not significantly different at $\alpha = 0.05$ (Fig. 1c; Fisher's Exact, $p = 0.141$).

No plants senesced during the first three months of the 2011 field experiment (Fig. 2). However, 70% of Scale stems senesced between mid-August and the end of the experiment. None of the No Scale stems senesced by the end of the experiment. If we conservatively assume that all No Scale plants senesced a week after the experiment ended, then No Scale stems senesced at least 27 days later than Scale stems (Fig. 2 inset; $t = 3.452$, $p = 0.003$). Importantly, the earlier senescence in Scale stems occurred during seed production.

Effects of scale removal and salinity reduction (2011 Mesocosm experiment)

Salinity of the pools in Freshwater treatments ranged between 0–3.5 ppt, with a mean of 0.3 ppt. In contrast, the salinity in Saltwater treatments was typically an order of magnitude larger (range = 27–59 ppt, mean = 39.8 ppt). *Spartina* in the high marsh can encounter salinities in excess of 80 ppt. Salinity tended to increase in pools after water changes because of evaporation from the pools. Similar to the 2011 field experiment, we saw a major dispersal event of scales in late June/early July (Fig. S2). However, we also saw a pronounced secondary dispersal event in mesocosms

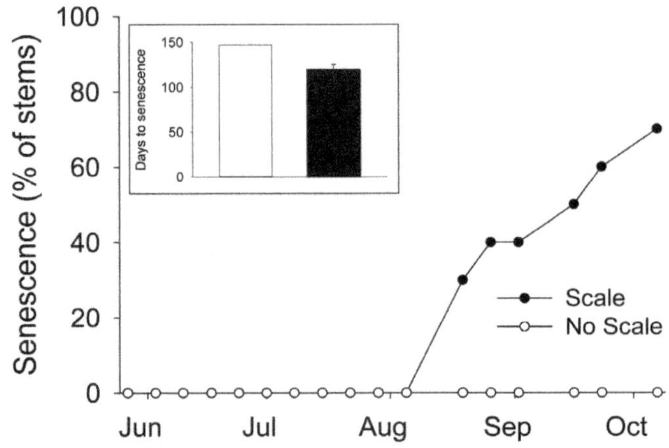

Figure 2. Scale effects on the timing of *Spartina* senescence during the 2011 field experiment. Senescence is reported as the cumulative percentage of stems senesced at each time point. Inset shows the number of days until *Spartina* senescence for No Scale (white bar) and Scale stems (black bar). Stems not senescing at the termination of the experiment were assigned a senescence date of 10/14/11. Because all 10 No Scale stems fell into this category but only 3 Scale stems did, the difference in days to senescence is conservative. This also explains the lack of error bars associated with the No Scale inset. Values in inset are means ± SE.

that started late in August 2011. Scale densities on Scale stems were higher in Freshwater than Saltwater treatments after the initial dispersal event (Fig. S2). This trend was not evident after the second dispersal event.

Plants senesced earlier in Saltwater treatments (days to senescence = 102±18 and 150±11 for Scale and No Scale stems, respectively) relative to Freshwater treatments (188±0 for both Scale treatments because no Freshwater plant had senesced by the end of the experiment; $F_{1,4} = 35.7$, $p = 0.004$). Although there was a trend for scales to cause earlier senescence in Saltwater treatments, neither Scales nor the interaction between Scales and Salinity influenced days to senescence ($F_{1,4} = 5.196$, $p = 0.085$).

Importantly, Salinity and Scales interacted to influence *Spartina* growth (Fig. 3, Table S1). Scales stimulated *Spartina* growth in Freshwater treatments but had no effect on growth in Saltwater treatments. As expected, *Spartina* growth increased during the experiment.

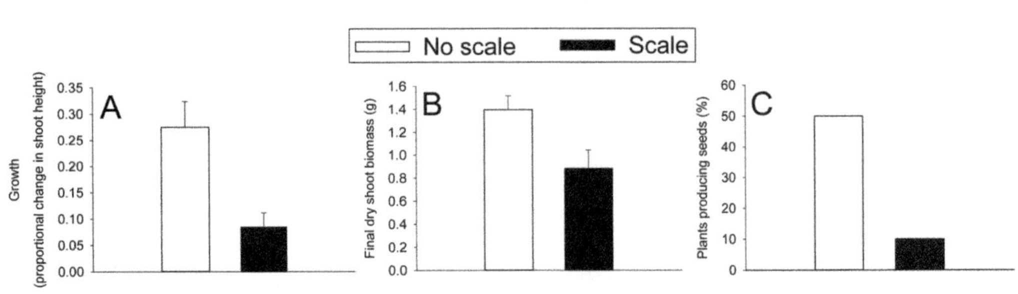

Figure 1. Scale effects on *Spartina* performance during the 2011 field experiment. The three performance measured included A) growth, B) final dry shoot biomass, and C) percentage of plants producing seeds. N = 10. Values are means ± SE.

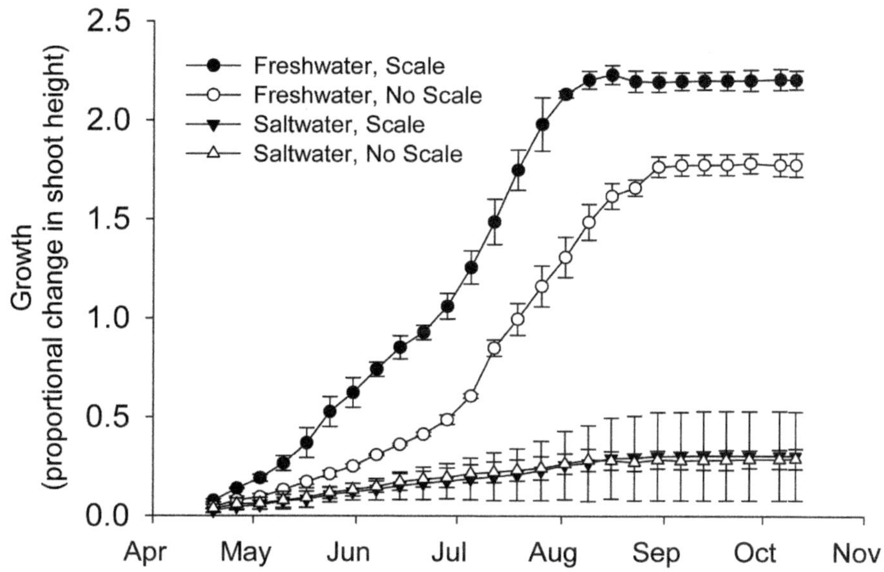

Figure 3. Effects of scales and reduced salinity on *Spartina* growth during the 2011 mesocosm experiment. Growth is reported as the proportional change in shoot height. Overcompensation to scale herbivory in freshwater treatments was absent in saltwater treatments. N = 7–8. Values are means ± SE.

Effects of scale removal and salinity addition (2012 Field experiment)

Plots with Elevated Salinity had higher porewater salinities than Ambient plots ($F_{1,54} = 21.2$, $p<0.001$; Fig. S3). The increase in soil salinity in Elevated plots was most pronounced three days after adding salt (9 July 2012) relative to two weeks after adding salt (all other dates, Fig. S3). Although overall soil salinities were similar at the North and South sites ($F_{1,54} = 0.663$, $p = 0.419$), Elevated Salinity plots contained 22% higher salinity on 9 July 2012 at the North versus the South site (84 vs. 69 ppt, respectively; $F_{1,28} = 4.20$, $p = 0.019$). Salinity did not differ with Scale treatment ($F_{1,54} = 0.117$, $p = 0.734$). All soil salinity measurements were within the range of soil salinities that can be currently experienced by *Spartina* at Sweetwater (we have measured a maximum soil salinity of 86 ppt).

Compared to the 2011 experiments (Figs. S1–S2), mean scale densities at the beginning of June were lower in 2012 (Fig. S4). For example, although our stems had ~100 scales per stem at the start of June 2011, mean scale densities in 2012 did not reach 100 scales per stem in any treatment until 6 July 2012. Thus, stems in the 2011 experiments may have experienced greater herbivore pressure than the 2012 experiment. The within-treatment variation of scale abundance increased with time, in part because of early plant senescence within this experiment. Because we removed scales from No Scale plants every two weeks, scale densities were always low on these treatments (Fig. S4).

For days to senescence, we observed a significant three-way interaction between Salt, Scale, and Site (Figure 4, Table S2). To better understand this interaction, we examined the interaction between Salt x Scale at each site separately using the appropriate error term. At the North site, the effect of Scales on senescence depended upon Salinity treatment ($F_{1,51} = 7.046$, $p = 0.011$). In contrast, we observed no interaction at the South site ($F_{1,51} = 0.437$, $p = 0.512$).

Discussion

The impact of scales on *Spartina* performance depended strongly upon environmental context. When grown under the lowest salinities (freshwater mesocosms), *Spartina* overcompensated for herbivory by growing taller shoots. However, increasing salinity stress switched the impact of scales from positive to neutral (2011 Mesocosm experiment). Additionally, scales negatively impacted *Spartina* by reducing growth and expediting senescence in the field. These data provide experimental evidence that *Haliaspis* can negatively impact *Spartina* and may have influenced the success of previous restoration efforts [33]. At the hypersaline field site where we observed negative effects of scales in 2011 and 2012, increasing soil salinity further returned the impact of scales to neutral (2012 Field experiment). Interestingly, scale densities and salinity were negatively related suggesting that scale impacts were not simply a linear function of scale abundance.

Previous studies examining the impact of environmental stress on plant compensatory responses largely focused on resource limitation as a stressor [8]. There, low nutrient conditions tended to weaken compensatory responses of monocots. Our study suggests that salinity stress acts similarly to weaken compensatory responses of *Spartina foliosa*. Reduced compensatory abilities with increasing salt stress might be particularly common in dominant salt marsh monocots like *Spartina* and *Juncus*. Salt stress decreases foliar nitrogen in *Spartina* [19], perhaps because of reduced uptake abilities [44,45]. Taken together, these observations suggest that salinity may alter the impact of marsh herbivores on marsh plants via reductions in compensatory abilities resulting from nutrient limitations. However, this hypothesis should be considered tentative, especially given the numerous exceptions to the general effect of resource limitation on monocots seen previously [8] and because salinity stress can directly impact other aspects of plant physiology [17].

While the interactive effects of environmental stress and herbivores can be complex and difficult to predict [7–9,46], there is some evidence that the impact of marsh herbivory might generally be more intense during intermediate salt stress because

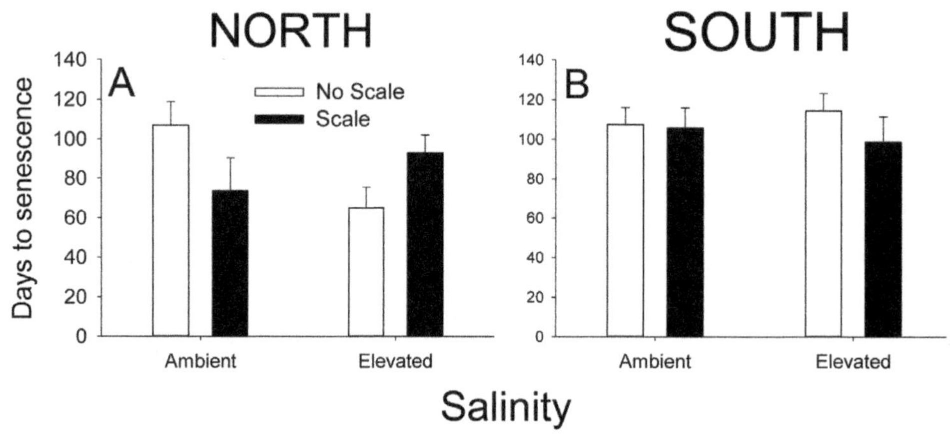

Figure 4. Effects of scales, elevated salinity, and site on *Spartina* senescence during the 2012 field experiment. Senescence is reported as the number of days after the start of the experiment that plants senesced at the North (A) and South Sites (B). N = 7–8. Values are means ± SE.

of reduced compensatory abilities. A study of a Northwest Atlantic Ocean salt marsh found that snail herbivory on *Spartina alterniflora* was much stronger in plots with experimental salt additions that increased salinity from ~35 to 57 ppt [27]. We observed positive or neutral effects of scale herbivory at low salinities (<4 to 40 ppt), negative effects at intermediate salinities (~50–55 ppt), and neutral effects at extremely high salinities (> 80 ppt). In contrast, previous research in a Southwest Atlantic Ocean salt marsh found that salt additions attempting to elevate ambient salinity to 60 ppt removed the negative effects of crab herbivores on marsh expansion seen at lower salinities [47]). In that case, crab impacts may have been low in salt addition plots because crabs avoided these areas [47]. Thus, salinity effects on marsh production may also depend upon how salinity directly affects herbivores.

Scales negatively influenced *Spartina* during the 2012 field experiment at only one of two sites (North site), suggesting that scale effects were site-specific. It is unlikely that salinity differences caused this specificity because soil salinity in Ambient Salinity plots showed no clear pattern with site. However, inundation can influence *Spartina* performance and two observations suggest that inundation differed between these sites. First, the North site was ~0.5 m lower in elevation. Second, soil salinities in Elevated Salinity plots three days after salt additions were much higher at the North site.

In addition to abiotic differences, at least two biotic factors may have influenced the site-specificity of scale effects. First, plant-plant competition may influence plant compensatory responses to herbivores [7,48]. Both sites have plant communities characteristic of high marsh but the makeup of these communities differed between the sites. *Jaumea carnosa* was more abundant and *Frankenia salina* and *Batis maritima* were less abundant at the site with negative scale effects. High marsh plants can suppress *Spartina* growth [49], and may influence *Spartina*'s ability to tolerate herbivory. Second, scale recruitment happened earlier at the North site, suggesting that scale pressure was greater at this site. Regardless of the mechanism, scales exerted a negative impact on *Spartina* at the site where overall plant senescence occurred earlier further supporting our contention that scale impacts depend upon plant stress.

Because we found the greatest impact of herbivores at intermediate salinities with reduced herbivore densities, our data suggest that the per capita effect of scales on *Spartina* was most negative at intermediate salinities. The realized impact of herbivory under stressful environmental conditions will also be determined by how the stressor impacts higher trophic levels, both directly and indirectly [50]. Unfortunately, the role of predators in this system is poorly known. Two observations suggest that predators may be important. First, potential predators such as beetles and parasitoids are present in these marshes [51,52]. Second, we observed a distinct secondary dispersal event in our mesocosms that was not present in the field in 2011. The lack of obvious predators in our mesocosms, but not the field, may have allowed for this dispersal event.

Understanding the complex interactions of salt stress and herbivory may help improve restoration strategies. In southern California, where loss of historic salt marshes has been extensive [53], a common restoration approach is to transplant *Spartina* into restoration sites. Perhaps because of the stress associated with transplantation, such approaches have been met with variable success [34,54,55]. One possible factor associated with restoration success has been heavy infestations of scale insects [34]. Our results suggest that scales may limit the success of restoration projects at intermediate soil salinities. Understanding this context-dependence may be particularly important given the anticipated changes in salinity with climate change that may alter the top-down control of marshes by herbivores.

Supporting Information

Figure S1 Densities of scales during the 2011 field experiment. Scales were removed from No Scale stems every 1–2 weeks. N = 10. Values are means ± SE.

Figure S2 Densities of scales during the 2011 mesocosm experiment. Scales were removed from No Scale stems every week. N = 7–8. Values are means ± SE.

Figure S3 Soil salinities of plots during the 2012 field experiment. Because salinity did not depend upon Scale treatment, data were combined for Scale and no Scale stems to facilitate comparisons of the North and South sites. Soil salinity was always measured two weeks after salt additions, except on July 9 when sampling occurred three days after salt additions. Values are means ± SE.

Figure S4 Densities of scales on stems during the 2012 field experiment. To highlight site and salt effects, data are shown separately for Scale (A) and No Scale (B) stems. Values are means ± SE.

Table S1 ANOVA table examining the effects of factors on *Spartina* growth during the 2011 Mesocosm experiment.

Table S2 ANOVA table examining the effects of factors on days to *Spartina* senescence during the 2012 Field experiment.

Acknowledgments

We thank Jeff Crooks and Brian Collins for providing access to Sweetwater Marsh. Steve Schroeter, Mark Page, and Luke Miller contributed thoughtful discussion on these topics. Heidi Long, Shelby Rinehart, and two anonymouns reviewers provided comments that improved the manuscript. Most importantly, data collection benefited from the significant efforts of the *Haliaspis* Chaps and *Spartina* Spartans (H. Ewertz, J. Hayward, P. Hitchcock, N. Nourn, and A. Warneke). This is Contribution No. 36 of the Coastal and Marine Institute Laboratory, San Diego State University.

Author Contributions

Conceived and designed the experiments: JL. Performed the experiments: JL LP. Analyzed the data: JL. Contributed reagents/materials/analysis tools: JL. Contributed to the writing of the manuscript: JL LP.

References

1. Menge BA, Sutherland JP (1987) Community regulation: variation in disturbance, competition, and predation in relation to environmental stress and recruitment. American Naturalist 130: 730–757.
2. Larsson S (1989) Stressful times for the plant stress-insect performance hypothesis. Oikos 56: 277–283.
3. White TCR (1969) An index to measure weather-induced stress of trees associated with outbreaks of psyllids in Australia. Ecology 50: 905–909.
4. Price PW (1991) Plant vigor and herbivore attack. Oikos 62: 244–251.
5. Huberty AF, Denno RF (2004) Plant water stress and its consequences for herbivorous insects: a new synthesis. Ecology 85: 1383–1398.
6. Rosenthal JP, Kotanen PM (1994) Terrestrial plant tolerance to herbivory. Trends in Ecology & Evolution 9: 145–148.
7. Maschinski J, Whitham TG (1989) The continuum of plant responses to herbivory: the influence of plant association, nutrient availability, and timing. American Naturalist 134: 1–19.
8. Hawkes CV, Sullivan JJ (2001) The impact of herbivory on plants in different resource conditions: a meta-analysis. Ecology 82: 2045–2058.
9. Wise MJ, Abrahamson WG (2005) Beyond the compensatory continuum: environmental resource levels and tolerance of herbivory. Oikos 109: 417–428.
10. Gonzales WL, Suarez LH, Molina-Montenegro MA, Gianoli E (2008) Water availability limits tolerance of apical damage in the Chilean tarweed *Madia sativa*. Acta Oecologica 34: 104–110.
11. Arthur AL, Hoffman AA, Umina PA (2013) Impact of *Halotydeus destructor* on crop seedlings at different plant developmental stages and levels of moisture stress. Environmental Entomology 42: 998–1012.
12. Louthan AM, Doak DF, Goheen JR, Palmer TM, Pringle RM (2013) Climatic stress mediates the impacts of herbivory on plant population structure and components of individual fitness. Journal of Ecology 101: 1074–1083.
13. Chase JM, Leibold MA, Downing AL, Shurin JB (2000) The effects of productivity, herbivory, and plant species turnover in grassland food webs. Ecology 81: 2485–2497.
14. Willis AJ, Ash JE, Groves RH (1993) Combined effects of two arthropod herbivores and water stress on growth of Hypericum species. Oecologia 96: 517–525.
15. Joern A, Mole S (2005) The plant stress hypothesis and variable responses by blue grama grass (*Bouteloua gracilis*) to water, mineral nitrogen, and insect herbivory. Journal of Chemical Ecology 31: 2069–2090.
16. Grinnan R, Carter TE Jr, Johnson MTJ (2013) The effects of drought and herbivory on plant-herbivore interactions across 16 soybean genotypes in a field experiment. Ecological Entomology 38: 290–302.
17. Munns R (2002) Comparative physiology of salt and water stress. Plant Cell and Environment 25: 239–250.
18. Stewart GR, Lahrer F (1980) Accumulation of amino acids and related compounds in relation to environmental stress. In: Miflin BJ, editor. Biochemistry of plants: a comprehensive treatise on amino acids and derivatives. New York: Academic Press. 609–635.
19. Bowdish TI, Stiling P (1998) The influence of salt and nitrogen on herbivore abundance: direct and indirect effects. Oecologia 113: 400–405.
20. Stiling P, Moon D (2005) Are trophodynamic models worth their salt? Top-down and bottom-up effects along a salinity gradient. Ecology 86: 1730–1736.
21. Bradley PM, Morris JT (1991) Relative importance of ion exclusion, secretion and accumulation in *Spartina alterniflora* Loisel. Journal of Experimental Botany 42: 1525–1532.
22. Pennings SC, Callaway RM (1992) Salt marsh plant zonation: the relative importance of competition and physical factors. Ecology 73: 681–690.
23. Zedler JB, Nordby CS, Kus BE (1992) The ecology of Tijuana Estuary: a National Estuarine Research Reserve. Washington, D.C.: NOAA Office of Coastal Resource Management, Sanctuaries and Research Division.
24. Moon DC, Stiling P (2002) Top-down, bottom-up, or side to side? Within-trophic-level interactions modify trophic dynamics of a salt marsh herbivore. Oikos 98: 480–490.
25. Jimenez JM, Wieski K, Marczak LB, Ho C-K (2012) Effects of an omnivorous katydid, salinity, and nutrients on a planthopper-*Spartina* food web. Estuaries and Coasts 35: 475–485.
26. Goranson CE, Ho CK, Pennings SC (2004) Environmental gradients and herbivore feeding preferences in coastal salt marshes. Oecologia 140: 591–600.
27. Silliman BR, van de Koppel J, Bertness MD, Stanton LE, Mendelssohn IA (2005) Drought, snails, and large-scale die-off of southern U.S. salt marshes. Science 310: 1803–1806.
28. Long JD, Mitchell JL, Sotka EE (2011) Local consumers induce resistance differentially between *Spartina alterniflora* populations in the field. Ecology 92: 180–188.
29. Adam P (2002) Saltmarshes in a time of change. Environmental Conservation 29: 39–61.
30. Hopkinson CS, Lugo AE, Alber M, Covich AP, Van Bloem SJ (2008) Forecasting effects of sea-level rise and windstorms on coastal and inland ecosystems. Frontiers in Ecology and the Environment 6: 255–263.
31. Cockfield SD, Potter DA (1986) Interaction of Euonymus scale (Homoptera: Diaspididae) feeding damage and severe water stress on leaf abscission and growth of *Euonymus fortunei*. Oecologia 71: 41–46.
32. Vranjic JA, Ash JE (1997) Scale insects consistently affect roots more than shoots: the impact of infestation size on growth of eucalypt seedlings. Journal of Ecology 85: 143–149.
33. Howard FW, Hamon A, McLaughlin M, Weissling T, Yang S-L (1999) *Aulacaspis yasumatsui* (Hemiptera: Sternorrhyncha: Diaspididae), a scale insect pest of cycads recently introduced into Florida. Florida Entomologist 82: 14–27.
34. Boyer KE, Zedler JB (1996) Damage to cordgrass by scale insects in a constructed salt marsh: effects of nitrogen additions. Estuaries 19: 1–12.
35. Zedler JB (1993) Canopy architecture of natural and planted cordgrass marshes: selecting habitat evaluation criteria. Ecological Applications 3: 123–138.
36. Crain CM, Silliman BR, Bertness SL, Bertness MD (2004) Physical and biotic drivers of plant distribution across estuarine salinity gradients. Ecology 85: 2539–2549.
37. Boyer KE (1994) Scale insect damage in constructed salt marshes: nitrogen and other factors. San Diego, CA: San Diego State University. 81 p.
38. Covin JD, Zedler JB (1988) Nitrogen effects on *Spartina foliosa* and *Salicornia virginica* in the salt marsh at Tijuana Estuary, California. Wetlands 8: 51–65.
39. Johnson KM (1991) The effects of host quality on a phytophagous insect (Homoptera: Delphacidae) and its predators in a California salt marsh system. San Diego, CA: San Diego State University. 85 p.
40. Spitzer BW (2004) Maternal effects in the soft scale insect Saissetia coffeae (Hemiptera: Coccidae). Evolution 58: 2452–2461.
41. Louda SM, Collinge SK (1992) Plant resistance to insect herbivores: a field test of the environmental stress hypothesis. Ecology 73: 153–169.
42. Trnka S, Zedler JB (2000) Site conditions, not parental phenotype, determine the height of *Spartina foliosa*. Estuaries 23: 572–582.
43. Pennings SC, Moore DJ (2001) Zonation of shrubs in western Atlantic salt marshes. Oecologia 126: 587–594.
44. Bandyopadhyay BK, Pezeshki SR, DeLaune RD, Lindau CW (1993) Influence of soil oxidation-reduction potential and salinity on nutrition, N^{15} uptake, and growth of *Spartina patens*. Wetlands 13: 10–15.
45. Brown CE, Pezeshki SR, DeLaune RD (2006) The effects of salinity and soil drying on nutrient uptake and growth of *Spartina alterniflora* in a simulated tidal system. Environmental and Experimental Botany 58: 140–148.
46. Wise MJ, Abrahamson WG (2007) Effects of resource availability on tolerance of herbivory: a review and assessment of three opposing models. American Naturalist 169: 443–454.
47. Canepuccia AD, Alberti J, Daleo P, Pascual J, Farina JL, et al. (2010) Ecosystem engineering by burrowing crabs increases cordgrass mortality caused by stem-boring insects. Marine Ecology Progress Series 404: 151–159.

48. Hjalten J, Danell K, Ericson L (1993) Effects of simulated herbivory and intraspecific competition on the compensatory ability of birches. Ecology 74: 1136–1142.
49. Pennings SC, Grant M-B, Bertness MD (2005) Zonation in low-latitude salt marshes: disentangling the roles of flooding, salinity, and competition. Journal of Ecology 93: 159–167.
50. Denno RF, Gratton C, Peterson MA, Langellotto GA, Finke DL, et al. (2002) Bottom-up forces mediate natural-enemy impact in a phytophagous insect community. Ecology 83: 1443–1458.
51. Williams K (1990) Control of scale insect infestation at Chula Vista Wildlife Reserve. San Diego, California.
52. Japoshvili G, Russell E (2012) A new parasitization record of *Haliaspis spartinae* (Diaspididae) and *Encarsia ellisvillensis* sp. nov. (Chalcidoidea: Aphelinidae) from the United States. Annals of the Entomological Society of America 105: 493–497.
53. Macdonald K (1990) South San Diego Bay Enhancement Plan. San Diego, California.
54. Langis R, Zalejko M, Zedler JB (1991) Nitrogen assessments in a consructed and a natural salt marsh of San Diego Bay. Ecological Applications 1: 40–51.
55. Zedler JB, Callaway JC (1999) Tracking wetland restoration: do mitigation sites follow desired trajectories. Restoration Ecology 7: 69–73.

Double-Stranded RNA Uptake through Topical Application, Mediates Silencing of Five CYP4 Genes and Suppresses Insecticide Resistance in *Diaphorina citri*

Nabil Killiny[1]*, Subhas Hajeri[2], Siddharth Tiwari[1], Siddarame Gowda[2], Lukasz L. Stelinski[1]

1 Department of Entomology and Nematology, Citrus Research and Education Center, IFAS, University of Florida, Lake Alfred, Florida, United States of America,
2 Department of Plant Pathology, Citrus Research and Education Center, IFAS, University of Florida, Lake Alfred, Florida, United States of America

Abstract

Silencing of genes through RNA interference (RNAi) in insects has gained momentum during the past few years. RNAi has been used to cause insect mortality, inhibit insect growth, increase insecticide susceptibility, and prevent the development of insecticide resistance. We investigated the efficacy of topically applied dsRNA to induce RNAi for five Cytochrome P_{450} genes family 4 (*CYP4*) in *Diaphorina citri*. We previously reported that these *CYP4* genes are associated with the development of insecticide resistance in *D. citri*. We targeted five *CYP4* genes that share a consensus sequence with one dsRNA construct. Quantitative PCR confirmed suppressed expression of the five *CYP4* genes as a result of dsRNA topically applied to the thoracic region of *D. citri* when compared to the expression levels in a control group. Western blot analysis indicated a reduced signal of cytochrome P_{450} proteins (45 kDa) in adult *D. citri* treated with the dsRNA. In addition, oxidase activity and insecticide resistance were reduced for *D. citri* treated with dsRNA that targeted specific *CYP4* genes. Mortality was significantly higher in adults treated with dsRNA than in adults treated with water. Our results indicate that topically applied dsRNA can penetrate the cuticle of *D. citri* and induce RNAi. These results broaden the scope of RNAi as a mechanism to manage pests by targeting a broad range of genes. The results also support the application of RNAi as a viable tool to overcome insecticide resistance development in *D. citri* populations. However, further research is needed to develop grower-friendly delivery systems for the application of dsRNA under field conditions. Considering the high specificity of dsRNA, this tool can also be used for management of *D. citri* by targeting physiologically critical genes involved in growth and development.

Editor: Immo A. Hansen, New Mexico State University, United States of America

Funding: This project was supported by a grant to LS from the Citrus Research and Development Foundation. The funders had no role in study design, data collection and analysis, decision to publish, or preparation of the manuscript.

Competing Interests: The authors have declared that no competing interests exist.

* Email: nabikilliny@ufl.edu

Introduction

RNA interference (RNAi) is a promising tool for studying functional genomics in eukaryotes and insects in particular [1,2]. Anti-sense (nonsense) RNA strand transcription has been used for over three decades to inhibit gene activity [3]. The efficacy of anti-sense silencing depends on hybridization between the injected RNA and an endogenous messenger. Since the discovery of double-stranded RNA (dsRNA) mediated gene-specific silencing in the nematode, *Caenorhabditis elegans* (Maupas) [4], dsRNA-mediated RNAi has been employed with various insects to silence specific genes [2]. RNAi has been widely used in various insect orders, including Coleoptera, Dictyoptera, Diptera, Hemiptera, Hymenoptera, Isoptera, Lepidoptera, Neuroptera, and Orthoptera [5,6,7]. The systemic nature of dsRNA-mediated RNAi has allowed this tool to be used in the management of various insect pests [8–13].

The cytochrome P_{450} monooxygenases are an important group of enzymes that are involved in the metabolism of xenobiotic compounds in insects. This group of enzymes is associated with insecticide resistance and metabolism of a wide range of endogenous and exogenous compounds that includes hormones, pheromones, insecticides, and plant secondary compounds in insects [14–18]. Overtranscription of families 4, 6, 9, and 12 has been frequently linked to insecticide metabolism and resistance [19–22].

The Asian citrus psyllid, *Diaphorina citri* Kuwayama (Hemiptera: Psyllidae), is perhaps the most destructive pest of citrus, mainly because it is a vector for the putative causal agent of huanglongbing (HLB), *Candidatus* Liberibacter asiaticus (*C*Las) [23]. HLB is a deadly citrus disease with no known cure [23,24]. Currently, the main tools that limit the spread of the disease are insecticides to manage the vector [18,25,26]. *D. citri* are susceptible to several insecticide classes, which includes the pyrethroids, organophosphates, carbamates, neonicotinoids, insect growth regulators, horticultural oils, and lipid synthesis inhibitors [27]. Foliar treatments may suppress populations for 3 weeks following application [27]. Broad-spectrum insecticides (pyrethroids, organophosphates, and neonicotinoids) are more effective against *D. citri* than IGRs or oils, and insecticide use against *D. citri* is most effective when populations are not actively reproducing [28]. Systemic soil-applied insecticides provide a much longer duration of population control (months) than foliar insecticides (weeks) [27]. The neonicitonoids have been the main class of effective systemic insecticides for *D. citri* control during the past decade [27]. Systemic neonicotinoids are particularly effective in protecting young trees as they mature into production [29].

Intense insecticide use has led to the development of varying levels of insecticide resistance in populations of *D. citri* in Florida, USA [26]. This is particularly concerning for the neonicotinoid class, since these are the main current tools for protecting young trees from *C*Las infection [29]. A metabolic mechanism for the evolution of insecticide resistance in populations of *D. citri*, particularly for neonicotinoids, is supported by increased activities of detoxifying enzymes and overexpression of Cytochrome P_{450} genes family 4 (*CYP4*) [17,21,26,30,31]. In the present study, we targeted the abovementioned *CYP4* genes for silencing by topical application of specific dsRNA to the thorax of newly emerged *D. citri* adults. Additionally, we tested the effect of dsRNA treatment on insecticide resistance by comparing mortality of known susceptible and resistant populations of *D. citri*.

Materials and Methods

Insect populations

A laboratory susceptible population (LS) of *D. citri* was maintained in a greenhouse at the Citrus Research and Education Center, Lake Alfred, Florida. The culture was established in 2000 using field populations from Polk County, Florida and maintained on sweet orange (*Citrus sinensis* (L.) Osbeck) without exposure to insecticides in a greenhouse at 27–28°C, with 60–65% relative humidity and a 14:10 (light:dark) photocycle hours. Three field populations of *D. citri* were collected from commercial citrus groves in Florida during 2013. The populations were collected with permissions from private groves. Name of groves, counties, and GPS coordinates are as the following: *i*) GapWay Groves (Private managed grove), Polk County (PL) (28° 05′ 40.14″ N; 81° 43′ 19.03″ W); *ii*) Winter Garden, Conserve II (Private managed grove), Lake County (LA) (28° 27′ 52.17″ N; 81° 39′ 31.69″ W); and *iii)* Uncle Matt's Organic (Organically managed grove), Lake County (OG) (28° 31′ 00.88″ N; 81° 40′ 01.90″ W). Adults were collected using sweep nets and aspirators, transferred to the laboratory, released onto citrus plants within Plexiglas cages (40×40×40 cm), and used in bioassays shortly thereafter.

Constructing dsRNA

A consensus sequence, derived from five previously published *CYP4* sequences [21], was used to design *CYP4*-specific primers (Table 1). The *CYP4*-specific primers were tailed with a T7 promoter sequence to generate sense and antisense transcripts separately.

Total RNA isolation was performed on groups of 40–50 psyllids using the SV total RNA isolation kit (Promega, Madison, WI, USA). One microgram of RNA was used to synthesize cDNA using the *CYP4*-specific reverse primers and iScript cDNA synthesis kit (Bio-Rad, Hercules, CA, USA). Sense and antisense PCR products were generated in separate PCR reaction using specific combination of primers (Table 1). To generate plus-sense

Table 1. Primers used in this investigation.

Purpose	Gene	Sequence	Reference
ds-RNA synthesis			
	CYP4	Forward CACG*ttaattaa*ACGTTCATGTTCGAGGGGCACGATACAACAAC	This
	Sense	Reverse GAGC*aggcct*GAAGGGTACATAGGAGTAAGGATGACGTTTCTG	investigation
	CYP4 Antisense	Forward GCAGCA*TAATACGACTCACTATAGGGAGA*GAAGGGTACATAGGAGTAAGGATGACGTTTCTG Reverse ACGTTCATGTTCGAGGGGCACGATACAACAAC	This investigation
	GFP	Forward GCGAACTAGTATGGCTAGCAAAGGAGAAGAACTTTTCACTG	El-Shesheny
		Reverse GAGACCGCGGCTACCCCTCGAGTTATTTGTAGAGCTCATC	et al. [32]
Gene expression			
	CYP4 genes		
	CYP4C67	Forward TGGAACGTGTCATCAAGGAG	Tiwari et al.
		Reverse CCGGATTGAAACTGTTAGGC	[21]
	CYP4DA1	Forward AGTGGTGTCGGAAATTGAGG	
		Reverse GTTCGAGCCACCTGGAGATA	
	CYPC68	Forward CTAGCCTGGACCCTCTTCCT	
		Reverse ACCCTCCCTATGAACGGAAC	
	CYPG70	Forward GCCGGAAGTTCTTTCTTCCT	
		Reverse TAACGGGTACTGGTGGGAAC	
	CYPDB1	Forward CTGTACGCTCTGGGACATCA	
		Reverse TTGAGCGGTGCATAGAGTTG	
	Reference genes		
	α-tubulin	Forward CAGGTCTTGTGTGGGACGTA	Tiwari et al. [21]
		Reverse GGCCACAGTTTGTTTCTTGC	
	Actin	Forward CCCTGGACTTTGAACAGGAA	
		Reverse CTCGTGGATACCGCAAGATT	

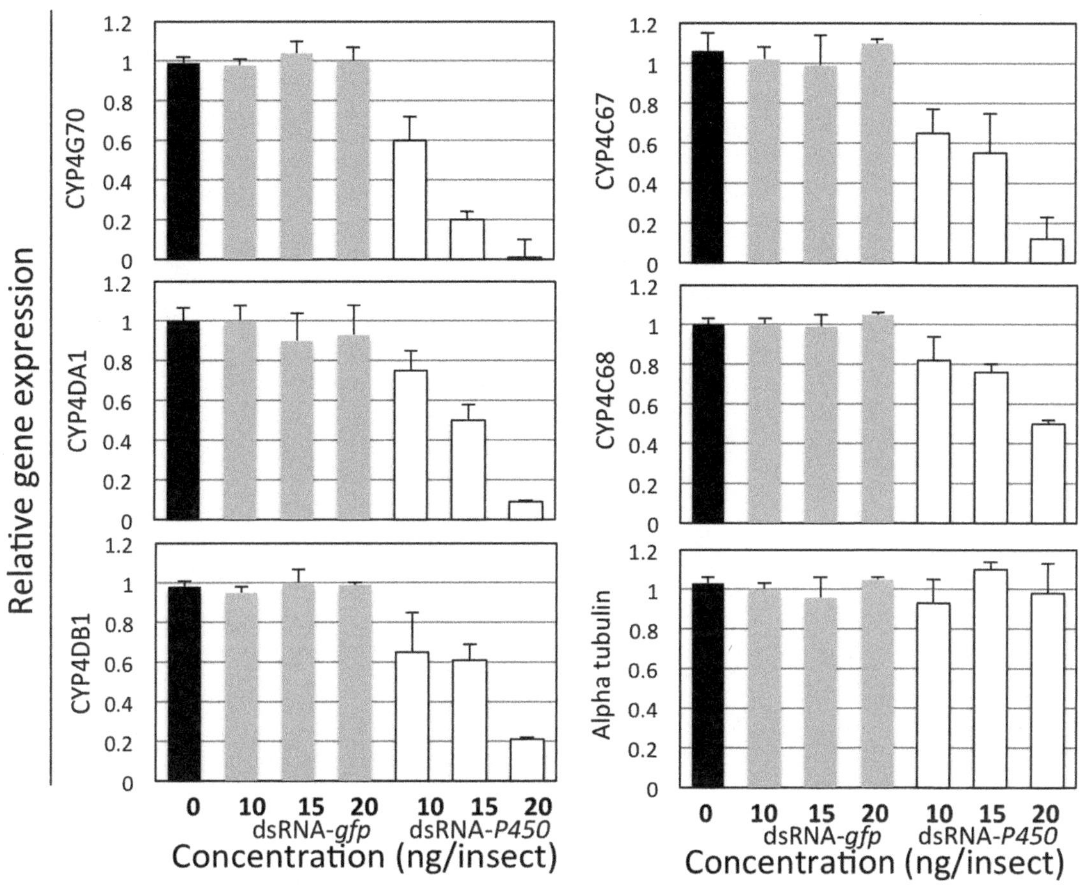

Figure 1. Relative expression levels of the five *CYP4* genes targeted by RNAi in *Diaphorina citri* adults 72 h after treatment with dsRNA. Ct values were first normalized to the endogenous control gene *Actin* followed by normalization to the treatment giving the lowest gene expression using the $2^{-\Delta\Delta CT}$ method. Standard deviations were calculated based on three independent experiments, each with three technical replicates. Alpha-tubulin was used as a non-target gene control. dsRNA-*gfp* treatment was used as a control targeting an irrelevant gene.

transcripts, sense primer with T7 promoter sequence and regular antisense primer were used; while to generate antisense transcripts, regular sense primer and antisense primer with T7 promoter sequence were used. Sense and antisense transcripts were annealed by denaturing at 70°C for 10 min, followed by slowly cooling to room temperature for 20 min. To eliminate DNA template and single-stranded RNA, dsRNA was treated with DNase I and RNase A. The dsRNA was then purified of proteins and free nucleotides using the phenol-chloroform purification method. The amount of purified dsRNA was measured with a NanoDrop Spectrophotometer. We used dsRNA-*gfp* as an irrelevant dsRNA (control). dsRNA-*gfp* was produced as described above. Green fluorescent protein (GFP) mRNA is 732 bp in length. Specific primers (Table 1) were used to amplify the full-length of GFP gene by using TMV-30BGFP according to El-Shesheny et al. [32].

D. citri treatment with dsRNA

Purified dsRNA was serially diluted using RNase-free water to obtain desired concentrations of dsRNA. Three concentrations of dsRNA (50, 75, and 100 ng/μl) and a control (0 ng/μl) were used to treat *D. citri* adults. *D. citri* adults were anaesthetized under CO_2 within a few hours of eclosion. A 0.2 μl droplet containing10, 15, or 20 ng of dsRNA was topically applied to the ventral side of the thorax using a 10 μl Hamilton syringe. To investigate the effect of dsRNA- P_{450} on gene and protein expression and enzymatic activity, treated adults were placed into 60 mm plastic disposable Petri dishes that were lined with citrus leaf disks, as a food source, over agar beds as described in Tiwari et al. [17]. Petri dishes with treated adults were kept at 25±1°C and 50±5% RH, with a 14:10 h light:dark photoperiod, in a growth chamber for 72 h. Insects were collected and stored in −20°C until use. dsRNA-*gfp* was used as a non-relevant dsRNA control.

Cytochrome P_{450} (general oxidase) assay

The activity of cytochrome P_{450} was quantified and expressed in terms of general oxidase level, which is an indirect measure of cytochrome P_{450} by using heme peroxidation as described in Tiwari et al. [31]. This method has been considered a reliable tool for comparing differences in general oxidase levels based on hemoprotein levels. Because heme constitutes the majority of cytochrome P_{450} in non-blood-fed insects, quantification of heme activity has been used to compare the levels of cytochrome P_{450} on the basis of general oxidase levels [33]. In brief, heme peroxidase activity was measured using 3,3′5,5′-tetra-methylbenzidine (TMBZ) (Sigma Aldrich) as the substrate. Five replicates, each consisting of three insects, were performed for each treatment.

Gene expression analysis

Live adult *D. citri* from each treatment were subjected to RNA isolation and cDNA synthesis. RNA isolations were performed in

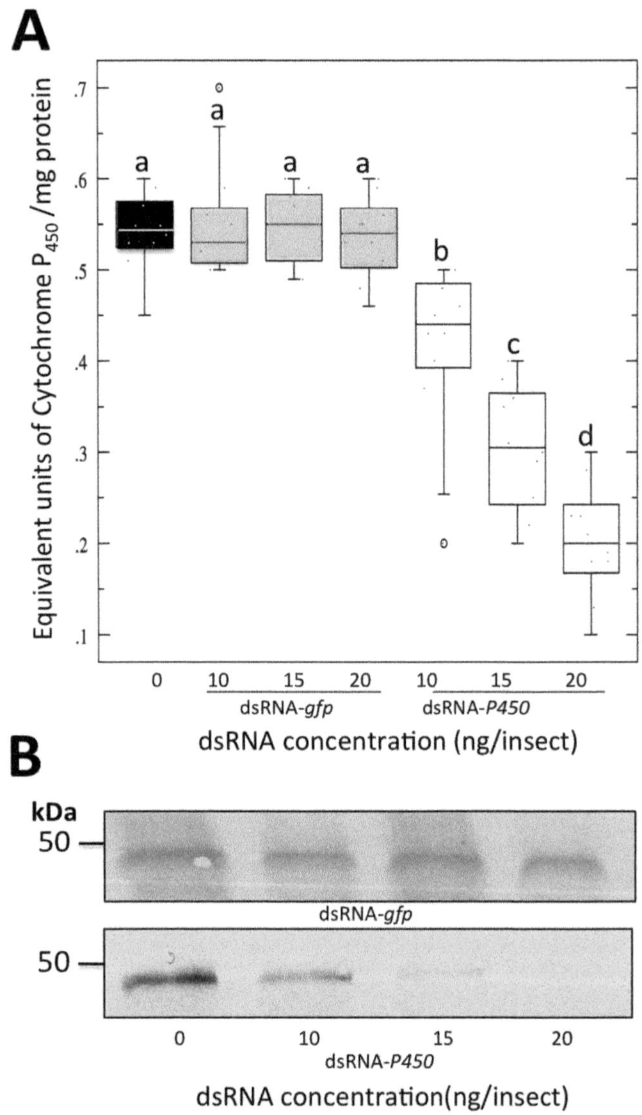

Figure 2. Cytochrome P_{450} (general oxidase) activity and protein expression in dsRNA-treated *D.citri*. A) Box blot representing general oxidase activity. Boxes indicate the interquartile range, including 50% of results, and the mid-horizontal lines represent the median; different letters above deviation error bars represent significant differences between treatments ($P<0.05$). B) Protein analysis of dsRNA-treated and control *D. citri* using Western blot. Western blot was performed using the microsomal proteins prepared from adults treated with three quantities of dsRNA and a control. dsRNA-*gfp* treatment was used as a control targeting an irrelevant gene.

three biological replicates using the SV total RNA isolation kit (Promega, Madison, WI, USA). The quantity and quality of RNA from each sample was measured on a NanoDrop 1000 Spectrophotometer using the absorbance at 260 nm and the A260/A280 ratio, respectively. Subsequently, cDNA was synthesized with the iScript cDNA synthesis kit (Bio-Rad, Hercules, CA, USA) for each replicate within each treatment. Quantitative real-time PCR (qPCR) was performed using iQ SYBR Green Supermix with an iCycler iQ real-time PCR detection system (Bio-Rad). Primers for five *CYP4* genes, *Alpha-tubulin* and the endogenous gene, *Actin*, were used to measure the gene expression of cytochrome P_{450}, as described in Tiwari et al. [21] (Table 1). Six biological replicates were performed for each treatment. The production of gene-specific products and absence of 'primer dimers' was verified by 1% agarose electrophoresis in TAE buffer with ethidium bromide staining.

The $2^{-\Delta\Delta CT}$ method was used to compare the relative expression of the consensus sequence among PCR products derived from the three dsRNA concentrations and control treatments [34]. This was done by first normalizing the expression level of dsRNA treated samples to *Actin* [21] gene expression, followed by normalization to the treatment giving the lowest gene expression. *Alpha-tubulin* was used as a non-targeted gene (control). Five biological replicates, each consisting of three insects and three technical replicates, were performed for each treatment.

Western blot assay

Subcellular protein fractions were extracted using the methods described by Wheeler et al. [35] from adults in each treatment. The protein concentration was determined by the Bradford method [36] using a protein assay kit (Bio-Rad Laboratories, Hercules, CA, USA) with ovalbumin as the standard. Since cytochrome P_{450} proteins were clearly detected in the microsomal protein [22], we used the microsomal fractions to perform the Western blot analysis, as described by Tiwari et al. [22].

Survival assay

The survival assay was carried out on *D. citri* treated with dsRNA-*gfp*, dsRNA- P_{450}, or RNase-free water as a control. Insects were placed on an autoclaved clear plant tissue culture container (75×75×100 mm) lined with 0.5 mm filter paper saturated with 20% sucrose. Fifty insects were placed per container and five replicates were performed for each treatment. Live insects were counted daily.

Residual activity of dsRNA- P_{450}

In order to assess the duration of the RNAi effect, 70 insects treated with RNase-free water or dsRNA-*P450* (20 ng/insect) were placed into plant tissue culture containers with filter paper saturated with 20% sucrose. Samples consisting of three insects were taken daily and kept at −20°C. Cytochrome P_{450} (general oxidase) activity was measured in all samples as described above. Five replicates were performed for both treatments.

Pesticide application

To investigate the effect of treatment with dsRNA- P_{450} on insecticide resistance, *D. citri* adults were treated with the dsRNAs (20 ng/insect) as described above. *D. citri* were initially maintained on Petri dishes with untreated citrus leaf disks for 72 h and thereafter transferred to new Petri dishes that contained leaf discs treated with insecticide solution. Briefly, the leaf disks (60 mm diameter) were excised, dipped in the insecticide solution made in acetone for 30 s, and allowed to air dry in a fume hood for 1 h prior to placement into the Petri dishes as described by Tiwari et al. [26]. We used analytical-grade imidacloprid at the LD_{50} dosage (0.02 ng AI/µl acetone) previously determined by Tiwari et al. [26]. The mortality of *D. citri* adults was assessed after 24 h. dsRNA-*gfp* was used as a negative control for dsRNA- P_{450}. Fives replicates (Petri dishes), with five insects each, were performed for each of the four *D. citri* populations tested. Each population was subjected to four different treatments: *D. citri* treated with RNase-free water on 1) RNase-free water- or 2) imidacloprid-treated disks, as well as *D. citri* treated with dsRNA- P_{450} on 3) RNase-free water- or 4) imidacloprid-treated leaf disks.

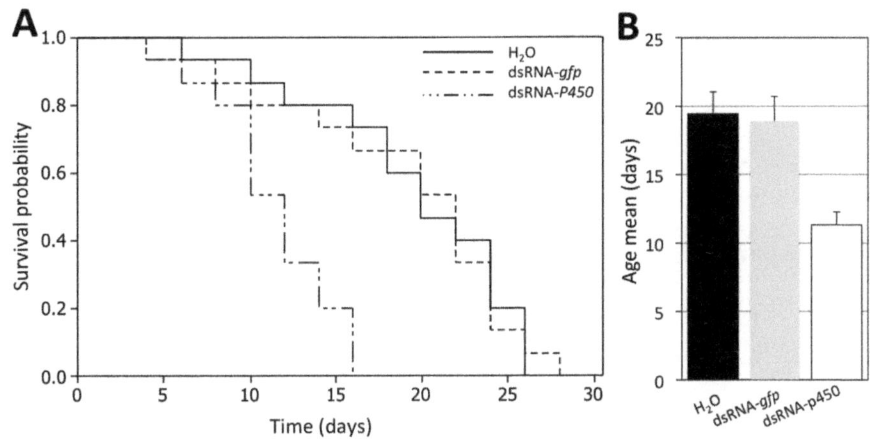

Figure 3. Effect of dsRNA- P_{450} on *D. citri* survival. A) Kaplan-Meier survival curve showing the effect of dsRNA- P_{450} treatment on lifespan of *D. citri*. B) Average life span of *D. citri*. Bars represent the standard deviations; different letters above error bars represent significant differences between treatments ($P<0.05$). dsRNA-*gfp* treatment was used as a control targeting an irrelevant gene.

Statistical analysis

All analyses were performed using SPSS version 19.0. Survival was calculated during the interval from initial treatment to when all insects died. Overall survival (OS) curves were obtained using the Kaplan-Meier method and comparisons were made using log rank and Wilcoxon tests. Analysis of variance (ANOVA) was used to compare: *i*) Calculated *D. citri* lifespans between various treatments, *ii*) The effect of dsRNA treatments on *CYP4* activity and insecticide resistance, and *iii*) The duration of the RNAi effect. Post hoc pairwise comparisons between treatments were performed with the Tukey honestly significant difference test. Statistical significance was established as $P<0.05$.

Results

Treatment with dsRNA-P_{450} causes down regulation of five *CYP4* genes

Relative expression levels for the five *CYP4* genes were compared between dsRNA- P_{450}, dsRNA-*gfp*-treated and control psyllids (Figure 1). Treatment with dsRNA- P_{450} caused reduced expression of the five *CYP4* genes. The effect of dsRNA- P_{450} was positively correlated with the quantity applied per treatment. The expression level of α-tubulin (non-target gene) remained constant among all treatments indicating the specificity of dsRNA- P_{450} to *CYP4* genes. In contrast, there was no effect of the dsRNA-*gfp* control treatments targeting irrelevant psyllid genes on expression levels of *CYP4* genes. The greatest reduction in expression level was found with *CYP4G70*, while the lowest effect was with *CYP4C68*. This observation may help in evaluating gene candidates for RNAi technology for *D. citri*.

Treatment with dsRNA-P_{450} reduces the protein expression and the enzymatic activity of *CYP4*

We investigated the effect of treating *D. citri* with dsRNA on general oxidase activity. The activities were similar for all doses of dsRNA-*gfp* treatment, while reduced when *D. citri* were treated with dsRNA-P_{450}. The activity decreased as the concentration of applied dsRNA- P_{450} was increased. Additionally, Western blots performed using the microsomal fractions revealed the presence of a band corresponding to a 45 kDa protein that cross-reacted with

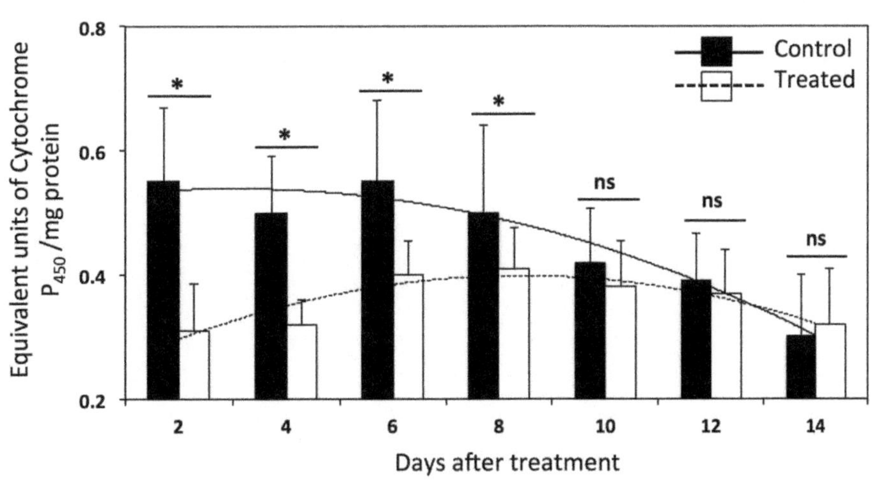

Figure 4. P_{450} as measured by general oxidase activity. Asterisks indicate significant differences between dsRNA- P_{450} treated and non-treated *D. citri*, while ns indicates no significant differences($P<0.05$).

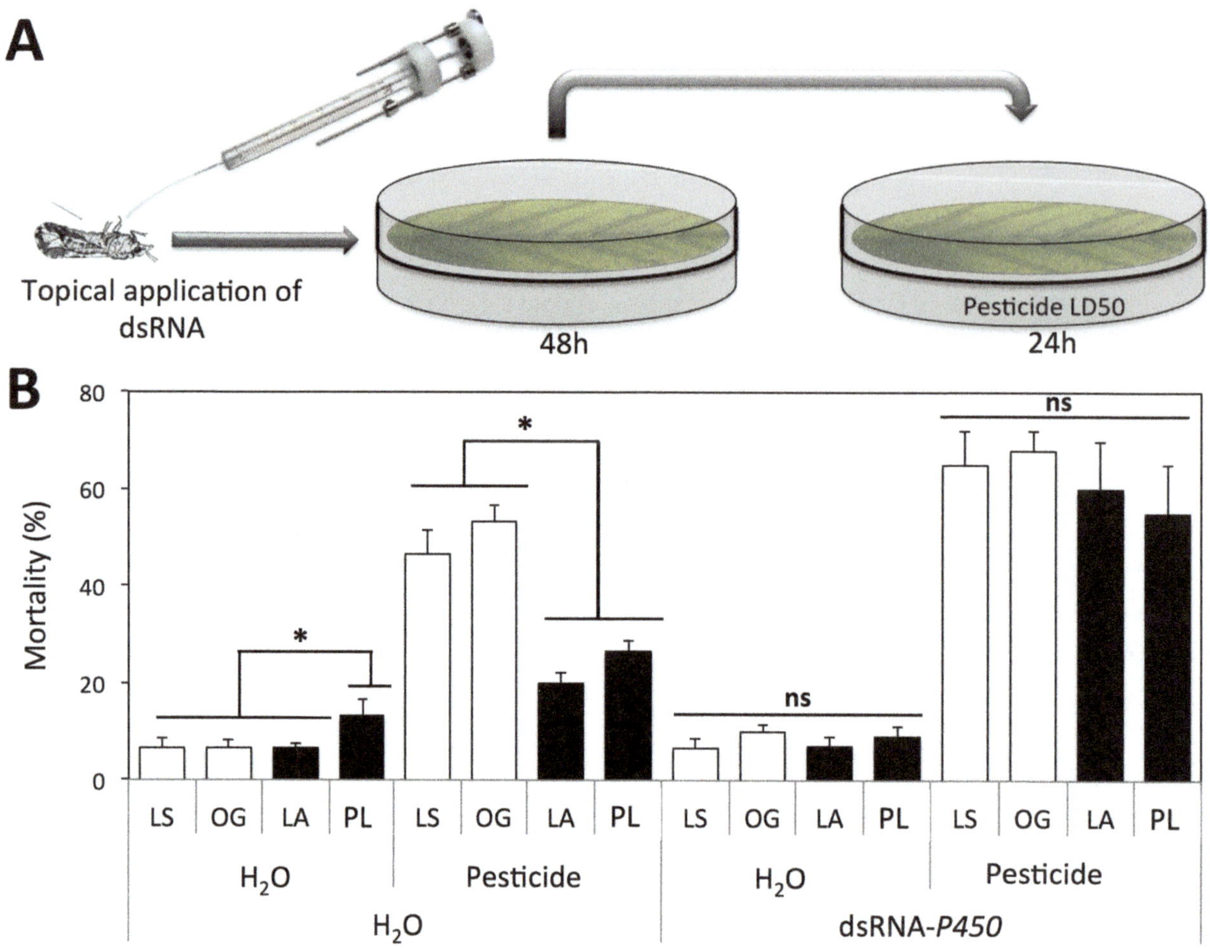

Figure 5. Effect of dsRNA- P_{450} treatment on *D. citri* susceptibility to imidacloprid using the LD_{50} dosage. A) Illustration of the protocol used to test the effect. B) Percent mortality of *D. citri* after exposure to imidacloprid for 24 h. Insects were maintained on non-treated leaf discs for 48 h after the dsRNA- P_{450} treatment prior to exposure to imidacloprid. LS: Laboratory susceptible population. OG: Susceptible population collected from an organic grove. PL: Resistant population collected from commercially managed Polk County, Florida. LA: Resistant population collected from commercially managed Lake County, Florida. Asterisks indicate significant differences, while ns indicates no significant differences ($P<0.05$).

the primary antibody of cytochrome P_{450} protein. Twenty-five micrograms of microsomal proteins for each treatment were used to perform the Western blot which indicated the highest amount of cytochrome P_{450} proteins (detoxifying enzymes) in *D. citri* treated with 0 ng/µl of dsRNA, followed by *D. citri* treated with 10, 15, and 20 ng/insect of dsRNA (Figure 2). There was no signal detected in *D. citri* treated with 20 ng/insect of dsRNA. Expression levels of target *CYP4* and oxidase activity, as a result of dsRNA treatment, directly correlated with the protein expression in adults treated with dsRNA.

Silencing of *CYP4* reduced the lifespan of *D. citri*

Survival of *D. citri* was quantified following treatment with dsRNA-*gfp*, dsRNA- P_{450}, and the control (water). The experiment was conducted under the conditions described earlier (Figure 3A). In this experiment, a 20 ng/insect concentration was used for both dsRNA-*gfp* and dsRNA- P_{450}. A Kaplan-Meier survival plot indicated significant differences among all treatments (log rank = 154.63, $P<0.001$). No significant differences in survival were found between the control and dsRNA-*gfp*-treated *D. citri* (log rank = 1.54, $P = 0.64$). The lifespan in dsRNA- P_{450}-treated *D. citri* was significantly shorter than that observed for other treatments. Mean lifespans for the treatments are presented in Figure 3B. This suggests that reduced expression of *CYP4* genes shortens the lifespan of *D. citri*.

Residual activity of dsRNA- P_{450} treatment

The residual activity of dsRNA- P_{450} was measured at 20 ng/insect concentration. We used general oxidase activity as an indicator for the residual of dsRNA- P_{450}. We compared oxidase activity between the control and dsRNA- P_{450}-treated *D. citri* daily after the treatment application. Oxidase activity was significantly reduced following application of dsRNA- P_{450} for up to 8 days (Figure 4).

Silencing of *CYP4* increased insecticide susceptibility

In order to determine the effect of dsRNA- P_{450} on insecticide susceptibility of *D. citri*, we used imidacloprid at the LD_{50} dosage. Two susceptible and two resistant populations were used in this experiment (Figure 5). *D. citri* that were treated with RNase-free water and then exposed to leaf discs treated with imidacloprid exhibited differing susceptibilities, depending on the population tested (Figure 5). Specifically, the two populations from commercially managed citrus groves that had received imidacloprid treatment over the previous several years (LA (Lake County) and PL (Polk County)) were less susceptible to imidacloprid at the LD_{50} dosage than *D. citri* collected from our Laboratory Susceptible culture (LS) and from the Organic Grove (OG) where imidacloprid had not been used previously (Figure 5). Mortality of *D. citri* exposed to imidacloprid after treatment with dsRNA- P_{450} was increased for each of the four populations as compared with the water control (Figure 5). Mortality of *D. citri* from the two resistant populations (LA, PL), at the LD_{50} dosage of imidacloprid, was significantly higher after treatment with dsRNA- P_{450} as compared with the water control (Figure 5). Given that dsRNA-P_{450} also increased mortality of *D. citri* from the two susceptible populations (LS and OG) further suggests that cytochrome P_{450} is implicated in imidacloprid resistance in *D. citri*.

Discussion

Most insect RNAi studies have relied on the delivery of specific dsRNA through either microinjections [4] or ingestion through feeding [10,13]. Each of these methods has advantages and disadvantages. The microinjection method requires intense training and is a notably time-consuming technique. In addition, optimization is required for volume selection, place of injection, and needle size for successful dsRNA injection into the insect body [7]. Delivery of dsRNA through ingestion also has limitations, such as reduced effectiveness for inducing RNAi [37], reduced efficacy of dsRNA due to the unfavorable gut environment [38], and difficulties in quantifying the amount of dsRNA ingested [39]. The current work highlights a novel method of dsRNA delivery through topical microapplication to the abdomen of adult *D. citri*. The ventral microapplication allows dsRNA uptake through the exoskeleton of insect. The uptake occurs via the intersegmental membranes. This investigation describes a relatively easy and efficient method for delivering and allowing the dsRNA to enter the insect's body to induce RNAi. This method has also been described for *D. citri* nymphs with high efficiency of activity [32]. Also, a similar delivery method was reported to induce RNAi in *Ostrinia nubilalis* larvae [11].

A dosage as low as 50 ng/μl of dsRNA down-regulated the expression of the consensus sequence derived from five *CYP4* genes from *D. citri,* as verified by qPCR and Western blot. The lifespan of *D. citri* following dsRNA treatment was statistically shorter as compared with untreated controls in the current investigation. In another example, mortality in *O. nubilalis* larvae ranged between approximately 40–50% following topical treatment with dsRNA [11]. The mortality, coupled with the lack of any other abnormality observed in the dsRNA-treated adult *D. citri*, suggests that the *CYP4* specific dsRNA are highly target specific. Target specificity of dsRNA is also useful considering the potential for dsRNA exposure to non-target organisms under field conditions. Designing target specific dsRNA is not uncommon; species-specific dsRNA has been shown to work like an insecticide by killing specifically targeted insect pests [40]. The low concentrations needed for induction of RNAi and the highly specific nature of dsRNA suggest it might be a tool for managing insecticide resistance in *D. citri*. Our results indicate that dsRNA-P_{450} reduced oxidase activity, which presumably increased insecticide susceptibility in both resistant and susceptible populations of *D. citri*. In comparison, dsRNA-*gfp* (our negative control treatment) did not affect *CYP4* gene expression or oxidase activity. These findings indicate specificity of RNAi for *D. citri* with the genes targeted in the present investigation.

An important challenge for the application of dsRNA for practical pest control is developing a delivery method for commercial field deployment. Another practical limitation of RNAi that needs to be addressed is that large quantities of dsRNA are expensive to produce. Currently, we are working on inserting the previously described dsRNA into citrus plants for direct ingestion by *D. citri* during feeding. Delivery of dsRNA through transgenic plants (Plant mediated RNAi) has been achieved in *Helicoverpa armigera* and *Diabrotica vergifera vergifera* [41,42]. The absence of interferon-regulated innate immunity pathways in insects allows the possibility of employing longer dsRNA for maximal RNAi [43]. Another potentially feasible way of delivering dsRNA would be to incorporate target-specific dsRNA into bacteria with an appropriate transfection reagent and then spraying the transformed bacteria onto citrus trees. However, future work is needed to evaluate the most efficient transfection reagents and bacterial formulations to prevent the breakdown of dsRNA under field conditions. Once the entire genome of *D. citri* is sequenced, this delivery method could be a convenient way to conduct high-throughput loss-of-function research for determining gene functions. In addition, the current results suggest that further work is needed to understand the mechanism of dsRNA entry into cells following topical application of dsRNA onto *D. citri* to induce RNAi.

Acknowledgments

We acknowledge Ian Jackson for *D. citri* collection and technical assistance.

Author Contributions

Conceived and designed the experiments: NK SH ST SG LLS. Performed the experiments: NK SH SG. Analyzed the data: NK SH ST SG LLS. Contributed reagents/materials/analysis tools: NK SH ST SG LLS. Contributed to the writing of the manuscript: NK SH ST SG LLS.

References

1. Scharf ME, Zhou X, Schwinghammer MA (2008) Application of RNA interference in functional genomics studies of a social insect. In: Barik S, editor. Methods in molecular biology, siRNA, shRNA and miRNA protocols. Vol. 442. Totowa, New Jersey: Humana Press. p. 205–229.
2. Huvenne H, Smagghe G (2010) Mechanisms of dsRNA uptake in insects and potential of RNAi for pest control: A review. J Insect Physiol 56: 227–235.
3. Izant JG, Weintraub H (1984) Inhibition of thymidine kinase gene expression by anti-sense RNA: A molecular approach to genetic analysis. Cell 36: 1007–1015.
4. Fire A, Xu S, Montgomery MK, Kostas SA, Driver SE, et al. (1998) Potent and specific genetic interference by double-stranded RNA in *Caenorhabditis elegans*. Nature 391: 806–811.
5. Tomoyasu Y, Miller SC, Tomita S, Schoppmeier M, Grossmann D, et al. (2008) Exploring systemic RNA interference in insects: A genome wide survey for RNAi genes in Tribolium. Genome Biol 9: R10.
6. Wuriyanghan H, Rosa C, Falk BW (2011) Oral delivery of double-stranded RNAs and siRNAs induces RNAi effects in the potato/tomato psyllid, *Bactericerca cockerelli*. PLoS ONE 6: e27736.
7. Yu N, Christiaens O, Liu J, Niu J, Cappelle K, et al. (2012) Delivery of dsRNA for RNAi in insect: An overview and future directions. Insect Sci 20: 4–14.
8. Gordon KH, Waterhouse PM (2007) RNAi for insect-proof plants. Nat Biotechnol 25: 1231–1232.
9. Price DRG, Gatehouse JA (2008) RNAi-mediated crop protection against insects. Trends Biotechnol 26: 393–400.
10. Zhou X, Wheeler MM, Oi FM, Scharf ME (2008) RNA interference in the termite *Reticulitermes flavipes* through ingestion of double-stranded RNA. Insect Biochem Molec Biol 38: 805–815.
11. Wang Y, Zhang H, Li H, Miao X (2011) Second-generation sequencing supply an effective way to screen RNAi targets in large scale for potential application in pest insect control. PLoS ONE 6(4): e18644.
12. Zhu F, Xu J, Palli R, Ferguson J, Palli SR (2011) Ingested RNA interference for managing the populations of the Colorado potato beetle, *Leptinotarsa decemlineata*. Pest Manag Sci 67: 175–182.

13. Rangasamy M, Siegfried BD (2012) Validation of RNA interference in western corn rootworm *Diabrotica virgifera virgifera* LeConte (Coleoptera: Chrysomelidae) adults. Pest Manag Sci 68: 587–591.
14. Hodgson E (1985) Microsomal monooxygenases. In: Kerkut GA, Gilbert LI, editors. Comprehensive insect physiology, biochemistry, and pharmacology. Vol. 11 Pharmacology. Elmsford, New York: Pergamon. p. 225–322.
15. Feyereisen R (1999) Insect P_{450} enzymes. Annu Rev Entomol 44: 507–533.
16. Scott JG (1999) Cytochromes P_{450} and insecticide resistance. Insect Biochem Mol Biol 29: 757–777.
17. Tiwari S, Stelinski LL, Rogers ME (2012) Biochemical basis of organophosphate and carbamate resistance in Asian citrus psyllid. J Econ Entomol 105: 540–548.
18. Tiwari S, Clayson PJ, Kuhns EH, Stelinski LL (2012) Effects of buprofezin and diflubenzuron on various developmental stages of Asian citrus psyllid, *Diaphorina citri*. Pest Manag Sci 68: 1405–1412.
19. Feyereisen R (2006) Evolution of insect P_{450}. Biochem Soc Trans 34: 1252–1255.
20. Li X, Schuler MA, Berenbaum MR (2007) Molecular mechanisms of metabolic resistance to synthetic and natural xenobiotics. Annu Rev Entomol 52: 231–253.
21. Tiwari S, Gondhalekar A, Mann RS, Scharf ME, Stelinski LL (2011) Characterization of five *CYP4* genes of Asian citrus psyllid and their expression levels in *Candidatus* Liberibacter asiaticus infected and uninfected adults. Insect Mol Biol 20: 733–744.
22. Tiwari S, Killiny N, Mann RS, Wenninger EJ, Stelinski LL (2012) Abdominal color of the Asian citrus psyllid, *Diaphorina citri*, is associated with susceptibility to various insecticides. Pest Manag Sci. 7 p. (wileyonlinelibrary.com) DOI 10.1002/ps.3407.
23. Halbert SE, Manjunath KL (2004) Asian citrus psyllids (Sternorrhyncha: Psyllidae) and greening disease of citrus: A literature review and assessment of risk in Florida. Fla Entomol 87: 330–353.
24. Manjunath KL, Halbert SE, Ramadugu C, Webb S, Lee RF (2008) Detection of '*Candidatus* Liberibacter asiaticus' in *Diaphorina citri* and its importance in the management of citrus Huanglongbing in Florida. Phytopathology 98: 387–396.
25. Sétamou M, Rodriguez D, Saldana R, Schwarzlose G, Parlang D, et al. (2010) Efficacy and uptake of soil-applied imidacloprid in the control of Asian citrus psyllid and a citrus leafminer, two foliar feeding citrus pests. J Econ Entomol 103: 1711–1719.
26. Tiwari S, Mann RS, Rogers ME, Stelinski LL (2011) Insecticide resistance in field populations of Asian citrus psyllid in Florida. Pest Manag Sci 67: 1258–1268.
27. Grafton-Cardwell E, Stelinski LL, Stansly PA (2013) Biology and management of Asian citrus psyllid, vector of huanglongbing pathogens. Annu Rev Entomol 58: 413–432.
28. Qureshi JA, Stansly PA (2010) Dormant season foliar sprays of broad spectrum insecticides: An effective component of integrated management for *Diaphorina citri* (Hemiptera: Psyllidae) in citrus orchards. Crop Prot 29: 860–866.
29. Serikawa RH, Backus EA, Rogers ME (2012) Effects of soil-applied imidacloprid on Asian citrus psyllid (Hemiptera: Psyllidae) feeding behavior. J Econ Entomol 105: 1492–1502.
30. Tiwari S, Pelz-Stelinski K, Stelinski LL (2011) Effect of *Candidatus* Liberibacter asiaticus infection on susceptibility of Asian citrus psyllid, *Diaphorina citri*, to selected insecticides. Pest Manag Sci 67: 94–99.
31. Tiwari S, Pelz-Stelinski K, Mann RS, Stelinski LL (2011) Glutathione S-transferase and cytochrome P_{450} (general oxidase) activity levels in *Candidatus* Liberibacter asiaticus-infected and uninfected Asian citrus psyllid (Hemiptera: Psyllidae). Ann Entomol Soc Am 104: 297–305.
32. El-Shesheny I, Hajeri S, El-Hawary I, Gowda S, Killiny N (2013) Silencing abnormal wing disc gene of the Asian Citrus Psyllid, *Diaphorina citri* disrupts adult wing development and increases nymph mortality. PLoS ONE 8(5): e65392.
33. Brogdon WG, McAllister JC, Vulule J (1997) Heme peroxidase activity measured in single mosquitoes identifies individuals expressing an elevated oxidase for insecticide resistance. J Am Mosquito Contr 13: 233–237.
34. Livak KJ, Schmittgen TD (2001) Analysis of relative gene expression data using realtime quantitative PCR and the $2^{-\Delta\Delta C}{}_{T}$ method. Methods 25: 402–408.
35. Wheeler MM, Tarver MR, Coy MR, Scharf ME (2010) Characterization of four esterase genes and esterase activity from the gut of the termite *Reticulitermes flavipes*. Arch Insect Biochem Physiol 73: 30–48.
36. Bradford MM (1976) A rapid and sensitive method for the quantitation of microgram quantities of protein utilizing the principle of protein-dye binding. Anal Biochem 72: 248–254.
37. Hunter CP (1999) Genetics: A touch of elegance with RNAi. Curr Biol 9: R440–R442.
38. Rajagopal R, Sivakumar S, Agrawal N, Malhotra P, Bhatnagar RK (2002) Silencing of midgut aminopeptidase N of *Spodoptera litura* by double-stranded RNA establishes its role as *Bacillus thuringiensis* toxin receptor. J Biol Chem 277: 46849–46851.
39. Surakasi VP, Mohamed AAM, Kim Y (2011) RNA interference of beta 1 integrin subunit impairs development and immune responses of the beet armyworm, *Spodoptera exigua*. J Insect Physiol 57: 1537–1544.
40. Whyard S, Singh AD, Wong S (2009) Ingested double-stranded RNAs can act as species-specific insecticides. Insect Biochem Molec Biol 39: 824–832.
41. Mao YB, Cai WJ, Wang JW, Hong GJ, Tao XY, et al. (2007) Silencing a cotton bollworm P_{450} monooxygenase gene by plant-mediated RNAi impairs larval tolerance of gossypol. Nature Biotechnol 25: 1307–1313.
42. Baum JA, Bogaert T, Clinton W, Heck GR, Feldmann P, et al. (2007) Control of coleopteran insect pests through RNA interference. Nature Biotechnol 25: 1322–1326.
43. Clemens MJ, Elia A (1997) The double-stranded RNA-dependent protein kinase PKR: Structure and function. J Interferon Cytokine Res 17: 503–524.

Biochemical Aspects of the Soybean Response to Herbivory Injury by the Brown Stink Bug *Euschistus heros* (Hemiptera: Pentatomidae)

Renata Velozo Timbó[1,2], Marcelo Hermes-Lima[1], Luciano Paulino Silva[2], Angela Mehta[2], Maria Carolina Blassioli Moraes[2], Débora Pires Paula[2]*

1 Department of Cell Biology, University of Brasília, Campus Universitário Darcy Ribeiro, Brasília, Distrito Federal, Brazil, **2** Department of Biological Control, Embrapa Genetic Resources and Biotechnology, Brasília, Distrito Federal, Brazil

Abstract

Plant defense response is an elaborate biochemical process shown to depend on the plant genetic background and on the biological stressor. This work evaluated the soybean biochemical foliar response to brown stink bug herbivory injury through an analysis of redox metabolism and proteomic 2DE profiles of susceptible (BRS Silvania RR) and resistant (IAC-100) varieties. The activity of lipoxygenase-3, guaiacol peroxidase, catalase and ascorbate peroxidase was monitored every 24 h up to 96 h. In the susceptible variety, injury caused an increase in the activities of lipoxygenase 3 and guaiacol peroxidase, no change in ascorbate peroxidase, and a decrease in catalase. In the resistant variety, injury did not cause an alteration of any of these enzymes. The proteomic profiles were evaluated after 24 h of injury and revealed to have a similar proportion (4–5%) of differential protein expression in both varieties. The differential proteins, identified by mass spectrometry, in the susceptible variety were related to general stress responses, to plant defense, and to fungal infections. However, in the resistant variety, the identified change in protein profile was related to Calvin cycle enzymes. While the susceptible variety showed adaptive changes in redox metabolism and expression of stress-responsive proteins, the resistant showed a defense response to circumvent the biological stressor.

Editor: Joseph Clifton Dickens, United States Department of Agriculture, Beltsville Agricultural Research Center, United States of America

Funding: This work was supported by Embrapa (www.embrapa.br) (DPP AM LPS), CNPq (www.cnpq.br) (LPS), and CAPES (www.capes.gov.br) (RVT). The funders had no role in study design, data collection and analysis, decision to publish, or preparation of the manuscript.

Competing Interests: The authors have declared that no competing interests exist.

* Email: debora.pires@embrapa.br

Introduction

Stink bugs (Hemiptera: Pentatomidae) are one of the most important pests of soybean in Brazil [1]. Pentatomids feed on various structures of the host plant, mainly fruits and immature seeds, affecting their quality, development and maturation [2]. The brown stink bug *Euschistus heros* is considered one of the main soybean pests [3,4]. Information about the soybean defense mechanisms to this herbivore when it colonizes in the late vegetative stages is important to support pest management.

Several varieties of soybean have been studied regarding their defense response to herbivores in order to identify herbivory resistant varieties [5,6,7,8]. These studies have mostly focused on primary and secondary metabolite production as parts of the plant defense strategy aiming to support environmentally friendly soybean production. Plant defense to herbivory injury involves two major systems: (i) the direct defense, involving structural components (e.g. thorns, trichomes), and production of primary metabolites (e.g. proteinase inhibitors, antioxidant enzymes) and non-volatile secondary metabolites (e.g. phenolic acids and isoflavonoids) [9,10]; and (ii) the indirect defense, which is related to the production of volatile organic compounds (VOCs), such as (*E*,*E*)-α-farnesene, methyl salicylate and *cis*-jasmone, which attract natural enemies (predators and parasitoids) of the herbivores that feed on the plants, or in some cases act as herbivore repellents [6,11].

Plant defense response varies according to the plant genetic background, and the type of stressor [10]. Therefore, it is valuable to study the response elicited by the key herbivores in different varieties. Biochemical studies on plant defense have focused on evaluating the activity of enzymes related to the redox metabolism (including antioxidant enzymes) [9,12–14], such as lipoxygenase (LOX), peroxidase (POD), glutathione reductase (GR), ascorbate peroxidase (APX), and catalase (CAT), as well as on the modification of protein profiles [15]. However, despite the significant economic impact of the damage caused by stink bugs to soybean [16,17], there are no reports regarding the soybean redox response and the regulation of protein expression induced by stink bug herbivory.

Plants differ in resistance to their key herbivorous pests according to tolerance, non-preference and antibiosis [18]. In this work we used two soybean varieties with large differences in antibiosis resistance to brown stinkbug: resistant IAC-100 and susceptible BRS Silvania RR. IAC-100 was developed by crossing the parental varieties IAC-12 and IAC 78-2318, showing resistance to chewing and sucking insects [19]. BRS Silvania RR was developed for glyphosate tolerance [20]. The goal was to characterize the biochemical effects of the brown stink bug herbivory injury on the foliar redox response, through the analysis of the activity of lipoxygenase-3 (LOX3), guaiacol peroxidase (GOPx), catalase (CAT) and ascorbate peroxidase (APX), as well as on the foliar protein expression, through two-dimensional electrophoresis and mass spectrometry.

Materials and Methods

Soybean cultivation and stink bug rearing

The seeds of IAC-100 and BRS Silvania RR were germinated in a neutral pH germination paper. Seedlings with equivalent development were transplanted to plastic pots of 350 mL containing mixture of soil and organic substrate (in a proportion of 3:1), maintained in a germination chamber at 25±2°C and 13 h photoperiod, and irrigated every two days. The bioassays were conducted with plants in vegetative stage 3 (V3, first three trifoliolate leaves fully expanded). Stink bugs colonize soybeans during the vegetative stage, during which soybean plants produce higher amount of volatiles compared to the reproductive stage, and are known to have their indirect defense elicited by injury of the brown stink bug [10,17,21]. Understanding the plant defense mechanisms during this stage could help control stink bug pest populations that result in the reproductive stage.

The brown stink bugs were obtained from a laboratory colony at Embrapa Genetic Resources and Biotechnology (Brasília, Brazil), reared according to Moraes *et al.* [18] in plastic cages (26×22 cm) containing cotton fiber soaked in water and fed on peanut seeds [*Arachis hypogaea* (L.)], soybean grains [*Glycine max* (L.) Merrill.], sunflower seeds [*Helianthus annus* (L.)] and green beans [*Phaseolus vulgaris* (L.)], at 26±2°C, 60±10% RH and 14 h photoperiod.

Herbivory bioassays

The herbivory bioassays consisted of no injury (control) or injury to individual V3 plants of the two soybean varieties by three male adult brown stink bugs. Plants were exposed to stink bugs for 24, 48, 72, or 96 h after which the plants were harvested. Simultaneously, plants were harvested from the control. There were six replicate plants per harvest per treatment in a completely randomized design. Only males were used to prevent possible interference of oviposition on the plants, which could trigger other plant responses [6,21]. The stink bugs were deprived of food for 24 h before the bioassays. The plants were covered with transparent plastic bags (with micro holes for aeration) to guarantee the constant presence of the stink bugs during the bioassay. All assays were performed in separate germination chambers (FANEM) for each treatment, so that the VOC released could not interfere with the other treatments. All leaves of the six plants per treatment were collected and transferred to 50 mL tubes and immediately frozen in liquid nitrogen and stored at −80°C for further biochemical analysis.

Quantification of enzyme activities

The enzymatic activity was measured for LOX3, GOPx, APX and CAT in the soybean foliar extracts of three plants from each treatment in the four 24 h sampling intervals. Foliar homogenates were prepared in a mortar containing liquid nitrogen, where the leaves were macerated until the formation of a fine powder. For every 100 mg of macerated leaf, 400 μL of 50 mM potassium phosphate buffer (KPi) containing 10 mM phenylmethanesulfonyl fluoride (PMSF), 5 mM ethylenediaminetetraacetic acid (EDTA) was added. After homogenization, the samples were centrifuged at 14,000×*g* for 15 min at 4°C. The supernatant was quantified by the Bradford method [22] and used in triplicate for each enzyme assay.

The conditions for the enzymatic activity assays are detailed in Table 1. The assays for GOPx, APX and CAT were performed in quartz cuvettes (10×10 mm) by adding 900 μL of KPi pH 7.0 and 1 to 20 μL of foliar supernatant. For LOX3, 2,975 μL Kpi pH 6.5 and 1 to 20 μL of foliar supernatant were used. The reactions were initiated by adding the substrate for each enzyme (100 μL for GOPx, APX and CAT and 25 μL for LOX3) and the absorbance was measured for 150 s at intervals of 10 s with a Smart Plus spectrophotometer (BioRad).

The enzymatic activities ($\mu mol.min^{-1}.mg^{-1}$) were compared by one-way ANOVA, and the paired comparisons among the treatments were performed by Tukey's HSD ($p<0.05$).

Total protein extraction for the proteomic analysis

The total foliar protein profiles of all treatments (three plants/each) were analyzed after 24 h of bioassay by 2-DE. Total protein extraction was performed adapting the TCA/acetone methodology [27]. According to Wildgruber *et al.* [28], this method inhibits the degradation of proteins by minimizing protease activity (due to low pH), and also improves cell lysis and subsequent precipitation of proteins at −20°C. The leaves from the same treatment were pooled and macerated in liquid nitrogen until the formation of a fine powder and chilled cold acetone solution, containing 10% (w/v) TCA and 0.07% (v/v) β-mercaptoethanol, was added in a proportion of 1:2 v/v. After vortex homogenization for 5 min and incubation for 24 h at −20°C, the samples were centrifuged at 10,000×*g* for 20 min at 4°C. The pellets were washed three times in cold acetone containing 0.07% (v/v) β-mercaptoethanol and centrifuged at 10,000×*g* for 10 min at 4°C. To each 12 mg of pellets 1 mL of solubilization buffer [8.5 M urea, 2.5 M thiourea, 2% (w/v) CHAPS, 1% (w/v) DTT, 1% (w/v) Serdolit, 2% (v/v) IPG buffer nonlinear 3–11, 1% (w/v) bromophenol blue] [29] was added and incubated at room temperature for 1 h under gentle stirring. After centrifugation at 14,000×*g* for 1 h at 25°C, the supernatants were collected and the proteins quantified by the Bradford method [22].

Two-dimensional electrophoresis (2-DE) and gel image analysis

The isoelectric focusing (IEF) was carried out in 13 cm Immobiline DryStrips pH 3–11 NL (GE Healthcare). The strips were hydrated overnight with 250 μL of solubilization buffer containing 500 μg of total protein extract (three strips for each treatment). The IEF was conducted at 20°C using the Multiphor II Electrophoresis System (GE Healthcare) coupled to a cooling system bath in one step of 300 V for 1 min and one step of 3,500 V for 5.5 h (all at 2 mA and 5 W). After IEF, the strips were reduced by the addition of 8 mL of equilibration buffer [50 mM Tris-HCl pH 8.8, 6 M urea, 30% (v/v) glycerol, 2% (w/v) SDS] containing 1% (w/v) DTT under gentle agitation for 15 min at room temperature; and alkylated by adding 8 mL of equilibration buffer containing 2.5% (w/v) iodoacetamide under gentle agitation for 15 min. The equilibrated strips were transferred to a 12.5% denaturaing polyacrylamide gel and electrophoresis was performed at 25°C in Tris/glycine/SDS buffer on a multiple Ettan DALTsix (GE Healthcare), coupled to a cooling bath system, according to manufacturer's recommendation. For molecular mass reference, the BenchMark Protein ladder (Invitrogen) was applied on a side of the gel using filter paper. The electrophoresis was carried out in two steps: first at 10 mA/80 V for 1 h and 40 mA/500 V for 4.3–6 h. The 2-DE gels were

Table 1. Enzymatic activity assay conditions in leaves of the soybean varieties BRS Silvania RR and IAC-100 in the V3 stage, with and without (control group) brown stink bug herbivory injury.

Enzyme	Substrate	ε ($mM^{-1}cm^{-1}$)	Wavelength (nm)	Final Volume (μL)	References
LOX3	80 mM sodium linoleate	22.00	234	3,000	[23]
GOPx	10 mM hydrogen peroxide +20 mM guaiacol	26.60	470	1,000	[24]
APX	10 mM hydrogen peroxide +40 mM ascorbic acid	14.50	265	1,000	[25]
CAT	10 mM hydrogen peroxide	0.04	240	1,000	[26]

stained for two days in a solution of 0.8% (w/v) Coomassie Brilliant Blue G-250, 20% (v/v) ethanol, 1.6% (v/v) phosphoric acid, 8% (w/v) ammonium sulfate, and rinsed five times in water to remove excess dye.

The 2-DE gel images were obtained by ImageScanner III (GE Healthcare). Spot detection and measurements, as well as spot matching in the three gel images per treatment, were performed using the ImageMaster 2D Platinum 7.0 software (GE Healthcare). A spot was considered differential when its intensity and area were different in the injured and uninjured gels. The comparison of the total average number of spots between the injured and uninjured treatments was performed by one-away ANOVA ($p<0.05$).

Protein identification

For the soybean variety BRS Silvania RR the differential spots were excised and sliced into small fragments. Spot destaining was conducted by successive washes in 30% ethanol and incubation in 50% (v/v) acetonitrile solution containing 25 mM ammonium bicarbonate under agitation for 15 min. The gel fragments were dehydrated with 100% acetonitrile for 10 min under agitation and dried in a centrifugal evaporator for 20 min. The spots were rehydrated with 50 μL of hydrolysis solution consisting of 50 mM Tris-HCl pH 7.6 added to 1 mM calcium chloride and 0.5 μL of trypsin at 1 μg/μL (Sequencing Grade Modified Trypsin, Promega), which was previously heated to 30°C for 15 min, and to 50 μL of 50 mM ammonium bicarbonate. The samples were incubated at 37°C for 14 to 18 h. For the soybean variety IAC-100, trypsin hydrolysis was carried out using the Trypsin Profile IGD Kit (Sigma), according to the manufacturer instructions.

The trypsin-hydrolyzed peptides were analyzed by MALDI-TOF mass spectrometry using an UltraFlex III MALDI TOF/TOF (Bruker Daltonics) controlled by the software FlexControl 3.3.108.0 using the following parameters: mass range 600-3,500 Da; RP_PEPMIX method; 200 automatic shooting at a frequency of 200 Hz, laser intensity between 20–45%; deflection mass of 600 Da; reflective mode; and positive external calibration by peptide calibration standard I. The method to obtain the MS/MS data was the LIFT; AUTO_LIFT method; using the following parameters: 2,000 shots for which, 400 and 1,600 shots for the precursor and the fragments, respectively; laser intensity between 22–55%; and the other parameters were the same as mentioned above. The peaks list files containing the m/z ratios of precursor ions and MS/MS fragmented ions were used to search similar oligopeptides in the National Center for Biotechnology Information (NCBI) database sequences using BLASTp and Mascot software. The parameters used in the Mascot program for MS/MS were: (i) NCBI-nr database, (ii) Viridiplantae taxonomy, (iii) fixed modifications: carbamidomethylation of cysteine, (iv) variable modification: oxidation of methionine, (v) monoisotopic mass spectra data, (vi) mass tolerance of 150 ppm, (vii) fragment tolerance of ±0.05 Da. Only significant hits, as defined by Mascot ($p<0.05$), were accepted.

Results and Discussion

Redox response

The herbivory injury by the stink bug affected the foliar enzymatic response ($p<0.05$) in the susceptible BRS Silvania RR variety (Figure 1). The LOX3 activity was significantly higher at 72 h of herbivory, with a decline at 96 h, but still significantly higher than in the first 48 h of injury and in the uninjured control. The GOPx activity did not differ over the time of injury, but was higher along the time evaluated compared to the uninjured plants. On the other hand, APX activity was not affected during the herbivory period, and CAT activity decreased after 48 h of injury compared to uninjured plants. In contrast, the herbivory injury by the brown stink bug did not affect the foliar activity of any of the enzymes evaluated even after 96 h of injury (Fig. 2) for the resistant IAC-100 variety, suggesting that the redox metabolism was not affected in this variety under *E. heros* feeding injury.

From the enzymes studied, LOX3 and GOPx have a clear relation to the plant defense response and an enhanced activity under herbivory injury has been reported [30–32]. LOX3 catalyzes the production of jasmonic acid (JA) from linolenic acid, which stimulates the expression of genes responsible for producing defense compounds and is related to plant defense against herbivorous insects, and GOPx activates defense genes [30–32]. Chen *et al.* [24] observed that herbivory by *Melanocallis caryaefoliae* (Hemiptera: Aphididae) induced higher activity of LOX3 and GOPx in susceptible pecans, although in moderate and high resistance pecans the GOPx activity did not change. Similarly, Ni *et al.* [33] reported increased activity of GOPx, CAT and polyphenol oxidase in wheat leaves after herbivory by *Diuraphis noxia* (Hemiptera: Aphididae), and that peroxidase activity in barley leaves was significantly higher in susceptible than in resistant plants after herbivory by the same aphid species.

Although uninjured BRS Silvania RR generally had higher activity of three enzymes related to the redox metabolism than IAC-100, LOX3 activity was higher in uninjured IAC-100 than in BRS Silvania RR. This result may be directly related to resistance, because one of the functions of the lipoxygenase pathways the production of JA. The constitutive high activity of lipoxygenase in plants is related to their resistance against insects [34], which may help to explain the resistance of IAC-100 against stink bugs [6,16] and why it produces high level of induced plant volatiles when injured by them, which were not detected in BRS Silvania RR [6].

APX catalyzes the redox reaction between hydrogen peroxide and ascorbate [35]. Bi & Felton [36] reported that APX activity in soybean leaves increased over time after herbivory by *Helicoverpa zea* (Lepidoptera: Noctuidae), with maximum activity at 96 h. In our results, the activity of APX was not affected by stink bug

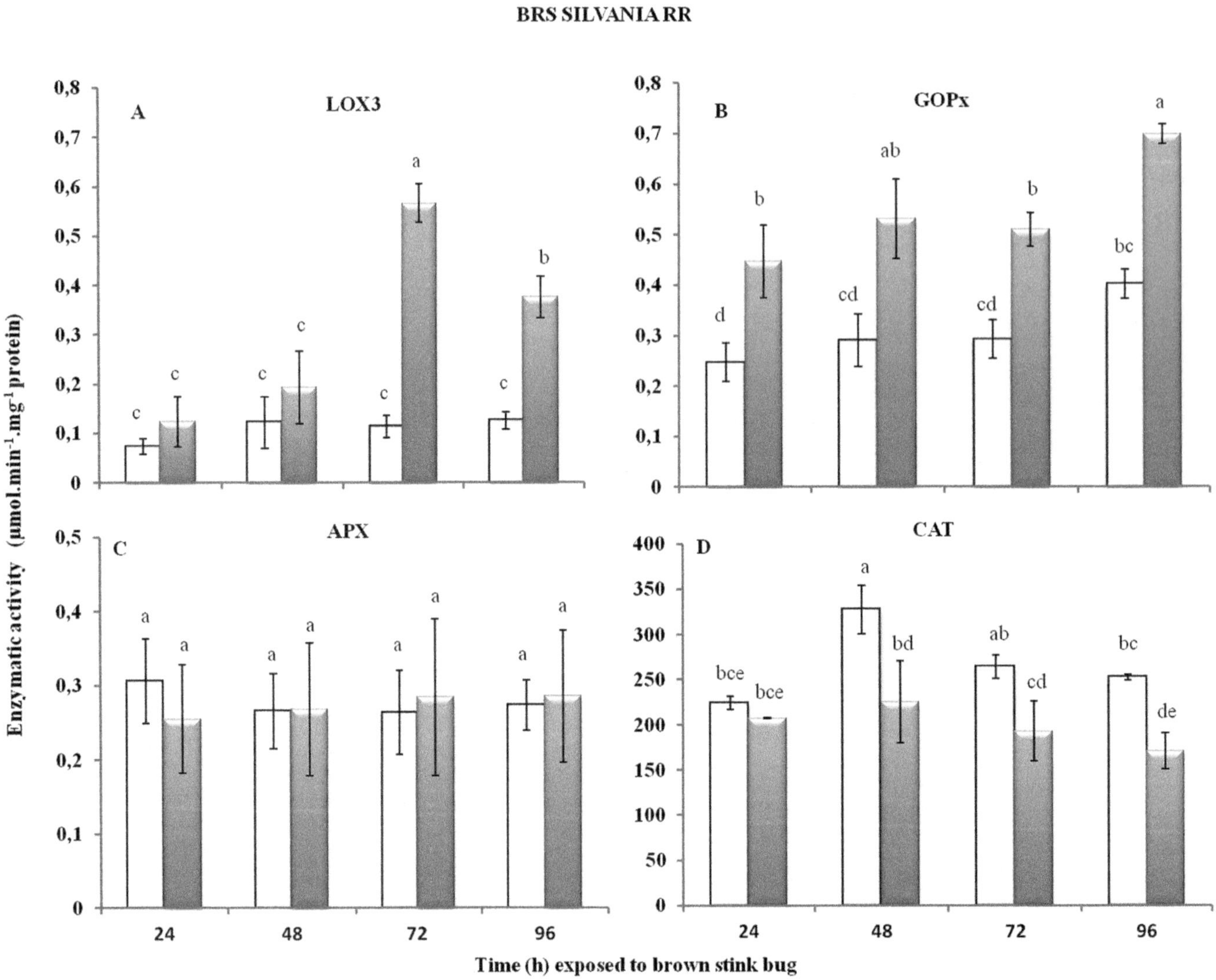

Figure 1. Enzymatic activity for (A) LOX3 ($F_{7,16}=2.32$; $p=0.0779$), (B) GOPx ($F_{7,16}=1.01$; $p=0.4629$), (C) APX ($F_{7,16}=3.35$; $p=0.0213$), and (D) CAT ($F_{7,16}=1.78$; $p=0.1609$) in leaves of BRS Silvania RR soybean variety (V3) with or without brown stink bug herbivory injury along different sampling periods. Injured is represented in grey and not injured in white. Bars with the same letter are not significantly different by Tukey's HSD ($p<0.05$).

herbivory on the susceptible soybean variety. Stink bugs are piercing-sucking insect, therefore provoking less damage to plants compared to chewing insects [37], and possibly inducing lower levels of hydrogen peroxide in the injury process.

CAT performs two major functions: (i) decomposition of hydrogen peroxide (a product of photorespiration and other plant metabolic activities) and (ii) oxidation of hydrogen donors such as methanol, ethanol, formic acid, phenols, with consumption of hydrogen peroxide [25,38]. An increase in CAT activity may be related to the control of hydrogen peroxide levels (which can be stressful at high concentrations) formed in response to herbivory injury [39]. In herbivory studies on various plant species, CAT activity remained unchanged [12,24,40], increased [33] or decreased [36,37]. In our study, a decrease in CAT activity was observed in response to stink bug injury only in the susceptible variety, showing that the varieties are managing the intracellular hydrogen peroxide levels differently. The observed increase in GOPx activity could be compensating the reduction in CAT levels in BRS Silvania RR leaves.

Overall, the susceptible soybean variety (BRS Silvania RR) showed redox gene expression changes, while the resistant variety (IAC-100) did not. The fluctuation in the activity of the redox metabolism enzymes might be associated with redox stress, which could be an indication of herbivory susceptibility. In the resistant variety, the redox genes might be constitutively expressed, so the stink bug injury might not produce enough reactive oxygen species (ROS) to elicit changed expression of these genes. Additional assessment of the redox alterations generated by herbivory across time would include measures such as protein carbonyls, lipid peroxidation and oxidized glutathione (GSSG), which are also classic indicators of redox oxidative stress, as well as other antioxidant enzymes, such as superoxide dismutases and peroxyredoxins [26].

Expression of foliar proteins

In both varieties, the foliar protein profile changed after 24 h of herbivore injury by the brown stink bug. Not all differentially expressed proteins could be identified, even when using the

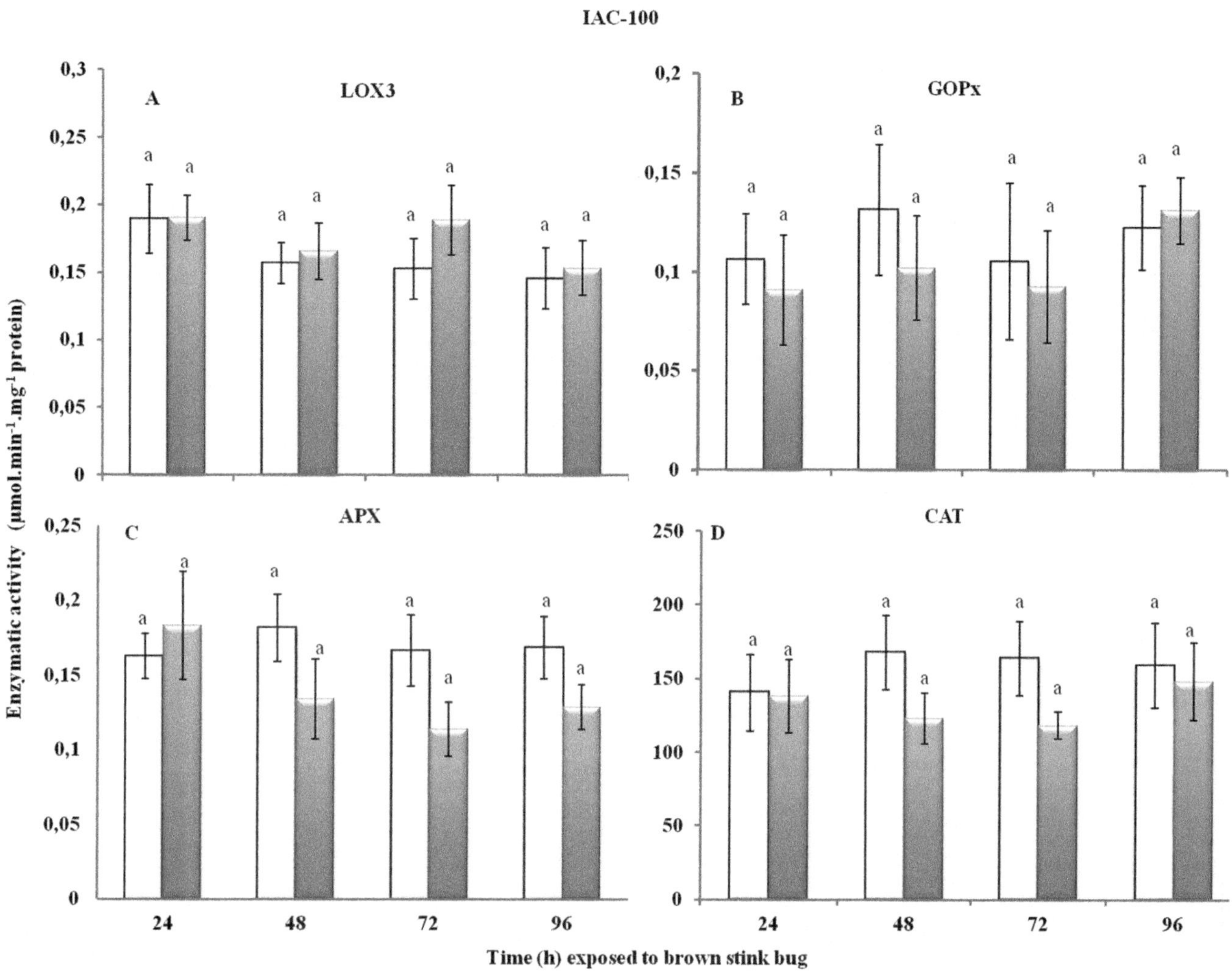

Figure 2. Enzymatic activity for (A) LOX3 ($F_{7,16}=34.52$; $p<0.0001$), (B) GOPx ($F_{7,16}=29.6$; $p<0.0001$), (C) APX ($F_{7,16}=0.16$; $p=0.9907$), and (D) CAT ($F_{7,16}=12.59$; $p<0.0001$) in leaves of IAC-100 soybean variety (V3) with and without brown stink bug herbivory injury along different sampling periods. Injured is represented in grey and not injured in white. Bars with the same letter are not significantly different by Tukey's HSD ($p<0.05$).

soybean genome reference database, possibly due to their low concentration or even absence in the database.

The 2-DE gel analyses of the BRS Silvania RR foliar protein profiles had an average of 386 protein spots per gel, while IAC-100 showed an average of 191 protein spots per gel, and most proteins were observed in the pH range from 4 to 8 and in the molecular mass range from 15 to 100 kDa (Fig. 3). The gel replicates of uninjured and injured plants had a high reproducibility: $r^2=0.95$ and 0.96, respectively for the susceptible; and $r^2=0.93$ and 0.97, respectively for the resistant.

When the foliar protein profile of the injured susceptible plants was compared to the uninjured profile, 17 proteins were differentially expressed, of which eight were up-regulated (Table 2). The identified up-regulated proteins were the chaperonin HSP20, glycine-rich 2b-like protein, both related to biotic and abiotic stress responses [41,42], and one hypothetical protein (gi|147800453).

The identified down-regulated proteins were CP4-EPSPS enzyme, a protease inhibitor, related to plant defense to herbivory, a hypothetical hydrolase function, related to plant defense against fungal infections [43], a heat-shock chaperonin 70 kDa protein (HSP), a glycine-rich RNA binding protein (GRP), related to general plant stress responses, a type 1 glutamine amidotransferase-like (GATase1) and three unknown hypothetical proteins (gi|40538952, gi|242054087, gi|147800453).

CP4-EPSPS participates in the shikimic acid pathway synthetizing the aromatic amino acids phenylalanine, tyrosine and tryptophan, which act as precursors for a wide range of products such as alkaloids, the plant hormone auxin, phytoalexins [44], flavonoids, and phenolic compounds such as lignin [45], and secondary volatile compounds such as methyl salicylate and indole, which are involved in indirect plant defense [6,20,21]. Michereff *et al.* [6] showed that the soybean variety BRS Silvania RR did not release methyl salicylate or indole when injured by the brown stink bug. Therefore the decrease of CP4-EPSPS in BRS Silvania RR might be related to its susceptibility to stink bugs [6,16]. HSPs are not exclusively related to heat stress, and some are induced by stresses such as injury, drought, and salinity [46]. In plants, GRPs are responsible for regulating cellular processes of genic regulation post-transcription, and translation [42,47], as well as cell elongation [48], protoxylem development [49,50], signal transduction [51], RNA chaperone activity [52], plant defense

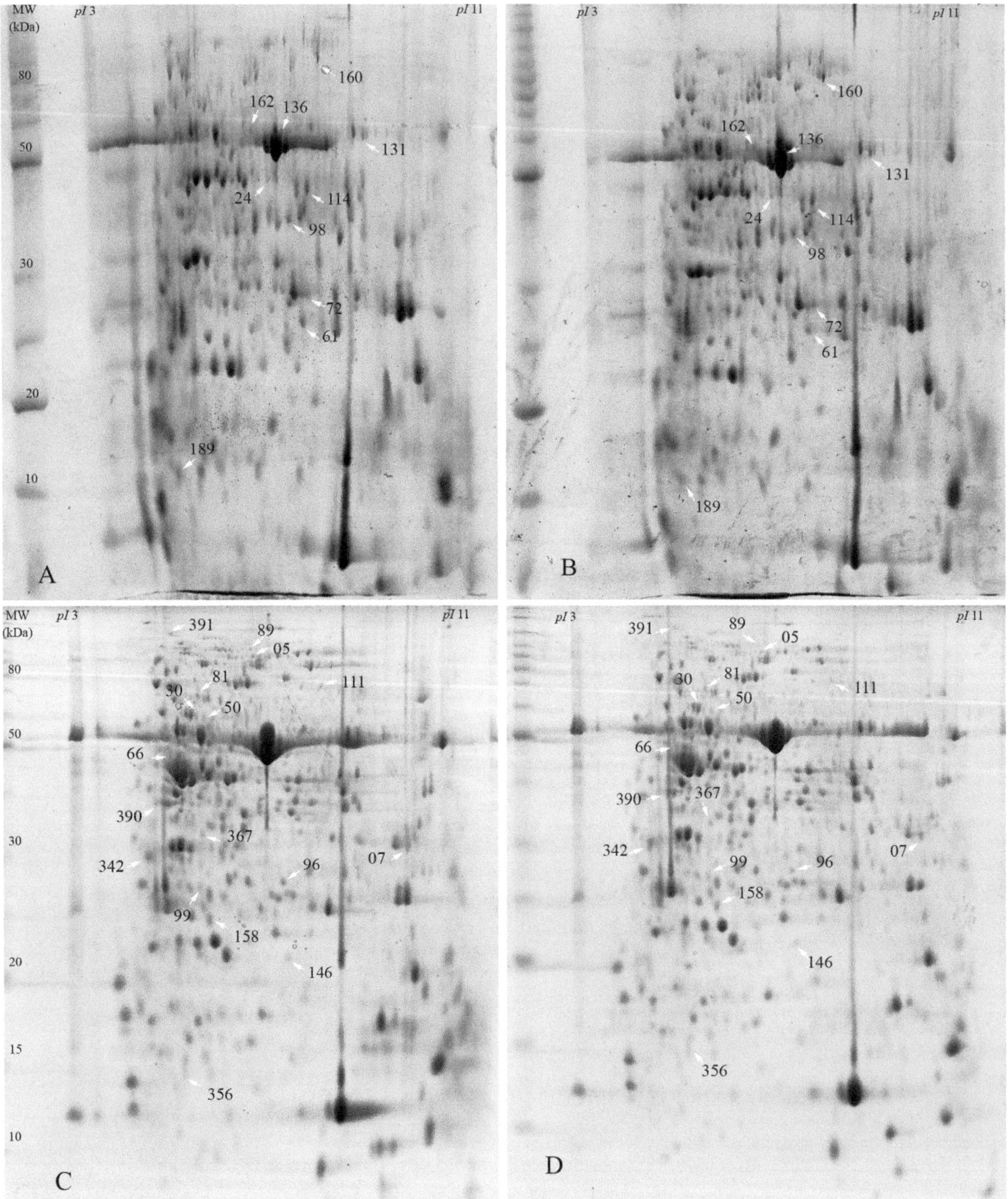

Figure 3. Foliar protein profiles (2-DE, 500 μg) of soybean varieties (V3). Variety IAC-100: (A) injured and (B) not injured, from the total average of 191 (±2) protein spots, 10 were differentially expressed and two identified (Table 2). Variety BRS Silvania RR: (C) injured and (D) not injured, from the total average of 386 (±4) protein spots, 17 were differentially expressed and 11 identified (Table 3). The differentially expressed proteins are indicated with white arrows and numbers, which correspond to identification numbers listed in the respective table.

Table 2. Identification of differentially expressed proteins by MS/MS in BRS Silvania RR soybean leaves (V3) ($p<0.05$).

Spot number	Protein identification	Accession (NCBI)	Molecular mass (kDa) theoretical/experimental	*p*I theoretical/experimental	Peptide sequence	Mascot/Blast Score	Search Type	RMS Error (ppm)	E-value	Expression level (%)
07	GATase1-like domain	gi|255639997	27.0/31.5	5.43/9	LSDFFEILATSEDR	117	MS/MS	44.40	4e-09	↓ 16.7
35*	Unknown	gi|40538952	-/57.8	-/5.6	GFIMKGATNATVDSMV	-	MS/MS	-	6e-13	↑ 30.4
66	CP4-EPSPS	gi|18266432	47.6/47.7	5.13/5	SAVLLAGLNTPGITTVIEPIMTR	57	MS/MS	12.80	2.2e-3	↓ 54.6
81*	70 kDa chaperonin-like	gi|473930207	-/64.6	-/5.6	AVITVPAYFNDAQR	-	MS/MS		1e-06	↓ 56.0
99*	STI-Soybean trypsin inhibitor	gi|351726331	19.8/26.8	-/5.5	GFPITISSPAR	-	MS/MS	-	0.003	↓ 24.0
111*	Unknown	gi|242054087	-/74.6	-/7.7	LSGIIPVHIGKLTK	-	MS/MS	-	5e-06	↓ 15.5
146	Glycine-rich protein 2b-like	gi|356538622	19.3/21.6	6.29/7.09	SLAEGESVEFAIESESDGR	132	MS/MS	11.60	1.6e-10	↑ 250.0
158	20 kDa chaperonin-like	gi|356556406	26.6/21.6	7.79/5.76	DKPSIGTVIAVGPGHLDEEGNR	146	MS/MS	0.86	3.2e-12	↑ 88.0
342	Unknown	gi|1478004532	50.0/31.0	5.43/4.7	SLLYVETADRPGLLVDLVK	90	MS/MS	4.51	1.6e-06	↓ 68.1
356	Glycine-rich RNA-binding	gi|351725469	15.8/15.0	6.58/5.3	AFSQYGEIVETK	77	MS/MS	20.90	2.9e-05	↓ 35.0
367	SGNH plant lipase-like	gi|255646252	40.0/33.4	5.83/5.6	AFFVFGDSLVDNGNNNFLATTAR	85	MS/MS	50.90	4.2e-06	↑ 58.9

* Identification by similarity search with BLASTp using peptide sequence obtained by manual *de novo* sequencing.

against pathogens [53], RNA binding activity [54], among others. Sachetto-Martins *et al.* [55] reported that the expression of GRP is regulated by a number of external stimuli, including stress by cold, water, high salinity, injury, and viral infection. The GATase1 belongs to a large group of enzymes capable of removing ammonia from glutamine and to transfer this group to other substrates to form amino acids, purine, pyrimidine nucleotides, and coenzymes [56]. Further studies are necessary to understand why these proteins were increased or decreased when the plants suffer injury by brown stink bug.

When the protein profiles of the injured and uninjured resistant plants were compared, 10 proteins had quantitative differential expression levels, of which four were increased in the injured plants. Two differentially expressed proteins were identified as ribulose bisphosphate carboxylase oxygenase (RuBisCO), which increased, and glyceraldehyde-3-phosphate dehydrogenase (GAPDH), which decreased (Table 3).

RuBisCO and GAPDH are enzymes in the Calvin cycle [57]. RuBisCO catalyzes the reaction of one carbon dioxide with ribulose-1,5-bisphosphate to form 3-phosphoglycerate (3-PGA), which in turn is catalyzed by phosphoglycerate kinase to 1,3-diphosphoglycerate. This compound, after catalysis by GAPDH produces glyceraldehyde-3-phosphate [43]. The increased expression of RuBisCO, and decreased expression of GAPDH, increases the level of the intermediate 3-phosphoglycerate. 3-Phosphoglycerate is metabolized during glycolysis into phosphoenolpyruvate, which in turn can be used in the shikimic acid pathway. Similar to what occurred in BRS Silvania RR, a protein involved in the shikimic pathway was affected by a piercing-sucking insect on IAC-100 variety. In general, sucking insects act on the shikimic acid pathway, and chewing insects act on the jasmonic acid pathway. Our results corroborate this hypothesis, because the main proteins identified here are involved in the shikimic acid pathway. However in contrast to BRS Silvania RR, the shikimic pathway was up-regulated in IAC-100. Michereff *et al.* [6] showed that IAC 100 injured by brown stink bugs produced higher amounts of methyl salicylate, and this volatile appears to be involved in the indirect defense of soybean.

Overall, the resistant variety (IAC-100) did not exhibit physiological stress to feeding injury by the brown stink bug, and seemed to shift its metabolism to produce secondary compounds for insect defense. In contrast, the susceptible variety (BRS Silvania RR) exhibited signs of physiological stress, both in its enzymes related to redox metabolism and in the protein profile response, which could affect its subsequent development in long term exposure to herbivory injury.

Conclusions

This work assessed the foliar redox response and protein expression profile of two soybean varieties (IAC-100, resistant, and BRS Silvania RR, susceptible, at stage V3) under herbivory injury by the brown stink bug. As could be expected from previous reports, the two soybean varieties responded differently at the biochemical level to herbivory by the brown stink bug. The susceptible variety BRS Silvania RR showed adaptive changes in the activity of enzymes related to redox metabolism and in the expression of stress-responsive proteins, probably to minimize biochemical damage caused by herbivory injury. On the other hand, in the resistant variety, the redox genes could be constitutively expressed, and the stink bug injury might not produce enough ROS to elicit changed expression of these genes. Overall, the susceptible variety responded to herbivory in ways that could compromise its development in the long term, while the

Table 3. Identification of differentially expressed proteins (by MS/MS) in IAC-100 soybean leaves (V3) ($p<0.05$).

Spot number	Protein identification	Accession (NCBI¹)	Molecular mass (kDa) theoretical/experimental	*p*I theoretical/experimental	Peptide sequence	Mascot Score	Search Type	RMS Error (ppm)	E-value	Expression level (%)
114	GAPDH	gi|66026	36.0/42.7	6.66/7.2	K.VIAWYDNEWGYSQR.V	64	MS/MS	61.7	8.4e-4	↓ 32.0
136	RuBisCO large subunit	gi|131898	52.0/53.9	6.23/6.24	K.TFQGPPHGIQVER.D	107	MS/MS	11.0	9.1e-08	↑ 33.0

resistant variety IAC-100 showed a defense response oriented to circumvent the biological stressor. Both varieties were injured by a piercing-sucking insect and the proteins regulated by this stress were linked to the shikimic acid pathway corroborating the hypothesis that sucking insects induce the shikimic acid route of defense more than the jasmonic acid route defense.

Acknowledgments

We would like to thank: Dra. Clara Beatriz Hoffmann-Campo for providing the soybean seeds; Dr. Miguel Borges and Dr. Raul A. Laumann for providing the brown stink bugs; Msc. Démerson André Polli for statistics consulting; MSc. Gabriela Villeth and Dr. Lilian Travassos for supporting the proteomic experiments and analysis; Dr. David A. Andow for English revisions.

Author Contributions

Conceived and designed the experiments: DPP RVT MCBM. Performed the experiments: RVT LPS. Analyzed the data: DPP MHL LPS RVT MCBM AM. Contributed reagents/materials/analysis tools: MHL DPP. Wrote the paper: DPP MHL LPS RVT MCBM AM.

References

1. Corrêa-Ferreira BS, Panizzi AR (1999) Percevejos da soja e seu manejo. Londrina: Embrapa - CNPSo, Circular Técnica 24.
2. Fernandes AP (2000) Análise comparativa da espermogênese normal e diapáusica em percevejos fitófagos (Hemiptera, Pentatomidae) – Revisão Universitas. Biociências (UNICEUB). Brasília-Brasil. 1(1): 87–95.
3. Todd JW, Herzog DC (1980) Sampling phytophagous Pentatomidae on soybean. In: Kogan M, Herzog D C (Ed) Sampling.
4. Companhia Nacional de Abastecimento (CONAB), Acompanhamento da Safra Brasileira de Grãos (2012) http://www.conab.gov.br/OlalaCMS/uploads/arquivos/12_10_17_16_09_58_boletim_graos_-_julho_2012.pdf.
5. Vieira CR, Moraes MCB, Borges M, Sujii ER, Laumann RA (2013) *cis*-Jasmone indirect action on egg parasitoids (Hymenoptera: Scelionidae) and its application in biological control of soybean stink bugs (Hemiptera: Pentatomidae). Biol. Control 64: 75–82.
6. Michereff MFF, Laumann RA, Borges RAM, Michereff-Filho M, Diniz IR, et al. (2011) Volatiles mediating a plant-herbivore-natural enemy interaction in resistant and susceptible soybean cultivars. J. Chem. Ecol. 37: 273–285.
7. Moraes MCB, Laumann RA, Pareja M, Sereno FTPS, Michereff MFF, et al. (2009) Attraction of the stink bug egg parasitoid *Telenomus podisi* to defence signals from soybean activated by treatment with *cis*-jasmone. Entomol. Exp. Appl. 131: 178–188.
8. Vieira SS, Bueno AF, Boff MIC, Bueno RC, Hoffman-Campo CB (2011) Resistance of soybean genotypes to *Bemisia tabaci* (Genn.) biotype B. Neo. Entomol. (Impress) 40: 117–122.
9. Dicke M, Hilker M (2003) Induced plants defence: from molecular biology to evolutionary ecology. Basic Appl. Ecol. 4: 3–14.
10. Courtois EA, Paine CE, Blandinieres PA, Stien D, Bessiere JM, et al. (2009) Diversity of the volatile organic compounds emitted by 55 species of tropical trees: a survey in French Guiana. J. Chem. Ecol. 35: 1349–1362.
11. Colazza S, Wajnberg E (2013) Chemical ecology of insect parasitoids: towards a new era. In: Wajnberg E, Colazza S, Chemical Ecology of Insect Parasitoids, 1st Edition, John Wiley & Sons (2013) 1–8.
12. McPherson RM, Buss GR (2007) Evaluation lepidopteran defoliation resistance in soybean breeding lines containing the stink bug (Hemiptera: Pentatomidae) resistance IAC-100 cultivar in their pedigrees. J. Econ. Entomol. 100: 962–968.
13. Panizzi AR, Mcpherson JE, James JG, Javabery M, McPherson RM (2000) Stink bugs (Pentatomidae). In: Schaefer CW, Panizzi AR (Eds.), Heteroptera of economic importance. Boca Raton: Florida, USA: CRC 432–434 pp.
14. Corrêa-Ferreira BS, Azevedo J (2002) Soybean seed damage by different species of stink bugs. Agr. Forest Entomol. 4: 145–150.
15. Bi JL, Murphy JB, Felton GW (1997) Antinutritive and oxidative components as mechanisms of induced resistance in cotton to *Helicoverpa zea*. J. Chem. Ecol. 23: 97–117.
16. Laumann RA, Farias Neto AL, Moraes MCB, Silva AP, Vieira CR, et al. (2008) Dinâmica populacional de percevejos (Hemiptera: Pentatomidae) em diferentes genótipos de soja. In: IX Simpósio Nacional Cerrado; II Simpósio Internacional Savanas Tropicais, Brasília, DF. Desafios e estratégias para o equilíbrio entre sociedade, agronegócio e recursos naturais: Anais. Planaltina, DF: Embrapa Cerrados, HTTP://www.cpac.embrapa.br/download/747t.
17. Rostas M, Eggert K (2008) Ontogenetic and spatio-temporal patterns of induced volatiles in *Glycine max* in the light of the optimal defense hypothesis. Chemoecol. 18: 29–38.
18. Moraes MCB, Laumann RA, Sujii ES, Pires CSS, Borges M (2005) Induced volatiles in soybean and pigeon pea plants artificially infested with the Neotropical brown stink bug, *Euschistus heros*, and their effect on the egg parasitoid, *Telenomus podisi*. Entomol. Exp. Appl. 115: 227–237.
19. Veiga RFA, Rossetto CJ, Razera LF, Galo PB, Bortoleto M (1999) Caracterização morfológica e agronômica da cultivar de soja 'IAC-100'. Instituto Agronômico de Campinas. Boletim Técnico, 175. 25p.
20. Gris CF, Pinho EVRV, Carvalho MLM, Diniz RP, Andrade T (2013) Lignificação da planta e qualidade de sementes de soja RR pulverizadas com herbicida glifosato. Rev. Ceres. Viçosa 60(2): 286–292.
21. Moraes MCB, Pareja M, Laumann RA, Hoffmann-Campo CB, Borges M (2008) Response of the parasitoid to induced volatiles from soybean damaged by stink bug herbivory and oviposition. J. Plant Interact. 3: 111–118.
22. Bradford MM (1976) A rapid and sensitive method for quantification of microgram quantities of protein utilizing the principle of protein-dye binding. Anal. Biochem. 72: 248–254.
23. Axelrod BT, Cheesbrough TM, Laakso S (1981) Lipoxygenase from soybeans. Methods Enzymol. 71: 441–445.
24. Chen Y, Ni X, Cottrell E, Wood BW, Buntin GD (2009) Changes of oxidase and hydrolase activities in pecan leaves elicited by black pecan aphid (Hemiptera: Aphididae) feeding, J. Econ. Entomol. 102 (3): 1262–1269.
25. Anderson JV, Chevone BI, Hess JL (1992) Seasonal variation in the antioxidant system of Eastern white pine needles. Plant Physiol. 98: 501–508.
26. Hermes-Lima M, Carreiro C, Moreira DC, Polcheira C, Machado DP, et al. (2012) Glutathione status and antioxidant enzymes in a crocodilian species from the swamps of the Brazilian Pantanal. Comp. Biochem. Physiol. Part A. 163: 189–198.
27. Méchin V, Damerval C, Zivy M (2007) Total protein extraction with TCA/acetone. In: Thiellement H, Zivy M, Damerval M, Méchin V (2007) Plant proteomics methods and protocols. Totowa, New Jersey, John M. Walker, 1–8.
28. Wildgruber R, Reil G, Drews O, Parlar H, Görg A (2002) Web-based two dimensional database of *Saccharomyces cerevisae* proteins using immobilized pH gradients from pH 6 to pH 12 and matrix-assisted laser desorption/ionization-time of flight mass spectrometry. Proteomics 2: 727–732.
29. Weiss W, Gorg A (2007) Two-dimensional electrophoresis for plant proteomics. Methods Mol. Biol. 355: 1421–143.
30. Farmer EF, Johnson RR, Ryan CA (1992) Regulation of expression of proteinase inhibitor genes by methyl jasmonate and jasmonic acid. Plant Physiol. 98: 995–1002.
31. Prost I, Dhondt S, Rothe G, Vicente J, Rodriguez MJ, et al. (2005) Evaluation of the antimicrobial activities of plant oxylipins supports their involvement in defense pathogens. Plant Physiol. 139: 1902–1913.
32. Ishige F, Mori H, Yamazaki K, Imaseki H (1993) Identification of a basic glycoprotein induced by ethylene in primary leaves of azuki bean as cationic peroxidase. Plant Physiol. 101: 193–199.
33. Ni X, Quisenberry SS, Heng-Moss T, Markwell J, Sarath G, et al. (2001) Oxidative responses of resistant and susceptible cereal leaves to symptomatic and nonsymptomatic cereal aphid (Hemiptera: Aphididae) feeding. Faculty Publications: Department of Entomology. Paper 113. http://digitalcommons.unl.edu/entomologyfacpub/113.
34. Devi PUM, Reddy PS, Rani NRU, Reddy KJ, Reddy MN, et al. (2000) Lipoxygenase metabolites of alpha-linolenic acid in the development of resistance in pigeonpea, *Cajanus cajan* (L.) Millsp, seedlings against *Fusarium udum* infection. European J. Plant Pathol. 106: 857–865.
35. Pérez FJ, Villegas D, Mejia N (2002) Ascorbic acid and flavonoid-peroxidase reactions as a detoxifying system of H_2O_2 in grapevine leaves. Phytochemistry 60: 573–580.
36. Bi L, Felton GW (1995) Foliar oxidative stress and insect herbivory: primary compounds, secondary metabolites, and reactive oxygen species as components of induced resistance. J. Chem. Ecol. 21: 1511–1530.
37. Ruuhola T, Yang S (2005) Wound-induced oxidative responses in mountain birch leaves. Ann. Botany 97: 29–37.
38. Aebi H (1984) Catalase *in vitro*. B. Chance. Act Chem. Scand. 1: 121–126.
39. Karuppanapandian T, Moon J, Kim C, Manoharan K, Kim W (2011) Reactive oxygen species in plants: their generation, signal transduction, and scavenging mechanisms. Australian J. Crop Sci. 5 (6): 709–725.
40. Gomez SK, Oosterhuis DM, Rajguru SN, Johnson DR (2004) Foliar antioxidant enzyme responses in cotton after aphid herbivory. J. Cotton Sci. 8: 99–104.
41. Li D-C, Yang F, Lu B, Chen DF, Yang WJ (2012) Thermotolerance and molecular chaperone function of the small heat shock protein HSP20 from hyperthermophilic aschaeon, *Sulfolobus solfataricus* P2. Cell Stress and Chaperon. 17: 103–108.
42. Mangeon A, Junqueira RM, Sachetto-Martins G (2010) Functional diversity of the plant glycine-rich proteins superfamily. Plant Signaling Behav. 5 (2): 99–104.
43. Taiz L, Zeiger E (2006) Secondary metabolites and plant defense. In: Taiz L, Zeiger E (Eds.) Plant Physiology, Sinauer Associates Inc., Sunderland, MA, 4th edition. 283–306 pp.
44. Kutchan TM (1995) Alkaloid biosynthesis: the basis for metabolic engineering of medicinal plants. Plant Cell 7: 1059–1070.
45. Vogt T (2010) Phenylpropanoid biosynthesis. Mol. Plant. 3: 2–20.

46. Lindquist S, Craig EA (1998) The heat-shock proteins. Annu. Rev. Genet. 22: 631–677.
47. Burd CG, Dreyfuss G (1994) Conserved structures and diversity of functions of RNA-binding proteins. Science 265: 615–21.
48. Mangeon A, Magioli C, Menezes-Salgueiro AD, Cardeal V, de Oliveira C, et al. (2009) AtGRP5, a vacuole-located glycine-rich protein involved in cell elongation. Planta 230: 253–265.
49. Ringle C, Hauf G, Keller B (2001) Hydrophobic interactions of the structural protein GRP1.8 in the cell wall of protoxylem elements. Plant Physiol. 125: 673–82.
50. Ryser U, Schorderet M, Guyot R, Keller B (2004) A new structural element containing glycine-rich proteins and rhamnogalacturonan I in the protoxylem of seed plants. J. Cell Sci. 117: 1179–90.
51. Park AR, Cho SK, Yun SJ, Jin MY, Lee SH, et al. (2001) Interaction of the arabidopsis receptor protein kinase Wak1 with a glycine-rich protein, AtGRP-3. J. Biol. Chem. 276: 26688–26693.
52. Kim JY, Park SJ, Jang B, Jung CH, Ahn SJ, et al. (2007) Functional characterization of a glycine-rich RNA-binding protein 2 in *Arabidopsis thaliana* under abiotic stress conditions. Plant J. 50: 439–451.
53. Fu ZQ, Guo M, Jeong BR, Tian F, Elthon TE, et al. (2007) A type III effector ADP-ribosylates RNA-binding proteins and quells plant immunity. Nature 447: 284–288.
54. Kim YO, Kim JS, Kang H (2005) Cold-inducible zinc finger-containing glycine-rich RNA-binding protein contributes to the enhancement of freezing tolerance in *Arabidopsis thaliana*. Plant J. 42: 890–900.
55. Sachetto-Martins G, Franco LO, de Oliveira DE (2000) Plant glycine-rich proteins: a family or just proteins with a common motif? Biochim Biophys Acta. 1492: 1–14.
56. Mäntsälä P, Zlakin H (1984) Replacement of the active-site cysteine in glutamine phosphoribosylpyrophosphate amidotranferase by site-directed mutagenesis. J. Soc. Biol. Chem. 259: 14230–14236.
57. Brynat DA, Friggard NU (2006) Porkaryotic photosynthesis and phototrophy illuminate. Trends Microbiol. 14: 488–496.

Biotic Interactions Overrule Plant Responses to Climate, Depending on the Species' Biogeography

Astrid Welk[1]*, Erik Welk[1,2], Helge Bruelheide[1,2]

1 Institute of Biology/Geobotany and Botanical Garden, Martin Luther University Halle-Wittenberg, Halle (Saale), Germany, **2** German Centre for Integrative Biodiversity Research (iDiv) Halle-Jena-Leipzig, Leipzig, Germany

Abstract

This study presents an experimental approach to assess the relative importance of climatic and biotic factors as determinants of species' geographical distributions. We asked to what extent responses of grassland plant species to biotic interactions vary with climate, and to what degree this variation depends on the species' biogeography. Using a gradient from oceanic to continental climate represented by nine common garden transplant sites in Germany, we experimentally tested whether congeneric grassland species of different geographic distribution (oceanic *vs.* continental plant range type) responded differently to combinations of climate, competition and mollusc herbivory. We found the relative importance of biotic interactions and climate to vary between the different components of plant performance. While survival and plant height increased with precipitation, temperature had no effect on plant performance. Additionally, species with continental plant range type increased their growth in more benign climatic conditions, while those with oceanic range type were largely unable to take a similar advantage of better climatic conditions. Competition generally caused strong reductions of aboveground biomass and growth. In contrast, herbivory had minor effects on survival and growth. Against expectation, these negative effects of competition and herbivory were not mitigated under more stressful continental climate conditions. In conclusion we suggest variation in relative importance of climate and biotic interactions on broader scales, mediated via species-specific sensitivities and factor-specific response patterns. Our results have important implications for species distribution models, as they emphasize the large-scale impact of biotic interactions on plant distribution patterns and the necessity to take plant range types into account.

Editor: Christopher J. Lortie, York U, Canada

Funding: The Graduate School of Saxony-Anhalt and the DFG Priority Programme 1374 "Infrastructure-Biodiversity-Exploratories" (BR 1698/14-1) supported the study financially. The funders had no role in study design, data collection and analysis, decision to publish, or preparation of the manuscript.

Competing Interests: The authors have declared that no competing interests exist.

* Email: astrid.welk@botanik.uni-halle.de

Introduction

Understanding the causes of species geographical distributions is a major research goal in ecology, often driven by the desire to model future species distributions in a world undergoing climate change. Soberón [1] summarized the main determinants of species distribution, which apart from a species' dispersal capacity comprise the physiological niche and biotic interactions. Depending on the species' physiological niche [2,3], a fundamental range of environmental, and in particular climatic conditions, defines the suitable range for growth, reproduction and establishment of populations [4]. This fundamental range of environmental conditions is usually modified by biotic interactions, for example owing to constricted tolerances in the presence of competitors or herbivores [1,5]. The result is the species' ecological [2,3]or realized niche [6], which is the environmental range of conditions under which a species does occur in nature.

Correlative species distribution models are based on the central assumption that on broad geographic scales, species' spatial distributions are in equilibrium with climate, while biotic interactions are of minor importance [7]. At the same time, is has been emphasized that biotic interactions are important at local scales, e.g. for presence and abundance in communities [1,8]. However, it has been recognized that detrimental biotic interactions, such as competition and herbivory, have the potential to limit plant distribution also on large spatial scales [9,10]. For instance, Bruelheide & Scheidel [11] demonstrated that the altitudinal distribution of a montane plant species is restricted to higher elevations because of increasing slug herbivory in the lowlands.

Impacts of biotic interactions on plant performance have been demonstrated to change along climate gradients as for example was proposed by the stress gradient hypothesis for competition [12,13]. Thereby, competition should gain in relative importance under benign climate conditions. For example, Loehle [14] suggested that the northern range limits of North American tree species are limited by cold tolerance, while competitive ability should determine the southern range limits. This has also been demonstrated with examples of increasing competition with decreasing altitude [15]. Similarly, other biotic interactions might vary with climate conditions. For example, slug herbivore pressure was shown to increase with decreasing altitude [16].

In addition to altitudinal and latitudinal climate gradients, environments in Europe are also structured by a distinct longitudinal differentiation in climate. A strong gradient exists from oceanic climate in Western Europe, with relatively narrow annual temperature ranges and constantly humid conditions, to continental climate in Eastern Europe, with large temperature seasonality and low annual precipitation. Along this gradient, the general physiological growth conditions for plants get harsher with increasing continentality. The strong impact of this continentality gradient on plant distribution is also evident in phytogeographical classification systems that classify plant species according to their geographic distribution along this gradient into oceanic and continental plant range types [17,18].

Species populations are expected to increase in performance, the closer their locations' growth conditions is to the species' climate optimum [19]. Under the general assumption that species' geographic distributions reflect their environmental requirements, species should do better under conditions which are climatically less peripheral. Testing this assumption for North American tree species, Purves [20] found distinct differences in growth and demographic rates between northern and southern peripheral range sections when compared with the core area. Accordingly, oceanic and continental plant species should perform better in oceanic and continental climates respectively. On the other hand, this view has recently been challenged as glasshouse experiments revealed no clear differences between responses of oceanic and continental species in respect to soil moisture levels and frost hardiness [21,22]. However, multiple species field tests on plant range type-specific adaptations are still missing.

If closely related plant species tend to be similar in their fundamental niche requirements, differences in geographical distribution patterns might be caused by dispersal limitation, biogeographical history and biotic interactions. Differences in biotic characteristics are probably responsible for the large amount of unexplained interspecific variation in periphery-core comparisons of species performance [20]. As predicted by the stress-gradient-hypothesis, plant-plant interactions can turn from competition to facilitation with increasing abiotic stress [15]. Thus, competition should be less intense in continental compared to oceanic regions. Similarly, mollusc herbivory might also be reduced in continental regions as slugs (e.g. Arionidae) show clear preferences to more benign climate conditions [16]. Hence, species distributed in continental regions may be less adapted to negative biotic interactions and more vulnerable to biotic stress. This mismatch might be disadvantageous in the climatically more favourable oceanic regions. In consequence, detrimental biotic interactions should have a different impact depending on the species' plant range type.

Based on these considerations a transplant experiment was set up in nine Botanic Gardens along a continentality gradient in Germany (Fig. 1), where the relative impact of biotic interactions (competition and mollusc herbivory) and climate was tested with congeneric plant species of contrasting plant range types. Such transplant experiments have been used before and demonstrated a strong climatic impact on the transplants' survival, growth and reproduction [23–25].

The following hypotheses were tested: H1) There is an interactive effect between climate and biotic treatments on plant performance. In particular, we expected the effect of competition and herbivory to become weaker with increasing climatic continentality. H2) At the oceanic end of the gradient, species with general oceanic distribution should perform better than species with general continental distribution range and vice versa, indicating range type-specific adaptation. H3) Depending on the plant range type, biotic treatments affect species differently, as the continental plant species should be more susceptible to competition and herbivory. Furthermore, assuming that the relative importance of herbivory and competition decreases with increasingly continental climate, the negative biotic effects on continental species should decrease with increasing continentality. Testing these hypotheses aims at improving the mechanistic understanding of species distribution patterns.

Material and Methods

Ethics statement

The authorities that issued the permit to use the Botanical Gardens were the scientific or technical directors of the gardens. As they also assigned the piece of land to us, carrying out the experiment without this permit would not have been possible. Seeds for the experiment were collected from public land and all regulations concerning protected or endangered species were respected.

Climate gradient

To establish a climatic gradient for common garden locations, we made use of the network of Botanical Gardens, as they provide excellent conditions for reproducibility, and generally have similar soil conditions, i.e. fertile garden soils (hortisols). Additionally, these gardens usually suffer from high mollusc densities, which motivated us to manipulate mollusc herbivory as a negative biotic impact. Using the geographic coordinates of 66 major Botanical Gardens in Germany we extracted the mean values for mean minimum temperatures of the coldest month and mean annual precipitation for the last 50 years, using the WORLDCLIM dataset [26]. We chose 12 gardens located along a gradient from oceanic (mild winter and high precipitation) to continental (cold winter and low precipitation) climate conditions. Out of these, nine gardens responded positively to our request for conducting an experiment (Fig. 1, Table 1).

For the study period (June 2008–October 2009), monthly mean data for temperature and precipitation were obtained from the nearest official meteorological stations. Table 1 shows mean temperatures as well as accumulated temperature and precipitation sums at the nine study sites for the investigation period. To analyze the weather conditions during the study period, the climate data of all nine Botanical Gardens were subjected to a principal component analysis (PCA, prcomp procedure; R 2.15.2, R Development Core Team 2012). In the PCA, the Botanical Gardens were ordered along a clear gradient related to temperature on the first, and precipitation variables on the second axis (explaining 53% and 28% of the overall variance in climate conditions, respectively). Summer temperature of 2009 had the highest loading and was positively correlated with the first PCA axis scores (Table 1). The remaining temperature variables for the vegetation period were also highly positively correlated with the first PCA axis while precipitation variables and minimum temperature were negatively correlated with this axis (Table 1). Regarding the second PCA axis, summer precipitation in 2009 had the highest loading and was negatively correlated with the respective axis scores (Table 1). The Botanical Gardens were arranged from Osnabrück (OSN) to Frankfurt (FRA) and from Frankfurt (FRA) to Braunschweig (BS) along PCA axis 1 and 2, respectively (Table 1).

Transplants and measurements

Fourteen herbaceous species were included in the experiment, two from each of seven genera. The two congeneric species are of similar growth form, have similar habitat preferences (Table S1), but differ in their geographical distribution range, in particular with respect to the longitudinal positions of their western range boundary (Figure S1). The following species were included in this transplant experiment (taxonomy according Jäger & Werner [27], plant range type as follows: oceanic-continental): *Carlina vulgaris*

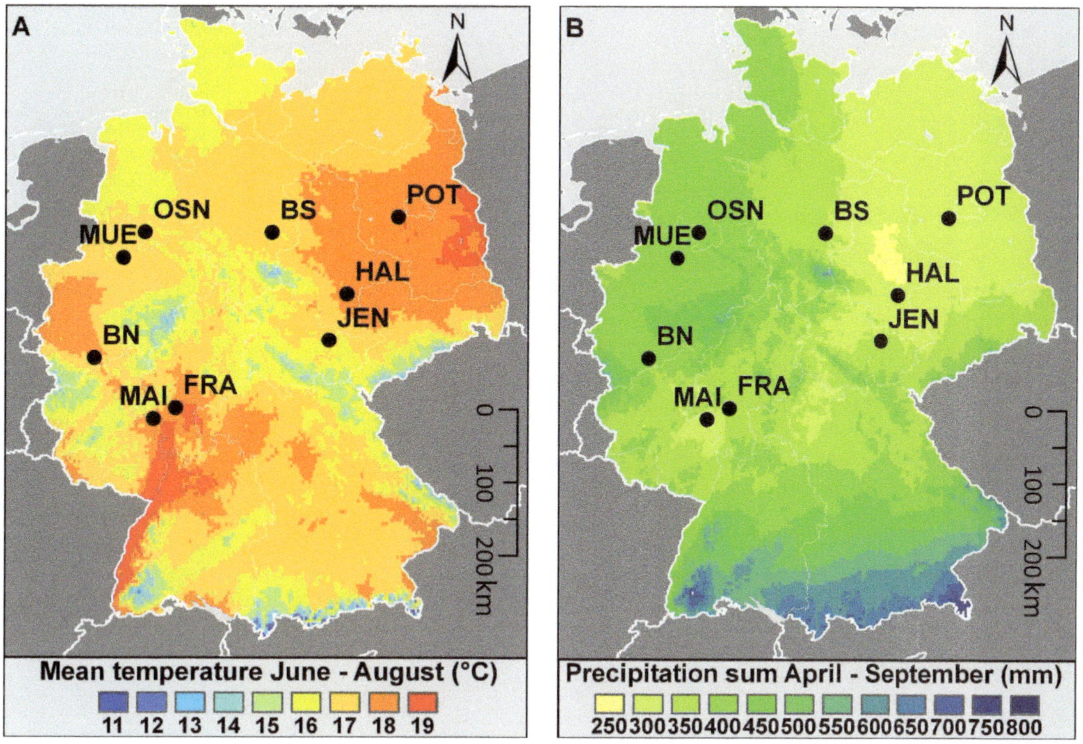

Figure 1. Locations of the Botanical Gardens in the transplant experiment showing the main gradients in climatic differences. A) Sum of the monthly mean temperature in summer (June-August), B) Precipitation of the vegetation period in mm. Climate data were obtainded from [26] and refer to the same periods as used in Table 1, but refer to long-term averages. For abbreviations of locations see Table 1.

- *Ca. biebersteinii*, *Centaurea scabiosa* - *Ce. stoebe*, *Dianthus deltoides* - *Di. carthusianorum*, *Inula conyzae* - *I. hirta*, *Koeleria pyramidata* - *K. macrantha*, *Scabiosa columbaria* -*Sc. ochroleuca* and *Silene nutans* - *Si. otites*. The species are perennial plants which are all native to Europe (see distribution maps in Figure S1) and occur mainly in dry to semidry open grasslands [27]. All species had already been investigated in other glasshouse [21,22] and field experiments [28].

Seeds of all species were collected in summer 2007 in Central Germany, using large populations to avoid negative effects of low genetic diversity (for geographical coordinates of the sampling localities see Table S1). Seedlings of all species were raised under controlled standardized glasshouse conditions in spring 2008. In June 2008, the young seedlings were transplanted into the plots. To ensure initial establishment, plants were watered regularly for one month. There were some species for which not enough seedlings were available, resulting in three species pairs that could not be planted in all gardens. These missing species pairs were randomly assigned to all gardens, except for those at the ends of the climate gradient which received all seven genera. We made sure that no more than one species pair was missing in any of the nine gardens.

At the beginning of the experiment, the leaf number of every transplant was counted to calculate relative growth rates according to Hunt [29]. At the end of the experiment, in October 2009, survival, flowering status, number of leaves, number of flowering units, plant height, specific leaf area (SLA) and the proportion of consumed leaf area (visually assessed) were recorded. Additionally, aboveground biomass was harvested and weighed after drying at 70°C for 48 hours.

All data are available at http://data.idiv.de/repo/data_Welk_etal_PlosOne.xls.

Experimental setup

In every garden one study plot consisting of 16 subplots of one by one metre area each was established (Figure S2). Each subplot was divided in four rows and four columns, resulting in 16 planting positions (Figure S2). One individual per plant species was planted in every subplot at randomly chosen positions resulting in a maximum of 14 transplants (and two empty positions) per subplot. For the competition treatment, seeds of *Festuca rubra* (cultivar *Wilma*, RUDLOFF Feldsaaten GmbH, Bad Schwartau, Germany) were sown ($5\ g/m^2$) on eight subplots at the time of planting of the transplants (Figure S2). All subplots were regularly weeded except for *Festuca rubra* in the competition subplots. For the mollusc exclosure we regularly applied a mollusc repellent (Ferramol, W. Neudorff GmbH KG, Emmerthal) on eight of the subplots. In spring 2009, these plots were additionally equipped with metal frame fences to exclude molluscs (IRKA, R+M Gartenbedarf, Rehling, www.der-schneckenzaun.de). The competition and mollusc exclosure treatments were fully crossed and all treatment combinations were replicated four times per Botanical Garden (Figure S2), resulting in a total of 144 subplots and 1824 transplant individuals.

Statistical analyses

All response variables (survival, incidence of flowering, RGR of leaf number, aboveground biomass, plant height, number of flowering units, SLA and proportion of consumed leaf area biomass) were analysed with separate generalized linear mixed effect models (GLMM, proc glimmix, type III SS, SAS 9.2, SAS Institute Inc. 2008). For survival and incidence of flowering, a logit-link function and binomial error distribution were used, while the GLMM for all other response variables had an identity-link

Table 1. Climatic conditions of the Botanical Gardens and results from the principal component analysis (PCA).

Climate variable	Botanical garden									*Loadings PCA1*	*Loadings PCA2*
	FRA	**MAI**	**HAL**	**POT**	**BS**	**JEN**	**BN**	**MUE**	**OSN**		
Sum of the monthly mean temperature in summer 2009 (June-August) in °C	57.9	56.6	54.4	54.4	53.6	53.2	53.5	52.9	52.7	***0.422***	*−0.227*
Sum of the monthly mean temperature of the vegetation period 2008 (August-October) in °C	42.8	42.0	41.6	41.0	41.4	40.6	40.9	41.0	40.5	***0.412***	*−0.186*
Sum of the monthly mean temperature of the vegetation period 2008 and 2009 in °C	147.0	144.4	138.5	138.2	137.1	135.5	136.8	136.1	135.2	***0.411***	*−0.258*
Sum of the monthly mean temperature of the vegetation period 2009 (April-September) in °C	104.2	102.4	96.9	97.2	95.7	94.9	95.9	95.1	94.7	*0.404*	*−0.269*
Precipitation in summer 2009 (June-August) in mm	205.5	163.2	151.1	152.1	139.2	173.3	182.9	209.3	198.1	*−0.140*	***−0.592***
Precipitation of the vegetation period 2009 (April-September) in mm	334.6	281.6	302.2	278.0	240.4	362.8	368.0	332.0	316.0	*−0.149*	***−0.399***
Minimum of monthly mean temperature 2008–2009	−1.9	−1.8	−2.1	−1.9	−0.7	−2.6	−0.8	−0.5	−0.2	*−0.279*	*−0.097*
Precipitation of the vegetation period 2008 (August–October) in mm	168.9	169.1	152.0	178.4	169.9	159.3	171.7	223.7	280.2	*−0.316*	*−0.244*
Precipitation of the vegetation period 2008 and 2009 in mm	503.5	450.7	454.2	456.4	410.3	522.1	539.7	555.7	596.2	*−0.317*	***−0.443***
First axis scores of the PCA (=PCA1)	*3.32*	*2.76*	*0.99*	*0.45*	*0.32*	*−0.98*	*−1.35*	*−2.20*	*−3.30*		
Second axis scores of the PCA (=PCA2)	*−2.66*	*−0.13*	*1.25*	*1.41*	*2.50*	*0.42*	*−0.54*	*−1.11*	*−1.13*		

Climate variables are arranged according to their loadings on the first PCA axis. Botanical gardens included in this study are sorted according to their scores of the first PCA axis. Scores and loadings of the Principal Component Analysis (PCA) are shown in italics. The three variables with highest absolute loadings per PCA axis are shown in bold letters. Botanical garden abbreviation: BN – Bonn, BS – Braunschweig, FRA – Frankfurt am Main, HAL – Halle (Saale), JEN – Jena Isserstedt, MAI – Mainz, MUE – Münster, OSN – Osnabrück, POT – Potsdam.

function and Gaussian error distribution. To identify the most relevant climatic drivers of the different response variables we included the scores of the first and second PCA axes as covariates in the models. Fixed categorical factors in all models were plant range type (oceanic, continental), competition (presence/absence) and herbivory (presence/absence). All possible two and three-way interactions of fixed factors and covariates were included. Subplot identity (nested in the interaction of garden identity, herbivory and competition treatment) and species identity (nested in plant range type) entered the models as crossed random factors.

Unbiased least square means (LS means) and standard errors were calculated using the LSMEANS statements in SAS 9.2 (SAS Institute Inc. 2008) and used to produce graphs. Tukey post-hoc tests were calculated for contrasts between treatment combinations. All graphs were produced with R 2.15.2 (R Development Core Team 2012).

Results

Main effects of climate, competition and herbivory

The climate gradient significantly affected survival, aboveground biomass, plant height, number of flowering units, SLA and proportion of leaf area consumed (Table 2). While temperature, which was captured by the first PCA axis, had no significant effect on any response variable, precipitation, with high loadings on the second PCA axis, played a major role. Survival ($p = 0.005$), plant height ($p = 0.013$) and number of flowering units ($p < 0.001$) decreased with decreasing precipitation (i.e. increasing PCA2 scores), while proportion of consumed leaf area increased ($p = 0.020$). Additionally, there were significant interactions of both PCA axes for survival ($p = 0.002$), aboveground biomass ($p = 0.005$), plant height ($p = 0.038$), number of flowering units ($p < 0.001$) and specific leaf area ($p = 0.042$, Table 2).

The competition and mollusc exclosure treatments had strong effects on all response variables except survival (Table 2). This demonstrates that our experimental subplot manipulation of biotic interactions was effective and that these two biotic factors were key determinants of plant performance across all species. Competition significantly reduced the incidence of flowering (by −7.1%, referring to LSmeans estimates from the GLMM, $p < 0.001$), RGR of leaf number (by −59.2%, $p < 0.001$), aboveground biomass (by −40.8%, $p < 0.001$) and number of flowering units (by −35.5%, $p < 0.001$). Furthermore, there was an increase in SLA (by +6.2%, $p = 0.022$) and proportion of consumed leaf area (by +50.1%, $p < 0.001$) when competitors were present. Mollusc exclosure resulted in increased plant height (by +3.5%, $p = 0.031$) and lower proportions of consumed leaf area (by −29.7%, $p < 0.001$). Additionally, proportions of consumed leaf area showed a significant interaction effect of competition with herbivory ($p = 0.004$): While the mollusc exclosures reduced the amount of consumed leaf area by −10.3% in absence of competition, the reduction was −40.5% in presence of competition.

Interaction of climate with impacts of competition and herbivory

In general, the climate gradient had no effects on the outcome of the biotic subplot manipulations for most response variables (Table 2). Along the first PCA axis (temperature) the competition effect on number of flowering units changed ($p = 0.043$, Table 2). With increasing temperature (i.e. increasing PCA1 scores), the number of flowering units increased when competitors were absent, while the number of flowering units remained unchanged when competitors were present. The effect of mollusc exclosure changed along the second PCA axis (precipitation) for incidence of flowering ($p = 0.027$, Table 2). With decreasing precipitation (i.e. increasing PCA2 scores) flowering was observed less frequently when molluscs were excluded, while flowering increased in frequency in the presence of mollusc herbivory.

Effects of plant range type and the interactions of plant range type with climate, competition and herbivory

No significant main effects of plant range type were observed for any of the response variables (Table 2). This also applied to SLA, showing that the two representatives for plant range types did not differ in basic functional traits. However, significant interactions indicated that species of different plant range types responded differently along the climate gradient and to the subplot treatments (Table 2).

Survival, incidence of flowering, aboveground biomass and number of flowering units were significantly affected by the interaction of plant range type with competition (Table 2). The plants of the oceanic range type survived slightly better in presence of competitors (+1.6%), whereas those of the continental range type had clearly lower survival rates (−9.0%) when competitors were present ($p = 0.025$, Fig. 2A). Similarly, plants with continental range type flowered less frequently in presence of competitors (−7.7%), whereas flowering of oceanic plants remained nearly unaffected by competition (−0.9%, $p = 0.030$, Fig. 2B). In contrast, aboveground biomass (oceanic −47.9%, continental −32.6%, $p = 0.008$) and number of flowering units (oceanic −40.5%, continental −27.0%, $p = 0.017$) were generally negatively affected by competition, with stronger negative effects on plants with oceanic than on continental range type (Fig. 2C,D).

With respect to the herbivory treatment, plant range type was important for the incidence of flowering ($p = 0.007$) and proportion of consumed leaf area ($p = 0.027$, Table 2). Plants of the continental range type showed a decrease in the proportion of flowering individuals (−4.8%) in the presence of molluscs while those of the oceanic range type displayed a slight increase (+0.6%, Fig. 2E). Without mollusc exclosure, plants of the oceanic range type were consumed more frequently than plants of the continental range type (Fig. 2F).

There were also threefold significant interactions of competition, herbivory and plant range type. The plant range types displayed contrasting response patterns to herbivory and competition in survival ($p = 0.028$, Fig. 3A,B), plant height ($p = 0.023$, Fig. 3C,D) and number of flowering units ($p = 0.022$, Fig. 3E,F, Table 2). Survival and plant height of continental plants were strongly negatively affected by the single effects of herbivory and competition (Fig. 3B,D). These effects had the same magnitude and were not additive when mollusc herbivory occurred in combination with competition. Survival and plant height of oceanic plants remained unaffected by the biotic treatments (Fig. 3A,C). The contrasting pattern was observed for the number of flowering units, where oceanic plants were strongly affected by biotic interactions (Fig. 3E), while continental plants were not (Fig. 3F). In oceanic plants, herbivory and competition had opposing effects on number of flowering units. While herbivory alone caused a significant increase in number of flowering units, competition alone had no significant effect. However, when competitors were present in addition to herbivores, the number of flowering units of oceanic plants strongly decreased (Fig. 3E). Neither RGR of leaf number nor the proportion of consumed leaf area showed significantly different effects for the contrasting range types in the presence or absence of herbivores or competition (Table 2).

The responses to the climatic variables of plants of contrasting range types differed for incidence of flowering (PCA2, $p = 0.018$,

Table 2. The effect of climatic covariates (PCA axes 1 and 2, representing temperature and precipitation effects, respectively), competition, herbivory and plant range type on the different variables of plant performance measured in the Botanical Gardens.

Effect		Survival rates			Incidence of flowering			RGR of leaf number			Aboveground biomass			Plant height			Number of flowering units			Specific leaf area (SLA)			Proportion of consumed leaf area		
	DF_{num}	DF_{den}	F		DF_{den}	F		DF_{den}	F		DF_{den}	F		DF_{den}	F		DF_{den}	F		DF_{den}	F		DF_{den}	F	
PCA1	1	1638	0		1500	2.78		1243	1.04		1225	0.13		1449	0.46		1499	2.5		1172	0.5		1237	0.16	
PCA2	1	**1638**	**7.84**	**	1500	0.15		1243	2.03		1225	2.04		**1449**	**6.20**	*	**1499**	**16.72**	***	1172	1.01		**1237**	**5.40**	*
PCA1×PCA2	1	**1638**	**10.06**	**	1500	1.41		1243	1.23		**1225**	**8.08**	**	**1449**	**4.30**	*	**1499**	**19.61**	***	**1172**	**4.13**	*	1237	0.26	
Competition (C)	1	129	0.28		**130**	**13.52**	***	**130**	**16.26**	***	**128**	**53.13**	***	130	3.02		**130**	**28.58**	***	**130**	**5.37**	*	**130**	**18.96**	***
C×PCA1	1	1638	0.25		1500	1.15		1243	0.01		1225	0		1449	0.65		**1499**	**4.12**	*	1172	0.07		1237	0.48	
C×PCA2	1	1638	2.21		1500	0.23		1243	0.47		1225	1.19		1449	0.04		1499	0.23		1172	0.60		1237	0.35	
C×PCA1×PCA2	1	1638	0.32		1500	1.37		1243	1.17		1225	0.77		1449	0.15		**1499**	**7.76**	**	1172	0.67		1237	0.88	
Herbivory (H)	1	129	0.87		130	1.72		130	0.66		128	0.04		**130**	**4.78**	*	130	0.44		130	2.41		**130**	**14.40**	***
H×PCA1	1	1638	0		1500	2.11		1243	0.09		1225	0.01		1449	0.68		1499	0.34		1172	0.11		1237	0	
H×PCA2	1	1638	0.26		**1500**	**4.90**	*	1243	0.39		1225	0.03		1449	0.72		1499	1.38		1172	0		1237	1.66	
H×PCA1×PCA2	1	1638	0.01		1500	0.47		1243	0.50		1225	0		1449	0.46		1499	1.29		1172	0.02		1237	0.38	
C×H	1	129	0.08		130	0.25		130	0.64		128	0.19		130	0.06		130	2.32		130	0.20		**130**	**8.47**	**
C×H×PCA1	1	1638	1.63		1500	0.17		1243	0.76		1225	0.07		1449	0.01		1499	0		1172	0.07		1237	0.05	
C×H×PCA2	1	1638	0.35		1500	2.47		1243	1.15		1225	1.11		1449	0.30		1499	0.17		1172	0.48		1237	2.74	
Range type (R)	1	12	1.49		12	0.50		12	2.50		12	0.02		12	0.07		12	0.63		12	1.13		12	0.37	
R×PCA1	1	1638	0.06		1500	1.67		1243	0.80		1225	0.02		1449	2.01		1499	0.80		1172	0.03		1237	0.45	
R×PCA2	1	1638	0.56		**1500**	**5.64**	*	1243	0		1225	0.54		**1449**	**6.65**	*	1499	0.51		1172	0.86		1237	0.88	
R×PCA1×PCA2	1	1638	2.29		1500	0.78		1243	0.40		1225	0.39		1449	0.53		1499	3.19		1172	2.42		1237	0.03	
C×R	1	**1638**	**5.07**	*	**1500**	**4.71**	*	1243	1.09		**1225**	**6.97**	**	1449	0.01		**1499**	**5.74**	*	1172	0.07		1237	1.59	
C×R×PCA1	1	1638	2.23		1500	0.01		1243	0.23		1225	0.44		1449	3.42		1499	0.62		1172	0.48		1237	0.11	
C×R×PCA2	1	1638	0.84		**1500**	**3.96**	*	1243	1.38		**1225**	**4.57**	*	1449	1.51		1499	3.20		1172	2.64		1237	1.27	
H×R	1	1638	0.44		**1500**	**7.76**	**	1243	0.03		1225	0.95		1449	1.26		1499	0.08		1172	0.02		**1237**	**4.91**	*
H×R×PCA1	1	1638	0.05		1500	2.36		1243	0.12		1225	0.50		1449	0.50		1499	1.22		1172	0.13		1237	0.01	
H×R×PCA2	1	1638	0.18		1500	0.02		1243	0.02		1225	0.01		1449	0.82		1499	0.99		1172	0.06		1237	0.03	
R×C×H	1	**1638**	**4.85**	*	1500	2.94		1243	3.31		1225	0.20		**1449**	**5.18**	*	**1499**	**5.29**	*	1172	0.50		1237	0.12	

The table shows degrees of freedom for the numerator and denominator (DF_{num} and DF_{den}, respectively as well as F values from a type III Anova based on generalized linear models. Bold F values indicate significant effects. ***$p < 0.001$, **$p < 0.01$, *$p < 0.05$.

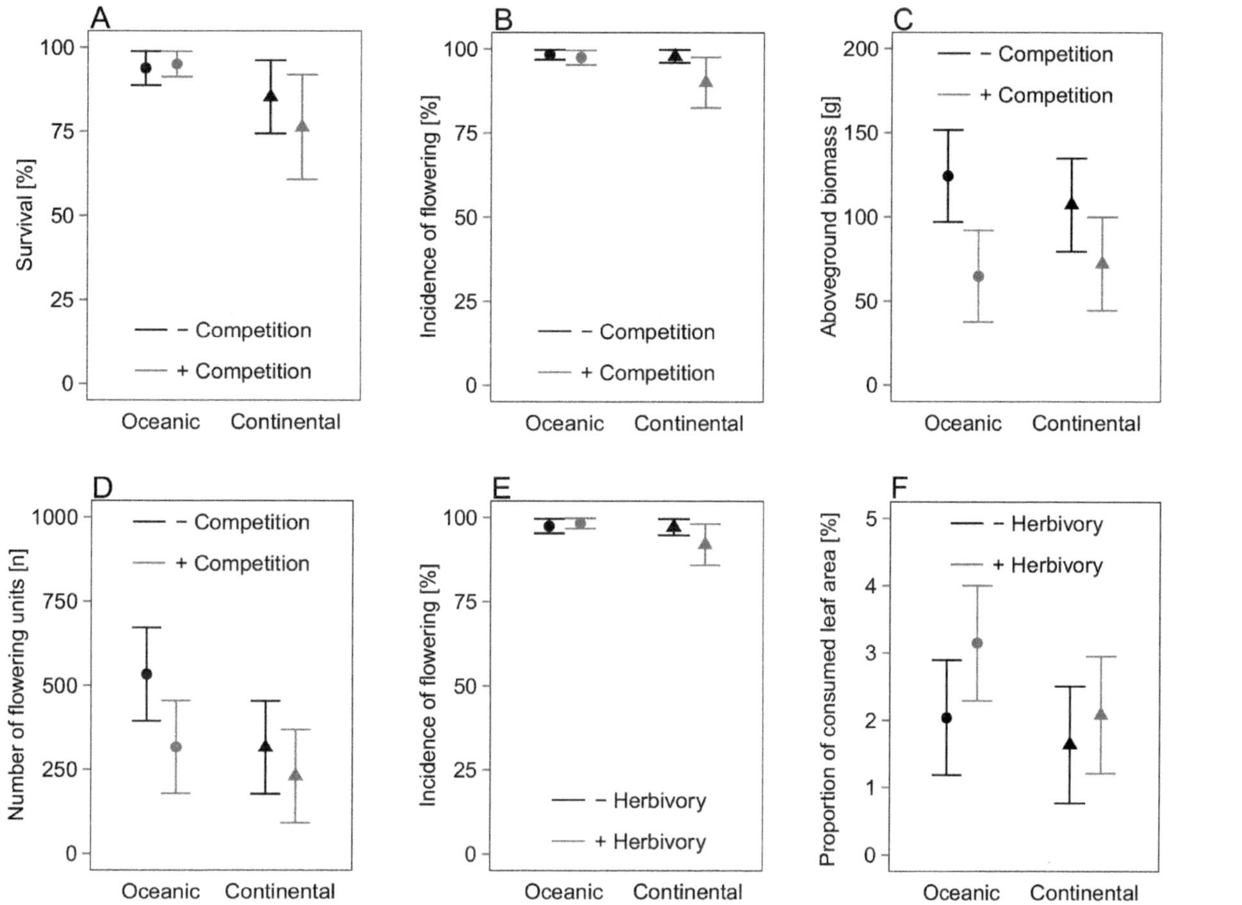

Figure 2. Significant interactive effects of plant range type with competition on A) survival, B) incidence of flowering, C) aboveground biomass and D) number of flowering units, and interactive effects of plant range type with herbivory on E) incidence of flowering and F) proportion of consumed leaf area. Graphs are plotted with LSmeans estimates and standard errors derived from the GLMM's. For statistical details see Table 2.

Fig. 4A) and plant height (PCA2, $p = 0.010$, Fig. 4B, Table 2). Incidence of flowering of continental plants decreased with decreasing precipitation (i.e. increasing PCA2 scores) while that of oceanic plants increased (Fig. 4A). Similarly, plant height of continental plants decreased with decreasing precipitation) but remained constant in oceanic plant species (Fig. 4B). At the moist end of the gradient, plants of the continental range type flowered more frequently and were taller than oceanic ones, while on the dry end the pattern was reversed (Fig. 4A,B).

Furthermore, along the precipitation gradient, the effect of competition differed between range types for incidence of flowering (PCA2, $p = 0.047$, Fig. 4C) and aboveground biomass (PCA2, $p = 0.033$, Fig. 4D, Table 2). For both response variables, oceanic plants showed changed climate responses when competition was manipulated. Incidence of flowering and aboveground biomass strongly increased with decreasing precipitation (i.e. increasing PCA2 scores) when competitors were absent (Fig. 4C). In the presence of competition, oceanic plants showed only a slight increase in incidence of flowering (Fig. 4C) and a decrease in aboveground biomass with decreasing precipitation (i.e. increasing PCA2 scores, Fig. 4D). In contrast, plants with continental plant range type showed almost no changes in their climate responses when competition was manipulated. With decreasing precipitation (i.e. increasing PCA2 scores) incidence of flowering and aboveground biomass of continental plants decreased, both in the absence and presence of competitors (Fig. 4C,D).

Discussion

The climate conditions in our transplant experiment were characterized by a distinct gradient from oceanic to continental climate. Across all species included in our study, temperature was less important for performance variation than precipitation. Plants benefited from higher precipitation in terms of higher survival, taller growth and increasing number of flowering units. Additionally, plant growth and reproduction were strongly negatively affected by competition and partly by mollusc herbivory, indicating that competition was more detrimental than herbivory. Biotic interactions affected all response variables except plant survival. This indicates that the relative importance of biotic interactions and climate differed among the different response variables.

Effects of climate on biotic interactions

In our first hypothesis, we expected that the negative effect of competition and herbivory is mitigated under the more stressful continental climate conditions. This was not the case for survival and plant growth. Biotic interactions only changed the response along the climate gradient in the case of variables related to

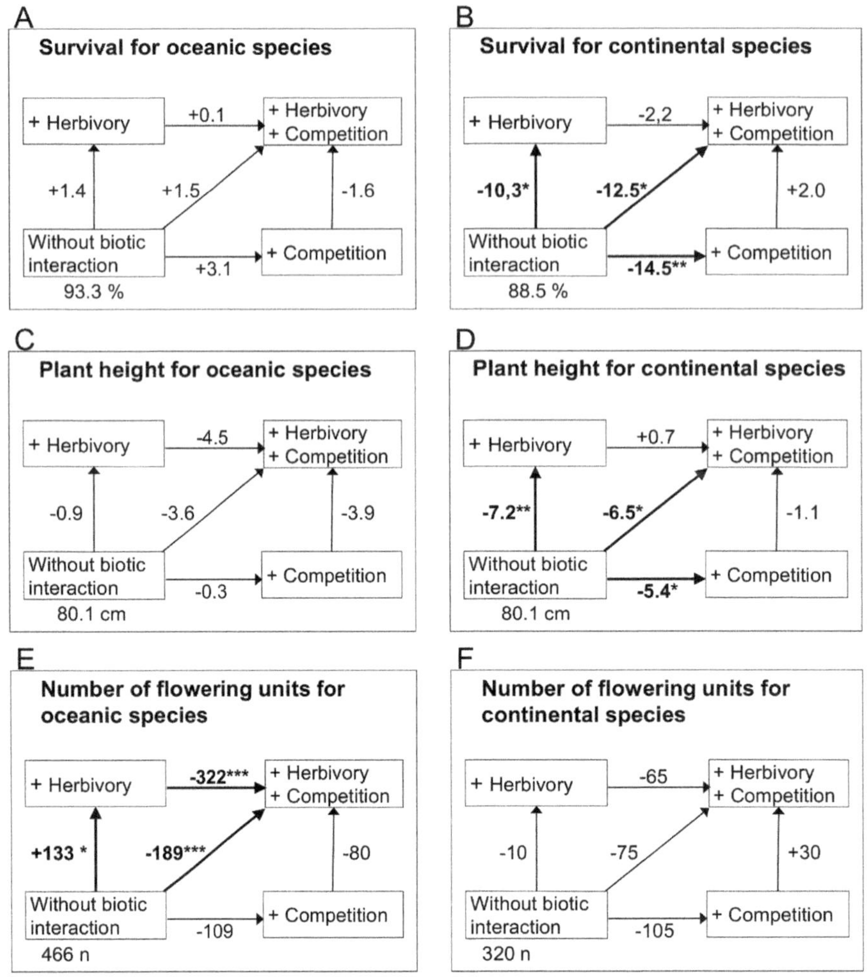

Figure 3. Summary of interactive effects of competition and herbivory on species with different plant range types for survival (A,B), plant height (C,D) and number of flowering units (E,F). Effects for the oceanic species are illustrated on the left side (A,C,E), for continental on the right side (B,D,F). The values in the bottom left corner are LSmeans estimates calculated from the GLMM for the treatment without biotic interactions. The values on the arrows are differences in the LSmeans estimates of this treatment-combination indicating the direction and strength of the relation. Units are percentage (A,B), cm (C,D) and numbers (E,F). Bold arrows indicate significant effects according to the Tukey post hoc-test. ***$p<0.001$, **$p<0.01$, *$p<0.05$.

flowering. Competition generally reduced aboveground biomass and growth, irrespective of climate. Restrictions in water supply at the dry end of the climatic gradient did not aggravate the effect of competition, probably because target plants and competitors were affected likewise. However, we also have to consider that the fertile garden soils used in our experiment have affected the interaction of climate and competition. While competition is expected to become more pronounced under fertile conditions, drought effects might have been reduced. In contrast to competition, mollusc herbivory had no impact on biomass production and was independent of the climate conditions in our experiment, except for incidence of flowering. This does not confirm the results of a recent global meta-analysis of Rodríguez-Castañeda [30], who found that the effect of general herbivory on plant performance increased in moister ecosystems. Bruelheide & Scheidel [11] described increased slug herbivory with decreasing altitude, which was reflected in increasing temperatures, and overall, more benign growth conditions. A potential explanation for this mismatch to our results might be that the climate gradient used in our study was not steep enough to evoke climate-dependent herbivory effects on plant performance. Additionally, the generally high slug abundances in Botanical gardens might have uncoupled the climate-herbivory relationship, which emphasises the need for assessing mollusc densities in future studies. Given the large distance between the experimental sites, we were not able to count molluscs during rain events or to quantify mollusc activity [11,16].

Interaction of climate with plant range type

We secondly hypothesised that the changing climate conditions in our study have a different impact on the transplants with respect to their range type. The expectation that species of the oceanic plant range type perform better at the oceanic end of the climate gradient and those of the continental plant range type at the continental end of the climate gradient was not confirmed. Instead, we even found opposing patterns with increased incidence of flowering, plant height and aboveground biomass for continental plants under moister conditions and for oceanic plants under drier conditions. These results support the outcome of previous experiments in the field and the greenhouse, where analyses with the same species set revealed similar responses to manipulated climate change [28] or altered soil moisture conditions [21,22]. Nevertheless, at least for the continental species, we expected a

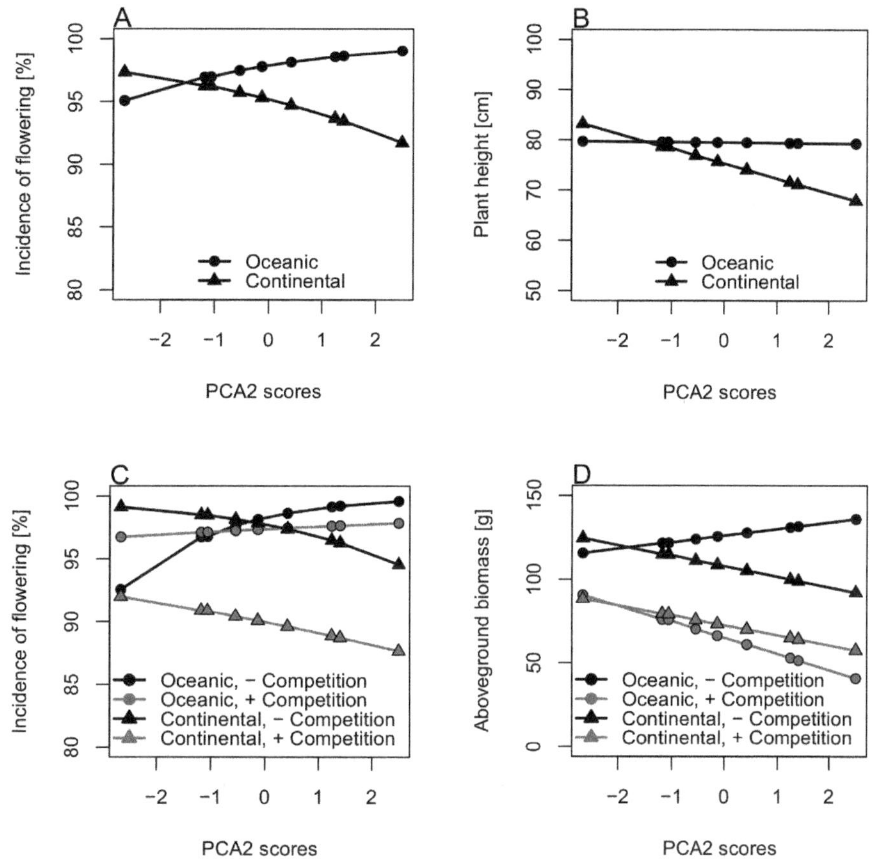

Figure 4. Significant effects of the climate gradient (climate covariate) on plants with different plant range types for A) incidence of flowering and B) plant height, and in interaction with competition for C) incidence of flowering and D) aboveground biomass. Climate covariates are PCA 2 scores which are negatively correlated with precipitation sums of the vegetation period during the experiment (Table 1). Graphs are plotted with LSmeans estimates derived from the GLMM's. For statistical details see Table 2.

poorer performance in the more oceanic climate since the experimental setup covered an area extending beyond their western distribution limit (Figure S1). However, the general impression was that plants of the continental range type responded stronger to the climatic gradient than those of the oceanic range type, especially with plant height. In contrast to our findings, numerous transplant experiments have described a decreased fitness beyond a species' current range [31]. However, most of these species' range boundaries were studied along latitudinal or altitudinal climate gradients [32,33], where temperature is the dominant driver. In contrast, the climate gradient covered in our experiment is less simple as higher temperatures that favour growth rates are counteracted by increasing drought risk. Consistent with our results, Stanton-Geddes et al. [34] observed fitness declines towards the northern range edge but not to the western range edge of *Chamaecrista fasciculata*, a widespread annual legume from North America. However, we also have to consider that short-term experiments, as presented here, might perhaps not be able to detect clear home-site advantages. For example, in a transplant experiment monitored over 30 years Bennington [35] encountered increasingly stronger home-site advantages for ecotypes of two arctic plant species with time. Our study design also did not allow for assessing the impact of climate on population demography as we did not focus on recruitment. Given that our species are almost all long-lived, although some of them are hapaxanth (Table S1), a 2 years-study period is certainly not sufficient to estimate the impact of climate on long-term survival and demography. This clearly underlines the need for long-term studies.

Interaction of competition and herbivory with plant range type

When climatic factors failed to explain species range limitation, biotic interactions have often been made accountable for directly limiting distribution ranges [31,36]. For example, Engels & Jensen [37] found that plant species from salt marshes performed similarly well in both freshwater and saltwater marshes, when competition was excluded. Similarly, in our experiment, plants with continental range type survived much better in absence of competition than those with oceanic plant range type. In addition, continental plants flowered more frequently when competitors or herbivores were removed or excluded, respectively. This sensitivity to competition and herbivory of continental plants was not modified by climate. This indicates that the geographic distribution of continental plants is not directly limited by climate conditions but by other factors, such as competition. However, biotic interactions might act on different spatial scales. In a tundra study system Le Roux et al. [38] found that small scale species distribution was shaped by horizontal biotic interactions (i.e. competition) rather than by vertical interactions (i.e. herbivory). Similarly, our experimental evidence on survival and incidence of flowering supports our third hypothesis stating that continental plants are more vulnerable to competition and herbivory than oceanic plants. However, oceanic plants were stronger negatively affected by competition in biomass and flower production than continental plants. Regarding these inconsistent results, the third hypothesis that continental plants are more susceptible to competition or herbivory has to be rejected.

Although individuals of both plant range types suffered from biotic stress, they showed different strategies to cope with that. Particularly, the combined effect of competition and herbivory caused different plant responses with respect to plant range type. Only plants of continental range type displayed lower survival rates and smaller plant height in presence of competitors and herbivores. In contrast, oceanic plants were significantly affected in their flower production by herbivory alone and in combination with competition. The reduced flower production of the oceanic plants might be an effect of the compensatory response to herbivory and demonstrates a shift in resource allocation to enhanced plant growth [39]. Further reproductive traits such as number of seeds, seed mass and seed dispersal mode, have also profound effects on plant persistence [40] but could not be assessed in our experiment. Nevertheless, survival of plant individuals is ecologically essential to maintain a population. A synthesis of all our results indicates, that continental plants run a higher risk to suffer from competition and herbivory than oceanic plants, particularly if they are subjected to the combination of both competition and herbivory.

Conclusion

From the results of our experiment we can conclude that, at the geographical scale of Germany, the continentality gradient is of minor direct importance for species of the considered plant range types compared to negative biotic interactions. Competition and herbivory affected particularly the performance of individuals of continental plant range type. There was only weak evidence for the assumption that the impact of competition and herbivory should vary with climate. In consequence, predictions of future geographic range dynamics of plants species have to be considered with caution, especially when based solely on occurrence data and climatic variables. Mechanistic models would be more suitable, but to devise such models a better understanding of plant ecology is needed [41]. Consequentially, Wisz et al [42] recommended targeted long-term field monitoring approaches. Such long-term field experiments should not only include species of different plant range types but also measure demographic rates, which requires the assessment of sexual and vegetative recruitment.

Supporting Information

Figure S1 Distribution maps of the congeneric species pairs used in the study. A – *Carlina*, B – *Centaurea*, C – *Dianthus*, D – *Inula*, E – *Koeleria*, F – *Scabiosa*, G - *Silene*. Species which were assigned to oceanic range types are coloured in blue, continental are coloured in red. Violet colour indicates range overlap of the two species The Botanical Gardens where the experimental sites were located are shown as black dots. Details on the compilation of the data for these distributions maps are given in Hofmann et al. (2013).

Figure S2 Plot scheme of the experimental design in every Botanical Garden. All treatments were randomly assigned to subplots and plants were randomly assigned to planting positions. All species were planted into subplots. Symbols: - C = absence of competitors (regular weeding), + C = presence of competitors (*Festuca rubra*), - H = slug herbivore exclusion (subplot with metal frame and slug repellents), + H = without slug herbivore exclusion (subplot without metal frame).

Table S1 Species characteristics (Jäger & Werner 2005) and coordinates of the localities where seeds were collected. Growth form: eg = evergreen, sg = summergreen, hc = hemicryptophyte, p = perennial.

Acknowledgments

We are very grateful to all the participating Botanical Gardens for providing space for the establishment and help for the maintenance of the experimental plots, especially (sorted by the Botanical Gardens): W. Lobin and J. Manner (Bonn), M. Kraft (Braunschweig), M. Wessel (Frankfurt a. M.), M. H. Hoffmann and A. Fläschendräger (Halle), S. Arndt (Jena), R. Omlor and B. Meyer (Mainz), H. Voigt (Münster), N. Friesen and H. Brüggemann (Osnabrück), M. Burkart and K. Kläring (Potsdam). We are also indebted to all technicians and student helpers who conducted field and lab work, particularly M. Baudis and S. Kosellek. Seed collection was assisted by V. Bremerich, J. Danihelka and J. Stolle. M. Hofmann was a great help in raising the plants. The manuscript was significantly improved by the comments of John Stanton-Geddes and another anonymous reviewer.

Author Contributions

Conceived and designed the experiments: AW EW HB. Performed the experiments: AW. Analyzed the data: AW EW HB. Contributed reagents/materials/analysis tools: AW EW HB. Wrote the paper: AW EW HB.

References

1. Soberón J (2007) Grinnellian and Eltonian niches and geographic distributions of species. Ecol Lett 10: 1115–1123.
2. Ellenberg H (1953) Physiological and ecological responses of the same plant species (in German). Physiologisches und ökologisches Verhalten derselben Pflanzenarten. Ber Dtsch bot Ges 65: 350–361.
3. Hector A, von Felten S, Hautier Y, Weilenmann M, Bruelheide H (2012) Effects of dominance and diversity on productivity along Ellenberg's experimental water table gradients. PLoS ONE 7: e43358.
4. Godsoe W (2010) I can't define the niche but I know it when I see it: a formal link between statistical theory and the ecological niche. Oikos 119: 53–60.
5. Colwell RK, Fuentes ER (1975) Experimental studies of niche. Annu Rev Ecol Syst 6: 281–310.
6. Hutchinson GE (1957) Concluding Remarks. Cold Spring Harb Symp Quant Biol 22: 415–427.
7. Dormann CF, Schymanski SJ, Cabral J, Chuine I, Graham C, et al. (2012) Correlation and process in species distribution models: bridging a dichotomy. J Biogeogr 39: 2119–2131.
8. Duffy JE, Cardinale BJ, France KE, McIntyre PB, Thébault E, et al. (2007) The functional role of biodiversity in ecosystems: incorporating trophic complexity. Ecol Lett 10: 522–538.
9. Louda SM, Keeler KH, Holt RD (1990) Herbivore influences on plant performance and competitive interactions. In: Grace JB, Tilman D, editors. Perspectives on plant competition. New York: Academic Press. pp. 413–444.
10. Salisbury EJ (1926) The Geographical Distribution of Plants in Relation to Climatic Factors. The Geographical Journal 67: 312–335.
11. Bruelheide H, Scheidel U (1999) Slug herbivory as a limiting factor for the geographical range of *Arnica montana*. J Ecol 87: 839–848.
12. Bertness MD, Callaway R (1994) Positive interactions in communities. Trends Ecol Evol 9: 191–193.
13. Maestre FT, Callaway RM, Valladares F, Lortie CJ (2009) Refining the stress-gradient hypothesis for competition and facilitation in plant communities. J Ecol 97: 199–205.
14. Loehle C (1998) Height growth rate tradeoffs determine northern and southern range limits for trees. J Biogeogr 25: 735–742.
15. Callaway RM, Brooker RW, Choler P, Kikvidze Z, Lortie CJ, et al. (2002) Positive interactions among alpine plants increase with stress. Nature 417: 844–848.
16. Scheidel U, Bruelheide H (2001) Altitudinal differences in herbivory on montane Compositae species. Oecologia 129: 75–86.

17. Jäger EJ (1968) The plant geographical oceanity-division of the Holarctis and the ocean affinity of plant distribution ranges (in German). Die pflanzengeographische Ozeanitätsgliederung der Holarktis und die Ozeanitätsbindung der Pflanzenareale. Fed Rep 79: 157–335.
18. Finnie TJR, Preston CD, Hill MO, Uotila P, Crawley MJ (2007) Floristic elements in European vascular plants: an analysis based on Atlas Florae Europaeae. J Biogeogr 34: 1848–1872.
19. Brown J, Mehlman D, Stevens G (1995) Spatial variation in abundance. Ecology 76: 2028–2043.
20. Purves DW (2009) The demography of range boundaries versus range cores in eastern US tree species. Proc R Soc B 276: 1477–1484.
21. Bütof A, Bruelheide H (2011) Effects of an unspecialized soil pathogen on congeneric plant species with different geographic distributions. Preslia 83: 205–217.
22. Hofmann M, Bütof A, Welk E, Bruelheide H (2013) Relationship between fundamental and realized niches with respect to frost and drought resistance. Preslia 85: 1–17.
23. Joshi J, Schmid B, Caldeira MC, Dimitrakopoulos PG, Good J, et al. (2001) Local adaptation enhances performance of common plant species. Ecol Lett 4: 536–544.
24. Bruelheide H, Heinemeyer A (2002) Climatic factors controlling the eastern and altitudinal distribution boundary of *Digitalis purpurea* L. in Germany. Flora 197: 475–490.
25. De Frenne P, Brunet J, Shevtsova A, Kolb A, Graae BJ, et al. (2011) Temperature effects on forest herbs assessed by warming and transplant experiments along a latitudinal gradient. Global Change Biol 17: 3240–3253.
26. Hijmans RJ, Cameron SE, Parra JL, Jones PG, Jarvis A (2005) Very high resolution interpolated climate surfaces for global land areas. Int J Climatol 25: 1965–1978.
27. Jäger EJ, Werner K, editors (2005) Werner Rothmaler: Exkursionsflora von Deutschland. 10. edition. München: Elsevier (Spektrum).
28. Bütof A, von Riedmatten LR, Dormann CF, Scherer-Lorenzen M, Welk E, et al. (2012) The responses of grassland plants to experimentally simulated climate change depend on land use and region. Global Change Biol 18: 127–137.
29. Hunt R (1990) Basic growth analyses. London: Unwin Hyman Ltd.
30. Rodríguez-Castañeda G (2013) The world and its shades of green: a meta-analysis on trophic cascades across temperature and precipitation gradients. Global Ecol Biogeogr 22: 118–130.
31. Sexton JP, McIntyre PJ, Angert AL, Rice KJ (2009) Evolution and ecology of species range limits. Annu Rev Ecol Evol S 40: 415–436.
32. Hennenberg KJ, Bruelheide H (2003) Ecological investigations on the northern distribution range of Hippocrepis comosa L. in Germany. Plant Ecol 166: 167–188.
33. Angert AL, Schemske DW (2005) The evolution of species' distributions: Reciprocal transplants across the elevation ranges of *Mimulus cardinalis* and *M. lewisii*. Evolution 59: 1671–1684.
34. Stanton-Geddes J, Tiffin P, Shaw RG (2012) Role of climate and competitors in limiting fitness across range edges of an annual plant. Ecology 93: 1604–1613.
35. Bennington CC, Fetcher N, Vavrek MC, Shaver GR, Cummings KJ, et al. (2012) Home site advantage in two long-lived arctic plant species: results from two 30-year reciprocal transplant studies. J Ecol 100: 841–851.
36. Gaston JK (2003) The structure and dynamics of geographic ranges. Oxford: Oxford University Press.
37. Engels JG, Jensen K (2010) Role of biotic interactions and physical factors in determining the distribution of marsh species along an estuarine salinity gradient. Oikos 119: 679–685.
38. Le Roux PC, Lenoir J, Pellissier L, Wisz MS, Luoto M (2013) Horizontal, but not vertical, biotic interactions affect fine-scale plant distribution patterns in a low-energy system. Ecology 94: 671–682.
39. Rosenthal JP, Kotanen PM (1994) Terrestrial plant tolerance to herbivory. Trends Ecol Evol 9: 145–148.
40. Maurer K, Durka W, Stöcklin J (2003) Frequency of plant species in remnants of calcareous grassland and their dispersal and persistence characteristics. Basic Appl Ecol 4: 307–316.
41. Buckley LB, Urban MC, Angilletta MJ, Crozier LG, Rissler LJ, et al. (2010) Can mechanism inform species' distribution models? Ecology Letters 13: 1041–1054.
42. Wisz MS, Pottier J, Kissling WD, Pellissier L, Lenoir J, et al. (2013) The role of biotic interactions in shaping distributions and realised assemblages of species: implications for species distribution modelling. Biol Rev Camb Philos Soc 88: 15–30.

Evaluating the Spatio-Temporal Factors that Structure Network Parameters of Plant-Herbivore Interactions

Antonio López-Carretero[1], Cecilia Díaz-Castelazo[1]*, Karina Boege[2], Víctor Rico-Gray[3]

1 Instituto de Ecología A.C., Xalapa, Veracruz, México, **2** Instituto de Ecología, Universidad Nacional Autónoma de México, México D.F., México, **3** Instituto de Neuroetología, Universidad Veracruzana, Xalapa, Veracruz, México

Abstract

Despite the dynamic nature of ecological interactions, most studies on species networks offer static representations of their structure, constraining our understanding of the ecological mechanisms involved in their spatio-temporal stability. This is the first study to evaluate plant-herbivore interaction networks on a small spatio-temporal scale. Specifically, we simultaneously assessed the effect of host plant availability, habitat complexity and seasonality on the structure of plant-herbivore networks in a coastal tropical ecosystem. Our results revealed that changes in the host plant community resulting from seasonality and habitat structure are reflected not only in the herbivore community, but also in the emergent properties (network parameters) of the plant-herbivore interaction network such as connectance, selectiveness and modularity. Habitat conditions and periods that are most stressful favored the presence of less selective and susceptible herbivore species, resulting in increased connectance within networks. In contrast, the high degree of selectivennes (i.e. interaction specialization) and modularity of the networks under less stressful conditions was promoted by the diversification in resource use by herbivores. By analyzing networks at a small spatio-temporal scale we identified the ecological factors structuring this network such as habitat complexity and seasonality. Our research offers new evidence on the role of abiotic and biotic factors in the variation of the properties of species interaction networks.

Editor: Anna R. Armitage, Texas A&M University at Galveston, United States of America

Funding: Research was supported by grant (project) number 106575 from CONACYT to Cecilia Díaz-Castelazo, CONACYT fellowship number 302184 to Antonio López-Carretero, and Universidad Veracruzana, I010/152/2014 C-133/2014 provided by CONACYT. We acknowledge the financial support provided by Instituto de Ecología A.C. (INECOL) through the Postgraduate Secretary to Antonio López-Carretero. CONACYT URL is http://www.conacyt.gob.mx/Paginas/InicioNueva.aspx, INECOL URL is http://www.inecol.mx/posgrado/. The funders had no role in study design, data collection and analysis, decision to publish, or preparation of the manuscript.

Competing Interests: The authors have declared that no competing interests exist.

* Email: diazcastelazogm@gmail.com

Introduction

Approximately 40% of terrestrial biodiversity is involved in plant-herbivore species interaction networks [1]; these interactions are one of the main conduits for upward energy flow to higher trophic levels [2], and are mainly found in tropical ecosystems [3], [4]. Plant-herbivore interactions are considered antagonistic since the loss of tissue of the host plants caused by herbivore feeding may have detrimental consequences in their fitness or survival [5]. Unlike mutualistic interactions and networks, antagonistic networks tend to be more specialized and have an essentially modular structure [6], [7], characterized by cohesive groups of species that interact intensively (more than expected by chance) and maintain few interactions with other groups of species [8], [9].

Despite the advances in the identification and description of structural patterns of ecological networks, relatively little is known about the ecological and phylogenetic mechanisms that give rise to such patterns [10]. For plant-herbivore trophic networks it has been recently demonstrated that their structural attributes are influenced by species body size, local abundance, phylogeny [7], and by disturbance, since deforestation, habitat fragmentation and secondary plant succession all reduce the specialization and structural complexity of plant-herbivore networks owing to the resulting alteration of the food resources of herbivorous insects [11], [12], [13].

Furthermore, both seasonal changes in precipitation and structural complexity of plant communities are intimately related with changes in microclimate (e.g. temperature and humidity) and with the availability and quality of host plants for herbivores [14], [15], [16], [17]. Thus, both factors can have an important effect on the structure, composition and specialization of herbivorous insect communities [15], [18], [19]. In general, plant communities with high heterogeneity and species diversity offer a wide variety of food and habitats for herbivores and can reduce stressing factors such as temperature and humidity, which affect the performance of invertebrate herbivores [20], [21]. Such conditions promote a more diverse and specialized community of insects than in less structured plant communities with fewer host species and more stressful conditions [4], [21]. Although the effects of seasonality and variation in the quality of food resources on herbivore communities have been well documented for different ecosystems [14], [18], [22], the influence of these factors on interaction networks has not been evaluated.

Addressing the spatio-temporal assessment of ecological networks represents a new frontier in community ecology [23]. This approach can provide evidence of the influence of biotic and abiotic factors in the variation of species interaction networks [24],

[25]. However, this information is particularly scarce for antagonistic interactions [7], [26]. In the present study, we analyzed plant-herbivore networks (lepidopteran larvae feeding on plants) at a small spatio-temporal scale, which has the advantage of identifying the ecological processes that influence the structure of such networks occurring within the same region (coastal habitats spaced as far as 0.4 ha) [27]. In particular we were interested in answering the following questions: i) What is the seasonal variation in the structural parameters of the plant-herbivore network in habitats that differ in their complexity? ii) How does resource availability vary across habitats and seasons? and, iii) Which is the simultaneous contribution of host availability, habitat structure and seasonality for the spatio-temporal structuring of this antagonistic network? In antagonistic systems coevolutionary processes will tend to reinforce constraints against grand generalists (only a small proportion of herbivorous animals are true generalists), thus compartments should often be apparent in such networks [9]. Since host use by phytophagous insects tend to specialization [28], compartmentation or modularity [8] and may be influenced by resource availability [6], [29] we hypothesized that: i) strong seasonal effects would be detectable at the structural parameters of plant-herbivore networks at the studied site, with higher selectiveness (network-level specialization) and modularity during the wet season and forest habitats, ii) host availability and interaction frequency would vary among habitats and temporal periods, thus we may expect iii) higher connectance, interaction evenness and lower selectiveness during dry season periods and open habitats.

This study is unique in the sense that, for the first time, the impacts of resource availability, habitat structural complexity and seasonality on the structure (parameters) of an antagonistic network were assessed.

Methods

Study site

Field work was carried out at La Mancha Coastal Research Center (Centro de Investigaciones Costeras La Mancha, CICOLMA) located on the coast of the state of Veracruz in Mexico (96°24′48″W, 19°40′33″N y 96°22′25″W, 19°31′49″N). Permit to perform the present research was issued by Instituto de Ecología, A.C. (INECOL), since CICOLMA is one of INECOL's field station and private conservation reserve. CICOLMA is a very small fraction of the 1336 Ramsar Site called "La Mancha el Llano" declared in 2005. Thus, the regulatory body concerned with protection of wildlife is given to INECOL by Mexican federal organisms in the matter and framed by The Ramsar Convention on Wetlands (intergovernmental treaty). The study site, the CICOLMA reserve area managed by INECOL covers an area of 70 ha, which includes several native vegetation types (including wetlands and dune vegetation), transformed forestry/agricultural sites, experimental areas and a field station. The major vegetation types here are tropical deciduous forest, tropical dry forest, sand dune scrub, mangrove forest, freshwater marsh and flooded deciduous forest [30], [31].

The climate of the study site is warm sub-humid and total annual precipitation ranges from 899 to 1829 mm, ca. 78% of the total annual precipitation falls during the rainy season (June - September). In this area, 837 plant species have been recorded, 50% of which are herbaceous plants and the rest are shrubs, trees and vines [30]. Currently, there is no list of lepidopteran species for CICOLMA. The field studies presented here did not involve endangered or protected species.

Recording plant-herbivore interactions

The study was carried out at five sites with vegetation types representative of the plant communities in the study region and with differing degrees of structural complexity: 1) pioneer dune vegetation (PIO), 2) coastal dune scrub (DUN), 3) recently established tropical lowland subdeciduous forest (SFY) 4) tropical lowland subdeciduous forest in an advanced stage of succession (SFO), 5) tropical lowland flood forest with a wetland ecotone (FFW).

Structural complexity and vegetation strata differ specially between the open (PIO, DUN) and the forest habitats (SFY, SFO, FFW). The open habitats have moderate vegetation cover due to increased sand movement at the dunes, with dominant herbaceous stratum (although two strata are common at these habitats).These habitats are exposed to abiotic stressful conditions because of high temperatures, radiation, salt water influence and mechanical damage of vegetation during storms and hurricanes [32], [33]. In contrast, the forested habitats (SFY, SFO, FFW) are more stable and with milder conditions because of shade; vegetation structure at these habitats is more complex since three distinct arboreal strata occur here, the taller elements may exceed 20 m; the medium strata with trees of 6 to 15 m high, and finally, a well-developed understory [30].

At each of these vegetation types we placed 14 quadrats (4×4 m) where plant-herbivore interactions were recorded, along with the attributes of each plant community. In monthly censuses during the dry (March, April, May, June) and the wet/rainy (July, August, September, and October) season of 2012, we collected all folivorous caterpillar species (excluding leaf miners) associated with host herbs, lianas, shrubs and trees. In order to search for caterpillars on adult trees, a subsample of four branches was taken per individual tree. All caterpillars were recorded and taken to the laboratory where they were fed until pupated. No ethical approval was required for manipulating lepidopteran specimens, since after caterpillars pupated and adults emerged, only unidentified species (at larval stage) were mounted as adults for taxonomical identification (and only one individual of the unidentified species was mounted). Data underlying the findings of the present study are available within this manuscript, but lepidopteran specimens found at the present study are being processed for subsequent molecular analysis (Instituto de Ecología UNAM, México, by K. Boege) after which specimen identity will be publicly available in reference to this manuscript. Host plants were identified and deposited in the XAL Herbarium of the Instituto de Ecología, A.C. (INECOL, Xalapa, México).

Structural network parameters

To assess spatio-temporal changes in the structural network parameters, for each of the five sampled vegetation types we constructed four matrices of the plant-caterpillar interactions throughout the dry and wet season: dry season 1 (DS1: March, April), dry season 2 (DS2: May, June), rainy season 1 (RS1: July, August) rainy season 2 (RS2: September, October). Previous work at the study site suggests that there is short-scale temporal variation in the structure of insect-plant –mutualistic- interaction networks derived from changes in abiotic properties (temperature and precipitation) [24]. In the present study, the monthly data of plant-herbivore interactions recorded at the field was fused bimestrially in order to have robust temporal information across vegetation types, since at some habitats, monthly census during dry season provide scarce interaction records; furthermore, these bimesters markedly reflect temporal biotic and abiotic changes at the study site: the beginning of the dry season period DS1 mark the end of cold fronts ("nortes", see below), DS2 months corresponds to peak maximum temperatures, RS1 months corresponds to peak rainfall and RS2 correspond to the end of the rainy season. We did not make census from November to February because these months correspond to "nortes", a season of winter cold fronts, characterized by strong winds and a considerable decrease in plant cover [34].

We obtained a total of 20 matrices of species interactions (that result from multiplying 5 vegetation types and 4 bimestrial censuses), which we refer to as sub-networks. We calculated the following structural parameters for each sub-network: connectance (C), Interaction evenness (IE), vulnerability (V), generality (G), modularity (M), number of modules (NM) and network specialization or selectiveness (selective use of the lower trophic level by the higher trophic level) of the entire bipartite network (H^2), given that H^2 describes to which extent observed interactions deviate from those that would be expected given the species marginal sums [27]. 1) Connectance was defined as the fraction of recorded interactions relative to the total number of possible interactions. 2) Interaction evenness indicates how homogeneously the plant-herbivore interactions are distributed throughout the network. It is calculated similarly to Shannon's Index, but does not take into account the absence of interactions (i.e., cells that contain a zero in the interaction matrix). Evenness approaches a value of 1 when the number of interactions between herbivore and host plant species is uniformly distributed. 3) Generality represents the proportion of herbivorous species to host species. 4) Vulnerability was calculated as the proportion of host species to herbivorous species. 5) The number of modules corresponds to the number of subgroups of the network that are not connected to other groups: modules are cohesive groups of highly connected nodes that are loosely connected to other nodes in the network [35], [36]. 6) Modularity is a measure of how host and herbivorous species tend to organize into subgroups of species that interact more frequently among themselves than with other members of the network. Because the degree of modularity determines how dense the connection between different groups of elements in an ecological system is, systems that largely differ in the degree of modularity often differ in their ecological and evolutionary [35]. This parameter was calculated with the M index (range: 0–1) in MODULAR [36] based on Newman and Girvan's algorithm. Its level of significance was calculated using Monte Carlo tests with 1000 randomizations. In order to carry out robust comparisons among sub-networks not affected by neither sampling effort nor network size, we provided a relative value of Modularity (Mr) estimated from the z-score values of the 1000 random replicates of the modularity analysis described above, as: Mr = M-Mz/Mz where M is the value of the real sub-network data matrix, while Mz refers to the average value of modularity of its random replicates, 7) Network specialization (H^2) is based on the deviation between the observed and expected number of interactions in each network, assuming that all species interact with other species in proportion to their total observed frequency. Given that this metric is not affected by neither sampling effort nor network size, it is possible to make robust, reliable comparisons between different networks [27]. H^2 values, ranging from 0 (low selectiveness) to 1 (high selectiveness). The network indices were calculated for each census and site using the BIPARTITE 3.0.1 package run in R [37].

Attributes of the plant community

In each vegetation type and census, the following plant community attributes were quantified. As a measure of vegetation structure, total plant species richness (TPR) was registered, along with an estimation of canopy cover provided by all plant species (included host and non-host species) found in each quadrat (total plant cover = TPC). As an indicator of food availability to herbivores, we measured plant cover of host species (i.e., exclusively those species in which caterpillars were found; host plant cover = HPC). Species richness of host plants used by herbivorous lepidopterans (host plant richness = HPR) was also registered. These variables were selected because they influence herbivore food availability, habitat structure and the abiotic factors that affect plant-herbivore interactions.

Data analysis

Effect of vegetation type and seasonality on network parameters and plant community attributes. The effect of vegetation type and census on plant community attributes (TPR, TPC, HPC) and structural parameters of each network (C, G, V, IE, M, Mr, NM, H^2) were evaluated using repeated measures ANOVAs, using the sub-network parameters as replicates for each vegetation type and census. Using a standard ANOVA in this case is not appropriate because it fails to model the correlation between the repeated measures (census and vegetation types): the data violate the ANOVA assumption of independence. Therefore the repeated measures ANOVA is suitable for analysis of auto correlated samples. To assess the influence of vegetation type, the model included census as the error term (within subject) and vegetation type as the explanatory variable (between subject). To evaluate the influence of census, the model included vegetation type as the error term (within subject) and census as the explanatory variable.

Spatio-temporal ordination of the sub-networks and their relation with plant community attributes. Plant-herbivore interactions were spatio-temporally analyzed at 20 sub-networks (i.e. 20 frequency matrices). Structural network parameters were subsequently estimated for each sub-network. In order to describe the dissimilarity among these matrices, we performed a NMDS analysis (Non-metric Multi-Dimensional Scaling). The resulting NMDS ordination (biplot) shows the structural parameters of all sub-networks: when sub-networks are close to a specific parameter, these have small dissimilarities in their values for that parameter. This is one of the most effective methods for the ordination of ecological data and the identification of underlying gradients, because it does not assume a linear relationship among the variables [38]. Because this analysis offers more than one solution, we carried out an iterative process to find the model with smallest stress value using the metaMDS function in the MASS routine run in R [39].

To evaluate if spatio-temporal dissimilarity among sub-network parameters was related to plant community attributes, we analyzed the effect of TPR, TPC, HPC (continuous variables) (see "Attributes of the plant community" within this section) on the ordination of plant-herbivore sub-networks using the envfit function from the VEGAN library of the statistical package R [39]. This function fits the vectors or continuous values of the environmental variables (TPR, TPC, HPC) to the NMDS ordination axes, generating a measure of correlation (r) and a significance value based on the probability that 1000 random permutations of environmental variables would have a better fit than the real environmental variables.

To test the null hypothesis of a similar composition of herbivore and host plants species among vegetation types and sampling censuses, we used a PERMANOVA (Permutational Multivariate Analysis of Variance), and the Bray-Curtis Index [40]. The PERMANOVA was performed using the *adonis* function of the VEGAN package run in R. This function runs a multivariate analysis using the distance between matrices (dissimilarities between objects or sampling sites), and the probability of significance P is obtained through permutations [40]. For these analyses, we used the following groups: season (censuses of the dry season vs. censuses of the rainy season) and structural complexity of the habitat (open habitats vs. forest).

To understand if spatio-temporal changes in herbivore species composition (ordination dissimilarity) could affect network structure, we analyzed the effect of parameters C, H^2, M y NM (quantitative variables or vectors) over the ordination (biplot) of an NMDS analysis of the community of lepidopteran herbivorous larvae found foraging on plants at the 20 sub-networks using the *envfit* function from the VEGAN package of the statistical software R. This function provides a correlation measure and is statistical significance (after comparing correlation values with 1000 random permutations).

Results

Effect of vegetation type and seasonality on network structure and plant community attributes

We recorded 654 larvae belonging to 176 morphospecies of Lepidoptera and associated with 56 host plant species. During the dry season there were 237 larvae belonging to 41 morphospecies of Lepidoptera feeding on 33 species of plants, whereas during the rainy season there were 417 larvae belonging to 66 species of Lepidoptera feeding on 47 species of host plants. The composition of the herbivore and host plant communities differed significantly between the studied open and forested habitats (PERMANOVA, herbivores: Pseudo-F = 2.667; P = 0.004, plants: Pseudo-F = 4.043; P = 0.004). For both herbivores (lepidopteran larvae) and hosts plants, species turnover was higher between groups (vegetation types) than within them (Tukey test, herbivores: P = 0.787, plants: P = 0.719). The mean values of C, M, NM and H^2 of the network varied significantly across censuses and among vegetation types (Table 1A). The network parameter V changed across time, while G and IE were not different between vegetation types or throughout the year (Tables 1A).

Post hoc tests indicated that the values for M, Mr, NM and H^2 of the forest habitats (SFY, SFO, FFW) were significantly higher than those with open vegetation (PIO, DUN) (Figure 1A, Table 2). Additionally, the networks of open habitats had the highest C, with values almost twice than those of forest habitats. The network indices did not vary significantly among the three types of forest vegetation, or between the two types of open vegetation (Figure 1A).

The first census of the dry season had the highest connectance of the entire study, with values twice as high as during the rest of the year (Figure 1B). In contrast, species networks of the last three censuses were characterized by high M, Mr (Table 2), NM and H^2 values that were twice as much of those estimated for the networks of the first census. None of these indices were different among the last three censuses (Figure 1B). It is important to notice that all sub-networks (but one) were significantly modular (Table 2).

The HPC and HPR varied significantly among vegetation types and across censuses, while total floristic richness differed only among vegetation types (Table 1B, Figure 1). The TPC did not differ statistically between vegetation types or censuses. In particular HPC in PIO was two times greater than in FFW and SFO and three times greater than in DUN and SFY. In contrast, the DUN had the lowest host plant cover of all the vegetation types studied, even with respect to the PIO (Figure 1C). The TPR and HPR in the forest habitats were twice than in open habitats (Figure 1C).

During the dry season, HPC was two times that of the wet season. Although TPR did not differ statistically across time, the HPR was about three times lower during DS1 relative to the rest of the year (Table 1B, Figure 1D). By comparing TPR and HPR values, we found that in all vegetation types and throughout all censuses, herbivores used approximately one quarter of the available plant species (Figure 1 C–D).

Spatio-temporal ordination of the sub-networks and their relation with plant community attributes

In the NMDS of plant-herbivore sub-networks, axis 1 was associated with the slope defined by seasonality (Figure 2). During the first census of the dry season the sub-networks clustered independently of the vegetation type because of their high C values. In addition sub-networks exhibited a clear separation between the dry and wet seasons, characterized by their high degree of H^2, M and NM (Figure 2). The sub-networks in the open vegetation habitats were defined by high values of C during the dry season. However, during the rainy season the sub-networks in these habitats were characterized by high values of IE, G and M (Figure 2). During the dry season, forest vegetation sub-networks had high values of IE and V, while the sub-networks of the rainy season had similarly high values of H^2, M and NM (Figure 2).

Some attributes of plant communities influenced the structure of plant-herbivore sub-networks: vector TPR (r = 0.485, P = 0.003) and vector HPC (r = 0.274, P = 0.046). The correlation of HPC was greater for the ordination of the sub-networks in habitats with open vegetation and during the dry season (Figure 2). TPR had the highest degree of correlation with sub-network ordination, and its correlation was stronger for the ordination of the sub-networks of the forest vegetation types and rainy season (Figure 2).

The NMDS exploring the relation among sub-network parameters and herbivore species composition showed an important spatio-temporal turnover of lepidopteran herbivore species (Figure 3). Independently of the bimester (seasonality), we found a clear dissimilarity among lepidopteran species foraging in plants at open habitats (PIO, DUN) and forest habitats (FFW, SFY, SFO). Vector fitting at NMDS showed that parameters NM, M and C were significantly correlated to dissimilarity in lepidopteran species composition (r = 0.464, P = 0.007; r = 0.389, P = 0.033; r = 0.364, P = 0.024, respectively). The correlation of M and NM was greater for the ordination of the sub-networks in habitats with forest vegetation, while the highest correlation of C occurred at open habitats (Figure 3). Thus, there are differences in lepidopteran species composition between sites that transcend to network structure (parameters): changes in lepidopteran species composition affects network general topology and connectance, but not specialization (selectiveness) (r = 0.236, P = 0.123).

Discussion

It has recently been demonstrated that the structure of plant-herbivore interaction networks is influenced by ecological, phylogenetic [7] and anthropogenic attributes [12], [13], [41]. However, although the patterns of plant-herbivore interactions can be influenced by finer and more immediate environmental factors such as temporal variation in the quality and availability of food [18], [42] or habitat structure [4], [43], the impact of these factors on the properties of species interaction networks had not been evaluated.

Our results show for the first time how the structural parameters of plant-herbivore networks are spatially and temporally dynamic, influenced by seasonality, habitat complexity and food resource availability. The spatio-temporal changes in host plant communities resulting from abiotic factors (climate and micro environmental variation) were not only reflected in the species composition of

Table 1. Results of the repeated measures analysis of variance (ANOVA) for the effect of census and vegetation type on structural network parameters (A) and the plant community attributes (B).

(A) Network parameters	*DF*	*SS*	*F*	*P*
Generality (G)				
Census	3	0.005	0.626	0.614
Vegetation type	4	0.004	0.363	0.831
Vulnerability (V)				
Census	3	1.961	4.167	***0.033***
Vegetation type	4	0.645	1.028	0.434
I. Evenness (IE)				
Census	3	0.034	1.808	0.203
Vegetation type	4	0.461	1.803	0.198
Connectance (C)				
Census	3	0.099	29.218	*<0.001*
Vegetation type	4	0.066	14.698	*<0.001*
Modularity (M)				
Census	3	0.125	11.568	***0.001***
Vegetation type	4	0.128	8.896	***0.009***
No. Modules (NM)				
Census	3	0.125	11.568	***0.001***
Vegetation type	4	0.128	8.896	***<0.001***
Network Specialization (H_2)				
Census	3	5.842	22.023	***<0.001***
Vegetation type	4	1.483	4.195	***0.026***
(B) Attributes of the plant community				
Total Plant Cover (TPC)				
Census	3	642.717	0.912	0.466
Vegetation type	4	1895.716	2.011	0.161
Total Plant Richness (TPR)				
Census	3	48.133	1.079	0.397
Vegetation type	4	1678.228	28.231	***<0.001***
Host Plant Cover (HPC)				
Census	3	0.342	10.462	***0.001***
Vegetation type	4	0.892	20.456	***<0.001***
Host Plant Richness (HPR)				
Census	3	222.64	11.478	***0.001***
Vegetation type	4	176.95	6.776	***0.005***

Significant values of P (≤0.05) are in bold and italics.

herbivore community as previously reported [14], but also in the emergent properties of the antagonistic network such as its connectance, specialization (selectiveness) and modularity.

In contrast with mutualistic plant-insect networks, which maintain their structural stability over time [44], [45], [46], [47], antagonistic networks (i.e., plant-herbivore) tend to maintain their modular structure [4], [8]. Indeed, in this study plant-herbivore networks were characterized by many specialist interactions and a marked compartmentalization, as has been reported for other plant-insect herbivore networks [6], [7], [8], [48]. To our knowledge this is the first study to report structural seasonal and spatial changes in a plant-herbivore interaction network. Our interaction sampling is robust; since our study habitats are relatively close to each other (spaced as far as 0.4 ha) and the area of each habitat is reduced, we could established permanent quadrats (n = 14) that covered most environmental gradient; thus we have a considerably good sampling effort in reference to the interactions studied. In general, in order to estimate sampling completeness within interaction networks, the number (richness) of interactions accumulated as sampling effort increased is frequently used [49]. Since in our study the assemblage of lepidopteran herbivores showed a high proportion of rare species, the result is similar to other tropical communities of herbivorous insects [3], [50] where asymptotic interaction richness is not reached [15], [29]. This non-asymptotic pattern of accumulation curves occurs as well for lepidopteran species richness hosted even by plant species of the same genus or family [3], [51].

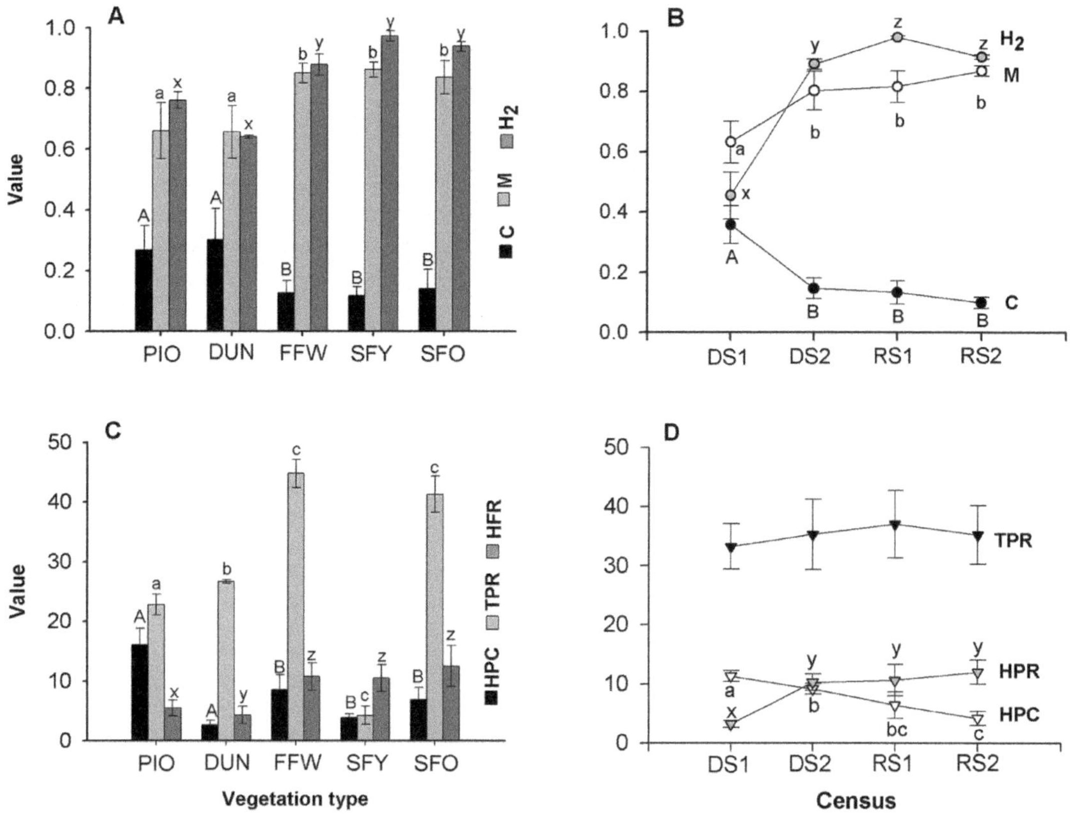

Figure 1. Vegetation type and seasonality effect on network parameters and plant community attributes. Vegetation types (PIO: pioneer dune vegetation, DUN: coastal dune scrub, FFW: flood forest with a wetland ecotone, SFY: recently established tropical lowland subdeciduous forest, SFO: tropical lowland subdeciduous forest in an advanced stage of succession). Census (DS1: Dry season 1, DS2: Dry season 2, RS1: Rainy season 1, RS2: Rainy season 2). Network parameters (H^2: network specialization, M: Modularity, C: Connectance). Plant community attributes (TPR: Total plant richness, HPR: Host plant richness, HPC: Host plant cover). (A) Network parameters by vegetation type. (B) Network parameters by census. (C) Plant community attributes by vegetation type. (D) Plant community attributes by census. Comparisons are valid only within each network parameter (panels A and B, for vegetation types and census, respectively) and within each plant community attribute (panels C and D, for vegetation types and census, respectively). Bars indicate standard error of the mean. Different letters represent significant differences resulting from post-hoc test ($P<0.05$).

Table 2. Relative modularity (Mr) based in z-score values estimated for the vegetation types and censuses (mean ±SD) in plant-herbivore sub-networks.

Vegetation type	Modularity (Mr)
PIO	1.887 ± 0.303^{a}
DUN	1.907 ± 0.194^{a}
FFW	2.611 ± 0.111^{b}
SFY	2.696 ± 0.191^{b}
SFO	2.669 ± 0.140^{b}
Census	
DS1	2.049 ± 0.227^{a}
DS2	2.325 ± 0.322^{ab}
RS1	2.440 ± 0.242^{bc}
RS2	2.686 ± 0.108^{c}

Different letters among vegetation types and censuses represent significant differences in post-hoc test at $P \leq 0.05$. See Methods section for vegetation types and censuses abbreviations.

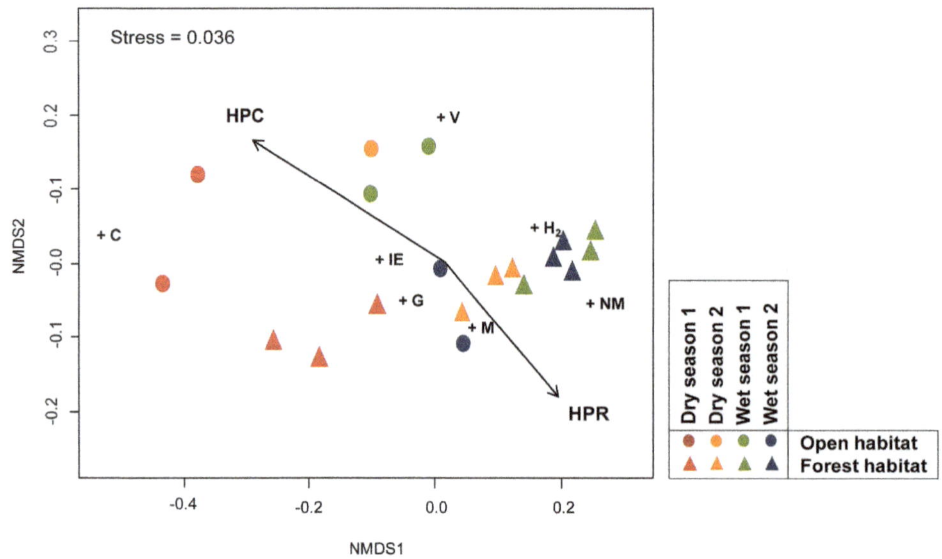

Figure 2. NMDS ordination of network parameters showing the spatio-temporal variation of plant-herbivore sub-networks. The plus signs (+) indicate the network parameters upon which the sub-network ordination was performed (H^2: network specialization, M: Modularity, C: Connectance, IE: interaction evenness, V: vulnerability, G: generality). The vectors on the ordination represent the gradient in HPR (Host plant richness) and HPC (Host plant cover) for all sub-networks. The arrow points to the direction in which the linear change in HPR and HPC was strongest, and the length of the arrow is proportional to the correlation between these variables and the sub-network ordination. Arrows were plotted only for variables with significance of $P \leq 0.05$. Factors fit in the ordination (season and habitat type) are not included. For habitats abbreviations see Figure 1.

Although connectance is a simple parameter sensitive to network size [52], the inverse relation trend generally found between connectance and network size stabilizes after a "threshold" above 50 species conforming the network [53]. Since at the present study our smaller sub-networks have much more than 50 nodes, we consider that connectance can be comparatively used to describe spatio-temporal changes in plant-herbivore interaction "saturation". In general, selectiveness (H^2) metric is not affected by neither sampling effort nor network size [27], and the relative modularity (Mr) provided by the z-score values of "M" metric make our spatio-temporal comparisons robust. In our study, temporal variation in the structure of plant-herbivore networks was associated with changes in climatic factors (e.g., precipitation and temperature) during the transition from the dry to the rainy season, and was accompanied by a decrease in connectance (lower interaction saturation in the community) and an increase in the selectiveness/modularity of networks. These temporal changes in the configuration of species networks could be associated with a high susceptibility of monophagous or oligophagous lepidopteran herbivores that need to synchronize their life cycles to plant phenology [14], [18], [22]. Given that at the present study, no traits of plant species are provided, the H^2 pattern among sub-networks suggest that, independently of the number of species constituting each sub-network, there is higher selectiveness of lepidopteran larvae (not necessarily trophic specialization) during the rainy season when offer of plant species (and foliage availability) is higher; this pattern may reflect changes in host choice by adult females, as a function of temporal changes in plant community [54], [55], [56]. Further studies including nutritional or defensive traits of plants used by herbivores at these networks or phylogenetic signaling of these interactions would shed light on the process generating these patterns.

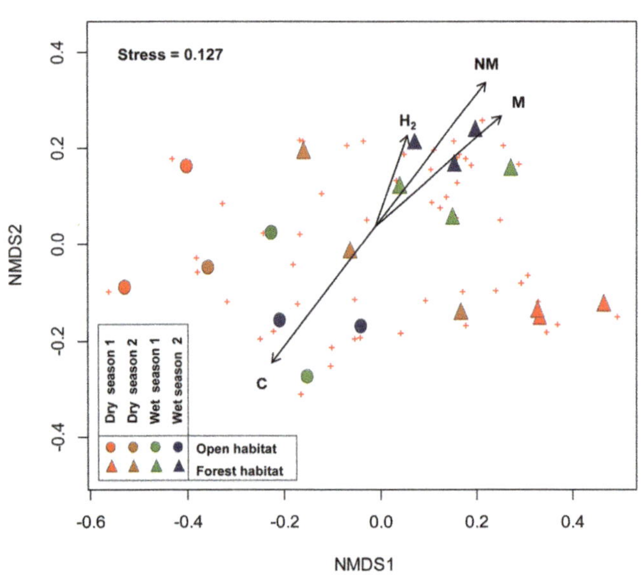

Figure 3. NMDS ordination of lepidopteran herbivore species composition showing the spatio-temporal variation in sub-networks. The plus signs (+) indicate different herbivore species upon which the sub-network ordination was performed. The vectors on the ordination represent the gradient in network parameter values across all sub-networks (for habitats abbreviations see Figure 1): H^2: network specialization, M: Modularity, NM: Number of modules, C: Connectance. The arrow points to the direction in which the linear change in network parameters was strongest, and the length of the arrow is proportional to the correlation between these variables and the ordination of herbivore species composition.

The ordination of the most specialized and highly modular sub-networks was highly correlated with floristic richness of forest habitats and the rainy season. The high degree of specialization and modularity of the networks in the richest floristic contexts could be related to a greater diversification by the herbivores in their use of plant resources: a greater availability of niches at the lower trophic level is likely to promote a greater niche differentiation (and therefore greater specialization) in the higher trophic level. The forest sites (tropical lowland subdeciduous forest) were not only floristically the richest of the studied sites, but also offered more canopy cover from trees and shrubs, as well as more plants with different life forms [30]. Greater plant richness and structural complexity of the vegetation provides a wide variety of microhabitats and food that can result in lower competition by food resources among herbivores [4], [57]. In turn, this could translate into greater specialization and modularity in the network patterns. There are two possible reasons that might promote specialization of plant-herbivore networks in more floristically diverse contexts: i) the advantages conferred by specialist foraging behavior in the face of the difficulties herbivores have choosing suitable hosts [58] and/or ii) specialization in host species that provide herbivores enemy-free space in more structurally complex habitats [59], [60]. These avenues remain unexplored in a plant-herbivore network context.

The highest sub-network connectance was found under the most stressful abiotic conditions: Open vegetation and the dry season create conditions that could cause greater water stress in plants as a result of high temperatures and solar radiation. In turn, this could affect their nutrition (water and nitrogen content) and a defense (toughness and phenolic compounds content) attributes rendering the plants unsuitable for many lepidopteran herbivores [14], [18], [19], [42], [61]. For example, the coastal dune scrub and pioneer dune vegetation have the most extreme temperatures (up to 60°C in the scrub) and solar radiation [31], [32] of all our study sites. Moreover, the pioneer dune vegetation is subject to continual disturbance (burial under moving sand, strong winds with a high saline concentration, abrasion from blown sand, flooding and mechanical damage caused by tropical storms and hurricanes) [33], which makes host plants in these habitats highly unpredictable resource for herbivores [62]. In this context, both the unpredictable availability and low nutritional quality of the host plants in open habitats and present during the dry season can impose strong selective pressure on the herbivores [42], favoring species that are less specialized and less susceptible to variations in host resources [63], [64], [65]. In fact, in open vegetation habitats and during the dry season we found a predominance of herbivore species with relatively low host specificity. A generalist habit could make these species less susceptible to fluctuations in resource availability and quality under stressing conditions given that they have more food options.

At the present study most generalist herbivores were associated with host plants with large plant cover, mainly pioneer species adapted to stressful natural habitat conditions such as *Chamaecrista chamaecristoides, Crotalaria incana* and *Porophyllum punctatum* [33], [66]. These host species may have provided protection and food under the stressful abiotic conditions and reduced nutritional quality of the plants. Greater plant cover is a structural attribute that provides greater food availability to herbivores that can take advantage of this cover, and would allow them to compensate for the decrease in leaf nutrition quality [67], [68], and reduce larval desiccation under the high temperatures that characterize these environments.

In antagonistic networks, connectance - whether promoted by the range of host plants eaten by herbivores or by resource availability- can mitigate the loss of species when environmental conditions are unstable or stressful [12]. Therefore, the increase in connectance found in our study may have buffered the negative effects of natural disturbance (in stressful habitats and conditions) in plant-herbivore networks. Furthermore, the high network specialization was a structural attribute characteristic of forest habitats with stable host plant population size, compared with open habitats where specialist insects exhibited less spatio-temporal variation [69]. These results are evidence that the structure of antagonistic networks, just as it occurs in mutualistic networks, tends to reduce species vulnerability to extinction, which promotes community stability [70], [71].

In the studied plant-herbivore sub-networks, higher connectance was detected during the dry season in the pioneer habitat, and this seems similar to the changes that other trophic networks undergo when subjected to anthropic disturbances. The way in which the loss of arboreal cover reduced network complexity has been reported as a process of spatio-temporal homogenization of species and their interactions [12], [41], [72]. Moreover, both habitat fragmentation and early secondary succession can decrease the size of the network, resulting in systems that are more connected and less modular owing to the possible loss of specialists [12], [13], [72], [73]. At the study site, the pioneer habitat and the dry season may, represent a significant decrease in resource availability, particularly in the abundance of rare plants and their specialist herbivores. Additionally, the lack of arboreal cover in this habitat increases stressful abiotic conditions for herbivore larvae.

An important property of our plant-herbivore network was the spatio-temporal prevalence of vulnerability and interaction evenness. Regardless of habitat type or season, the ratio of herbivore to plant species was the lowest and no dominant interactions were recorded. Our results are similar to those of other trophic networks for habitats with differing degrees of fragmentation and where there was no evidence of variation in the structural attributes of the network [11]. In contrast, in parasitoid-herbivore trophic networks the spatio-temporal variation in interaction evenness is greater where deforested habitats are dominated by a reduced number of interactions [72]. In our study, the spatio-temporal prevalence of these indices could be a consequence of the high specialization plant-herbivore interactions [12], since the great majority of lepidopteran herbivores in tropical ecosystems are rare species that are low in abundance [14], even in disturbed tropical habitats [74]. Low vulnerability could also be a consequence of the loss of higher trophic levels as occurs in very small habitat fragments [12] [75].

In recent years, there have been efforts to assess the spatio-temporal variation of ecological networks [44], [45], [47], [76]. To our knowledge, this is the first study to address how plant-herbivore trophic networks vary in space and time at small scales, and to evaluate simultaneously the role of habitat complexity, resource availability and seasonality in such variation. This approach opens a new frontier in the study of plant-herbivore interactions in an effort to understand the processes that drive the modular and specialized patterns observed in antagonistic interaction networks.

Acknowledgments

We especially wish to acknowledge Michelle Ramos-Robles for logistical support in the field and caterpillars rearing. We would like to thank also Arturo Bonet Ceballos, María de Jesús Peralta Méndez, Gibrán Pérez Toledo for their support during fieldwork. The suggestions made by Víctor Parra-Tabla at several stages of our work enriched greatly this manuscript. We thank Wesley Dáttilo for his support during modularity analysis. We thank Carlos Durán and Adolfo Ibarra for taxonomic identification of plants and lepidopterans, respectively.

Author Contributions

Conceived and designed the experiments: ALC CDC KB VRG. Performed the experiments: ALC. Analyzed the data: ALC CDC. Contributed reagents/materials/analysis tools: CDC VRG KB. Wrote the paper: ALC CDC KB VRG.

References

1. Price PW (2002) Resource-driven terrestrial interaction webs. Ecol Res 17: 241–247.
2. Futuyma D, Agrawal AA (2009) Macroevolution and the biological diversity of plants and herbivores. Proc Natl Acad Sci USA 84: 8054–18061.
3. Novotny V, Basset Y (2005) Host specificity of insect herbivores in tropical forests. Proc Biol Sci 272: 1083–1090.
4. Lewinsohn TM, Novotny V, Basset Y (2005) Insects on plants: Diversity of herbivore assemblages revisited. Annu Rev Ecol Evol Syst 36: 597–620.
5. Coley PD, Barone JA (1996) Herbivory and plant defenses in tropical forests. Annu Rev Ecol Syst 27: 305–335.
6. Thébault E, Fontaine C (2010) Stability of ecological communities and the architecture of mutualistic and trophic networks. Science 329: 853–856.
7. Cagnolo L, Salvo A, Valladares G (2011) Network topology: patterns and mechanisms in plant-herbivore and host-parasitoid food webs. J AnimEcol 12: 1–10.
8. Prado PI, Lewinsohn TM (2004) Compartments in insect-plant associations and their consequences for community structure. J Anim Ecol 73: 1168–1178.
9. Lewinsohn TM, Prado PI, Jordano P, Bascompte J, Olesen JM (2006) Structure in plant-animal interaction assemblages. Oikos 113: 174–84.
10. Vazquez DP, Bluthgen N, Cagnolo L, Chacoff NP (2009) The determinants of patterns in mutualistic networks: a review. Ann Bot 103: 1445–1457.
11. Kaartinen R, Roslin T (2011) Shrinking by numbers: landscape context affects the species composition but not the quantitative structure of local food webs. J Anim Ecol 80: 622–631.
12. Valladares G, Cagnolo L, Salvo A (2012) Forest fragmentation leads to food web contraction. Oikos 121: 299–305.
13. Villa-Galaviz E, Boege K, del-Val E (2012) Resilience in plant-herbivore networks during secondary succession. PLoS ONE 7(12): e53009.
14. Janzen DH (1993) Caterpillar seasonality in a Costa Rican dry forest. In: Stamp NE, Casey TM, editors. Caterpillars. Ecological and evolutionary constraints on foraging. Chapman and Hall, New York. pp. 448–477.
15. Basset Y (1996) Local communities of arboreal herbivores in Papua New Guinea: predictors of insect variables. Ecology 77: 1909–1916.
16. Poorter L, de Plassche MV, Willems S, Boot RGA (2004) Leaf traits and herbivory rates of tropical tree species differing in successional status. Plant Biol (Stuttg) 6: 746–754.
17. Boege K (2005) Herbivore attack in *Casearia nitida* influenced by plant ontogenetic variation in foliage quality and plant architecture. Oecologia 143: 117–125.
18. Murakami M, Ichie T, Hirao T (2008) Beta-diversity of lepidopteran larval communities in a Japanese temperate forest: effects of phenology and tree species. Ecol Res 23: 179–187.
19. Trotter III, Cobb RT, Whitham TG (2008) Arthropod community diversity and trophic structure: a comparison between extremes of plant stress. Ecol Entomol 33: 1–11.
20. Brehm G, Homeier J, Fiedler K (2003) Beta diversity of geometrid moths (Lepidoptera: Geometridae) in an Andean montane rainforest. Divers Distrib 9: 351–366.
21. Jeffries JM, Marquis JR, Forkner RE (2006) Forest age influences oak insect herbivore community structure, richness and density. Ecol Appl 16: 901–912.
22. Janzen DH (1988) Ecological Characterization of a Costa Rican Dry Forest Caterpillar Fauna. Biotropica 20: 120–135.
23. Bascompte J, Jordano P (2007) Plant-animal mutualistic networks: the architecture of biodiversity. Annu Rev Ecol Evol Syst 38: 567–593.
24. Rico-Gray V, Díaz-Castelazo C, Ramirez-Hernandez A, Guimaraes PR, Holland JN (2012) Abiotic factors shape temporal variation in the structure of an ant-plant network. Arthropod-Plant Interact 6: 289–295.
25. Sánchez-Galván IR, Díaz-Castelazo C, Rico-Gray V (2012) Effect of Hurricane Karl on a plant–ant network occurring in coastal Veracruz, Mexico. J Trop Ecol 28: 603–609.
26. de Ruiter PC, Wolters V, Moore JC, Winemiller KO (2005) Food web ecology: playing Jenga and beyond. Science 309: 68–71.
27. Blüthgen N, Menzel F, Bluthgen N (2006) Measuring specialization in species interaction networks. BMC Ecology 6: 9.
28. Thompson JN (1999) What we know and do not know about coevolution: insect herbivores and plants as a test case. In: Herbivores: Between Plants and Predators, H Olff, V.K Brown and R.H Drent, editors. Oxford, Blackwell Science. pp. 7–30.
29. Novotny V, Miller SE, Leps J, Bito D, Janda M, et al. (2004) No tree an island: the plant-caterpillar food web of secondary rainforest in New Guinea. Ecol Lett 7: 1090–1100.
30. Castillo-Campos G, Travieso-Bello AC (2006) La flora. In: Moreno-Casasola P, editor. Entornos Veracruzanos: la costa de La Mancha. Instituto de Ecología AC, Xalapa. pp. 171–204.
31. Moreno-Casasola P, Travieso AC (2006) Las playas y dunas. In: Moreno-Casasola P, editors. Entornos Veracruzanos: la costa de La Mancha. Instituto de Ecología AC, Xalapa. pp. 205–220.
32. Moreno-Casasola P (1982) Ecología de la vegetación de dunas costeras: factores físicos. Biotica 7: 577–602.
33. Pérez-Maqueo O (1995) Análisis del efecto de los disturbios en la dinámica de la playa de El Morro de La Mancha, Ver. México DF: UNAM. 65 p.
34. Travieso-Bello AC, Campos (2006) Los componentes del paisaje. In: Moreno-Casasola P, editors. Entornos Veracruzanos: la costa de La Mancha. Instituto de Ecología AC, Xalapa. pp. 140–150.
35. Olesen JM, Bascompte J, Dupont YL, Jordano P (2007) The modularity of pollination networks. Proc Natl Acad Sci USA 104 (50): 19891–19896.
36. Marquitti FMD, Guimaraes PR, Pires MM, Bittencourt LF (2013) MODULAR: Software for the Autonomous Computation of Modularity in Large Network Sets. Ecography arXiv: 1304.2917.
37. R Core Team (2013). R: A language and environment for statistical computing. R Foundation for Statistical Computing, Vienna, Austria. URL: http://www.R-project.org/.
38. Quinn GP, Keough MJ (2002) Experimental design and data analysis for biologists. Cambridge University Press. 537 pp.
39. Dixon P (2009) VEGAN, a package of R functions for community ecology. J Veg Sci: 927–930.
40. Anderson MJ (2001) A new method for non-parametric multivariate analysis of variance. Austral Ecol 26: 32–46.
41. Laliberté E, Tylianakis JM (2010) Deforestation homogenizes tropical parasitoid-host networks. Ecology 91: 1740–1747.
42. Barber NA, Marquis RJ (2011) Leaf quality, predators, and stochastic processes in the assembly of a diverse herbivore community. Ecology 92: 699–708.
43. Lepš J, Novotny V, Basset Y (2001) Habitat and successional status of plants in relation to the communities of their leaf-chewing herbivores in Papua New Guinea. J Ecol 89: 186–199.
44. Alarcón R, Waser NM, Ollerton J (2008) Year-to-year variation in the topology of a plant-pollinator interaction network. Oikos 117: 1796–1807.
45. Olesen JM, Bascompte J, Elberling H, Jordano P (2008) Temporal dynamics in a pollination network. Ecology 89: 1573–1582.
46. Dupont YL, Olesen JM (2009) Ecological modules and roles of species in heathland plant–insect flower visitor networks. J Anim Ecol 78: 346–353.
47. Dupont YL, Padrón B, Olesen JM, Petanidou T (2009) Spatio-temporal variation in the structure of pollination networks. Oikos 118: 1261–1269.
48. Fontaine C, Thébault E, Dajoz I (2009) Are insect pollinators more generalist than insect herbivores? Proc Biol Sci 276: 3027–3030.
49. Chacoff NP, Vázquez DP, Lomáscolo SB, Stevani EL, Dorado J, et al. (2012) Evaluating sampling completeness in a desert plant–pollinator network J An Ecol 81(1): 190–200.
50. Price PW, Abrahamson WG, Hunter MD, Melika G (2004) Using gall wasps on oaks to test broad ecological concepts. Conserv Biol 18: 1405–1416.
51. Novotny V, Basset Y (2000) Rare species in communities of tropical insect herbivores: pondering the mystery of singletons. Oikos 89: 564–572.
52. Heleno R, Devoto M, Pocok M (2012). Connectance of species interaction networks and conservation value: is it any good to be well connected? Ecol Indic 14: 7–10.
53. Dormann CF, Fründ J, Blüthgen N, Gruber B (2009) Indices, graphs and null models: analyzing bipartite ecological networks. The Open Ecology J 2: 7–24.
54. Rausher DM, Papaj DR (1983) Host plant selection by *Battus philenor* butterflies: evidence for individual differences in foraging behavior. Anim Behav 31: 341–334.
55. Rausher MD, Odendaal FJ (1987) Switching and the pattern of host use by *Battus philenor* butterflies. Ecology 68: 869–877.
56. Nylin S (1988) Host plant specialization and seasonality in a polyphagous butterfly, *Polygonia C-Album* (Nymphalidae). Oikos 53: 381–386.
57. Lewinsohn TM, Roslin T (2008) Four ways towards tropical herbivore megadiversity. Ecol Letts 11: 398–416.
58. Bernays EA (2001) Neural limitations in phytophagous insects: implications for diet breadth and evolution of host affiliation. Annu Rev Entomol 46: 703–727.
59. Jeffries MJ, Lawton JH (1984) Enemy free space and the structure of ecological communities. Biol J Linn Soc Lond 23: 269–286.
60. Bernays E, Graham M (1988) On the evolution of host specificity in phytophagous arthropods. Ecology 69: 886–892.
61. Staley JT, Mortimer SR, Masters GJ, Morecroft MD, Brown VK, et al. (2006) Drought stress differentially affects leaf-mining species. Ecol Entomol 31: 460–469.
62. Rico-Gray V, Castro G (1996) Effect of an ant-aphid-plant interaction on the reproductive fitness of *Paullinia fuscescens* (Sapindaceae). Southwest Nat 41: 434–440.
63. Gaston KJ, Lawton JH (1988) Patterns in body size, population dynamics, and regional distribution of bracken herbivores. Am Nat 132: 662–680.
64. Redfearn A, Pimm SL (1988) Population variability and polyphagy in herbivorous insect communities. Ecol Monogr 58: 39–55.
65. Hughes JB (2000) The scale of resource specialization and the distribution and abundance of lycaenid butterflies. Oecologia 123: 375–383.
66. Martínez ML (1994) Sobrevivencia y establecimiento de plántulas de una especie colonizadora de dunas costeras: *Chamaecrista chamaecristoides*. México DF: UNAM. México. 64 p.
67. Bezemer TM, Jones TH (1998) Plant-insect herbivore interactions in elevated atmospheric CO2: quantitative analyses and guild effects. Oikos 82: 212–222.
68. Whittaker JB (2001) Insects and plants in a changing atmosphere. J Ecol 89: 507–518.

69. Lepš J, Spitzer K, Jaroš J (1998) Food plants, species composition and variability of the moth community in undisturbed forest. Oikos 81: 538–548.
70. Dunne JA, Williams RJ, Martinez ND (2002) Food-web structure and network theory: the role of connectance and size. Proc Natl Acad Sci USA 99: 12917–12922.
71. Thébault E, Huber V, Loreau M (2007) Cascading extinctions and ecosystem functioning: contrasting effects of diversity depending on food web structure. Oikos 116: 163–173.
72. Tylianakis JM, Tscharntke T, Lewis OT (2007) Habitat modification alters the structure of tropical host-parasitoid food webs. Nature 445: 202–205.
73. Melián C, Bascompte J (2002) Complex networks: two ways to be robust? Ecol Lett 5: 705–708.
74. López-Carretero A (2010) Composición y diversidad de lepidópteros en la cronosecuencia sucesional del bosque tropical caducifolio, consecuencias sobre la herbivoría de *Casearia nitida*. México DF: UNAM. 86 p.
75. Holt RD, Lawton JH, Polis GA, Martinez ND (1999) Trophic rank and the species-area relationship. Ecology 80: 1495–1504.
76. Díaz-Castelazo C, Guimarães PR, Jordano P, Thompson JN, Marquis RJ, et al. (2010) Changes of a mutualistic network over time: reanalysis over a 10-year period. Ecology 9: 793–801.

PERMISSIONS

All chapters in this book were first published in PLOS ONE, by The Public Library of Science; hereby published with permission under the Creative Commons Attribution License or equivalent. Every chapter published in this book has been scrutinized by our experts. Their significance has been extensively debated. The topics covered herein carry significant findings which will fuel the growth of the discipline. They may even be implemented as practical applications or may be referred to as a beginning point for another development.

The contributors of this book come from diverse backgrounds, making this book a truly international effort. This book will bring forth new frontiers with its revolutionizing research information and detailed analysis of the nascent developments around the world.

We would like to thank all the contributing authors for lending their expertise to make the book truly unique. They have played a crucial role in the development of this book. Without their invaluable contributions this book wouldn't have been possible. They have made vital efforts to compile up to date information on the varied aspects of this subject to make this book a valuable addition to the collection of many professionals and students.

This book was conceptualized with the vision of imparting up-to-date information and advanced data in this field. To ensure the same, a matchless editorial board was set up. Every individual on the board went through rigorous rounds of assessment to prove their worth. After which they invested a large part of their time researching and compiling the most relevant data for our readers.

The editorial board has been involved in producing this book since its inception. They have spent rigorous hours researching and exploring the diverse topics which have resulted in the successful publishing of this book. They have passed on their knowledge of decades through this book. To expedite this challenging task, the publisher supported the team at every step. A small team of assistant editors was also appointed to further simplify the editing procedure and attain best results for the readers.

Apart from the editorial board, the designing team has also invested a significant amount of their time in understanding the subject and creating the most relevant covers. They scrutinized every image to scout for the most suitable representation of the subject and create an appropriate cover for the book.

The publishing team has been an ardent support to the editorial, designing and production team. Their endless efforts to recruit the best for this project, has resulted in the accomplishment of this book. They are a veteran in the field of academics and their pool of knowledge is as vast as their experience in printing. Their expertise and guidance has proved useful at every step. Their uncompromising quality standards have made this book an exceptional effort. Their encouragement from time to time has been an inspiration for everyone.

The publisher and the editorial board hope that this book will prove to be a valuable piece of knowledge for researchers, students, practitioners and scholars across the globe.

LIST OF CONTRIBUTORS

Denise Lange and Kleber Del-Claro
Instituto de Biologia, Universidade Federal de Uberla^ndia, Uberla^ndia, Minas Gerais, Brazil

Tomoko Mitsunami and Gen-ichiro Arimura
Department of Biological Science & Technology, Faculty of Industrial Science & Technology, Tokyo University of Science, Tokyo, Japan

Kohei Fujita, Masahiro Nishihara and Nobuhiro Sasaki
Iwate Biotechnology Research Center, Kitakami, Japan

Ivan Galis, Kabir Md Alamgir and Yuko Hojo
Institute of Plant Science and Resources, Okayama University, Kurashiki, Japan

Keichiro Nemoto and Tatsuya Sawasaki
Proteo-Science Center, Ehime University, Matsuyama, Japan

Nadja K. Simons, Martin M. Gossner, Wolfgang W. Weisser, Esther Pašalić and Manfred Türke
Terrestrial Ecology Research Group, Department of Ecology and Ecosystem Management, School of Life Sciences Weihenstephan, Technische Universita¨t Mu¨ nchen, Freising, Germany

Thomas M. Lewinsohn
Department of Animal Biology, Institute of Biology, University of Campinas, Campinas, Sao Paulo, Brazil

Steffen Boch and Stephanie A. Socher
Institute of Plant Sciences, University of Bern, Bern, Switzerland

Markus Lange
Max-Planck-Institute for Biogeochemistry, Jena, Germany

Jörg Müller
Institute of Biochemistry and Biology, University of Potsdam, Potsdam, Germany

Markus Fischer
Biodiversity and Climate Research Centre, Senckenberg Gesellschaft für Naturforschung, Frankfurt/Main, Germany

Steffen Hagenbucher and Jörg Romeis
Agroscope, Institute for Sustainability Sciences (ISS), Zurich, Switzerland

Felix L. Wäckers
Centre for Sustainable Agriculture, Lancaster University, Lancaster, United Kingdom

Mitra Shariatinajafabadi, Tiejun Wang, Andrew K. Skidmore and Albertus G. Toxopeus1
Faculty of Geo-Information Science and Earth Observation (ITC), University of Twente, Enschede, The Netherlands

Andrea Kölzsch
Max Planck Institute for Ornithology, Department of Migration and Immuno-Ecology, Vogelwarte Radolfzell, Radolfzell, Germany
Department of Animal Ecology and Project group Movement Ecology, Netherlands Institute of Ecology (NIOO-KNAW), Wageningen, The Netherlands

Bart A. Nolet
Department of Animal Ecology and Project group Movement Ecology, Netherlands Institute of Ecology (NIOO-KNAW), Wageningen, The Netherlands

Klaus-Michael Exo
Institute of Avian Research, Wilhelmshaven, Germany

Larry Griffin
Wildfowl & Wetlands Trust, Slimbridge, Gloucestershire, United Kingdom

Julia Stahl
Sovon Dutch Centre for Field Ornithology, Nijmegen, The Netherlands

David Cabot
Environmental Consultancy Services, Carrigskeewaun, Carrowniskey, Westport, Co. Mayo, Ireland

Justin S. Strong
Department of Biology, Memorial University of Newfoundland, St. John's, NL, Canada
Biology Department, Concordia University, Montreal, QC, Canada

Shawn J. Leroux
Department of Biology, Memorial University of Newfoundland, St. John's, NL, Canada
Kimberly A. Peyton and Lauren M. Valentino
Hawaii Institute of Marine Biology, University of Hawaii - Manoa, Kaneohe, Hawaii, United States of America

Karen P. Maruska
Department of Biological Sciences, Louisiana State University, Baton Rouge, Louisiana, United States of America

Jitka Horáčková
Department of Ecology, Charles University in Prague, Faculty of Science, Prague 2, Czech Republic
Department of Zoology, Charles University in Prague, Faculty of Science, Prague 2, Czech Republic

Lucie Juřičková
Department of Zoology, Charles University in Prague, Faculty of Science, Prague 2, Czech Republic

Arnošt L.šizling
Center for Theoretical Study, Charles University and the Academy of Sciences of the Czech Republic, Prague 1, Czech Republic

Vojtěch Jarošík and Petr Pyšek
Department of Ecology, Charles University in Prague, Faculty of Science, Prague 2, Czech Republic
Department of Invasion Ecology, Institute of Botany, Academy of Sciences of the Czech Republic, Průhonice, Czech Republic

Keenan Stears and Adrian M. Shrader
School of Life Sciences, University of KwaZulu-Natal, Pietermaritzburg, South Africa

Graham I. H. Kerley
Centre for African Conservation Ecology, Department of Zoology, Nelson Mandela
Metropolitan University, Port Elizabeth, South Africa

Ken Keefover-Ring
Department of Plant Physiology, Umeå Plant Science Centre, Umeå University, Umeå, Sweden
Department of Entomology, University of Wisconsin-Madison, Madison, Wisconsin, United States of America
Maria Ahnlund
Department of Forest Genetics and Plant Physiology, Swedish University of Agricultural Sciences, Umeå Plant Science Centre, Umeå, Sweden

Ilka Nacif Abreu
Department of Plant Physiology, Umeå Plant Science Centre, Umeå University, Umeå, Sweden
Department of Forest Genetics and Plant Physiology, Swedish University of Agricultural Sciences, Umeå Plant Science Centre, Umeå, Sweden

Stefan Jansson
Department of Plant Physiology, Umeå Plant Science Centre, Umeå University, Umeå, Sweden

Thomas Moritz
Department of Forest Genetics and Plant Physiology, Swedish University of Agricultural Sciences, Umeå Plant Science Centre, Umeå, Sweden

Benedicte Riber Albrectsen
Department of Plant Physiology, Umeå Plant Science Centre, Umeå University, Umeå, Sweden
Department of Plant and Environmental Sciences, University of Copenhagen, Frederiksberg, Denmark

Peter Houk
University of Guam Marine Laboratory, UOG Station, Mangilao, Guam
Pacific Marine Resources Institute, Saipan, Northern Mariana Islands

David Benavente, John Iguel, Steven Johnson and Ryan Okano
CNMI Bureau of Environmental and Coastal Quality, Saipan, Northern Mariana Islands

Hugh W. McGregor
School of Biological Sciences, University of Tasmania, Hobart, Tasmania, Australia
Australian Wildlife Conservancy, Derby, Western Australia, Australia

Sarah Legge
Australian Wildlife Conservancy, Mornington Wildlife Sanctuary, Derby, Western Australia, Australia

Menna E. Jones and Christopher N. Johnson
School of Biological Sciences, University of Tasmania, Hobart, Tasmania, Australia

Regino Zamora
Grupo de Ecología Terrestre, Departamento de Ecología, Facultad de Ciencias, Universidad de Granada, Granada, Spain

Luis Matías
Grupo de Ecología Terrestre, Departamento de Ecología, Facultad de Ciencias, Universidad de Granada, Granada, Spain
Biological and Environmental Sciences, School of Natural Sciences, University of Stirling, Stirling, United Kingdom

Anne Ebeling, Nico Eisenhauer and Anja Vogel
Institute of Ecology, University of Jena, Jena, Germany

Sebastian T. Meyer
Department of Ecology and Ecosystem Management, Center for Food and Life Sciences Weihenstephan, Technische Universität München, Freising, Germany

Maike Abbas and Helmut Hillebrand
Institute for Chemistry and Biology of the Marine Environment, Carl-von-Ossietzky-University Oldenburg, Wilhelmshaven, Germany

Markus Lange
Max Planck Institute for Biogeochemistry, Jena, Germany

Christoph Scherber
DNPW, Agroecology, Georg-August University Göttingen, Göttingen, Germany

Alexandra Weigelt
Department for Systematic Botany and Functional Biodiversity, University of Leipzig, Leipzig, Germany

Wolfgang W. Weisser
Institute of Ecology, University of Jena, Jena, Germany
Department of Ecology and Ecosystem Management, Center for Food and Life Sciences Weihenstephan, Technische Universität München, Freising, Germany

Jian Sun and Shuli Niu
Synthesis Research Centre of Chinese Ecosystem Research Network, Key Laboratory of Ecosystem Network Observation and Modelling, Institute of Geographic Sciences and Natural Resources Research, Chinese Academy of Sciences, Beijing, China

Xiaodan Wang, Genwei Cheng, Jianbo Wu and Jiangtao Hong
The key laboratory of mountain surface processes and eco-regulation, Institute of Mountain Hazard and Environment, Chinese Academy of Sciences, Chengdu, China

Nele Ilmberger, Simon Güllert, Joana Dannenberg, Ulrich Rabausch, Jeremy Torres, Christel Schmeisser, Wolfgang R. Streit and Jennifer Chow
Universität Hamburg, Biozentrum Klein Flottbek, Abteilung für Mikrobiologie & Biotechnologie, Hamburg, Germany

Bernd Wemheuer, Anja Poehlein and Rolf Daniel
Georg-August-University Göttingen, Institute of Microbiology and Genetics, Göttingen, Germany

Malik Alawi
University Medical Center Hamburg-Eppendorf, Bioinformatics Service Facility, Hamburg, Germany

Dimitrij Turaev and Thomas Rattei
University of Vienna, CUBE - Division for Computational Systems Biology, Department of Microbiology and Ecosystem Science, Vienna, Austria

Jesper Salomon, Peter B. Olsen and Martin S. Borchert
Novozymes A/S, Microbial Discovery, Bagsværd, Denmark

Adam Grundhoff
Heinrich Pette Institute, Leibniz Institute for Experimental Virology, Hamburg, Germany

Carla R. Flöthe and Uwe John
Section Ecological Chemistry, Alfred-Wegener-Institut, Helmholtz-Zentrum für Polar- und Meeresforschung, Bremerhaven, Germany

Markus Molis
Section Functional Ecology,Alfred-Wegener-Institut, Helmholtz-Zentrum für Polar- und Meeresforschung, Bremerhaven, Germany

James P. Gibbs
Department of Environmental and Forest Biology, State University of New York College of Environmental Science and Forestry, Syracuse, New York, United States of America

Elizabeth A. Hunter
Warnell School of Forestry and Natural Resources, University of Georgia, Athens, Georgia, United States of America

Kevin T. Shoemaker
Department of Ecology and Evolution, Stony Brook University, Stony Brook, New York, United States of America

Washington H. Tapia
Department of Applied Research, Galapagos National Park Service, Puerto Ayora, Galapagos, Ecuador

Linda J. Cayot
Galapagos Conservancy, Fairfax, Virginia, United States of America

Rachel M. Mitchell
School of Environmental and Forest Sciences, University of Washington, Seattle, Washington, United States of America

Jonathan D. Bakker
School of Environmental and Forest Sciences, University of Washington, Seattle, Washington, United States of America
Smithsonian Environmental Research Center, Edgewater, Maryland, United States of America

Jeremy D. Long and Laura D. Porturas
Biology Department and Coastal & Marine Institute Laboratory, San Diego State University, San Diego, California, United States of America

Nabil Killiny, Lukasz L. Stelinski and Siddharth Tiwari
Department of Entomology and Nematology, Citrus Research and Education Center, IFAS, University of Florida, Lake Alfred, Florida, United States of America

Subhas Hajeri and Siddarame Gowda
Department of Plant Pathology, Citrus Research and Education Center, IFAS, University of Florida, Lake Alfred, Florida, United States of America

Renata Velozo Timbó
Department of Cell Biology, University of Brasília, Campus Universitário Darcy Ribeiro, Brasília, Distrito Federal, Brazil
Department of Biological Control, Embrapa Genetic Resources and Biotechnology, Brasília, Distrito Federal, Brazil

Marcelo Hermes-Lima
Department of Cell Biology, University of Brasília, Campus Universitário Darcy Ribeiro, Brasília, Distrito Federal, Brazil

Luciano Paulino Silva, Angela Mehta, Maria Carolina Blassioli Moraes and Débora Pires Paula
Department of Biological Control, Embrapa Genetic Resources and Biotechnology, Brasília, Distrito Federal, Brazil

Astrid Welk
Institute of Biology/Geobotany and Botanical Garden, Martin Luther University Halle-Wittenberg, Halle (Saale), Germany

Erik Welk and Helge Bruelheide
Institute of Biology/Geobotany and Botanical Garden, Martin Luther University Halle-Wittenberg, Halle (Saale), Germany
German Centre for Integrative Biodiversity Research (iDiv) Halle-Jena-Leipzig, Leipzig, Germany

Antonio López-Carretero and Cecilia Díaz-Castelazo
Instituto de Ecología A.C., Xalapa, Veracruz, México

Karina Boege
Instituto de Ecología, Universidad Nacional Auto´noma de México, México D.F., México

Víctor Rico-Gray
Instituto de Neuroetología, Universidad Veracruzana, Xalapa, Veracruz, México

Index